Advances in Intelligent Systems and Computing

Volume 898

Series editor

Janusz Kacprzyk, Systems Research Institute, Polish Academy of Sciences,
Warsaw, Poland
e-mail: kacprzyk@ibspan.waw.pl

The series "Advances in Intelligent Systems and Computing" contains publications on theory, applications, and design methods of Intelligent Systems and Intelligent Computing. Virtually all disciplines such as engineering, natural sciences, computer and information science, ICT, economics, business, e-commerce, environment, healthcare, life science are covered. The list of topics spans all the areas of modern intelligent systems and computing such as: computational intelligence, soft computing including neural networks, fuzzy systems, evolutionary computing and the fusion of these paradigms, social intelligence, ambient intelligence, computational neuroscience, artificial life, virtual worlds and society, cognitive science and systems, Perception and Vision, DNA and immune based systems, self-organizing and adaptive systems, e-Learning and teaching, human-centered and human-centric computing, recommender systems, intelligent control, robotics and mechatronics including human-machine teaming, knowledge-based paradigms, learning paradigms, machine ethics, intelligent data analysis, knowledge management, intelligent agents, intelligent decision making and support, intelligent network security, trust management, interactive entertainment, Web intelligence and multimedia.

The publications within "Advances in Intelligent Systems and Computing" are primarily proceedings of important conferences, symposia and congresses. They cover significant recent developments in the field, both of a foundational and applicable character. An important characteristic feature of the series is the short publication time and world-wide distribution. This permits a rapid and broad dissemination of research results.

More information about this series at http://www.springer.com/series/11156

Jiacun Wang · G. Ram Mohana Reddy
V. Kamakshi Prasad · V. Sivakumar Reddy
Editors

Soft Computing and Signal Processing

Proceedings of ICSCSP 2018, Volume 2

 Springer

Editors
Jiacun Wang
Department of Computer Science
and Software Engineering
Monmouth University
West Long Branch, NJ, USA

V. Kamakshi Prasad
Department of Computer Science
and Engineering
JNTUH College of Engineering Hyderabad
Hyderabad, Telangana, India

G. Ram Mohana Reddy
Department of Information Technology
National Institute of Technology Karnataka
Surathkal, Mangaluru, Karnataka, India

V. Sivakumar Reddy
Department of Electronics and
Communication Engineering
Malla Reddy College of Engineering &
Technology
Secunderabad, Telangana, India

ISSN 2194-5357   ISSN 2194-5365   (electronic)
Advances in Intelligent Systems and Computing
ISBN 978-981-13-3392-7   ISBN 978-981-13-3393-4   (eBook)
https://doi.org/10.1007/978-981-13-3393-4

Library of Congress Control Number: 2018962132

This Springer imprint is published by the registered company Springer Nature Singapore Pte Ltd.
The registered company address is: 152 Beach Road, #21-01/04 Gateway East, Singapore 189721, Singapore

Organizing Committee

Chief Patron

Sri. Ch. Malla Reddy
Hon'ble MP, Government of India
Founder Chairman, MRGI

Patrons

Sri. Ch. Mahendar Reddy, Secretary, MRGI
Sri. Ch. Bhadra Reddy, President, MRGI

Conference Chair

Dr. V. S. K. Reddy, Principal

Publication Chair

Dr. Suresh Chandra Satapathy, Professor, KIIT, Bhubaneswar

Convener

Prof. P. Sanjeeva Reddy, Director, ECE and EEE

Organizing Chair

Dr. M. Murali Krishna, Dean, Academics

Organizing Secretaries

Dr. S. Srinivasa Rao, HOD, ECE
Dr. D. Sujatha, HOD, CSE
Dr. G. Sharada, HOD, IT

Session Chairs

Dr. C. Suchismita, Professor, NIT Rourkela
Dr. Ram Murthy Garimella, Professor, IIIT Hyderabad
Dr. Chandra Sekhar, Professor, Osmania University
Dr. Mohammed Arifuddin Sohel, Professor, Muffakham Jah CET
Dr. Samrat Lagnajeet Sabat, Professor, HCU
Dr. Malla Rama Krishna Murty, Professor, ANITS, Visakhapatnam
Dr. Mohana Sundaram, Professor, VIT, Vellore
Dr. Suresh Kumar Nagarajan, Professor, VIT, Vellore

Coordinators

Dr. S. Shanthi, Professor, CSE
Dr. N. S. Gowri Ganesh, Professor, IT
Dr. V. Chandrasekar, Professor, CSE
Mr. G. S. Naveen Kumar, Associate Professor, ECE
Mr. K. Mallikarjuna Lingam, Associate Professor, ECE
Mr. M. Vazralu, Associate Professor, IT

Organizing Committee

Prof. K. Kailasa Rao, Director, CSE and IT
Prof. K. Subhas, Professor and Head, EEE
Dr. B. Jyothi, Associate Professor, ECE
Dr. Pujari Lakshmi Devi, Professor, ECE

Dr. C. Ravishankar Reddy, Professor, ECE
Dr. Ajeet Kumar Pandey, Professor, CSE
Dr. A. Mummoorthy, Professor, IT
Dr. V. M. Senthil Kumar, Professor, ECE
Dr. Murugeshan Rajamanickam, Professor, ECE
Sri. B. Rajeswar Reddy, Administrative Officer

Web Developer

Mr. K. Sudhakar Reddy, Assistant Professor, IT

Proceedings Committee

Dr. Sucharitha Manikandan, Associate Professor, ECE
Ms. P. Anitha, Associate Professor, ECE
Ms. M. Gayatri, Associate Professor, CSE
Ms. D. Asha, Assistant Professor, ECE
Mr. T. Vinay Simha Reddy, Assistant Professor, ECE
Mr. N. Sivakumar, Assistant Professor, CSE

Technical Program Committee

Dr. E. Venkateshwar Reddy, Professor, CSE
Dr. R. Roopa Chandrika, Professor, IT
Dr. A. Mummoorthy, Professor, IT
Mr. M. Sandeep, Associate Professor, CSE
Mr. M. Ramanjaneyulu, Associate Professor, ECE
Mr. K. Murali Krishna, Associate Professor, ECE
Mr. N. Ramesh, Associate Professor, EEE
Mr. K. Srikanth, Associate Professor, CSE
Mr. P. Bikshapathy, Associate Professor, CSE
Mr. D. Chandrasekhar Reddy, Associate Professor, CSE
Mr. M. Sambasivudu, Associate Professor, CSE
Mr. M. Jaypal, Associate Professor, CSE
Ms. J. Suneetha, Associate Professor, IT

Publicity Committee

Ms. D. Radha, Associate Professor, CSE
Mr. Ch. Kiran Kumar, Assistant Professor, ECE
Ms. P. Swetha, Associate Professor, ECE
Mr. R. Chinna Rao, Assistant Professor, ECE
Ms. Arthi Jeyakumari, Assistant Professor, CSE
Mr. P. Raji Reddy, Assistant Professor, EEE
Mr. K. D. K. Ajay, Assistant Professor, ECE
Ms. Renju Panicker, Assistant Professor, ECE
Mr. K. L. N. Prasad, Assistant Professor, ECE
Mr. T. Srinivas, Assistant Professor, ECE
Ms. R. Sujatha, Assistant Professor, CSE
Mr. A. Yogananda, Assistant Professor, IT

Registration Committee

Ms. M. Anusha, Assistant Professor, ECE
Mr. K. Suresh, Assistant Professor, ECE
Mr. V. Shiva Raja Kumar, Assistant Professor, ECE
Ms. B. Srujana, Assistant Professor, ECE
Ms. D. Kalpana, Assistant Professor, CSE
Mr. S. Vishwanath Reddy, Assistant Professor, CSE
Mr. Naresh, Assistant Professor, CSE

Hospitality Committee

Mr. A. Syam Prasad, Associate Professor, CSE
Mr. G. Ravi, Associate Professor, CSE
Mr. P. Srinivas Rao, Associate Professor, IT
Mr. M. Venu, Assistant Professor, CSE
Ms. Novy Jacob, Assistant Professor, IT
Mr. M. Anantha Gupta, Assistant Professor, ECE
Mr. G. Sekhar Babu, Assistant Professor, EEE
Mr. S. Rakesh, Assistant Professor, EEE
Mr. B. Mahendar, Assistant Professor, IT
Mr. P. Harikrishna, Assistant Professor, IT
Ms. W. Nirmala, Assistant Professor, CSE
Ms. Shruthi Rani Yadav, Assistant Professor, CSE
Ms. V. Alekya, Assistant Professor, CSE

Ms. G. Shamini, Assistant Professor, CSE
Mr. Naveen, Assistant Professor, CSE

Certificate Committee

Mr. M. Sreedhar Reddy, Associate Professor, ECE
Ms. S. Rajani, Assistant Professor, ECE
Ms. M. Hima Bindu, Assistant Professor, ECE
Mr. Manoj Kumar, Assistant Professor, CSE
Mr. K. Srinivas, Assistant Professor, CSE
Ms. Srilakshmi, Assistant Professor, IT

Decoration Committee

Mr. M. Anantha Gupta, Assistant Professor, ECE
Ms. N. Saritha, Assistant Professor, ECE
Mr. O. Saidulu Reddy, Assistant Professor, EEE
Mr. B. Srinivasa Rao, Assistant Professor, EEE
Ms. M. Nagma, Assistant Professor, ECE
Ms. D. Kavitha, Assistant Professor, ECE
Mr. Maheswari, Assistant Professor, ECE
Ms. Honey Diana, Assistant Professor, CSE
Ms. K. Swetha, Assistant Professor, IT
Ms. Sireesha, Assistant Professor, CSE
Mr. Y. Dileep Babu, Assistant Professor, CSE

Transportation Committee

Mr. V. Kamal, Associate Professor, CSE
Mr. P. Dileep, Associate Professor, CSE
Mr. G. Ravi, Associate Professor, CSE
Mr. M. Arun Kumar, Assistant Professor, ECE
Mr. E. Mahender Reddy, Assistant Professor, ECE
Mr. Saleem, Assistant Professor, CSE

International and National Advisory Committee

Dr. Heggere Ranganath, Chair of CS, The University of Alabama in Huntsville, USA

Dr. Someswar Kesh, Professor, Department of CISA, University of Central Missouri, USA

Mr. Alex Wong, Senior Technical Analyst, Diligent Inc., USA

Dr. Bhaskar Kura, Professor, University of New Orleans, USA

Dr. Ch. Narayana Rao, Scientist, Denver, Colorado, USA

Dr. Arun Kulkarni, Professor, University of Texas at Tyler, USA

Dr. Sam Ramanujan, Professor, Department of CIS and IT, University of Central Missouri, USA

Dr. Richard H. Nader, Associate Vice President, Mississippi State University, USA

Prof. Peter Walsh, Head of the Department, Vancouver Film School, Canada

Dr. Ram Balalachandar, Professor, University of Windsor, Canada

Dr. Asoke K. Nandi, Professor, Department of EEE, University of Liverpool, UK

Dr. Vinod Chandran, Professor, Queensland University of Technology, Australia

Dr. Amiya Bhaumik, Vice Chancellor, Lincoln University College, Malaysia

Prof. Soubarethinasamy, UNIMAS International, Malaysia

Dr. Sinin Hamdan, Professor, UNIMAS

Dr. Hushairi bin Zen, Professor, ECE, UNIMAS

Dr. Bhanu Bhaskara, Professor, Majmaah University, Saudi Arabia

Dr. Narayanan, Director, ISITI, CSE, UNIMAS

Dr. Koteswararao Kondepu, Research Fellow, Scuola Superiore Sant' Anna, Pisa, Italy

Shri B. H. V. S. Narayana Murthy, Director, RCI, Hyderabad

Prof. P. K. Biswas, Head, Department of E & ECE, IIT Kharagpur

Dr. M. Ramasubba Reddy, Professor, IIT Madras

Prof. N. C. Shiva Prakash, Professor, IISc, Bangalore

Dr. B. Lakshmi, Professor, Department of ECE, NIT Warangal

Dr. Y. Madhavee Latha, Professor, Department of ECE, MRECW, Hyderabad

Preface

The International Conference on Soft Computing and Signal Processing (ICSCSP 2018) was successfully organized by Malla Reddy College of Engineering & Technology, an UGC autonomous institution, during June 22–23, 2018, at Hyderabad. The objective of this conference was to provide opportunities for the researchers, academicians, and industry persons to interact and exchange the ideas, experience, and gain expertise in the cutting-edge technologies pertaining to soft computing and signal processing. Research papers in the above-mentioned technology areas were received and subjected to a rigorous peer review process with the help of program committee members and external reviewers. ICSCSP 2018 received a total of 574 papers, each paper was reviewed by more than two reviewers, and finally, 156 papers were accepted for publication in two separate volumes in Springer AISC series.

We would like to express our sincere thanks to Chief Guest Dr. S. B. Gadgil, Outstanding Scientist, Associate Director, RCI, DRDO, and keynote speakers Mr. Aninda Bose, Senior Editor, Springer Nature; Dr. C. Suchismita, Professor, NIT Rourkela; and Dr. Rishu Gupta, Senior Application Engineer, MathWorks, India.

We would like to express our gratitude to all the session chairs, viz., Dr. Ram Murthy Garimella, IIIT Hyderabad; Dr. Chandra Sekhar, Osmania University; Dr. Mohammed Arifuddin Sohel, Muffakham Jah College of Engineering and Technology; Dr. Samrat Lagnajeet Sabat, HCU; Dr. Malla Rama Krishna Murty, ANITS, Visakhapatnam; Dr. Mohana Sundaram, VIT, Vellore; and Dr. Suresh Kumar Nagarajan, VIT, Vellore, for extending their support and cooperation.

We are indebted to the program committee members and external reviewers who have produced critical reviews in a short time. We would like to express our special gratitude to Publication Chair Dr. Suresh Chandra Satapathy, KIIT, Bhubaneswar, for his valuable support and encouragement till the successful conclusion of the conference.

We express our heartfelt thanks to our Chief Patron Sri. Ch. Malla Reddy, Founder Chairman, MRGI; Patrons Sri. Ch. Mahendar Reddy, Secretary, MRGI,

Sri. Ch. Bhadra Reddy, President, MRGI; Convener Prof. P. Sanjeeva Reddy, Director, ECE and EEE; and Organizing Chair Dr. M. Murali Krishna, Dean.

We would also like to thank the organizing secretaries, viz., Dr. S. Srinivasa Rao, HOD, ECE; Dr. D. Sujatha, HOD, CSE; and Dr. G. Sharada, HOD, IT, for their valuable contribution. Our thanks also to all the coordinators and the organizing committee as well as all the other committee members for their contribution in the successful conduct of the conference.

Last but not least, our special thanks to all the authors without whom the conference would not have taken place. Their technical contributions have made our proceedings rich and praiseworthy.

New Jersey, USA	Jiacun Wang
Surathkal, India	G. Ram Mohana Reddy
Hyderabad, India	V. Kamakshi Prasad
Hyderabad, India	V. Sivakumar Reddy

Contents

About the Editors

Dr. Jiacun Wang received his Ph.D. in computer engineering from Nanjing University of Science and Technology (NJUST), China, in 1991. He is currently Professor in the Department of Computer Science and Software Engineering at Monmouth University, West Long Branch, New Jersey. From 2001 to 2004, he was a member of scientific staff with Nortel Networks in Richardson, Texas. Prior to joining Nortel, he was Research Associate of the School of Computer Science, Florida International University (FIU), Miami. Prior to joining FIU, he was Associate Professor at NJUST.

He has published two books and about 80 papers. He is Secretary of the Organizing and Planning Committee of the IEEE SMC Society. He is an associate editor of several international journals and has served as program chair, program co-chair, special session chair, or program committee member for many international conferences. He has been a senior member of IEEE since 2000.

His research interests include software engineering, discrete event systems, formal methods, wireless networking, and real-time distributed systems. Most recently, he has been focusing his research on workflow modeling and analysis for emergency management systems.

Dr. G. Ram Mohana Reddy is Professor and Head in the Department of IT at NITK Surathkal, Mangalore. He completed his B.Tech. (1987) in electronics and communication engineering from Sri Venkateswara University, Tirupati, Andhra Pradesh, India, and M.Tech. (1993) in telecommunication systems engineering from Indian Institute of Technology Kharagpur, India. He received his Ph.D. (2005) in cognitive hearing science from the University of Edinburgh, Edinburgh, UK. His areas of interests include affective human-centered computing, big data and cognitive analytics, cognitive hearing and speech science, cloud/edge/fog computing and QoS-aware green datacenters, Internet of things and smart city/home/office applications, social multimedia, and social network analysis. He has 75 publications in various reputed international/national journals and conferences. To his credit, he has authored many textbooks and chapters and he has worked on many significant funded projects and received many awards and honors.

Dr. V. Kamakshi Prasad is currently working as Professor in the Department of Computer Science and Engineering at Jawaharlal Nehru Technological University, Hyderabad. He completed his B.Tech. in civil engineering from K.L. College of Engineering during 1985–1989 and his M.Tech. in computer science and technology from Andhra University, Visakhapatnam, during 1990–1992. He received his Ph.D. in speech recognition from IIT Madras, Chennai, during 1998–2002. His areas of interest include speech recognition and processing, image processing, pattern recognition, data mining, ad hoc networks, and computer graphics. He has guided 11 Ph.D. students and is having a vast experience of 21 years in the teaching field. His research interests include speech recognition and processing, image processing, neural networks, data mining, and ad hoc networks. To his credit, he authored many textbooks and 95 publications in various reputed international/ national journals and conferences. He is acting as the TEQIP coordinator and a member in various academic bodies.

Dr. V. Sivakumar Reddy is working as Principal and Professor in the Department of Electronics and Communication Engineering at Malla Reddy College of Engineering & Technology (Autonomous). He completed his B.Tech. in electronics and communication engineering from S.V. University, Tirupati, India, and M.Tech. (first class) in digital systems from JNTU, Hyderabad, India. He received his Ph.D. in electronics and communication engineering from IIT Kharagpur, India. His areas of research interest include computer networks and communications, video processing, multimedia system design, operating systems, TCP/IP networks and protocols, system architecture, and advanced microprocessors. He is having a total of 21 years of experience at various levels of academic and administrative positions. To his credit, he is having 105 publications in various reputed international/national journals and conferences and received many awards and honors and he is the reviewer for many international journals.

A Graphical Computational Tool for Computerized Ventricular Extraction in Magnetic Resonance Cardiac Imaging

Ayush Goyal, Disha Bathla,
Sai Durga Prasad Matla Leela Venkata Manikanta, Gahangir Hossain,
Rajab Challoo, Ashwani K. Dubey, Anupama Bhan and Priya Ranjan

Abstract In this research, a graphical computational tool for segmenting the ventricles (both left ventricle and right ventricle) using images that are taken from cardiac MRI has been developed and tested. The purpose of this research is to develop a tool to aid cardiologists in the extraction of clinically relevant medical information such as ejection fraction and stroke volume from the patient's cardiac MRI images. The tool has been developed to allow the user to load any cardiac MRI image and performs segmentation upon the click of a button. Moreover, along with all other above-mentioned features, it will provide a cardiac disease prediction framework for extracting clinically relevant medical information and clinical parameters from the patient's cardiac MRI images and for assisting cardiologists and cardiac

A. Goyal (✉) · S. D. P. M. L. V. Manikanta · G. Hossain · R. Challoo
Texas A&M University - Kingsville, Kingsville, TX, USA
e-mail: ayush.goyal@tamuk.edu

S. D. P. M. L. V. Manikanta
e-mail: sai_durga_prasad.matla_leela_venkata_manikanta@students.tamuk.edu

G. Hossain
e-mail: gahangir.hossain@tamuk.edu

R. Challoo
e-mail: rajab.challoo@tamuk.edu

D. Bathla · A. K. Dubey · A. Bhan · P. Ranjan
Amity University Uttar Pradesh, Noida, UP, India
e-mail: dishabathla@gmail.com

A. K. Dubey
e-mail: akdubey@amity.edu

A. Bhan
e-mail: abhan@amity.edu

P. Ranjan
e-mail: pranjan@amity.edu

© Springer Nature Singapore Pte Ltd. 2019
J. Wang et al. (eds.), *Soft Computing and Signal Processing* ,
Advances in Intelligent Systems and Computing 898,
https://doi.org/10.1007/978-981-13-3393-4_1

researchers for creating patient-specific personalized cardiac treatment plans based on the extracted cardiac parameters such as left ventricular ejection fraction and stroke volume.

Keywords Segmentation · Cardiac MRI · Left ventricle · Right ventricle

1 Introduction

Across the globe, cardiac diseases are one of the major reasons for deaths. In 2011, the United Nations generated a death chart in which people dying of cardiovascular diseases were 33.7%, cancer deaths were 12.5%, and tumor-related deaths were 12.5%. The estimation predicts that by 2020, cardiovascular deaths would reach the top of the death chart. In order to understand the anatomy behind cardiac ischemia, coronary arterial network provides effective opportunities. In parallel, research is being done in cardiology to evaluate the functioning of a patient's heart. This circumstance makes cardiologists for distinguishing cardiac ailments precisely by taking very less amount of time. For assisting cardiologists and doctors, medical image processing is being used to facilitate the extraction of information from patient images [1–4]. In this work, a graphical computational tool has been developed in order to segment the cardiac MRI images, which can help provide clinically relevant medical parameters such as ejection fraction and stroke volume.

With a rapid increment of heart anomalies in patients, investigating patient's heart through the extraction of MRI images and through using other data had become the tough task for neurologists or specialized doctors who are in biomedical field. Magnetic resonance imaging (MRI) image segmentation is being utilized as a part of numerous biomedical applications in order to visualize and to measure the patient's heart functioning [5]. Proper usage of MRI by performing segmentation provides clear-cut information about the position, type, and size of a ventricle or tumor or other objects of interest. Proper treatment can be given in a very less time, considering the complete details of a ventricle or tumor or other objects of interest [6].

In the early stages, segmentation is usually performed manually by medical specialists, which is tedious and had a major possibility of error occurrence [7]. In order to prevent these, many techniques were discovered in order to segment MRI images properly, which are error-free or the ones that possess few errors [8]. The techniques that are introduced for proper segmentation of MRI images got classified as pixel-based, model-based, threshold-based, and region-based [9]. The complete procedure is guided using the computer without any scope for human errors [10].

For the purpose of segmenting the images that are generated from MRI, fuzzy c-means clustering technique had been used [11]. Along with fuzzy c-means clustering, connected component analysis, noise removal from the segmented image is done [12]. It had been the dominating and most widely used approach that gives efficient results when segmentation of MRI images is performed [13]. It deals with combining similar components or objects of same cluster or not similar objects that are taken within other clusters [14]. This technique is used in my research to segment the images that got generated from MRI [15].

This graphical cardiac MRI segmentation tool will be free of cost and accessible to all cardiac researchers, cardiologists, and doctors in any part of the world [16]. It will be a stand-alone independent GUI-based software platform for easy-to-use user-friendly approach which will help a lot to facilitate cardiologists in order to minimize the need for manual tracing or user intervention [17]. However, it will support all medical image data types (such as NIfTI, DICOM, PNG).

2 Methodology

In general, there are three categories of segmentation: high-level segmentation, low-level segmentation, and medium-level segmentation [18, 19]. MRI is presently a common modality used by cardiologists for diagnosing heart sickness noninvasively. Many imaging techniques, for example, computed tomography (CT), are generally utilized as modalities or technologies in order to extract clinical information and cardiac parameters for gaining an understanding of a patient's heart disease. Magnetic resonance imaging (MRI) is most generally utilized since this single methodology is non-obtrusive, free of radiation, and approved by radiologists and specialists other than CT scans, ultrasonography, PETs. MRI can satisfy an extensive variety of patient's information requirements [20–29].

2.1 MRI Frames Segmentation

A graphical user interface is developed which allows uploading the cardiac MRI images (which are produced from the cardiac MRI machine) into it and performing particular image segmentation operations in order to obtain clinically relevant medical information, e.g., EDV, ESV, and SV.

After uploading an image, following operations take place in order to segment the uploaded cardiac MRI image. Figure 1 provides the complete information about the methodology involved.

2.2 MRI Frames Processing

In general, MRI machine resembles a long, narrow tube. When you are put within the tube, you are encompassed by a magnetic field. The human body is comprised of various elements, a large portion of which is magnetic. The magnetic field encompassing

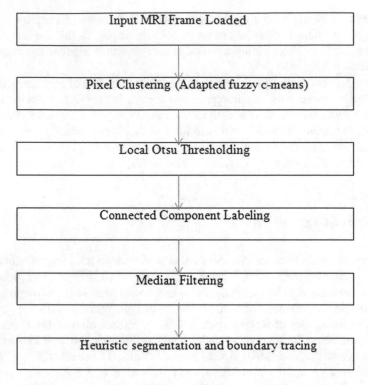

Fig. 1 Methodology for segmenting both the ventricles (left and right ventricles)

your body responds with the magnetic components inside your body to transmit a radio signal. For instance, your body contains a lot of hydrogen atoms, and those particles are exceptionally magnetic. The MRI machine's magnetic field energizes the hydrogen atoms, in your body, which thus generates a small radio signal. A computer reads the radio signal that got generated and transforms it into an image that can be observed on a computer monitor. In MRI, each and every image is called a slice. One MRI exam produces several image slices, which can be stored on a computer called a slice.

This section presents the algorithms used to segment the ventricles in the cardiac MRI images. The entire procedure is described sequentially step by step. From earlier research, the three methods explored—max flow, adaptive k, and fuzzy c-means clustering—are compared for segmenting left and right ventricles. Pixels are grouped into clusters in fuzzy c-means clustering. This groups the pixels in terms of connectivity, intensity, or distance, which brings about pixel classification. The pixels that are in separate clusters are not the same as another. In this clustering, n number of pixels are clustered following a particular measure. The center points are determined in each and every cluster. It can be observed in Eq. (1):

$$w_k = \sum_j \frac{d(center_k, x)^{\frac{2}{m-1}}}{d(center_j, x)}. \tag{1}$$

The centers of the cluster can be observed in Eq. (2):

$$c_k = \frac{\sum_x w_k(x^m)x}{\sum_x w_k(x^m)}. \tag{2}$$

After the clustering, the obtained image is binarized by utilizing thresholding technique for automatic segmentation of foreground from background clusters. As soon as thresholding gets completed, connected component labeling is done [30]. Based on the connectivity of pixels, it is used for grouping of pixels to get components like background, left ventricle, right ventricle. Through the usage of median filtering, an image gets filtered in order to effectively eliminate the regions which are very minute and do not deserve to be called as noise. In this work, heuristics get utilized to choose which region belongs to right ventricle and which region belongs to left ventricle based on calculating the distance, which is exactly from the center of the image and least in eccentricity. Pixels that are present at left and right ventricle boundary regions are made to delineate for defining left and right ventricle boundary points. After the segmentation, the clinical cardiac parameters are measured. The parameters measured are described as follows: end-systolic volume—the amount of blood that is present inside the left ventricle during systole (end of contraction) and end-diastolic volume—the amount of blood which is present inside the left ventricle during diastole (during the beginning of expansion). Stroke volume can be obtained by taking the difference between the two. It can be found in Eq. (3):

$$Stroke\ Volume = End\ Diastolic\ Volume - End\ Systolic\ Volume. \tag{3}$$

The measure of the ejection fraction parameter is calculated as the ratio between the parameters (stroke volume and end-diastolic volume). The formula for ejection fraction can be found defined below in Eq. (4).

$$EjectionFraction = \frac{ED\ Volume - ES\ Volume}{ED\ Volume}. \tag{4}$$

3 Graphical Computational Tool

The entire process of the cardiac ventricle segmentation, which includes feature extraction, classification, and segmentation, was developed as a graphical computer interface (GUI) in this work. The objective was to develop a stand-alone graphical user application where the user can upload the cardiac MRI images upon the click of a button. The generated GUI framework redraws the GUI by uploading the cardiac MRI image, drawing it, and grayscale and binary versions obtained with MATLAB functions. Medical specialists can use this graphical computational tool and segment images of heart patients generated from cardiac MRI. This would be helpful to segment the image and detect the ventricles automatically, which takes an ample amount of time manually. The following are the characteristics of the GUI:

1. It is easy to use and user-friendly.
2. It is accurate, and cardiac experts have verified its accuracy against manual segmentation.
3. Stand-alone software, especially useful for those who are in the biomedical imaging field, is made available to reduce the complexity of their work.
4. The MRI image that will get uploaded in this software will be automatically segmented, and this aids automated detection of the ventricles.
5. It will support all medical image data types (such as NIFTI, DICOM, PNG).
6. There is no need of manual segmentation, which eliminates human-made errors.

The computerized extraction of cardiac ventricles was done using the graphical user interface of a software platform that was developed in this work. While segmenting left ventricle, images generated from MRI are taken. MRI images are segmented in GUI, and both the ventricles (left and right ventricle endocardiums along with their boundaries) can be observed in Fig. 2.

For the ventricle segmentation and area calculation, in general, ventricular segmentation takes place through taking information of the entire cardiac cycle of each patient that gets generated from MRI [30]. In order to segment ventricles, a graphical user interface was developed. While segmenting ventricles, images generated from MRI were taken. MRI images are segmented in the GUI performing fuzzy c-means clustering, followed by connected component labeling and eliminating noisy regions. Figure 3 shows the extraction of the left and right ventricle endocardium regions and their calculated areas.

4 Validation

The validation gives a complete confirmation to show why the proposed cardiac image processing algorithm is more accurate when compared to previous methods. Automatic versus manual segmentation of both ventricles can be observed in Fig. 4, and accuracy results can be observed in Table 1.

Fig. 2 Stepwise procedure for segmenting left and right ventricles

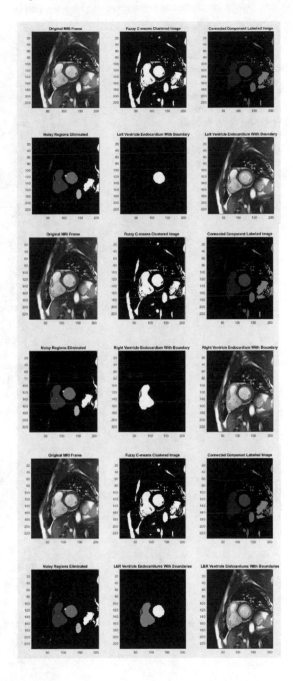

Fig. 3 Screenshots of the stand-alone computational software GUI for cardiac disease prognosis and analysis. This tool (developed in this work) segments both ventricles (left and right) in MRI images (cardiac MRI) along with calculating the areas of these regions of interest

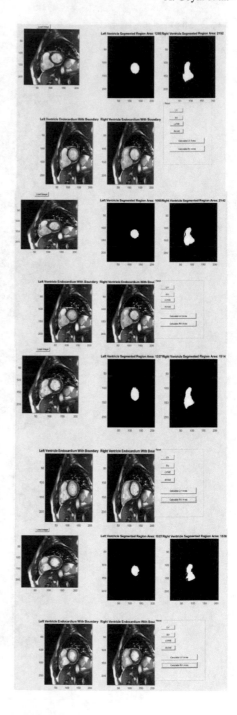

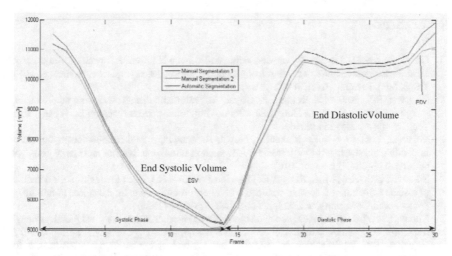

Fig. 4 Area plot calculated from the automatic and manual segmentations of cardiac ventricles

Table 1 Accuracy of the clinical parameter calculation with the automatic cardiac image processing algorithm compared against manual segmentation

Clinical parameter calculation	ED area (MM^2)	ES area (MM^2)	ED volume (MM^3)	ES volume (MM^3)	Stroke volume (MM^3)	Ejection fraction
Manual segmentation 1	1924.4169	869.9055	11546.5014	5219.433	6327.0684	0.547964113
Manual segmentation 2	1824.8094	847.3278	10948.8564	5083.9668	5864.8896	0.5356623
Max flow graph cut method	1925.745	982.794	11554.47	5896.764	5657.706	0.489655172

5 Conclusion

The objective of this paper was the development of a graphical user interface and computational software platform for assisting cardiologists and cardiac researchers. The software and associated GUI developed in this project have an easy-to-use interface in which left and right ventricles can be segmented by processing the cardiac images generated from MRI automatically with the graphical computational tool with minimum user intervention. Ejection fraction (EF) and stroke volume (SV) can be calculated after segmenting the ventricles (both left ventricle and right ventricle) from the images that get generated from cardiac MRI to assist cardiologists in cardiac disease detection and treatment planning.

References

1. Goyal, A., Lee, J., Lamata, P., van den Wijngaard, J., van Horssen, P., Spaan, J., Smith, N.P.: Model-based vasculature extraction from optical fluorescence cryomicrotome images. IEEE Trans. Med. Imaging **32**(1), 56–72 (2013)
2. Sikarwar, B.S., Roy, M.K., Ranjan, P., Goyal, A.: Automatic disease screening method using image processing for dried blood microfluidic drop stain pattern recognition. J. Med. Eng. Technol. **40**(5), 245–254 (2016)
3. Sikarwar, B.S., Roy, M.K., Ranjan, P., Goyal, A.: Imaging-based method for precursors of impending disease from blood traces. In: Advances in Intelligent Systems and Computing, vol. 468, pp. 411–424. Springer (2016)
4. Sikarwar, B.S., Roy, M.K., Ranjan, P., Goyal, A.: Automatic pattern recognition for detection of disease from blood drop stain obtained with microfluidic device. In: Advances in Intelligent Systems and Computing, vol. 425, pp. 655–667. Springer (2015)
5. Chhabra, M., Goyal, A.: Accurate and robust iris recognition using modified classical hough transform. In: Lecture Notes in Networks and Systems, vol. 10, pp. 493–507. Springer (2017)
6. Goyal, A., Ray, V.: Belongingness clustering and region labeling based pixel classification for automatic left ventricle segmentation in cardiac MRI images. Transl. Biomed. **6**(3) (2015)
7. Goyal, A., Roy, M., Gupta, P., Dutta, M.K., Singh, S., Garg, V.: Automatic detection of mycobacterium tuberculosis in stained sputum and urine smear images. Arch. Clin. Microbiol. **6**(3) (2015)
8. Goyal, A., van den Wijngaard, J., van Horssen, P., Grau, V., Spaan, J., Smith, N.: Intramural spatial variation of optical tissue properties measured with fluorescence microsphere images of porcine cardiac tissue. In: Annual International Conference of the IEEE Engineering in Medicine and Biology Society (EMBC), pp. 1408–1411 (2009)
9. Duta, M., Thiyagalingam, J., Trefethen, A., Goyal, A., Grau, V., Smith, N.: Parallel simulation for parameter estimation of optical tissue properties. In: Euro-Par2010-Parallel Processing, pp. 51–62 (2010)
10. Bhan, A., Goyal, A., Chauhan, N., Wang, C.W.: Feature line profile based automatic detection of dental caries in bitewing radiography. In: International Conference on Micro-Electronics and Telecommunication Engineering (ICMETE), pp. 635–640. IEEE (2016)
11. Bhan, A., Goyal, A., Ray, V.: Fast fully automatic multiframe segmentation of left ventricle in cardiac MRI images using local adaptive k-means clustering and connected component labeling. In: 2nd International Conference on Signal Processing and Integrated Networks (SPIN), pp. 114–119. IEEE (2015)
12. Bhan, A., Goyal, A., Dutta, M.K., Sankhla, D., Khanna, P., Travieso, C.M., Hernandez, J.B.A.: Left ventricle wall extraction in cardiac MRI using region based level sets and vector field convolution. In: 4th International Work Conference on Bioinspired Intelligence (IWOBI), pp. 133–138. IEEE (2015)
13. Bhan, A., Goyal, A., Dutta, M.K., Riha, K., Omran, Y.: Image-based pixel clustering and connected component labeling in left ventricle segmentation of cardiac MR images. In: 7th International Congress on Ultra Modern Telecommunications and Control Systems and Workshops (ICUMT), pp. 339–342. IEEE (2015)
14. Goyal, A., Bathla, D., Sharma, P., Sahay, M., Sood, S.: MRI image based patient specific computational model reconstruction of the left ventricle cavity and myocardium. In: 2016 International Conference on Computing, Communication and Automation (ICCCA), pp. 1065–1068. IEEE (2016)
15. Ray, V., Goyal, A.: Image based sub-second fast fully automatic complete cardiac cycle left ventricle segmentation in multi frame cardiac MRI images using pixel clustering and labelling. In: 8th International Conference on Contemporary Computing (IC3), pp. 248–252. IEEE (2015)
16. Ray, V., Goyal, A.: Image-based fuzzy c-means clustering and connected component labeling subsecond fast fully automatic complete cardiac cycle left ventricle segmentation in multi frame cardiac Mri images. In: International Conference on Systems in Medicine and Biology (ICSMB). IEEE (2015)

17. Clerk Maxwell, J.: A Treatise on Electricity and Magnetism, vol. 2, pp. 68–73, 3rd edn. Clarendon, Oxford (1892); Jacobs, I.S., Bean, C.P.: Fine particles, thin films and exchange anisotropy. In: Rado, G.T., Suhl, H. (eds.) Magnetism, vol. III, pp. 271–350. Academic, New York (1963)
18. von Schutthess, G.K.: The effects of motion and flow on magnetic resonance imaging. In: Morphology and Function in MRI, Ch. 3, pp. 43–62, 1989
19. Goshtasby, A., Turner, D.A.: Segmentation of cardiac cine MR images for extraction of right and left ventricular chambers. IEEE Trans. Med. Imaging 14(1), 56–64 (1995)
20. Paragios, N.: A level set approach for shape-driven segmentation and tracking of the left ventricle. IEEE Trans. Med. Imaging 22(6), 773–776 (2003)
21. Chen, C., Luo, J., Parker, K.: Image segmentation via adaptive Kmean clustering and knowledge-based morphological operations with biomedical applications. IEEE Trans. Image Process. 7(12), 1673–1683 (1998)
22. Mhlenbruch, G., Das, M., Hohl, C., Wildberger, J., Rinck, D., Flohr, T., Koos, R., Knackstedt, C., Gnther, R., Mahnken, A.: Global left ventricular function in cardiac CT. Evaluation of an automated 3D region-growing segmentation algorithm. Eur. Radiol. 16(5), 1117–1123 (2005)
23. Vandenberg, B., Rath, L., Stuhlmuller, P., Melton, H., Skorton, D.: Estimation of left ventricular cavity area with an on-line, semiautomated echocardiographic edge detection system. Circulation 86(1), 159–166 (1992)
24. Petitjean, C., Dacher, J.: A review of segmentation methods in short axis cardiac MR images. Med. Image Anal. 15(2), 169–184 (2011)
25. Lynch, M., Ghita, O., Whelan, P.: Automatic segmentation of the left ventricle cavity and myocardium in MRI data. Comput. Biol. Med. 36(4), 389–407 (2006)
26. Kaus, M., Berg, J., Weese, J., Niessen, W., Pekar, V.: Automated segmentation of the left ventricle in cardiac MRI. Med. Image Anal. 8(3), 245–254 (2004)
27. Suri, J.: Computer vision, pattern recognition and image processing in left ventricle segmentation: the last 50 years. Pattern Anal. Appl. 3(3), 209–242 (2000)
28. Cootes, T.F., Edwards, G.J., Taylor, C.J.: Active appearance models In: Proceedings of the European Conference on Computer Vision, vol. 2, pp. 484–498. Springer (1998)
29. Cootes, T.F., Edwards, G.J., Taylor, C.J.: Active appearance models. IEEE Trans. Pattern Recogn. Mach. Intell. 23(6), 681–685 (2001)
30. Bhan, A., Bathla, D., Goyal, A.: Patient-specific cardiac computational modeling based on left ventricle segmentation from magnetic resonance images. In: Advances in Intelligent Systems and Computing, vol. 469, pp. 179–187. Springer (2016)

Speckle Noise Suppression in Ultrasound Images by Using an Improved Non-local Mean Filter

Aditi Gupta, Vikrant Bhateja, Avantika Srivastava, Ananya Gupta
and Suresh Chandra Satapathy

Abstract In medical field, ultrasound imaging technique is a major medium for the doctors and radiologists to diagnose the patient's problem. The presence of speckle noise degrades the visuals of the ultrasound image and making it difficult for the doctors to diagnose properly. Hence, denoising and restoration of these images came up as a challenge to enhance the quality of the image. In this paper, an improved non-local mean filter is proposed for suppression of speckle noise in ultrasound image. This improved filter is the combination of two conventional filters, i.e., non-local mean filter and bilateral filter. The newly modified method which is emerged by combining the functions of both the conventional filters results in the improvement of quality of image. The three image quality assessment parameters are being used to analyze the performance of improved non-local mean filter, i.e., peak signal-to-noise-ratio (PSNR) and structural similarity (SSIM).

Keywords Speckle suppression · Non-local mean filter · Bilateral filter
Improved non-local mean filter · PSNR · SSIM

A. Gupta · V. Bhateja (✉) · A. Srivastava · A. Gupta
Department of Electronics and Communication Engineering, Shri Ramswaroop Memorial Group
of Professional Colleges (SRMGPC), Lucknow 226028, Uttar Pradesh, India
e-mail: bhateja.vikrant@gmail.com

A. Gupta
e-mail: agaditi12@gmail.com

A. Srivastava
e-mail: avantika.srivastava1996@gmail.com

A. Gupta
e-mail: ananyag530@gmail.com

S. C. Satapathy
School of Computer Engineering, KIIT (Deemed to be University), Bhubaneswar, Odisha, India
e-mail: sureshsatapathy@gmail.com

© Springer Nature Singapore Pte Ltd. 2019
J. Wang et al. (eds.), *Soft Computing and Signal Processing* ,
Advances in Intelligent Systems and Computing 898,
https://doi.org/10.1007/978-981-13-3393-4_2

1 Introduction

In ultrasound imaging technique, denoising and restoration appear as a big challenge since suppression of speckle noise is quite difficult because it is a multiplicative noise [1]. The speckle noise is undesirable noise which generally appears during the acquisition process of ultrasound image. During the acquisition process, waves sent from the transducer are scattered back and results in collision of ultrasonic waves which forms the granular pattern of speckle noise [2]. This uneven pattern of noise ruins the major details of the image; hence, there is a need for restoration for bringing the clarity in the visuals of ultrasound image. The previous conventional filters can be broadly classified as local statistics filters which are mean [3], median [4], Lee [5], Kuan [6, 7], Frost [7], Wiener [8], LSMV [9], anisotropic diffusion [10], and SRAD [11]. They cannot differentiate between edges and noise; therefore, they result in over smoothening, causing the loss of fine structures and details of the image. The improved non-local mean filter is a newly modified filter which works on the function of two conventional filters that are non-local mean and bilateral filter. These two filters are the successful one that improves the quality by removing the noise and keeping the edges. Baudes et al. proposed non-local mean filter which involves working of two search windows simultaneously, and it assigns weight to each pixel according to the similarity with the target pixel and eventually takes the average of all similar pixels. After taking the average, it replaces the target pixel with that average weight [12]. Tomasi et al. proposed bilateral filter which is a non-iterative and simple method. It works on both spatial and intensity domains which result in better improvement of filtered image without losing any important detail [13, 14]. The performance analysis of improved non-local mean filter is done by applying it to the ultrasound images for different values of noise variance through two image quality assessment parameters that are PSNR and SSIM [15, 16]. Finally, the result of the improved non-local mean filter is obtained for different values of noise variance and comparison is done with the existing conventional filters that are non-local mean and bilateral filter. The remaining sections of the paper are organized as follows. In the next section, description on the working of improved non-local mean filter is briefly reviewed which is followed by its algorithm. Third and the fourth sections of the paper are divided into results and discussions and conclusion, respectively.

2 Improved Non-local Mean Filter for Suppression of Speckle Noise in Ultrasound Images

Conventional non-local mean filter takes a pixel value and calculates the average values of all the pixels around it or the center pixel is replaced by that mean value. In conventional non-local mean filter, two types of windows are used which include search window and similarity window. Non-local mean filter is one of the spatial domain filters. Spatial domain filtering is linear method of filtering which is given in the paper Srivastava et al. [17]. A conventional bilateral filter is based on non-linear method, and the intensity of every single pixel is exchanged by a weighted average intensity of nearby pixels. In conventional bilateral filter, two types of weights are computed, that is, spatial domain weight and intensity domain weight. In spatial domain, the value of any given pixel in the output image is computed by values of the pixels in the vicinity of the corresponding input pixel which is given in the paper Gupta et al. [18]. An improved non-local mean filter follows combined approach of both non-local mean filtering and bilateral filtering. This filter uses search window and similarity window concept like non-local mean filter. This filter uses the domain filtering and range filtering method just like bilateral filter [19]. Then, this filter processes on both spatial domain weight and intensity domain weight same as bilateral filter does. Finally, combined non-local bilateral weight is computed, and normalize weight is computed like non-local mean filter. This new improved non-local mean filter reduces speckle noise more effectively in ultrasound image as compared to existing conventional non-local mean filter and bilateral filter. Finally, the performance of the improved non-local mean filter is compared with the conventional non-local mean filter and bilateral filter by using image quality assessment parameter. The proposed algorithm for improved non-local mean filter is as follows.

3 Results and Discussions

The improved non-local mean filter method involves certain parameters which are fixed such as window size, domain value σ_d, and range value σ_r. The window size is fixed to 3. The domain value is fixed to 0.06, and range is 0.16. The original ultrasound image which is used here is of a fetal spine shown in Fig. 1(i) [20]. The various simulation results with different noise variance are shown in Table 1. The improved non-local mean filter is applied to the ultrasound image which contains the different levels of noise variance ranges from 0.001, 0.02, 0.04, 0.06, and 0.08.

The value of 0.001 implies the very low value of noise existing in the ultrasound image, and the result for the low variance of noise in the image is shown in Fig. 1(ii), (iii). Here the value of noise variance is increasing up to the maximum level noise variance = 0.08, and the result for the maximum variance of noise is shown in Fig. 1(x), (xi). The values of noise variance 0.02, 0.04 are shown in Fig. 1(iv)–(vii). It is known that for the greater value of PSNR, the quality of image is enhanced. PSNR denotes peak signal-to-noise ratio and it means that if the value of peak signal is more, then it shows the image which is more clearer; that is, it contains less noise. It is shown in Table 2, and the PSNR value is higher for image with lowest amount of noise, and as the content of noise in the image is increased, the PSNR value is decreased. Here decrease in PSNR value implies more noise in the image. Here SSIM shows the structural similarity, higher the value of SSIM more clear is the image.

Fig. 1 (i) Original ultrasound image, (ii) noisy image with variance = 0.001, (iii) filtered image, (iv) noisy image with variance = 0.02, (v) filtered image, (vi) noisy image with variance = 0.04, (vii) filtered image, (viii) noisy image with variance = 0.06, (ix) filtered image, (x) noisy image with variance = 0.08, (xi) filtered image

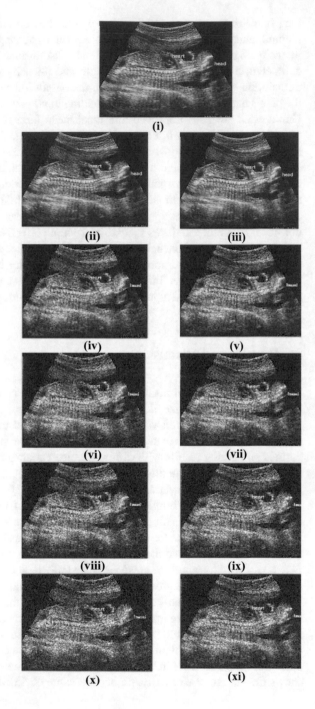

Table 1 Algorithm of improved non-local mean filter

BEGIN	**Step 1: *Input*** Ultrasound image containing some speckle noise.
	Step 2: *Convert* Colour image to grayscale image (Y).
	Step 3: *Initialize* Search window (t), similarity window (f), and filtering parameter h = 10.
	Step 4: *Initialize* Kernel size $n = 3$, filter scale value, i.e., σ_{range}, $\sigma_1 = 0.6$ and σ_{domain}, $\sigma_2 = 0.15$.
	Step 5: *Initialize* Kernel size for window (i) and Kernel size for image (j).
	Step 6: *Compute* Spatial domain weight $$d(i, j, k, l) = \exp\left(-\frac{(i-k)^2+(j-l)^2}{2\sigma_d^2}\right) \quad (1)$$
	Step 7: *Compute* Intensity domain weight $$r(i, j, k, l) = \exp\left(-\frac{\|f(i,j)-f(k,l)\|^2}{2\sigma_r^2}\right) \quad (2)$$
	Step 6: *Compute* Combined non-local bilateral filter weight $$w(i, j, k, l) = \exp\left(-\frac{(i-k)^2+(j-l)^2}{2\sigma_d^2} - \frac{\|f(i,j)-f(k,l)\|^2}{2\sigma_r^2}\right) \quad (3)$$
	Step 7: *Compute* Normalize weight $Z(i) = \sum \exp \frac{\|Y(N_i)-Y(N_j)\|_2^2}{h^2}$ (4)
	Step 8: *Compute* Weighted average $NLMB(Y(i)) = \sum w(i, j, k, l)Y(j)$
	Step 7: *Repeat* Step 3 to Step 8, until we reach the last pixel of the image.
	Step 8: *Display* Denoised ultrasound image (Y).
END	

Table 2 Simulation result of PSNR and SSIM for ultrasound image

Noise variance	PSNR (in dB)	SSIM
0.001	68.7862	0.9844
0.02	67.7613	0.7808
0.04	61.8018	0.6528
0.06	54.6902	0.5648
0.08	53.0518	0.5012

The simulation results of improved non-local mean filter for PSNR and SSIM are shown in Table 2. Here the comparison between conventional non-local mean filter, bilateral filter, and improved non-local mean filter is shown by evaluating the value of PSNR for each filter. The various results are shown in Table 3. The value of PSNR for improved non-local mean filter is higher; hence from Table 3, it can be concluded that the performance of improved non-local mean filter is better in comparison with other conventional filters.

Table 3 Simulation result of PSNR for conventional non-local mean, bilateral, and improved non-local mean filter

Noise variance	Conventional non-local mean filter	Conventional bilateral filter	Improved non-local mean filter
0.001	58.5468	46.6179	68.7862
0.02	48.8228	33.7892	67.7613
0.04	46.8574	31.0187	61.8018
0.06	44.6197	29.3202	54.6902
0.08	42.9808	28.1777	53.0518

4 Conclusion

In this paper, simulation of improved non-local mean filter is performed by adding different amount of noise variance to the original ultrasound image. The results are being obtained by calculating image quality assessment parameters such as SSIM and PSNR as shown in Table 2. The foremost intention of this paper is to implement improved non-local mean filter for speckle noise suppression which should be capable of retaining the fine structure of the original image. The improved filter which is developed is better in comparison with other conventional existing filters.

References

1. Bhateja, V., Tripathi, A., Gupta, A., Lay-Ekuakille, A.: Speckle suppression in SAR images employing modified anisotropic diffusion filtering in wavelet domain for environment monitoring. Measurement **74**, 246–254 (2015)
2. Bhateja, V., Singh, G., Srivastava, A., Singh, J.: Speckle reduction in ultrasound images using an improved conductance function based on anisotropic diffusion. In: Proceedings of International Conference of Computing for Sustainable Global Development (INDIACom), pp. 619–624. IEEE, March 2014
3. Bhateja, V., Tiwari, H., Srivastava, A.: A non local means filtering algorithm for restoration of Rician distributed MRI. In: Emerging ICT for Bridging the Future-Proceeding of the 49th Annual Convention of the Computer Society of India CSI, Vol. 2, pp. 1—8, Cham, Springer (2015)
4. Zhang, P., Li, F.: A new adaptive weighted mean filter for removing salt and pepper noise. IEEE J. Sig. Process. Lett. **21**(10), 1280–1283 (2014) (IEEE)
5. Loupas, T., McDicken, N.W., Allan, L.P.: An adaptive weighted median filter for speckle suppression in medical ultrasound images. IEEE Trans. Circuits Syst. **36** (1), 129–135 (1989) (IEEE)
6. Lee, S.J.: Digital image enhancement and noise filtering by use of local statistics. IEEE Trans Pattern Anal. Mach. Intell. **2**(2), 165–168 (1980) (IEEE)
7. Loizou, P.C., Pattichis, S.C., Christodoulou, I.C., Istepanian, S.R., Pantziaris, M., Nicolaides, A.: Comparitive evaluation of despeckle filtering in ultrasound imaging of the cartoid artery. IEEE Trans. Ultrason. Ferroelectr. Freq. **52**(10), 1653–1669 (2005) (IEEE)
8. Singh, S., Jain, A., Bhateja, V.: A comparative evaluation of various de-speckling algorithms for medical images. In: Proceedings of the CUBE International Information Technology Conference, pp. 32–37. ACM (2012)

9. Sivakumar, J., Thangavel, K., Saravanan, P.: Computed radiography skull image, enhancement using Wiener filter. In: Proceedings of International Conference on Pattern Recognition, Informatics and Medical Engineering, Henan, China, pp. 307–311. IEEE, March 2012
10. Tripathi, A., Bhateja, V., Sharma, A.: Kuan modified anisotropic diffusion approach for speckle filtering. In: Proceedings of the First International Conference on Intelligent Computing and Communication, pp. 537–545. Springer, Singapore (2017)
11. Baudes, A., Coll, B., More, M.J.: A non-local algorithm for image denoising. In: IEEE Computer Society Conference on Computer Vision and Pattern Recognition, vol. 2, pp. 60–65, June 2005
12. Tomasi, C., Manduchi, R.: Bilateral filter for gray and color images. In: Proceedings of IEEE International Conference on Computer Vision, pp. 1–8. IEEE, Bombay, India, July 1988
13. Bhateja, V., Sharma, A., Tripathi, A., Satapathy, S.C., Le, D.N.: An optimized anisotropic diffusion approach for despeckling of SAR images. In: Digital Connectivity Social Impact, pp. 134–140. Springer, Singapore (2016)
14. Baudes, A., Coll, B., Morel, J.: Neighborhood filters and pde's. In: Technical report (2005-04)
15. Perona, P., Malik, J.: Scale-space and edge detection using anisotropic diffusion. IEEE Trans. Pattern Anal. Mach. Intell. **12**(7), 629–639 (1990)
16. Bai, J., Feng, X.: Fractional-order anisotropic diffusion for image denoising. IEEE Trans. Image Process. **16**(9) (2007)
17. Srivastava, A., Bhateja, V., Gupta, A., Gupta, A.: Non local mean filter for suppression of speckle noise in ultrasound images. In: Proceedings of 2nd International Conference on Smart Computing & Informatics (SCI-2018), pp. 1–7, Vijayawada, India, January 2018
18. Gupta, A., Bhateja, V., Srivastava, A., Gupta, A., Satapathy, C.S.: Suppression of speckle noise in ultrasound images using bilateral filter. In: Proceedings of International Conference on Informatics & Communication Technology for Intelligent System (ICTIS), pp. 1–7, Ahemdabad, India, April 2018
19. Bhateja, V., Mishra, M., Urooj, S., Lay-Ekuakille, A.: Bilateral despeckling filter in homogeneity domain for breast ultrasound images. In: Proceedings of International Conference on Advances in Computing, Communications and Informatics (ICACCI), pp. 1027–1032. IEEE, September 2014
20. Ultrasound Pictures, www.ob-ultrasound.net/frames.htm

Internet of Things: Illuminating and Study of Protection and Justifying Potential Countermeasures

Nitesh Chouhan, Hemant Kumar Saini and S. C. Jain

Abstract Presently, Internet of things (IoTs) contains environs with explore in innumerable mixed plans are linked by the classes of Internet. That decides the plans in novel capability. Such having a important responsibility in the privacy of informatics. Conservatively, security method would not be directly sensible to IoT knowledge due to the different philosophy and communiqué alarmed. Here crucial study is complete in divulge the safety and solitude with IoT plans and quarrel diverse contradict events alleviate that supply the potential extent for subsequently investigate with IoT coerce.

Keywords IOT · Eavesdrop · Sinkhole attack · Sniffer · MITM · CPS · IDS
IOTCube · RFID · UNB · RC4

1 Preface

In everyday life the IoT the stage a strange place the incidence of IoT alleviate quite a lot of every daylight proceedings, supplement the technique that assist with the flexible environs, and addition the community transportation with additional public and substance. This idea picks up quite a few concerns, as IoT proffer security at that echelon? And the seclusion of such users is how secluded by IoT?

Vulnerabilities aim the corporal boundary of IoT plans, wireless protocols, and addict boundary. It is urgent that invader understand how to charge the invasion

N. Chouhan (✉)
Department of IT, Manikya Lal Verma Textile & Engineering College, Bhilwara 311001,
Rajasthan, India
e-mail: niteshchouhan_9@yahoo.com

H. K. Saini
Department of CSE, Modern Institute of Technology & Research Center, Alwar 301001,
Rajasthan, India
e-mail: hemantrhce@rediffmail.com

S. C. Jain
University Department, Rajasthan Technical University, Kota 324010, Rajasthan, India

© Springer Nature Singapore Pte Ltd. 2019
J. Wang et al. (eds.), *Soft Computing and Signal Processing* ,
Advances in Intelligent Systems and Computing 898,
https://doi.org/10.1007/978-981-13-3393-4_3

facade, observe intimidation, and develop the probable to sagacity stabbing in IoT location.

1.1 IOT Model

To recognize policies, IoT decides about building and communication in diversity with plans plus systems [1].

Following the IoT depiction subsequently confers a diversity of credible intimidation on the IoT scheme.

1.2 Threats on Each Layer

As IoT processes, the huge quantity of data is composed by users. This roadmaps for numerous invaders [2]. in addition, IoT rudiments, to assault the third party feature [3] can be negotiation. It is temporarily argued in Fig. 1.

Sinkhole Attack: To boom the system, the energy is augmented expensing the terminals. This attack incongruity where the center of nasty terminal with adjoining terminals that amplify communication installment.

Man-in-the-Middle Attack (MITM): Here, invader monitors communication among consumers. So identity not only shams but also converse like the sender.

Data Tampering: Client can fiddle the in sequence by excavate the IoT statistics.

Denial-of-Service Attack (DoS): It accesses the exterior with effort by closet system since engaged in world.

Sniffer: The statistics composed in order; aggressor implant masquerader that beat records, credentials with message patterns.

Session Hijacking: It exploits defects wherever, that unveil consumer independence Captivating eccentricity.

2 Related Works Done in IOT

This section gives the better understanding in a variety of current researches in IoT. The following Table 1 summarizes current mechanism inside a variety of features.

Fig. 1 Representation on different attacks on each layer in IoT model

Application Layer-DDoS hit
Support Layer-Dos, Fabrication
Network Layer-MITM hit
Perception Layer-Spoofing, spy

Table 1 Summary of IoT-related works

Reference divulge	Scheme	Approach	Comprehension	Deal with
[4]	Todays confront in IoT	Security algorithms	Perfection in security prospect defending devices shield tidy policy	Virtual attacks Recurrently modernize the firmware
[5]	Diminish the safety risks in IoT system	Business security policy	Recognize rogue devices frustrate possible IP passage	Undeveloped IoT devices Watch and prevailing
[6]	Diverse viable IoT smart devices	Charge quite a lot of viable devices used like separate sensors, microcontroller component, printed circuit boards	Satisfying the gap in conditions of size, lifetime, and cost to activate the expected exponential enlargement of the IoT	Quality energy efficiency in IoT nodes
[7]	Security events tackle of IoT by secluding patient inspect system	Frame the analyses sanctuary provisions and later on approximation the realizable intimidation	Device-specific security algorithms identified	Particular algorithm is not adequate
[8]	Discriminate the safety and privacy in IoT system	IoT model	Layer-based bullying and pedals	One threat risk many layers
[9]	Protection menace and awareness toward overpower	Damaged protected IoT device	Coating stand security challenges analyzed	Confirmation DDoS attacks
[10]	IoT analysis expertise, representation, and illustrate near	Separate on foundation of IoT inundated security facial exterior like authentication, privacy, access control	Device safekeeping, ACL	Authentication API security Middleware support
[11]	Safety connected premium bear process in IoT tactics	Bi-IoT directory investigate Web based-crawlings	Genuine IoT strategy within a enormous system	Internet material assault
[12]	Overexcited connectivity and IoT procedures	Safety practice plus infrastructure fulfill by typical attention on sinking the vulnerabilities	Execute vulnerabilities with verification	CWE database updates device to understands the appropriate risk

(continued)

Table 1 (continued)

Reference divulge	Scheme	Approach	Comprehension	Deal with
[13]	Consequence of the community accessibility of IoT datasets in manipulative the security	Blockchain expertise in IoT to make certain the isolation and reliability of IoT allied figures sets	Identify the cooperation firmware	Isolation of datasets Existence of datasets
[14]	The noise problem in sensor stream data	Discrete Wavelet Transform (DWT) applied	Applied technique decrease the noise during broadcast by retreating the information	Faulty in sequence Time and assets route pointless figures
[15]	Safety of cyber-physical system (CPS) in IoT	Runtime safekeeping	Run time CPS decrease	Untrusted principles
[16]	Convalesce IoTCube, to interrelate through a Web-based border	IoTCube clients analyzed	IoT implementation needs the genuine sphere component exploitation	Connectivity susceptibility location covert safety Harms
[17]	IoT in the hardware viewpoint	Periphery compute, multimode compute, with artificial intelligence	Inexpensive tempo by IoT structure	Dependability Accessibility Presentation
[18]	IDS automata	Model-based input/output ticket development method comparable to automaton	Review the imitation on Raspberry Pi mechanism in finding of pseudo-assault, replay-assault, and squeeze-assault	Circumstance freedom ignition trouble For imperfect attacks
[19]	Community locked IoT	Investigations done	Supplementary protection of IoT	Enlargement of IoT plans
[20]	IoT with cloud computing give issues	A multiplicity of mechanism were planned	Application in farming, medicare, etc.	Mobility context-based tuning QoS
[21]	IoT merged in cloud and big data	IoT reference model	Smart home, smart engineering, accuracy farming	Incorporation Scalability

(continued)

Table 1 (continued)

Reference divulge	Scheme	Approach	Comprehension	Deal with
[22]	The elegant city scheme by IoT situation	Semtech and SigFox's technology called LoRa ultra-narrowband (UNB)	Arrange IoT campaign in a low cost	Networking transportation policy concern
[23]	Clean development in formation virtual city	Review on smart city mechanism	Method used by elegant areas, houses	Deprived consistency and recognition Scalability energy restriction
[24]	IoT in therapeutic applications	Map plinth on smart treatment arrangement	Isolated ensure of fitness care system	Data security

3 Countermeasures to Mitigate Vulnerability

Ever since the dissimilar invasions transaction in diverse in Table 2 such lacunae evaded and such confront prevent to next investigate in diverse refuge era that is depicted in each issue.

Table 2 Various countermeasures to defend the vulnerabilities

Protection issue	Treaty in	Vulnerability	Countermeasures
Insecure net frontier	GUI built into IoT strategy to collaborate amid the equipment	1. Banking details 2. Certificate for scrawny 3. Certificate transfer the set up 4. XSS	i. Default credential must be distorted throughout preliminary company ii. Ensure key resurgence mechanism are strong
Not satisfactory endorse-ment/agreement	Poor mechanisms satisfy	1. Low password complexity 2. Low protected Credentials 3. Good password improvement	i. Strong passwords mandatory ii. Access control necessary
Poor network forces	Poor in the complex services admission the IoT device appliance	1. Vulnerable forces 2. Shield excess 3. Unravel point via UPnP 4. Exposed UDP services 5. Denial-of-Service	i. Ports mandatory are showing and existing ii. Forcibly caring overload and fuzzing assault
Inefficient carrying encryption	It show with statistics forward by the IoT device in an deencrypted way. Such make vulnerablity	1. Deencrypted Services 2. Poorly SSL/TLS	i. SSL/TLS encodes during the transmission of networks

4 Scope

By means of augmentation and adulthood, the investigation on IOT will extend out to varied field such as rule, direction, operations finances.

5 Conclusion

This paper explains 'Internet of Things', state persist to plant at a speedy velocity. This entails safekeeping that cut back from advance and enlarges to accomplishment. The insufficiency of evenness accompanying aggravates the predicament and employ security as it gives the impression to pay. Thereby unlike countermeasures invade the vulnerability to secure confer that would expand the scholars the occasion for the future research.

References

1. Arseni, S., Halunga, S., Fratu, O., Vulpe, A., Suciu, G.: Analysis of the security solutions implemented in current Internet of Things platforms. In: 2015 Conference Grid, Cloud & High Performance Computing in Science (ROLCG) (2015)
2. Mosenia, A., Jha, N.: A comprehensive study of security of Internet-of-Things. IEEE Trans. Emerg. Topics Comput. **4**, 586–602 (2017)
3. Okul, S., Ali Aydin, M.: Security attacks on IoT. In: 2017 International Conference on Computer Science and Engineering (UBMK) (2017)
4. Daniel, M.: Hidden dangers of Internet of Things. Women Secur. 69–75 (2017)
5. Haber, M., Hibbert, B.: Internet of Things (IoT). In: Privileged Attack Vectors, pp. 139–142 (2017)
6. Alioto, M., Shahghasemi, M.: The Internet of Things on its edge: trends toward its tipping point. IEEE Consum. Electron. Mag. **7**(1), 77–87 (2018)
7. Jaiswal, S., Gupta, D.: Security requirements for Internet of Things (IoT). Adv. Intel. Syst. Comput. 419–427 (2017)
8. Al-Gburi, A., Al-Hasnawi, A., Lilien, L.: Differentiating security from privacy in Internet of Things: a survey of selected threats and controls. In: Computer and Network Security Essentials, pp. 153–172 (2017)
9. Adat, V., Gupta, B.: Security in Internet of Things: issues, challenges, taxonomy, and architecture. Telecommun. Syst. (2017)
10. Sain, M., Kang, Y., Lee, H.: Survey on security in Internet of Things: state of the art and challenges. In: 2017 19th International Conference on Advanced Communication Technology (ICACT) (2017)
11. Payne, B., Abegaz, T.: Securing the Internet of Things: best practices for deploying IoT devices. Comput. Netw. Secur. Essent. 493–506 (2017)
12. Dawson, M.: Cyber security policies for hyperconnectivity and Internet of Things: a process for managing connectivity. Adv. Intell. Syst. Comput. 911–914 (2017)
13. Banerjee, M., Lee, J., Choo, K.: A blockchain future to Internet of Things security: a position paper. Dig. Commun. Netw. (2017)
14. Lopes De Faria, M., Cugnasca, C., Amazonas, J.: Insights into IoT data and an innovative DWT-based technique to denoise sensor signals. IEEE Sens. J. **18**(1), 237–247 (2018)

15. Wolf, M., Serpanos, D.: Safety and security in cyber-physical systems and Internet-of-Things systems. Proc. IEEE **106**(1), 9–20 (2018)
16. Hong, S., Kim, Y., Kim, G.: Developing usable interface for Internet of Things (IoT) security analysis software. In: Human Aspects of Information Security, Privacy and Trust, pp. 322–328 (2017)
17. Mohanty, S., Huebner, M., Xue, C., Li, X., Li, H.: Guest editorial circuit and system design automation for Internet of Things. IEEE Trans. Comput. Aided Des. Integr. Circuits Syst. **37**(1), 3–6 (2018)
18. Fu, Y., Yan, Z., Cao, J., Koné, O., Cao, X.: An automata based intrusion detection method for Internet of Things. Mob. Inf. Syst. **2017**, 1–13 (2017)
19. Dahabiyeh, L.: The security of Internet of Things: current state and future directions. Inf. Syst. 414–420 (2017)
20. Malik, A., Om, H.: Cloud computing and Internet of Things integration: architecture, applications, issues, and challenges. In: Sustainable Cloud and Energy Services, pp. 1–24 (2017)
21. Oppitz, M., Tomsu, P.: Internet of Things. In: Inventing the Cloud Century, pp. 435–469 (2017)
22. Hammi, B., Khatoun, R., Zeadally, S., Fayad, A., Khoukhi, L.: IoT technologies for smart cities. IET Netw. **7**(1), 1–13 (2018)
23. Shahid, A., Khalid, B., Shaukat, S., Ali, H., Qadri, M.: Internet of Things shaping smart cities: a survey. In: Studies in Big Data, pp. 335–358 (2017)
24. Javdani, H., Kashanian, H.: Internet of Things in medical applications with a service-oriented and security approach: a survey. Health Technol. (2017)

GA-Based Feature Selection for Squid's Classification

K. Himabindu, S. Jyothi and D. M. Mamatha

Abstract In this work, twenty features are extracted from Squid species that is from their shape, color, and texture features. The extracted features are fin width, fin length, head length, head width, mantle length, mantle width, total length, contrast, correlation, homogeneity, entropy, R mean, R standard deviation, R skewness, G mean, G standard deviation, G skewness, B mean, B standard deviation, B skewness. These too many extracted features may contain a lot of redundancy, increases the time complexity, and hence automatically degrade the accuracy. Hence, we adopted genetic algorithm for feature selection. Feature selection enhances the performance of concerned classifiers. Selected features using GA are validated with fuzzy system (FS), and it gives the better accuracy.

Keywords Squid species · Feature selection · Genetic algorithm · Fuzzy system Species classification

1 Introduction

Squid is successful class cephalopods of the phylum mollusca. For identifying Squid species, we need to train taxonomists who are experts in such a field and also assign taxonomic labels to them. Hence, such skilled subject experts are rare to available. This has lead to an increasing interest in automating the process of species identification and related tasks. Among Squid species, morphological characters such as shape, texture, and color are particularly used for species to differentiate their phenotypic

K. Himabindu (✉) · S. Jyothi
Department of Computer Science, Sri Padmavati Mahila Visvavidyalayam, Tirupati, India
e-mail: bindukar.karamala@gmail.com

S. Jyothi
e-mail: jyothi.spmvv@gmail.com

D. M. Mamatha
Department of Sericulture, Sri Padmavati Mahila Visvavidyalayam, Tirupati, India
e-mail: drdmmamatha@gmail.com

© Springer Nature Singapore Pte Ltd. 2019   29
J. Wang et al. (eds.), *Soft Computing and Signal Processing* ,
Advances in Intelligent Systems and Computing 898,
https://doi.org/10.1007/978-981-13-3393-4_4

differentiation and to the fact that their industrial processing [1, 2]. For automatic identification of Squid species, color, texture, and shape are common visual features. In this, some non-informative features reduce the classification accuracy and may require high computational cost which decreases the effectiveness of feature representation. There are different feature selection methods that are available for obtaining a subset of relevant features; in that one of the most familiar methods is Wrapper method.

In Wrapper method, particle swarm optimization (PSO) approach does not employ crossover and mutation operators, and ant colony optimization (ACO) approaches suffer from inadequate rules of pheromone update and heuristic information. Comparing GA with PSO and ACO algorithms, genetic algorithm (GA) has to improve image classification performance using feature selection due to its simplicity and powerful search capability upon the exponential search spaces [3, 4]. For the present study, GA is considered for feature selection and then it is implemented in MATLAB software.

2 Feature Selection

For classifying, the species features are very important. Feature selection is the process in which relevant feature subset is selected and removes the irrelevant and noisy information into original features for classification and also it can automatically speed up the learning process [5]. The selection of inappropriate and irrelevant features may give the incorrect results. Hence, to solve this problem, feature selection is considered and also it can increase the classification accuracy. Feature selection is applied in various areas such as classification, clustering, and regression [6]. The main advantage of the feature selection can improve the better understanding of dataset, and it helps to increase the classification accuracy.

In our work, features of Squid species such as morphometric features (i.e., shape), color, and texture are used for the classification. We extracted the shape, color, and texture features of Squid species done in our previous paper [7]. Totally 20 features are used, i.e., for shape seven features, nine features for color, and four features for texture. The extracted features of shape, texture, and color are fin width, fin length, head length, head width, mantle length, mantle width, total length, contrast, correlation, homogeneity, entropy, R mean, R standard deviation, R skewness, G mean, G standard deviation, G skewness, B mean, B standard deviation, B skewness are shown in Table 1. More features of Squid species may reduce accuracy.

In this paper, feature selection method is chosen for dimensionality reduction. Different feature selection techniques such as Wrapper methods and embedded methods are available to obtain relevant features. Every feature selection algorithm uses any one of the three feature selection techniques. We adopted Wrapper method, one of the most important heuristic search algorithms [8]. Heuristic algorithm usually finds solution, and it is very close to the best solution. In heuristic algorithm, one of the best feature selection algorithms is genetic algorithm (GA) which follows the process

Table 1 Extracted features of shape, texture, and color of Squid species

Features	Descriptor	No.
Shape	Fin width, fin length, head length, head width, mantle length, mantle width, total length	7
Texture	Contrast, correlation, homogeneity, entropy	4
Color	R mean, R standard deviation, R skewness, G mean, G standard deviation, G skewness, B means, B standard deviation, B skewness	9

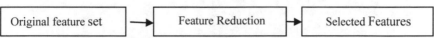

Fig. 1 Process of feature selection

of natural evolution. The GA generates useful solutions to optimization and search problems [9]. For this problem, GA-based feature selection algorithm is adopted for the selection of necessary features [10]. The processing of feature selection is shown in Fig. 1.

3 GA-Based Feature Selection

Genetic algorithm (GA) is a search-based heuristic algorithm, and it is mainly used for natural feature selection. The GA has set of candidate solutions that is call it as population. The GA gives the optimal solutions by using the survival of the fittest principle. The generated high fitness populations are represented as chromosomes. The chromosomes provide the acceptable solution to the particular problem [11].

The k-nearest neighbor approach has to be used for feature subset selection using genetic algorithm as in Fig. 2. The main components of GA are initial population of chromosomes (i.e., chromosome encoding), fitness evaluation, selection, recombination, and mutation. The first chromosome in initial population which is indicated by "1" depicts the feature selected and "0" indicates not selected. Each chromosome has its value "0" or "1" which represents the features at particular location is selected or not. The selected features are indicated by "1", and the selected features all are having the rankings; based on these rankings, it can produce new generation kids [12].

3.1 Generation of Initial Population

This work begins with generating the initial population and then initial population of chromosomes is represented in bit string format. In this study, 20 features are generated from 15 categories of Squid species. Therefore, it is difficult to represent all

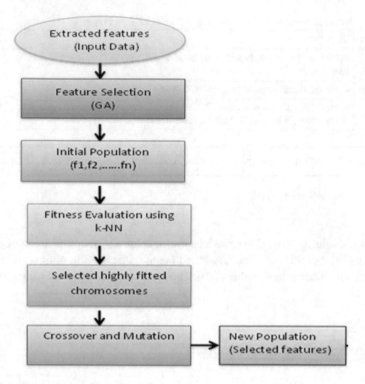

Fig. 2 Feature selection based on GA

features of chromosomes in the search space. So we have generated the chromosomes randomly of fixed length related to our population size. Based on Squid species, maximum number of generations of color, texture, and shape features are produced. The optimal solution may be produced after the maximum number of generations. This feature selection problem, individual population fitness are evaluated by using objective or fitness function with respect to some criteria [13].

3.2 Fitness Evaluated Function

In GA system, to evaluate the proficient feature subset, feature selection is necessary. To evaluate fitness of each chromosome using fitness function, here k-NN-based fitness function is used for each chromosome fitness for GA because of k-NN algorithm get to solve classification problems more accurately [14]. Here k depends upon the k number of features and k-NN accuracy calculated by following formula.

$$k - NN \text{ accuracy} = x/d \tag{1}$$

where x indicates species correctly classified and d represents number of species images in the set. Given a set of f label pairs (X_i, Y_j) where $i = 1, 2, ..., f$, where X_i represents feature set and Y_j represents class of species. The similarity measure between X_i and Y_j computed by using Euclidean distance is expressed in Eq. 2.

$$F(X_i, Y_j) = \sqrt{\sum_{i=1}^{20} \sum_{j=1}^{15} (X_{i,j} - Y_{i,j})^2} \tag{2}$$

After fitness assignment is evaluated, new populations are generated using crossover and mutation operators. It chooses the individuals that will combine for creating next new generation. These operators select the individuals depending upon the fitness level.

3.3 Crossover and Mutation

The genetic algorithm mainly composed of three operators to produce the new feature subsets that are crossover, reproduction, and mutation by using the Darwin evolution principle. The probability of survival and chromosomes fitness values are proportional to each other. Chromosomes with higher fitness value have more chances to involve in the processes such as mutation, crossover. And two chromosomes will be created by exchanging their parts at crossover point. The chromosomes in GA are represented by bit strings. So that GA can be applied on binary search space. After that, chromosomes will be ranked on the top n ranks of elicit children to survive in the next generation. Elitism identifier is Elite Count; it is always fixed by 2 [15, 16]. Mutation operator disturbs the genes in each chromosome, and it can be performed through flipping of bits randomly. For this process to increase the diversity of the population, we have implemented GA with k-NN that can automatically find out the parameters of Squid species using crossover rate and mutation rate.

4 Parameter Settings for GA Experiment

We have used Squid images dataset for feature selection. Here 50 Squid species with 15 classes are used. The parameter values that are applied for the algorithms are as the following: N (population size) = 50, population type is bit string, m (the length of chromosome) = 50, maximum iteration = 100, crossover rate = 0.8, crossover factor = 0.7, mutation rate = 0.1, mutation parameter = 0.6, two-point crossover, elitist preserved model. For k-NN classifier, k value is set to 50.

5   Experimental Results

Based on the experimental parameters, configuration of GA implementation provides the following results and also evidenced by GA simulation diagram shown in Fig. 3.

Based on fitness value, the GA reaches the termination condition. The features are selected by GA approach based on Squid features, that is, shape, texture, and color datasets. The selected features of Squid dataset are 2, 4, 6, 12. The best fitness value of GA for dataset is 0.01538 and mean of fitness value of GA is 0.019533. Then GA-based selected features classification accuracy is evaluated by using fuzzy system (FS) with four features that are from shape, color, and texture of Squid database. Generally, while comparing extracted features of Squid species using GA with FS classifier approach gives accuracy performance is good with fewer features. The

Table 2 Performance accuracy using GA and fuzzy system

Fuzzy system classifier	Number of selected features	Class accuracy (%)	Recall	Precision	Specificity
Without GA feature selection	1, 2, 3, 4, 5, 6, 7, 8, 9, 10, 11, 12, 13, 14, 15, 16, 17, 18, 19, 20	85.7	79.64	81.25	90.2
With GA feature selection	2, 4, 6, 12	91.59	87.65	80.56	94.97

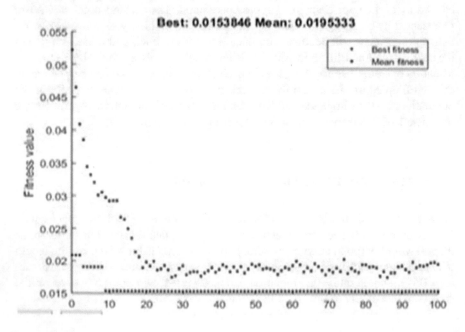

Fig. 3 GA simulation diagram

experimental results showed GA feature selection using FS and without using GA feature selection accuracy, recall, precision, and specificity computed result as shown in Table 2.

6 Conclusion

In this study, we have considered Squid's features dataset for classification. These too many features provide the low classification accuracy and time-consuming process. To increase the accuracy, we selected some important features from dataset by using GA feature selection. While using feature selection, it can give good accuracy than without using GA feature selection.

Acknowledgements This work is carried out under DBT-MRP, New Delhi.

References

1. Emam, W.M., Saad, A.A., Riad, R., et al.: Morphometric study and length—weight relationship on the squid Loligo forbesi (Cephalopoda: Loliginidae) from the Egyptian Mediterranean waters. Int. J. Environ. Sci. Eng. (IJESE) **5**, 1–13 (2014)
2. Chakraborty, S.K., Biradar, R.S., Jaiswar, A.K.: Growth, mortality and population parameters of three cephalopod species, Loligo duvauceli (Orbigny), Sepia aculeata (Orbigny) and Sepiella inermis (Orbigny) from north-west coast of India. Indian J. Fish. **60**(3), 1–7 (2013)
3. Hassan, R., Cohanim, B., Weck, O., et al.: A comparison of particle swarm optimization and genetic algorithm (2005)
4. Jianjiang, L., Zhao, T., Zhang, Y.: Feature selection based-on genetic algorithm for image annotation. Knowl.-Based Syst. **21**, 887–891 (2008)
5. Lu, H., Chen, J., Yan, K., et al.: A hybrid feature selection algorithm for gene expression data classification. Neurocomputing (2017), 2016.07.080 0925-2312/© 2017, Elsevier
6. Li, B., Lai, Y.K., Rosin, P.L.: Example-based image colorization via automatic feature selection and fusion. Neurocomputing **266**, 687–698 (2017)
7. Himabindu, K., Jyothi, S., Mamatha, D.M.: Squid species clustering based on color, shape and texture features. Int. J. Inf. Technol. (2017) [submitted paper waiting for further process]
8. Kumbhar, P., Mali, M.: A survey on feature selection techniques and classification algorithms for efficient text classification. Int. J. Sci. Res. (IJSR) **5**(5) (2016). ISSN (Online) 2319-7064
9. Agrawal, N., Gonnade, S.: An approach for unsupervised feature selection using genetic algorithm. Int. J. Eng. Sci. Res. Technol. (2016). ISSN 2277-9655
10. Huang, C.-L., Wang, C.J.: A GA-based feature selection and parameters optimization for support vector machines. Expert Syst. Appl. **31**, 231–240 (2006)
11. Melanie, M.: An introduction to genetic algorithms. In: A Bradford Book. The MIT Press (1999)
12. Chatterjee, S., Bhattacherjee, A.: Genetic algorithms for feature selection of image analysis-based quality monitoring model: an application to an iron mine. Eng. Appl. Artif. Intell. **24**, 786–795 (2011)
13. Zeng, D., Wang, S., Shen, Y., et al.: A GA based feature selection and parameter optimization for support tucker machine **111**, 17–23 (2017)
14. Sangari Devi, S., Dhinakaran, S.: Crossover and mutation operations in GA-genetic algorithm. Int. J. Comput. Organ. Trends **3**(4), 157–159 (2013). ISSN 2249-2593

15. Bhanu, B., Lin, Y.: Genetic algorithm based feature selection for target detection in SAR images. Elsevier Sci. Image Vis. Comput. **21**, 591–608 (2003). https://doi.org/10.1016/s0262-8856(03)00057-x
16. Khare, P., Burse, K.: Feature selection using genetic algorithm and classification using weka for ovari an cancer. Int. J. Comput. Sci. Inf. Technol. **7**(1), 194–196 (2016). ISSN 0975-9646

Enabling Cognitive Predictive Maintenance Using Machine Learning: Approaches and Design Methodologies

Vijayaramaraju Poosapati, Vedavathi Katneni, Vijaya Killu Manda and T. L. V. Ramesh

Abstract Asset reliability and 100% availability of machines are a competitive business advantage in complex industrial environment as they play a vital role in improving productivity. Preventive maintenance models help to identify the performance degradation or failure of machines well ahead to prevent unscheduled breakdown of machines. However, lack of knowledge in identifying the root cause or the lack of knowledge to fix the problem may delay the corrective actions, which in turn impacts the productivity. To overcome this problem, cognitive predictive maintenance model is proposed which helps in classifying and recommending corrective actions along with predicting time to failure of machine. We discussed in detail about building a cognitive system using rule-based bottom-up approaches. We also presented the high-level design of a system to build a software solution using open-source technologies.

Keywords Industrial Internet of things · Machine learning · Time to failure
Cognitive predictive model

1 Introduction

Cognitive predictive maintenance helps to enhance the predictive maintenance performance by leveraging the latest advances in technologies and abilities of hardware to store and process huge volume of data. Productivity is a key performance measure

V. Poosapati (✉) · V. Katneni · V. K. Manda · T. L. V. Ramesh
GITAM (Deemed to be University), Visakhapatnam 530045, Andra Pradesh, India
e-mail: vijayaramaraju.poosapati@gmail.com

V. Katneni
e-mail: vedavathi.katneni@gitam.edu

V. K. Manda
e-mail: mvkillu@gmail.com

T. L. V. Ramesh
e-mail: tlv.ramesh@gmail.com

© Springer Nature Singapore Pte Ltd. 2019
J. Wang et al. (eds.), *Soft Computing and Signal Processing* ,
Advances in Intelligent Systems and Computing 898,
https://doi.org/10.1007/978-981-13-3393-4_5

of an industry and it will have a significant impact if any machine breaks down. Machine health can be monitored by tracking its parameters like vibration, noise, temperature, pressure [1], and they rarely break without a change in these parameters (warnings). These warnings can be monitored by analyzing the sensors' data attached to the machines using machine learning techniques [2] which helps to predict the root cause that may result in failure of the machine. Fixing the root cause in advance will help in preventing a failure. Further, the actions taken and root cause of the failure are categorized appropriately to assist in future for similar maintenance issues. The process of categorizing and suggesting appropriate recommendations can be achieved through cognitive predictive maintenance system [3]. These systems are self-improving and update themselves by processing the different patterns received, and save the accomplished action. They enable machine-assisted decision making and help industries to automate most of the repetitive tasks during maintenance [4].

Cognitive predictive systems store and process data from multiple internal and external sources to improve the equipment quality, uptime of machines, and reduce the service time. The data collected from various sources are processed using unsupervised and supervised machine learning models which act as an input to a cognitive system. In an unsupervised learning, the relations, patterns, and actions taken are identified without mentioning the parameters to classify the results using clustering technique [5]. In a clustering technique, machine run multiple combinations and perform grouping of similar outputs called clusters. Machine-generated clusters are used in segmentation to identify the possible root cause of a problem for machine failure. In supervised learning, for each input the corresponding segmentation is provided, and the data is grouped. Classification and regression are the techniques used to handle data in supervised learning [6]. The data classified by machine learning algorithms is an input to the cognitive systems. The actions taken for each combination of clusters is saved and processed by a cognitive system.

2 Cognitive Predictive Maintenance Process

It is a process to identify and prevent the cause of failure by automatically performing actions which help to prevent the failure that may occur [7]. For example, a conveyor belt is running too fast and it might break in next 3 h if it runs at same speed. In an ordinary predictive maintenance model, an operator gets an alert, but prevention of failure depends on human action [8], whereas in a cognitive predictive maintenance process, the model automatically reduces the speed of belt and informs the operator, thereby ensuring to avoid the problem to occur. The building of such cognitive models involves processing a huge volume of internal (industry data, sensors data) and external (relevant data outside Industry) data to detect anomalies and continuously learning from the actions taken [9]. It is a six-step iterative process as described in Fig. 1.

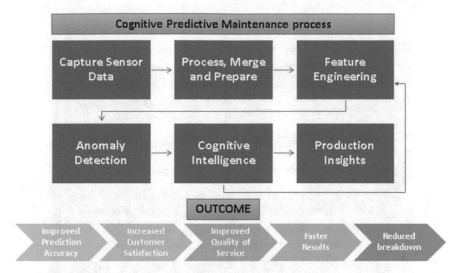

Fig. 1 Cognitive predictive maintenance process

Capturing Sensor Data: In this process, a huge volume of data generated from connected devices is collected in real time.

Process, Merge, and Prepare data: Data collected from multiple sources is processed for usage, and here, data-driven business questions are formalized to categorize the data collected from devices and it is merged with data from other sources.

Feature Engineering: In this, segmentation and grouping of data based on patterns and historic information are done.

Anomaly Detection: In this step, patterns are analyzed to identify the root cause and the process is trained to classify information, recognize patterns, behavior, trends, and sentiments.

Cognitive Intelligence: Learnings from the patterns received are recorded and possible actions to be taken are suggested in this step [10]. The context for predicting outcomes and suggesting recommendations based on context is made in this step.

Production Insights: Possible actions to be taken and best practices based on the real-time situations are suggested. The system is trained to take actions automatically on pre-occurred patterns.

3 Cognitive Framework and Widely Used Cognitive Technologies

Cognitive framework is categorized into four steps; they are foundation, learn, interact, and expand. Each step is a composition of multiple processes which figure out the context and reasoning behind each and every action taken. These are summarised as shown in Fig. 2.

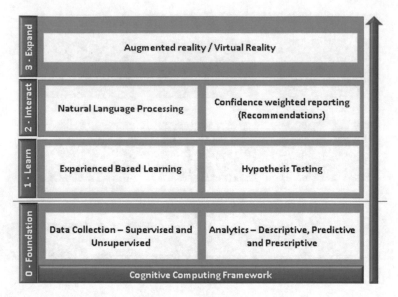

Fig. 2 Cognitive computing framework

Rule-Based Analytics

Rule-based analytics are also known as complex event processing (CEP) which are widely used in building cognitive solutions. They work based on predefined rules or a set of instructions that guide in decision making [11]. They are formalized by a domain expert or a functional consultant by considering the cause-and-effect relationships. Rule-based analytics work often in conjunction with other cognitive technologies and are executed using the CPU of the system.

Machine Learning (ML)

Commonly used algorithms like linear regression, logistic regression, support vector machine (SVM), naïve Bayes, and random forest are typically used for predictions, and they are run either by a central processing unit (CPU) or graphics processing unit (GPU) depending upon the prediction or classification [12]. They use statistical algorithms to find a complex function which will predict the result with accuracy.

Deep Learning (DL)

It is a process of determining a data feature from a given data set by using multi-level and complex neural network. They process a huge volume of data crunching and use matrix multiplications. GPUs are best suited as they used a huge volume of computational power. Convolutional neural networks (CNNs), recurrent neural networks (RNNs), region-based convolutional neural networks (R-CNNs) are few of the neural networks widely used in deep learning [13].

Reinforcement Learning (RL)

Reinforcement learning algorithms seek to improve its rewards based on the algorithm behavior. A lot of trial and error happens to test various permutations and combinations. Q-learning, Double Q-learning, state–action–reward–state–action (sarsa) are few reinforcement learning algorithms which are widely used [14].

4 Cognitive System Architecture

Various methodologies of learning and memory organization techniques have evolved during last few decades, among those few vastly used are iCub [15], Adaptive components of thought rational (ACT-R) which used a top-down learning approach [16]. Connectionist Learning Adaptive Rule Induction on Line (CLARION) is a hybrid model which used both top-down and bottom-up learning approaches. Learning Intelligent Distribution Agent (LIDA) uses IDA framework which was designed to automate assignment of sailors to new tours. They use perceptual, episodic, and procedural learning which are all part of bottom-up learning approaches. DUAL architecture is a hybrid, multi-agent, and general-purpose architecture inspired by Minsky's "society of mind."

In the present study for industrial cognitive predictive maintenance, we have used bottom-up rule-based cognitive techniques which suggest possible actions to prevent the failure and assist the operator to identify the root cause by considering the industry compliance norms and historic data. The data from sensors is continuously fed to machine learning models and the outcomes are sent as an input to the cognitive system [17]. The cognitive systems integrate the data from external sources, equipment asset management (EAM), process the information received, and take appropriate actions [18].

The sematic architecture of industrial cognitive predictive system [19] is presented in Fig. 3.

The proposed cognitive model is a progressive model. This model used bottom-up approach and has six stages (Fig. 4).

Data Capture: It helps gather relevant data from multiple sources.

Understand: The captured data is understood by the system based on rules and learnings from previous iterations.

Reason: The processed data is categorized into multiple groups based on the predefined parameters and executes reasoning.

Learn: Continuously learn from the interaction and sharpen the expertise of the system.

Ensure: The hypothesis generated is evaluated by comparing with real-time data and actions taken in real scenario.

Review: The model continuously reviews the rules and the actions taken by self-learning algorithms for identifying any deviations. All hypothesis changes are reviewed for better understanding.

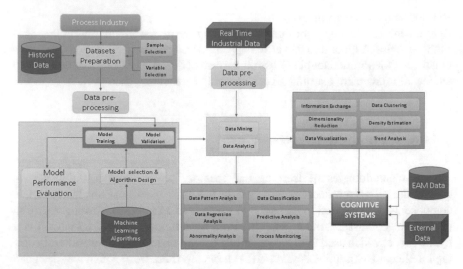

Fig. 3 Industrial cognitive predictive system architecture

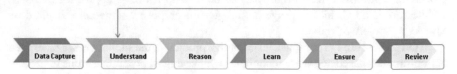

Fig. 4 Proposed cognitive model stages

5 Anomaly Detection Approach and System Design

Top-down and bottom-up approaches are the most widely used learning methods [20]. In a top-down approach, data is captured from multiple sensors for every asset and features are generated for each of the sensors. The features' dimensionality is reduced using principal component analysis or nonnegative matrix factorization, and then, clustering is applied to group the data. Anomalies are detected using the center of mass analysis. The top-down approach starts by considering all the data, thereby making it difficult to start without capturing complete data. Bottom-up approach resolves this problem by considering data of each sensor independently and builds the cognitive capabilities in a progressive way.

5.1 Proposed Rule-Based Bottom-up Approach for Building Cognitive Model

Predictive actions taken prior are recorded in rules' criterion; the derivation of it can be based on experience or equipment specific data. An elaborate way of explanation:

If the sequence or distance variation of the sensor data satisfies the predefined criteria say pressure is 20% more than 1000 pa, then increase the speed of conveyor belt, and algorithm is run to modify or create new rules and to recommend new actions. This is an iterative process, and the algorithms are self-learning, wherein it captures the human decisions and builds recommendations based on various parameters.

Algorithm:

1. Collect data for all the sensors independently. $X = \{x_1, x_2, x_3, ... x_n\}$.
2. Import or define rule criteria based on best practices and industry benchmarks (functional data to be used).
3. Identify the stages in time series in each sensor.
4. Compare the stages with baseline data (data within same asset).
5. Generate distance-based variance and sequence-based variance.
6. Generate machine-level anomalies by grouping variance of all sensors of a machine.
7. Perform following actions based on the variance.

   a. If variance is not in the defined limit of rules criteria, revisit all the rules with the matching criteria (extraction or addition of rule).
   b. If variance is found and there is no rule defined for that variance, then add rule and record the action taken (generalization).
   c. If variance is found and rule is defined, generalize that rule criteria (specialization).

8. Repeat the process.

At each step, the parameter (x, y, v, a) is checked, where 'x' is the state in which it is there before the action 'a' is taken, the changed state is denoted by 'y', and v is the state after action is taken. 'c' denotes the count.

PoM(c) is for positive match and NeM(c) is for negative match; then, positivity can be defined as γmax $Q(y, b) + r - Q(x-a) > \text{threshold}$ and negativity can be defined as γmax $Q(y, b) + r - Q(x-a) < \text{threshold}$.

For positive update, the count of c is given by PoM(c) = PoM(c) + 1, and for negative update, the count of c is given by NeM(c) = NeM(c) + 1.

At the end, the count is discounted based on the probability PoM(c) = PoM(c) * 0.9 and NeM(c) = NeM(c) * 0.9 (assuming probability as 90%) and the information gained (IG) is calculated by

IG(A, B) = Log2 [(PoM(A) + C1)/(PoM(A) + NeM(A) + C2] - Log2 [(PoM(B) + C1) / (PoM(B) + NeM(B) + C2]

where A and B are two different rule conditions which result in the same action (example: pressure more than 20% increase the speed of belt by 10%, temperature less than 10%, increase the speed by 10%).

6 Open-Source S/W to Build a Cognitive System

The objective of this study is to develop an open-source cognitive predictive system which can integrate data from multiple sources and help to prevent the occurrence of failure to happen. The data captured from multiple sources is stored in a MySQL server 5.6 database and accessed using a Web application build using Angular 5, Node JS, HTML, and CSS. APACHE 2.4.29 a Web server is installed on an Ubuntu server 16.04.3. Python scikit-learn, a free machine learning library for Python is used for predictive analysis. Sensors are connected using Arduino, and data is transferred using Socket I/O which is a JavaScript-based library. D3 JS visualization tool is used to monitor the data generated by sensors.

To handle huge volume of data, Apache Spark5 is used. It is scalable with the growing needs of the data acquisition. Apache Spark comes with an inbuilt machine learning library which can be effectively used for predictive analysis instead of scikit-learn.

7 Conclusion

The predictive maintenance techniques combined with cognitive methods can help to solve most of the maintenance problem; however, these are progressively developed and require a huge volume of data which is distributed across and challenging to get the accurate data [21]. In this paper, we outlined the cognitive predictive maintenance process and proposed the open-source software combination on which a prototype can be built. In future, a cognitive predictive model can be built on open-source s/w and results can be validated.

References

1. Akbar, A., Khan, A., Carrez, F., Moessner, K.: Predictive analytics for complex IoT data streams. IEEE Internet Things J. **4**, 1571–1582 (2017)
2. Ayoubi, S., Limam, N., Salahuddin, M.A., Shahriar, N., Boutaba, R., Estrada-Solano, F., Caicedo, O.M.: Machine Learning for Cognitive Network Management. IEEE Commun. Mag. **56**(1), 158–165 (2018)
3. Wang, J., Li, C., Han, S., Sarkar, S., Zhou, X.: Predictive maintenance based on event-log analysis: a case study. IBM J. Res. Dev. **61**, 11:121–11:132 (2017)
4. Goyal, A., Aprilia, E., Janssen, G., Kim, Y., Kumar, T., Mueller, R., Phan, D., Raman, A., Schuddebeurs, J., Xiong, J., Zhang, R.: Asset health management using predictive and pre-scriptive analytics for the electric power grid. IBM J. Res. Dev. **60**, 4.1–4.14 (2016)
5. Susto, G.A., Schirru, A., Pampuri, S., McLoone, S., Beghi, A.: Machine learning for predictive maintenance, a multiple classifier approach. IEEE Trans. Ind. Inform. **11**(3), 812–820 (2015)
6. Chang, C.-C., Lin Libsvm, C.J.: A library for support vector machines. ACM Trans. Intell. Syst. Technol. **2**, 20–29 (2013)

7. Ren, L., Sun, Y., Wang, H., Zhang, L.: Prediction of bearing remaining useful life with deep convolution neural network. IEEE Access **6**, 13041–13049 (2018)
8. Deutsch, J., He, D.: Using deep learning-based approach to predict remaining useful life of rotating components. IEEE Trans. Syst. Man Cybern. Syst. **48**, 11–20 (2017)
9. Dhoolia, P., Chugh, P., Costa, P., Gantayat, N., Gupta, M., Kambhatla, N., Kumar, R., Mani, S., Mitra, P., Rogerson, C., Saxena, M.: A cognitive system for business and technical support: a case study. IBM J. Res. Dev. **61**, 7:74–7:85 (2017)
10. Damerow, F., Knoblauch, A., Korner, U., Eggert, J.: Towards self-referential autonomous learning of object and situation models. Springer Cognit. Comput. **8**, 703–719 (2016)
11. Taher, A.: Rule mining and prediction using the Flek machine—a new machine learning engine. In: Bagheri, E., Cheung, J. (eds.) Advances in Artificial Intelligence. Canadian AI Lecture Notes in Computer Science, vol. 10832. Springer, Cham (2018)
12. Ahmad, I., Basheri, M., Iqbal, M.J., Rahim, A.: Performance comparison of support vector machine, random forest, and extreme learning machine for intrusion detection. IEEE Access **6**, 33789–33795 (2018)
13. Zhu, J., Song, Y., Jiang, D., Song, H.: A new deep-Q-Learning-based transmission scheduling mechanism for the cognitive internet of things. IEEE Internet Things J. (2017)
14. Xie, Z., Jin, Y.: An extended reinforcement learning framework to model cognitive development with enactive pattern representation. IEEE Trans. Cogn. Dev. Syst. (2018)
15. Vernon D., von Hofsten C., Fadiga L: The iCub cognitive architecture, a roadmap for cognitive development in humanoid robots. In: Cognitive Systems Monographs, vol. 11. Springer, Berlin, Heidelberg (2017)
16. Oltramari, A., Lebiere, C.: Pursuing artificial general intelligence by leveraging the knowledge capabilities of act-r. In: Artificial General Intelligence, vol. 7716, pp. 199–208. Springer (2012)
17. Ge, Z., Song, Z., Ding, S.X., Huang, B.: Data mining and analytics in the process industry—the role of machine learning. IEEE Access **5**, 20590–20616 (2017)
18. Chen, M., Tian, Y., Fortino, G., Zhang, J., Humar, I.: Cognitive internet of vehicles. Elsevier Comput. Commun. **120**, 58–70 (2018)
19. Poosapati, V., Katneni, V., Manda, V.K.: Super SCADA systems: a prototype for next gen SCADA system. IAETSD J. Adv. Res. Appl. Sci. **3**, 107–115 (2018)
20. Suna, R., Zhangb, X.: Top-down versus bottom-up learning in cognitive skill acquisition. Elsevier Cogn. Syst. Res. **5**, 63–89 (2004)
21. Antonio, L., Lebiere, A., Oltramarid, A.: The knowledge level in cognitive architectures, current limitations and possible developments. Elsevier Cogn. Syst. Res. **48**, 39–55 (2018)

Agricultural Monitoring and Controlling System Using Wireless Sensor Network

Karthik Chunduri and R. Menaka

Abstract In recent years, the automation in agricultural has become a significant issue. The parameters like soil pH, soil moisture, light intensity, temperature, and humidity play a crucial role to increase the productivity of the crops. The continuous monitoring of these environmental and soil parameters helps in taking profitable decisions. For effective utilization of resources like irrigation water and fertilizers, sprinkling of water can be done according to the soil moisture levels in the soil and spraying of the fertilizers can be done according to the soil pH value. Health monitoring of the plants is crucial for sustainable agriculture. Camera module is used to acquire the images of the plants, and image processing techniques can be applied on the acquired images for pest's detection and disease detection. The system gathers the camera data and sensor data and analyzes continuously in a feedback loop which actuates the control devices according to the threshold values.

Keywords Agricultural automation · Disease detection
Computer vision for automation · Precision agriculture · Wireless sensor network
Whiteflies detection · Image processing · Segmentation · Feature extraction
Leaf segmentation · IoT · Raspberry Pi

1 Introduction

Today energy reserves are becoming deficient and therefore become more expensive. In concurrence with the growth of the population over last century, the demand for finding new, more economical, and feasible methods of agricultural farming and food production has turned out more critical. To provide the useful data about water supply management, soil temperature, soil humidity, and the general condition of their

K. Chunduri (✉) · R. Menaka
School of Electronics Engineering, VIT Chennai, Chennai, India
e-mail: karthikchunduri98@gmail.com

R. Menaka
e-mail: menaka.r@vit.ac.in

© Springer Nature Singapore Pte Ltd. 2019
J. Wang et al. (eds.), *Soft Computing and Signal Processing* ,
Advances in Intelligent Systems and Computing 898,
https://doi.org/10.1007/978-981-13-3393-4_6

fields in a user understandable and easily accessible manner, a process facilitation is done by designing, building, and analyzing a system for agricultural monitoring and controlling operations. Our intention is to make a system, which is useful for irrigation and cultivation in efficient manner and that would help the farmer to take the decisions using the desirable information for time-saving and effective resource utilization.

2 Background

To make energy usage more efficient and low cost, CoT network collects data analytics that provide the practical information and are used with the combination of the data acquired by sensors. This allows the farmers to utilize pesticides, fertilizers, water in precise quantities with better scheduling of time to increase the production of the crops [1–4]. To estimate the growth and status of the plants, a set of numerical models of soil has taken into consideration and a fuzzy theory-based inference design engine is used to mimic the farmer's requirements [5]. Mascarello et al. [6], Kapoor and Bhat [7], Pooja and Uday [8], and Prathibha et al. [9] deal with harmless tiltrotor for farming applications and the configurations of remotely piloted aircraft system (RPAS) regarding the performance. Kapoor and Bhat [7] use an IoT network for acquisition of the readings of the environmental factors and the leaf lattice images, and the system uses MATLAB software with the help of histogram analysis to achieve conclusive results. Zhang and Zhang [10] use Internet of things and big data technologies for real-time monitoring and collection of data regarding the crop growth. The system uses duplex communication link based on a cellular–Internet interface [11–13].

3 Proposed System

The agricultural monitoring and controlling system uses temperature sensor, humidity sensor, light sensor, soil moisture sensor, soil pH sensor, and camera module for data acquisition. LCD display and mobile application are used for monitoring the field parameters. Solenoid valve is used for controlling the amount of water sprinkled on the plants in Fig. 1.

4 Methodology

IoT helps farmers to monitor their farms from anywhere and anytime. With the help of cameras, the field can be monitored in the form of images or videos remotely. Using smart phone and IoT empowers the farmer to keep monitoring the real-time

Fig. 1 Proposed system

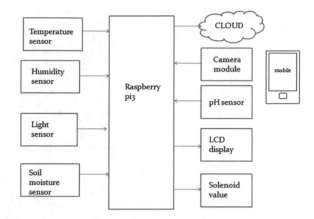

conditions of the agricultural field. The achievement of automation in agriculture, to a degree, is predicated on the availability of smart phones as they can be easily connected to different devices via the Internet and are exceptionally portable.

4.1 Field Monitoring System

The field parameters play a key role in the growth of the plant. So, a sensor node is placed in the field that contains temperature sensor to detect the temperature in the field and humidity sensor for detecting humidity levels. The pH sensor measures the pH level of soil [14–17]. The pH value of the soil gives the information about the nutrients present in the soil. If the pH value is between 5.5 and 6.8, it gives optimum yield. The light-dependent resistor determines the light intensity available to the plants. The flowchart for the soil monitoring system is shown in Fig. 2.

4.2 Pest Detection

The Raspberry Pi camera module captures the crop images and converts the captured images into grayscale images. To detect the whiteflies in the images, threshold operation has been implemented on the grayscale images. The resulting image contains noise due to water droplets and dust, so noise reduction can be done by smoothing operation using median filter following with erosion and dilation operations. Then, the resulted image is free from the noise. Whiteflies-presented areas on the leaves can be detected by finding their contours, and then, the number of whiteflies on the leaves

Fig. 2 Flowchart of soil
monitoring system

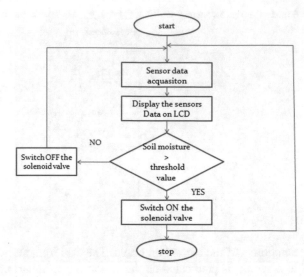

Fig. 3 Methodology to
detect whiteflies on the plant

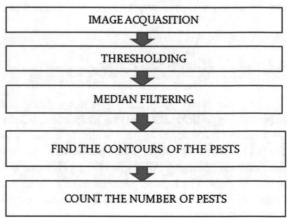

is counted [18]. Based on population of the insects that affect the plant, intimation
will be given to the farmers regarding how much amount of pesticide concentration
is required for spraying in the field. Figure 3 shows the methodology for detecting
whiteflies on the plant.

4.3 Septoria Leaf Spot Disease Detection

The BGR images acquired from the camera module are converted into HSV images,
and with the help of green pixels, a mask is created that represents the healthy plant.
Logical AND operation between the mask and acquired image can find the amount of

Fig. 4 Methodology for detecting Septoria leaf spot disease on the plant

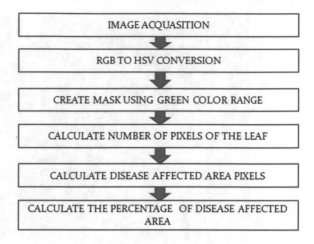

infected region on the leaves [19, 20]. By calculating the number of pixels belonging to healthy region and number of pixels belonging to the infected region, the amount of plant infected by the disease will be obtained. Figure 4 shows the methodology for detecting Septoria leaf spot on the plant.

5 Experimental Setup

The proposed system uses Raspberry Pi 3 as microcontroller and Raspberry Pi camera module v1 for capturing the images of the plants. Temperature sensor (LM35), soil moisture sensor (SKU-SEN0114), soil pH sensor (YXC 8.0SDX), LDR sensor, and humidity sensor (DHT11) are used to monitor the environment and soil parameters. Solenoid valve is used to release water to the plants. Figure 5 shows the experimental setup for the proposed system.

6 Results and Discussion

The proposed system delivers the results regarding the pest population, density of the area affected by the disease, the sensor outputs, and controlling of irrigation water supply using solenoid valve. Messages are sent regarding the soil properties using mobile app, e-mail, and twitter.

Figure 6 shows the pest detection on the leaves. Figure 6a shows the original image, Fig. 6b shows the result after the implementation of threshold on the grayscale image of the original image, and Fig. 6c shows the resultant image after noise removal of Fig. 6b.

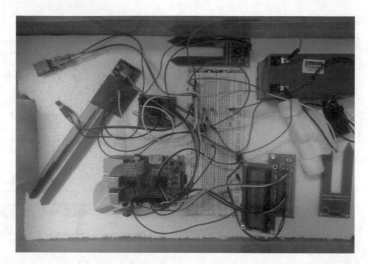

Fig. 5 Experimental setup

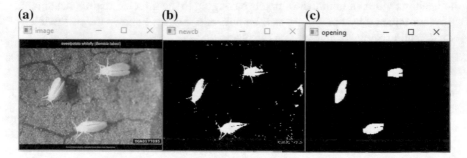

Fig. 6 Pest detection

The image shown in Fig. 7a is the Septoria leaf spot disease-affected tomato plant leaf; the HSV conversion of the image is shown in Fig. 7b; the image shown in Fig. 7c is the healthy region of the plant leaf.

The image shown in Fig. 7d contains the disease-affected pixel value assigned to 0, the image shown in Fig. 7e is the grayscale image of Fig. 7d, and the image shown in Fig. 7f represents the leaf area pixel value with 255. Figure 8 shows the disease-affected area on the leaf in the output terminal.

Figure 9 shows the field parameters like temperature, humidity, soil moisture, soil pH, and light intensity in the field with help of the mobile app. The proposed system used the soil moisture sensor to measure the soil moisture value. The Raspberry Pi continuously checks the soil moisture level with the threshold value. When the soil moisture level is less than the threshold value, Raspberry Pi opens the solenoid valve to supply the irrigation water to the plants.

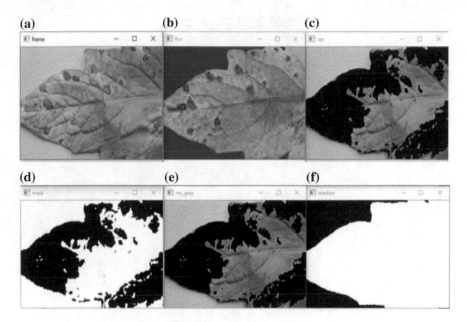

Fig. 7 Septoria leaf spot disease detection

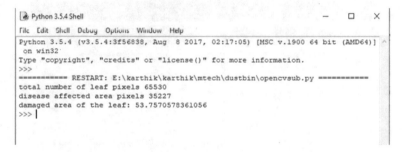

Fig. 8 Terminal output to display damaged area of the leaf

Figures 10 and 11 show the data plots using the sensor parameters uploaded to the ThingSpeak acquired from the sensor nodes. Field 1 Chart shows the temperature data, and Field 2 Chart shows the soil moisture data.

Field 3 Chart shows the pH data of the soil, Field 4 Chart shows the humidity data, and Field 5 Chart shows the light intensity data. Field 6 Chart shows the solenoid valve status in the field. When the soil is dry, the solenoid valve is in ON status, and when the soil is wet, the solenoid valve is in OFF position.

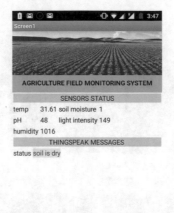

Fig. 9 Sensor output shown in mobile app

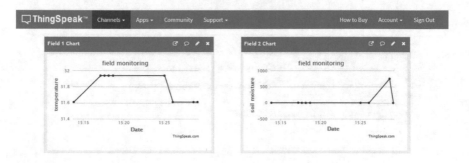

Fig. 10 Data plots of temperature and humidity in ThingSpeak

7 Conclusion

To increase the productivity of the crops and for the efficient usage of resources, Internet of things (IoT) plays an important role in the agriculture crop monitoring and controlling system. For acquisition of crop health conditions and changes in environment, camera module and various sensors are placed at the sensor node of wireless sensor network. The information about the plant health condition will be uploaded to the cloud platforms and also transmitted to the farmers through mobile

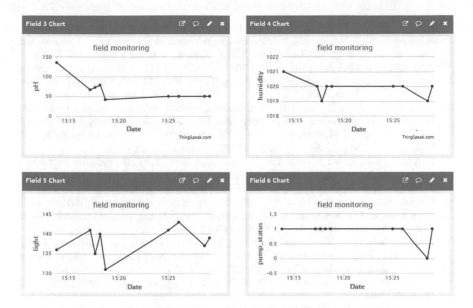

Fig. 11 Data plots of pH, humidity, light intensity, and solenoid valve status in ThingSpeak

app and e-mail for taking control actions. Real-time monitoring of the environment conditions can be helpful to increase the production of the crop yield and to reduce the wastage of irrigation water, fertilizers, and pesticides.

References

1. Roopaei, M., Rad, P., Choo, K.-K.R.: Cloud of things in smart agriculture: intelligent irrigation monitoring by thermal imaging. In: IEEE Cloud computing, vol. 4, pp. 127–131. IEEE Computing Society (2017)
2. Oksanen, T., Linkolehto, R., Seilonen, I.: Adapting an industrial automation protocol to remote monitoring of mobile agricultural machinery: a combine harvester with IoT. IFAC-PapersOnLine **49**, 127–131 (2016)
3. Potamitis, I., Rigakis, I.: Aperture optoelectronic devices to record and time-stamp insects wingbeats. IEEE Sens. J. **16**, 6053–6061 (2016)
4. Srbinovska, M.: Environmental parameters monitoring in precision agriculture using wireless sensor networks. J. Clean. Prod. **88**, 297–307 (2015)
5. Viani, F., Bertolli, M., Salucci, M.: Low-cost wireless monitoring and decision support for water saving in agriculture. IEEE Sens. J. **17**, 4299–4309 (2017)
6. Mascarello, L.N., Quagliotti, F., Ristorto, G.: A feasibility study of an harmless tiltrotor for smart farming applications. In: International Conference on Unmanned Aircraft Systems (ICUAS), pp. 1631–1639. IEEE Press, Miami, FL, USA (2017)
7. Kapoor, A., Bhat, S.I.: Implementation Of IoT (internet of things) and image processing in smart agriculture. In: International Conference on Computational Systems and Information Systems for Sustainable Solutions, pp. 21–26. IEEE Press, India (2016)

8. Pooja, S., Uday, D.V.: Application of MQTT protocol for real time weather monitoring and precision farming. In: International Conference on Electrical, Electronics, Communication, Computer, and Optimization Techniques (ICEECCOT), pp. 1–6. IEEE Press, India (2017)

9. Prathibha, S., Hongal, A., Jyothi, M.: IOT Based monitoring system in smart agriculture. In: International Conference on Recent Advances in Electronics and Communication Technology, pp. 81–84. IEEE Press, India (2015)

10. Zhang, P., Zhang, Q.: The construction of the integration of water and fertilizer smart water saving irrigation system based on big data. In: IEEE International Conference on Computational Science and Engineering (CSE) and IEEE International Conference on Embedded and Ubiquitous Computing (EUC), pp. 392–397. IEEE Press, China (2017)

11. Roy, S., Rajarshi, R.: IoT, big data science & analytics, cloud computing and mobile app based hybrid system for smart agriculture. In: 8th Annual Industrial Automation and Electromechanical Engineering Conference, pp. 303–304. IEEE Press Thailand (2017)

12. Mohanraj, I., Ashokumar, K., Naren, J.: Field monitoring and automation using IOT in agriculture domain. In: 6th International Conference on Advances in Computing & Communications, Procedia Computer Science, vol. 93, pp. 931–939 (2016)

13. Imteaj, A., Rahman, T., Hossain, M.K., Zaman, S.: IoT based autonomous percipient irrigation system using Raspberry pi. In: 19th International Conference on Computer and Information Technology, pp. 563–568. IEEE Press, Bangladesh (2016)

14. Ivanov, S., Bhargava, K., Donnelly, W.: Precision farming sensor analytic. IEEE Intell. Syst. **30**, 76–80 (2015)

15. Ojha, T., Misra, S., Raghuwanshi, N.S.: Sensing-cloud: leveraging the benefits for agricultural applications. Comput. Electron. Agric. **135**, 96–107 (2017)

16. Gutierrez, J., Villa-Medina, J.F., Nieto-Garibay, A.: Automated irrigation system using a wireless sensor network and GPRS module. IEEE Trans. Instrum. Meas. **63**, 166–176 (2014)

17. Gevaert, C.M., Suomalainen, J., Tang, J., Kooistra, L.: Generation of spectral–temporal response surfaces by combining multispectral satellite and hyperspectral UAV imagery for precision agriculture applications. IEEE J. Sel. Top. Appl. Earth Obs. Remote Sens. **8**, 3140–3146 (2015)

18. Duy, N.T.K., Tu, N.D., Son, T.H., Khanh, L.H.D.: Automated monitoring and control system for shrimp farms based on embedded system and wireless sensor network. In: IEEE International Conference on Electrical, Computer and Communication Technologies, pp. 1–5. IEEE Press, India (2015)

19. Balamurali, R., Kathiravan, K.: An analysis of various routing protocols for precision agriculture using wireless sensor network. In: IEEE International Conference on Technological Innovations in ICT for Agriculture and Rural Development, pp. 156–159. IEEE Press, India (2015)

20. Mahmoud, R., Yousuf, T., Aloul, F.: Internet of things (IoT) security: current status, challenges and prospective measures. In: Internet Technology and Secured Transactions (ICITST), pp. 336–341. IEEE Press, UK (2015)

A Modified Firefly Swarm Optimization Technique to Improve the Efficiency of Underwater Wireless Sensor Networks

A. M. Viswa Bharathy, V. Chandrasekar and D. Sujatha

Abstract The sensor nodes in UWSNs do not remain within the confined zone due to water currents. The sensor nodes have to be clustered in order to enable them to send and receive the collected data in their deployed environment. Clustering of sensor nodes helps the network in reducing transmission time. A grouping algorithm derived from the firefly swarm optimization (FSO) is tested to improve the stability and proximity of the underwater wireless sensor networks (UWSNs). The firefly algorithm helps in keeping the sensor nodes intact and produces fewer failures in network connectivity. The simulation results are convincing, and the same has been given at the end.

Keywords Clustering · Firefly · Stability · Sensor networks · Optimization

1 Introduction

The sensor networks in general can be categorized as wired and wireless. Wired sensor networks are used in places where the sensor nodes are closely installed to each other. WSN is used where the distance between sensor nodes is longer. The wireless sensor networks when deployed underwater behave like mobile ad hoc networks. This deployment is called underwater wireless sensor networks. The sensor nodes installed underwater tend to move dynamically due to wave currents and easily move

A. M. Viswa Bharathy
Department of CSE, Jyothishmathi Institute of Technology
and Science, Karimnagar, Telangana, India
e-mail: viswabharathy86@yahoo.co.in

V. Chandrasekar (✉) · D. Sujatha
Department of CSE, Malla Reddy College of Engineering
and Technology, Hyderabad, Telangana, India
e-mail: drchandru86@gmail.com

D. Sujatha
e-mail: sujatha.dandu@gmail.com

© Springer Nature Singapore Pte Ltd. 2019
J. Wang et al. (eds.), *Soft Computing and Signal Processing* ,
Advances in Intelligent Systems and Computing 898,
https://doi.org/10.1007/978-981-13-3393-4_7

from the network [1, 2]. The sensor nodes detach themselves from the network due to high currents in the water which results in loss of connectivity between the nodes. In the past, many researches have been done to improve the throughput and packet delivery ratio (PDR) of underwater wireless sensor networks (UWSNs) [3, 4].

But the less interpreted fact is that the throughput and PDR increase only when the stability of the sensor nodes is high. The high stability helps in improving the connectivity among the nodes which in turn optimizes the routing options available within the network. The sensed data from the sensor nodes are transmitted to base station through other intermediate nodes. Apart from stability and connectivity, there are other factors such as energy, memory, location and topology that affect the performance efficiency of the underwater wireless sensor networks (UWSNs). The factors need to be optimized for a better performance of UWSN are energy, memory, location and topology [5]. Energy required to transmit a data to the base station from the nodes is higher [6]. In general, the memory in a sensor node is very limited, so that they transfer the data as soon as possible after it is collected and processed [7]. They cannot run a complex algorithm to process the data [8]. Usually, the nodes are employed in a remote area where regular maintenance is impossible. The nodes get physically damaged due to environmental conditions [9–11]. The topology of the UWSN is highly dynamic in nature such that they are easily prone to wear and tear. When a node moves out of place, it takes a lot of human effort and money to install a new node in that place [12].

2 Related Work

Many studies and researches have been conducted and presented their findings about the issues associated with the wireless sensor networks. The findings of the paper suggested various problems to be rectified in alleviating the performance of the sensor networks [13, 14]. In another work which extended the lifetime of the wireless sensor networks, added some extra nodes to the network by proposing a modified form of LEACH—Spare Management LEACH-SM) protocol to manage those spare nodes [15]. Debasmita Sengupta and Alak Roy studied about the various physical structure control approaches in the wireless sensor networks and issues associated with them [16, 17]. In another attempt of the experiment, the idea of a clustering protocol derived from genetic algorithm for improving the lifetime of the sensor networks and the stability among the sensor nodes was tested [18–20]. The approach genetic algorithm-based energy-efficient hierarchy protocol with adaptive clustering (GAEEP) is executed in phases, namely set-up phase, steady-state phase and data transfer phase [21].

Sachin Gajjar et al. designed a fuzzy-based ant colony optimization routing protocol for WSNs (FAMACRO) which focused on clustering. The heads of the clusters were elected on the following factors, namely average remaining energy, number of adjacent nodes and the quality of the transmission path. The ant colony optimization (ACO) is used for routing of packets from the cluster heads to the base station. In this inter-clustering routing protocol, the relay nodes are selected based on the average remaining energy, average distance from the cluster head and base station and receiving rate of the node [22]. Zhi Ren et al. proposed the idea ring-based multi-hop clustering routing (RBMC). In this approach instead of clustering in each round, multi-round clustering is done [23]. Hicham et al. proposed an clustering algorithm based on multi-objective weights. The proposed method divides the sensor network into varying clusters and selects an efficient sensor based on its residual energy [2].

Chirihane et al. submitted their energy balanced adaptive clustering protocol in a distributed fashion which yielded better energy efficiency and overall lifetime. It was a load and energy balancing self-adaptive clustering. The aim of this work is to extend the lifetime of the large-scale wireless sensor networks. The method also reduced the communication overhead and optimized the time complexity. The heads of the clusters were elected on the following factors, namely average remaining energy, inter-cluster head distances, distance between nodes and their weights. The selection of cluster head is done either locally or distantly. The results of the experiment were compared with DEEAC and PEGASIS. The DEACP yielded better results in terms of energy consumption and lifetime [18]. Hari et al. proposed K-means algorithm to form clusters among the sensor nodes and a relative neighbourhood graph (RNG) to control the structure of the individual clusters. The connectivity of the clusters is made strong by an inter-cluster topology. The simulation experiment in OMNet++ had two factors, node power depletion and node lifetime. The K-means algorithm worked in three phases. First, it formed clusters randomly, then intra-clustering is done, and third, inter-clustering is done. The integrity among the sensor nodes was also increased, and the topology was in control [24]. Vrinda Gupta and Rajoo Pandey presented the power efficient non-equal clustering, in which the head of the cluster is decided by the location of the base station, energy level of the node and the number of adjacent nodes. The method proved to increase the lifetime of the network by a varying margin. The next hop node is selected based on distance and energy information. This cluster head and next hop node selection strategy yielded better performance efficiency. The protocol was tested under three conditions [4].

3 Firefly Algorithm

This metaheuristics approach was given by Xin-She Yang, and it is a mathematical optimization technique. The algorithm was inspired from the lighting behaviour of the fireflies [25]. Generally, the fireflies flashlight signals to attract other fireflies. The basic firefly algorithm is based upon the following considerations shown in Figs. 1 and 2.

1. Fireflies are unisexual in nature.
2. A firefly's flash attracts all the other neighbourhood fireflies.
3. The brightness of the firefly is directly proportional to the attraction level of the firefly. The less glowing one is attracted towards the high glowing one.
4. In a given neighbourhood area, if no firefly is glowing beyond a threshold, all fireflies move randomly.

Fig. 1 Process flow of
firefly algorithm

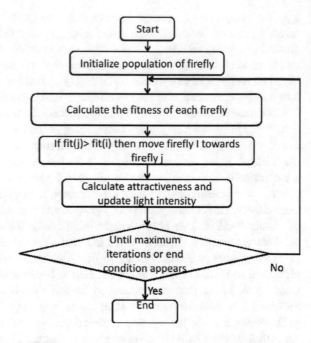

Fig. 2 Picture showing the
firefly attraction

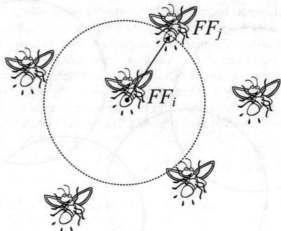

The algorithm of the firefly-based technique is given below

Step 1: Generate an objective function $f(x)$ and I denoting light intensity linked to
        $f(x)$, where I is either directly proportional or equal to $f(x)$.
Step 2: Distribute the n fireflies in the space.
Step 3: Define the absorption coefficient γ for the population.
Step 4: Vary the light intensity with respect to the distance.
Step 5: Depending upon the intensity, the fireflies are attracted to each other.

Step 6: The fireflies form clusters among themselves.
Step 7: After several iterations, the fireflies are clustered and form a strong bond based on the light intensity.

4 Firefly Swarm Intelligence Algorithm for UWSNs

The altered FSO for UWSN is given in this section. The flash in a firefly is calculated based on the following factors like energy level, mobility and distance from the base station. These factors are, respectively, denoted by e, m and d.

$$F = \sqrt{\frac{e}{m + d}}$$

where F is denoted as the flash of the firefly. Here, it denotes the power of the sensor nodes. The factors e need to be high and m and d need to be low. The optimum flash value for a sensor node is estimated at close to 2.23. The nodes which have values plus or minus very close to this flash value act as cluster head, and other nodes are attracted to this node within a close distance. Figure 3 below shows the deployment of sensor nodes in a two-dimensional plane.

The step-by-step process of forming a cluster in UWSNs using firefly-based swarm intelligence technique is given below.

Step 1: Determine the population space of the sensor nodes.
Step 2: Calculate the flash value of the sensor nodes using the factors—average energy, mobility and distance from the base station.
Step 3: The flash value is displayed to all the sensor nodes.
Step 4: The nodes which have optimum flash value will attract other nodes around and form a cluster.

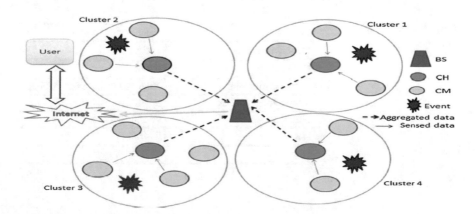

Fig. 3 Clusters formed in UWSN using FSO

Step 5:   The process is repeated until all nodes are into a cluster.
Step 6:   The moment the cluster formed is found to be stable and close to the base station, the sensed data are sent to the base station.
Step 7:   Likewise in future iterations, the data are sent once the cluster formed is stable.

The pseudocode of the technique is

```
INITIALIZE all the sensor nodes with equal e, m and d.
INITIALIZE the position of sensor nodes p
CALCULATE the flash value F of the nodes
FOR all nodes (where F~2.23)
          INFORM neighbouring nodes as HEAD
          nodes other than HEAD join one or more clusters
REPEAT
          Nodes ∈ ((F~2.23) AND (P_w > P_5))
          REMOVE from respective clusters
UNTILL stable cluster is formed
SEND data once a stable cluster is formed
```

P denotes the alpha position of the nodes. (it can keep track of up to ten positions).

5   Experimental Set-up

The algorithm is tested in MATLAB simulation tool. The method is tested with the following factors in Table 1.

Table 1 Simulation parameters

Parameters	Value
Deployment area	120 * 120
Base energy	0.49 J
Server station node	49 m * 49 m and 98 m * 98 m
Transmitter/receiver	50 nJ/bit
No. of nodes	99 and 299
$\varepsilon f\, s$	9 pJ/bit/m
εmp	0.0014 pJ/bit/m

6   Results and Comparative Analysis

This section presents the test results of the approach and has been compared with the different existing methods to prove the superiority of our method. Figures 4, 5 and 6 below display the initial placement of sensor nodes in the network in the water body.

The proposed clustering model was compared with various techniques like FAMACRO, distributed energy-efficient adaptive clustering protocol (DEACP) and multi-objective weighted clustering algorithm (IMOWCA) in Table 2 and Fig. 7.

Fig. 4 Initial placement of sensor nodes

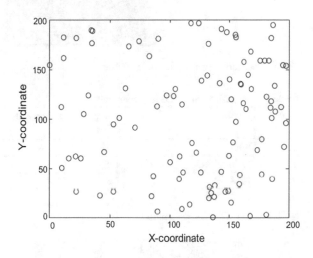

Fig. 5 Clustered groups in network

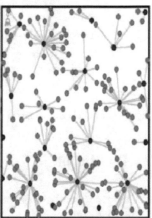

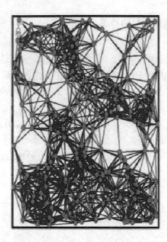

Fig. 6 Connectivity among nodes in cluster

Table 2 Comparison of clustering techniques

S. no.	Model	Energy Exhaustion Rate (EER)	Stability
1.	FSO	0.66	99.31
2.	FAMACRO	1.62	97.64
3.	DEACP	1.81	98.23
4.	IMOWCA	2.13	99.07

*EER in terms of joules per second

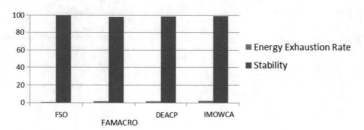

Fig. 7 Graphical representation of Table 2

7 Conclusion

In this paper, we have proposed a firefly swarm intelligence-based clustering technique for the underwater wireless sensor networks (UWSNs). The results of the proposed technique were tested and compared with other approaches, and the method yielded good results in terms of stability and energy exhaustion rate. In future, more stable techniques can be used for optimization.

References

1. Salehian, S., Subraminiam, S.K.: Unequal clustering by improved particle swarm optimization (IPSO) in wireless sensor network. In: The 2015 International Conference on Soft Computing and Software Engineering (SCSE 2015). Procedia Comput. Sci. **62**, 403–409
2. Ouchitachen, H., Hair, A., Idrissi, N.: Improved multi-objective weighted clustering algorithm in wireless sensor network. Egypt. Inform. J. http://dx.doi.org/10.1016/j.eij.2016.06.001
3. Amine, D., Nasr-Eddine, B., Abdelhamid, L.: A distributed and safe weighted clustering algorithm for mobile wireless sensor networks. Procedia Comput. Sci. **52**, 641–646 (2015)
4. Gupta, V., Pandey, R.: An improved energy aware distributed unequal clustering protocol for heterogeneous wireless sensor networks. Eng. Sci. Technol. Int. J. **19**, 1050–1058 (2016)
5. Perrig, A., Szewczyk, R., Wen, V., Culler, D.E., Tygar, J.D.: SPINS: security protocols for sensor networks. Wirel. Netw. **8**(5), 521–534 (2002)
6. Hill, J., Szewczyk, R., Woo, A., Hollar, S., Culler, D.E., Pister, K.: System architecture directions for networked sensors. In: Proceedings of the 9th International Conference on Architectural Support for Programming Languages and Operating Systems. ACM Press, New York, pp. 93–104 (2000)
7. Carman, D.W., Krus, P.S., Matt, B.J.: Constraints and approaches for distributed sensor network security. Technical Report 00-010, NAI Labs, Network Associates Inc., Glenwood, MD (2000)
8. Mahajan, S., Malhotra, J., Sharma, S.: An energy balanced QoS based cluster head selection strategy for WSN. Egypt. Inform. J. **15**, 189–199 (2014)
9. Viswa Bharathy, A.M., Basha, A.M.: A multi-class classification MCLP model with particle swarm optimization for network intrusion detection. In: Sadhana: Academy Proceedings in Engineering Science, vol. 42, no. 5, pp. 631–640 (2017)
10. Viswa Bharathy, A.M., Basha, A.M.: A hybrid intrusion detection system cascading support vector machine and fuzzy logic. World Appl. Sci. J. **35**(1), 104–109 (2016)
11. Viswa Bharathy, A.M., Basha, A.M.: A hybrid network intrusion detection technique using variable multiplicative K-means with self-organising PSO. Middle East J. Sci. Res. **24**(12), 3812–3819 (2016)
12. Kingsly, S.R., Viswa Bharathy, A.M.: Secure neighbor discovery scheme for dynamic clustering in manet. Int. J. Sci. Eng. Res. **3**(4) (2015)
13. Sen, J.: A survey on wireless sensor network security. Int. J. Commun. Netw. Inf. Secur. (IJCNIS) **1**(2), 55–78 (2009)
14. Gowrishankar, S., Basavaraju, T.G., Manjaiah, D.H., Sarkar, S.K.: Issues in wireless sensor networks. In: Proceedings of the World Congress on Engineering 2008, vol I
15. Bakr, B.A., Lilien, L.T.: Extending lifetime of wireless sensor networks by management of spare nodes. Procedia Comput. Sci. **34**, 493–498 (2014)
16. Sengupta, D., Roy, A.: A literature survey of topology control and its related issues in wireless sensor networks. Int. J. Inf. Technol. Comput. Sci. **10**, 19–27 (2014)
17. Taherian, M., Karimi, H., Kashkooli, A.M., Esfahanimehr, A., Jafta, T., Jafarabad, M.: The design of an optimal and secure routing model in wireless sensor networks by using PSO algorithm. Procedia Comput. Sci. **73**, 468–473 (2015)
18. Gherbi, C., Aliouat, Z., Benmohammed, M.: A load-balancing and self-adaptation clustering for lifetime prolonging in large scale wireless sensor networks. Procedia Comput. Sci. **73**, 66–75 (2015)
19. Krishna, K.H., Babu, Y.S., Kumar, T.: Wireless sensor network topology control using clustering. Procedia Comput. Sci. **79**, 893–902 (2016)
20. Fan, Z., Jin, Z.: A multi-weight based clustering algorithm for wireless sensor network. PRZEGLĄD ELEKTROTECHNICZNY (Electrical Review). ISSN 0033-2097, R. 88 NR 1b/2012, 19–21
21. Abo-Zahhad, M., Ahmed, S.M., Sabor, N., Sasaki, S.: A new energy-efficient adaptive clustering protocol based on genetic algorithm for improving the lifetime and the stable period of wireless sensor networks. Int. J. Energy Inf. Commun. **5**(3), 47–72 (2014)

22. Gajjar, S., Sarkar, M., Dasgupta, K.: FAMACRO: fuzzy and ant colony optimization based MAC/routing cross-layer protocol for wireless sensor networks. Procedia Comput. Sci. **46**, 1014–1021 (2015)
23. Ren, Z., Chen, Y., Yao, Y., Li, Q.: Energy-efficient ring-based multi-hop clustering routing for WSNs. In: IEEE Fifth International Symposium on Computational Intelligence and Design, pp. 14–17 (2012)
24. Krishna, H., Babu, Y.S., Kumar, T.: Wireless sensor network topology control using clustering. Procedia Comput. Sci. **79**, 893–902 (2016)
25. Amine, D., Nassreddine, B., Bouabdellah, K.: Energy efficient and safe weighted clustering algorithm for mobile wireless sensor networks. In: The 9th International Conference on Future Networks and Communications (FNC 2014). Procedia Comput. Sci. **34**, 63–70

Using Hierarchies in Online Social Networks to Determine Link Prediction

Ravinder Ahuja, Vipul Singhal and Alisha Banga

Abstract Social hierarchy is one of the most basic concepts in sociology and analysis of networks. In today's world, online social networks are very popular, and thus, there is a need to analyze social hierarchies in the form of complex networks. Social hierarchies can help in improving link prediction, optimize page rank, provide efficient query results, etc. These networks can provide us with important information about social interactions and how to efficiently use it. Social network will be modeled as directed graph, and then, social hierarchies are explored by converting directed graph to directed acyclic graph. By converting to directed acyclic graph (DAG), we can partition the graph into different levels which will divide the network into multiple hierarchies. The topmost level signifies highest social ranking, while bottom ones represent the least ranking. Most algorithms widely used calculate DAG in $O(n^3)$, which is fine for small graphs, but when dealing with large social networks, it is not practically feasible. Thus here, we devise a new method to compute DAG. In this method, we will remove all the cycles from graph G'. In addition to this, we use a two-phase algorithm—compute and refine. The result is used to improve link prediction. The algorithm used two orders faster than the basic algorithm.

Keywords Directed acyclic graph · Eulerian graph · Link prediction
Social hierarchy · Social network

R. Ahuja (✉) · V. Singhal
Jaypee Institute of Information Technology, Noida, India
e-mail: ahujaravinder022@gmail.com

V. Singhal
e-mail: vipulsinghal2903@gmail.com

A. Banga
Satyug Darshan Institute of Engineering and Technology, Faridabad, Haryana, India
e-mail: alishabanga47@gmail.com

© Springer Nature Singapore Pte Ltd. 2019
J. Wang et al. (eds.), *Soft Computing and Signal Processing* ,
Advances in Intelligent Systems and Computing 898,
https://doi.org/10.1007/978-981-13-3393-4_8

67

1   Introduction

Social hierarchy is a pyramid-like structure where the minorities (high social ranking) are at top, while the majorities (low social ranking) are at bottom. This is a
universal feature which is seen everywhere. Social hierarchy is a well-researched
topic in sociology, psychology, etc. It is a common practice seen on various social
networks like Twitter and Instagram. Apart from this, social hierarchy can be used
in link prediction, efficient query results, and complex networks. Social networks
are very dynamic in nature; i.e., they change very quickly. Link prediction [1, 2]
can be used to predict edges (links) which could be added in the near future. This
is also called as the link prediction problem, i.e., at a particular time (t), predicting
which edge will be added between time t and t'. Furthermore, social networks like
Facebook need to predict the content users will like so as to increase their popularity.
The most important use of link prediction in social network is collaboration and
collision prediction. Here, we discuss ways to determine link prediction using social
hierarchies. We focus on some social networks like Google+, Weibo, Twitter, which
can be represented as directed graphs. For a given social network, it can be converted
to a DAG. In DAG, all the nodes in a graph are presented into various levels. It shows
the ranking of a node in the social hierarchy. For a given graph G', there are various
methods to compute a DAG. A random DAG can be formed by contracting the nodes
in a strongly connected component (SCC). But it does not solve the purpose. Furthermore, determining the maximum DAG is NP-hard problem [3], hence not feasible.
We thus compute a maximum Eulerian subgraph, which will give us $G' = \text{ч}(G') \cup$
G'_d, where G'_d is the subgraph with no cycles [4]. Here, we will first compute the
DAG with a simple algorithm which computes in $O(n^2)$. Then, we use a two-phase
algorithm—compute and refine. The compute function uses a greedy approach to
find maximum Eulerian subgraph, and the refine algorithm refines it to give us the
DAG.

2   Literature Survey

Gould [5] developed a model to represent the emergence of social hierarchy which
could precisely predict the complex network structure. Maiya and Berger-Wolf [6]
inferred social hierarchy from weighted social networks on the basis of maximum
likelihood assuming that hierarchy is the primary factor. In paper [7], they model
the advisors–advisees mining problem using temporal networks utilizing a joint
likelihood objective function. Here, we consider an unweighted graph $G' = (V, E)$,
where $V(G')$ is a set of nodes and $E(G')$ represents the set of directed edge (as shown
in Fig. 1). $|V(G')|$ represents the number of nodes, while $|E(G')|$ represents the number
of edges. Let us consider a path $p(v_1, v_2 \ldots v_n)$ as a sequence of edges from v_1 to
v_n, where len (p) denotes the length of the path p. A path is considered as a cycle
if a particular node appears more than once. For any node v_i in graph G', the in

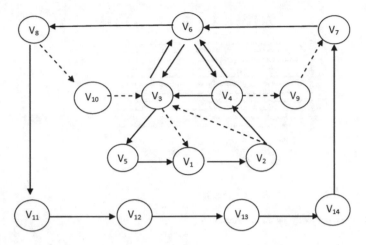

Fig. 1 Reference graph G′

neighbors of v_i (in degree) is denoted as $d_i(v_i)$ and the out neighbors of v_i (out-degree) is denoted as $d_o(v_i)$. The analysis done on some databases is shown below in Table 1. By analyzing the nodes as shown in Fig. 2, G′ of $d_i(u) - d_o(u)$ shows that some datasets show pyramid structure (which have low ratio) while some datasets (like Slashdot) do not show hierarchical pyramid structure as shown in Table 2.

Table 1 Real dataset results

Graph G′	|V|	|E|	|V(ɥ(G′))|	|E(ɥ(G′))|	|V(G′D)|
Wiki-vote	7115	103680	1287	17679	7106
Epinions	75879	508893	33580	265007	67790
Slashdot0902	82168	870151	71811	748511	44600
Pokec	1632803	30622544	1297378	20911978	1562647
G′plus2	84690	2867782	51825	770814	82366
Weibo0	97906	2431511	73582	850154	96704

Table 2 Datasets with corresponding pyramid characteristics

Graph G′	|V|	|V(G′D)|	|V| − |V(G′D)|	((|V| − |V(G′D)|)/|V|)
Wiki-vote	7115	7106	9	0.00126
Epinions	75879	67790	8089	0.1066
Slashdot0902	82168	44600	37568	0.4572
Pokec	1632803	1562647	70156	0.0429
G′plus2	84690	82366	2324	0.0274
Weibo0	97906	96704	1204	0.01229

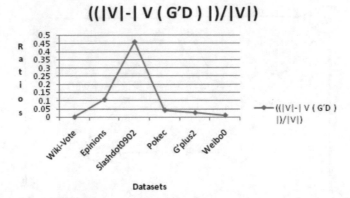

Fig. 2 Datasets showing pyramid structure

3   Proposed Algorithms

The maximum Eulerian subgraph is our final result [8, 9] (denoted by $ч(G')$). G'_D can now be easily computed using $G' = ч(G') \cup G'_d$, and $E(ч(G')) \cup E(g'_d) = \emptyset$. It is to be noted that the maximum Eulerian subgraph is not unique [10]. Let S is a strongly connected component, such that $S = (G'_1, G'_2 \dots g'_n)$.

Algorithm 1: Base (G')
- ➢ Compute SCC (G'), S=(G'₁,G'₂..g'ₙ)
- ➢ For every G'ᵢ in S, do
    - ○ $\bar{υ}(g'_i) <-$ Greedy(g'ᵢ);
    - ○ Move cycles in G'ᵢ - $\bar{υ}(g'_i)$ to $\bar{υ}(g'_i)$
    - ○ $ч(g'_i) <-$ Compute ($\bar{υ}(g'_i)$, g'ᵢ)
- ➢ Finish for

The above algorithm first computes all the SCC, and for each SCC a Eulerian subgraph is computed which is passed to the computer function to calculate maximum Eulerian subgraph.

Algorithm 2: Greedy (G')
- ➢ L' <- 1 ; G'' <-- G';
- ➢ While node u in G'' with lab(u) > 0 do
    - ○ G'' <- ppath(G'',l);
    - ○ L' <- l'+1
- ➢ Finish
- ➢ Ret G''

This function deletes edge from G' to make $d_i(v) = d_o(v)$ for every node. Node label: lab $(u) = d_i(v) - d_o(v)$. If lab (u) is not equal to zero, it needs to delete some nodes to make the above condition true.

Ppath: ppath stands for positive negative path between two nodes. Here, $p = (p_1, p_2 \dots p_n)$ is a ppath if lab(u) $>$ 0, lab(v)$<$0 and all lab(v_i)$= 0$ for i in 1 to l.

For l$=$1, initially greedy algorithm deletes ppath($v2$, $v3$), making lab(v_2)$=$ lab(v_3)$=0$. For l$=$2, ppath(v_4, v_1) is deleted. Finally l$=$5, ppath(v_8, v_7) is deleted, as shown in Fig. 2.

Algorithm 3: ppath (G',l)

- ➤ G'_1 <- l-subg(G',l);
- ➤ Enqueue nodes u in V(G'), with lab(u)>0 into queue q';
- ➤ While q' not in Ø do
  - ○ U <-q'.top();
  - ○ Do DFS in g'_l,
  - ○ Traverse not visited edge,
  - ○ Select them as traversed;
  - ○ Let path v be ppath(u,v), with level(v)=l;
  - ○ If ppath(u,v) not in Ø do
    - ▪ Delete edge in ppath(u,v) from G';
    - ▪ Lab(u) <- lab(u)-1;
    - ▪ Lab(v) <- lab(v)-1;
    - ▪ If lab(u)=0 then q'.dequeue()
  - ○ Else
    - ▪ Q'. dequeue();
  - ○ Finish if
- ➤ Finish loop
- ➤ Ret G'

Ppath algorithm computes subgraphs of a particular length l that can be detected using the lsub algorithm. It means that all edge in E(G') can be in ppath p. Now, ppath deletes ppaths from graph G'. Algorithm returns a subgraph which is passed to the greedy algorithm.

Algorithm 4: lsub (G',l)
  ➢ Consider Every node u do
      ○ Level(u) <- infinity;
  ➢ Finish for
  ➢ Add node s and an edge to every node in G' if lab(u)>0
  ➢ Level(s) <- -1;
  ➢ Level(u) <- level(parent(u))+1,
  ➢ Start BFS from s;
  ➢ Construct G'^{T},
  ➢ $E(G'^{T})$={u,v in E(G')};
  ➢ node t and an edge to be added to every node in G'^{T} if lab(u)<0
  ➢ Rlevel(t) <- -1;
  ➢ Rlevel(u) <- rlevel(parent(u))+1,
  ➢ Start BFS from t;
  ➢ Extract subgraph g'_{l};

The lsub algorithm extracts G'_{l} from G' by calling BFS twice. In first BFS, a node s is added to every node in G' with lab (u)>0. In the BFS, a level is assigned to the level (parent).

Algorithm 5: Compute ($\bar{v}(G')$,G')
  ➢ For each edge (v_i,v_j) in E(G') do
      ○ If(v_i,v_j) in ʯ(G') then
          ▪ Reverse edge in (v_i,v_j) in G',
          ▪ $W(v_i,v_j)$ <- +1;
      ○ Else
          ▪ $W(v_i,v_j)$ <- -1;
      ○ Finish if
  ➢ Finish for
  ➢ Assign dst(u) for node u in V(G') do
  ➢ Relax(u) <- true;
  ➢ Pos(u) <- 0;
  ➢ Enqueue every node u in V(G') in q';
  ➢ U <- q'.top()
  ➢ While q' not in Ø do
      ○ S_V <-Ø;
      ○ S_E <-Ø;
      ○ NV <-Ø;
      ○ If relax(u) == 1 then,
          ▪ Cycle will be reversed
          ▪ Q' = q' U S_V
      ○ Else
          ▪ Q'.pop();
          ▪ U <- q'.top();
      ○ Finish if
  ➢ Finish while

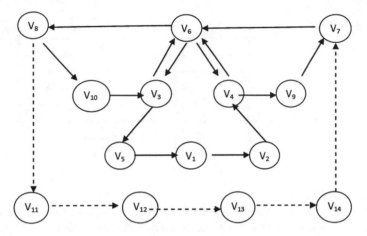

Fig. 3 Graph at l=3

Here, the input is Eulerian subgraph and output is a maximum Eulerian subgraph. In the initial phase, all the edges of $\bar{\upsilon}(G')$ are reversed. In Fig. 3, dst(v_1)$=-2$, dst(v_3)$=-1$, dst(v_7)$=-5$, dst(v_{11})$=-1$, dst(v_{12})$=-2$, dst(v_{13})$=-3$, dst(v_{14})$=-4$, and others have dst(v)$=0$. Find (G', v_1) loop dst(v_5)$=-1$, false is returned, making relax(v_1)$=$relax(v_5)$=$false. Find (G', v_7) finds a negative cycle (v_7, v_9, v_4, v_2, v_3, v_{10}, v_8, v_{11}, v_{12}, v_{13}, v_{14}, v_7). Find (G', v_2) finds a negative cycle (v_2, v_4, v_3, v_2), thus giving us the desired result which is a DAG [4, 11].

Now, we need to form a hierarchy from the DAG. To find the optimum value of our hierarchy, we need to assign a score to each node and lab them so as to form a hierarchy. The input of our new algorithm will be the DAG calculated above.

Algorithm 6: Hierarchy (DAG)
- ➤ Set lab (l_v) <- 0, for each node v in DAG
- ➤ While ∃ edge (u,v) such that l(v) < l(u) – w(u,v) do
  - ○ L(v) <- l(u) – w(u,v)
- ➤ Finish
- ➤ X(u,v) <-0, for edge (u,v) in DAG
- ➤ X(u,v) <- l(u) – l(v) + 1, for edge (u,v) in DAG

This algorithm will label all the nodes thus forming an effective hierarchy.

3.1 Using Hierarchies for Link Prediction

Given a DAG and a hierarchy of the graph, there are various methods for link prediction.

Table 3 Katz output

Path length	$|\text{path} v_4, v_3|$	$|\text{path} v_4, v_2|$	Dampling (βl)
1	1	0	1/2
2	1	0	1/4
3	0	1	1/8
4	2	2	1/16
5	0	0	0
6	1	1	1/32
7	0	1	1/64

Method 1: Shortest Path:
This method works on the fact that there is the shortest distance between two nodes. This estimates the likelihood that a link exists between two nodes.

$$S_{x,y}^{sp} = -d_{x,y}$$

This method can be implemented by running a BFS from each node. Assuming that every node has n neighbors, the complexity of this algorithm is $O\ (V^l\ n^l)$. In Fig. 1, both v_1 and v_7 are at a distance of 2 from v_4, thus both having the same probability.

Method 2: Common Neighbor:
This method is based on common neighbors of various nodes. That is to say, if there are two people x and y, then x has friends a, b, c, while y has friends a, b, c, d. It is to be seen that x and y both have common friends. Thus, it can be concluded that there is a high likelihood that x and y are also friends.

$$S_{x,y}^{cn} = |\Gamma(x) \cup \Gamma(y)|$$

Algorithm:

- Store graph in the adjacency list
- Sort the list of nodes
- Compare the lists

This can be done in $O(n \log n)$. From Fig. 1, v_4 has v_6 and v_3 as neighbors, while v_2 has v_3 as neighbors. It is found that there is an edge between v_2 and v_4, thus proving the hypothesis.

Method 3: Katz:
Assuming that if one short path exists between two nodes, then many paths exist between two nodes indicate a stronger likelihood that a path exists between the two nodes.

Taking our starting node as v_4 and destination nodes as v_3 and v_2, Table 3 has been created.

$$\{v_4, v_3\} = 1/2 * 1 + 1/4 * 1 + 1/8 * 0 + 1/16 * 2 + 1/32 * 1 + 1/64 * 0 = 0.90675$$

$\{v_4, v_2\} = 1/2 * 0 + 1/4 * 0 + 1/8 * 1 + 1/16 * 2 + 1/32 * 1 + 1/64 * 1 = 0.29687$

Here, the formula by which the above is calculated is:

$$S_{x,y}^k = \beta \cdot |path_{x,y}| + \beta^2 \cdot |path_{x,\,y}| + \ldots$$

Here, β is chosen such that it does not affect longer chains.

This method helps to predict links. It can be implemented using DFS and has a time complexity of $O(V^l\, n^l)$.

4 Conclusion

This paper finds how to compute DAG with less time complexity (time complexity of $O(cn^2)$), where c is a very small quantity, thus reducing the time complexity effectively. To find DAG, we use greedy and compute approach. Labeling method is used to find the hierarchy. Then, several techniques are used like shortest distance, nearest neighbor, Katz for link prediction. Shortest distance has a complexity of $O(V^l\, n^l)$, nearest neighbor $O(n \log n)$, and Katz has a complexity of $O(V^l\, n^l)$. The Katz algorithm is the best among all others, as it offers the best complexity and is easily applicable on complex networks. Katz algorithm gives us the probability which tells us the chance that an edge exists between two nodes. This algorithm is very effective for link prediction as we can guess (with high precision) the presence of links between multiple users.

References

1. Zhang, H., Quan, W., Song, J., Jiang, Z., Yu, S.: Link state prediction-based reliable transmission for high-speed railway networks. IEEE Trans. Veh. Technol. **65**(12), 9617–9629 (2016)
2. Nagappan V.K., Elango, P.: Agent-based weighted page ranking algorithm for web content information retrieval, pp. 31–36 (2015)
3. Li, D., Li, D., Mao, J.: On a maximum number of an edge in a spanning Eulerian subgraph. Discr. Math. **274**(1/3), 299–302 (2004)
4. Gupta, M., Shankar, P., Li, J., Muthukrishnan, S., Iftode, L.: Finding hierarchy in directed online social networks. In: Proceedings of 20th International Conference on World Wide Web, pp. 557–566 (2011)
5. Gould, R.V.: The origins of status hierarchies: a formal theory and empirical test. Am. J. Soc. **107**(5), 1143–1178 (2002)
6. Maiya, A.S., Berger-Wolf, T.Y.: Inferring the maximum likelihood hierarchy in social networks. In: Proceedings of International Conference Computer Science Engineering, pp. 245–250 (2009)
7. Zhang, J., Wang, C., Wang, J.: Who proposed the relationship?: recovering the hidden directions of undirected social networks. In: Proceedings of 23rd International Conference on World Wide Web, pp. 807–818 (2014)
8. Catlin, P.A.: Supereulerian graphs: a survey. J. Graph Theory **16**(2), 177–196 (1992)

9. Chen, Z., Lai, H.: Reduction techniques for supereulerian graphs and related topics: a survey. Combinatorics Graph Theory, vol. 1. World Sci. Pub., River Edge, NY, USA (1995)
10. Fleischner, H.: Eulerian Graphs and Related Topics, vol. 1. North-Holland, Amsterdam, The Netherlands (1990)
11. Doreian, P., Batagelj, V., Ferligoj, A.: Symmetric-acyclic decompositions of networks. J. Classif. **17**(1), 3–28 (2000)

Data Analysis, Visualization, and Leak Size Modeling for Water Distribution Network

Shikha Pranesh Gupta and Umesh Kumar Pandey

Abstract Data analysis and visualization are greatly used in ubiquitous computing environments. Contemporary water distribution networks (WDN) have become extremely adaptive, dynamic, heterogeneous, and large scaled. Management of such system is not trivial to fulfill these features, leading to more and more complex management. Along with encompassing state of art and novel techniques for such diversely dynamic system, in this paper two hydraulic parameters namely pressure and flow is measured. Epanet is used for such system's setup. This technique has their own pros and cons which makes them suitable according to the requirements and contextual situations. Detail analysis of hydraulic parameter of WDN is done in this paper, and it can be a benchmark to lead toward most appropriate solution to model leak size. This paper incorporates a practical approach to collect the information from the network setup and after the visualization, analysis and computation of that data, leak size is modeled.

Keywords Water distribution network · Analysis · Visualization · Modeling
Leakage

1 Introduction

Water distribution network is playing very important role across the world to fulfill the basic demand of the human being. For live beings, water is a primary necessity. It is a very large physical network to deal with across the world. Smooth and safe water transmission through the pipeline which will be capable of fulfilling the demand of public is the expectation from this WDN. In most of the countries, this water distribution sector is handled by the government of that country. Most common kind

S. P. Gupta (✉) · U. K. Pandey
MATS University, Raipur, India
e-mail: 09.shikha@gmail.com

U. K. Pandey
e-mail: umesh6326@gmail.com

© Springer Nature Singapore Pte Ltd. 2019
J. Wang et al. (eds.), *Soft Computing and Signal Processing* ,
Advances in Intelligent Systems and Computing 898,
https://doi.org/10.1007/978-981-13-3393-4_9

of failure in the network is the leakage problem in the WDN which degrades the quality of water also. So the government has to take initiative to resolve the problem as soon as possible.

According to Times of India news paper 15 March 2016, Mumbai city is losing 25% of its water due to leakage and theft, despite of multicrore water augmentation projects by BMCs [1]. In Mumbai, growing demand of water is 4,500 million liters daily (MLD) and Mumbai is getting 3,650 MLD of water. Mumbai loses 900 MLD of water due to leakage and theft. In five years, BMC has got 40,000 leakage complaints.

According to UN report, it has been predicted till 2025 almost 3.4 billion people will be living in "water-scarce" country. People of all over the world start suffering water crisis due to drastic change in the environment in terms of pollution and growing population as well as climate change. In future about 2050, so many human conflicts will grow across the globe which will become worse situation for the entire world. "The report, published on the eve of the World Water Day [2], indicated that the Indian subcontinent may face the brunt of the crisis where India would be at the center of this conflict due to its unique geographical position in South Asia" [2].

Flooding during rainy season, water loss due to leakage of pipeline, and less awareness of total ground water resources are the major challenges [3].

Researchers started work in this area long ago, but due to the large physical WDN and it's heterogeneous nature unable to provide efficient and accurate solution to the leakage detection problem for WDN. Initially, the leakage detection is based on the acoustic characteristic of water [4]. Afterward, inverse transient method, transient damping method, inverse resonance method [5], and genetic algorithm [6] were used by the researchers for the purpose of leakage detection. With the technological enhancement, artificial neural networks (ANNs)-based method and neurofuzzy technique were also introduced by the researchers for the monitoring purpose of pipes in WDN [7, 8]. Mashford J. et al. describe a method in which data mining technique is applied for finding the leak size and location with the help of pressure sensor monitored value using SVM [9].

In order to come out from this water crisis, it becomes mandatory for this generation researchers to plan, design, and build suitable water distribution network with proper water supply schemes. In this paper for the leak size modeling, SVM is used. SVM worked on smaller training set, and its generalization characteristic is also better in comparison with ANNs [10]. Leakage detection in water distribution network is a very important domain that can ensure a reliable water supply to the people for which the network has been designed.

2 Modeled Network Description

A water distribution network is designed for the sprinkler network. The hypothetical water distribution network has been designed by considering certain assumptions like elevation, length, and diameter along with pipe properties such as pipe material. A total of 24 nodes and 23 pipes with 1 reservoir and 1 pump are designed and shown in Fig. 1.

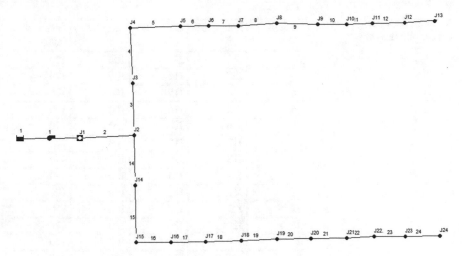

Fig. 1 Sprinkler network with pipes and junction

Table 1 Modeled network properties

Sr. no.	Properties	Value
1	No. of nodes	24
2	No. of pipes	23
3	Average pipe length (m)	100
4	Pipe diameter (mm)	170
5	Max. allowable velocity	Approx. 2 m/s
6	Roughness coefficient	100

The basic purpose of this network is to provide a hypothetical network for the analysis and visualization of leak size modeling within hydraulic constraints of the system. Hydraulic constraints for the modeled network are maximum allowable flow velocity. The network properties are summarized in Table 1.

2.1 Network Design

The network has 20 hydrants in sprinkler network. Hydrants consider node from J4 to J13 and J15 to J24. Each of the hydrants has a base demand of 2.25 l/s. Basic design data for sprinkler network is shown in Table 2.

After running the program with no leakage condition in EPANET, various relations can be found among flow, pressure, demand, velocity, elevation, diameter, length, head, demand, contour, etc. [11]. Here, the main hydraulic relation of interest is flow and pressure.

The simulation is done for 120 trials of various leakage conditions to calculate the pressure and flow of the whole network shown in Fig. 2. As the sprinkler network can

Table 2 Basic design data
for sprinkler network

Link Id	Start node	End node	Length (m)	Diameter (mm)
2	J1	J2	100	170
14	J2	J14	100	170
15	J14	J15	100	170
16	J15	J16	100	170
17	J16	J17	100	170
18	J17	J18	100	170
19	J18	J19	100	170
20	J19	J20	100	170
21	J20	J21	100	170
22	J21	J22	100	170
23	J22	J23	100	170
24	J23	J24	100	170
3	J2	J3	100	170
4	J3	J4	100	170
5	J4	J5	100	170
6	J5	J6	100	170
7	J6	J7	100	170
8	J7	J8	100	170
9	J8	J9	100	170
10	J9	J10	100	170
11	J10	J11	100	170
12	J11	J12	100	170
13	J12	J13	100	170

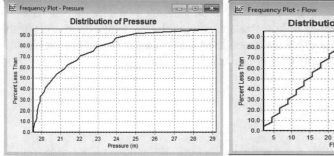
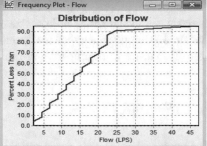

Fig. 2 Pressure and flow frequency distribution of sprinkler network

be seen as combination of two similar structures, leaky node is modeled for upper
structure. The model is studied basically for three various leak sizes, i.e., small,
medium, and large leak.

Table 3 Various test conditions considered for the study

Scenario no.	Explanation
Condition 0	No leakage
Condition 1	Small leak = 3.25 l/s
Condition 2	Medium leak = 4.5 l/s
Condition 3	Large leak = 7.25 l/s
Condition 4	Small leak = 3 l/s
Condition 5	Medium leak = 4.75 l/s
Condition 6	Large leak = 7.5 l/s
Condition 7	Small leak = 2.5 l/s
Condition 8	Medium leak = 5 l/s
Condition 9	Large leak = 8 l/s

2.2 Leakage Modeling for Sprinkler Network

For modeling the leakage, discharge of water is added to various nodes which shows the additional water coming out from the pipe at particular junction. The simulation is done to measure the pressure and flow changes in the network. According to Wu et. al. [12], 5 l/s discharge from the mains is considered as the large size of leak, as well as 1 l/s discharge is considered as small leak size. In this work, 0.1–1 l/s discharge load condition is considered for the small leak, 5 l/s and greater than this is considered for the large size leak and in between 1 and 5 l/s is considered for the medium leak size. Test conditions considered for the study are described in Table 3.

The first experiment carried out for the test condition 0, which is no leak condition. Leakage is considered one by one for all the hydrant nodes. In this scenario, pressure and flow measured at every junction is tabulated and average pressure and flow was calculated. Afterward one by one the experiment was done for all the 9 test conditions and similar to test condition 0, average pressure and flow is calculated. This data is used for the computation of pressure difference and flow difference. In comparison with the test case scenario 0, percentage pressure difference and percentage flow difference are tabulated, which generates the 120 observations.

3 Data Analysis, Visualization, Classification, and Prediction Using Support Vector Machine (SVM)

Vapnik has introduced the term support vector machine, which gets high popularity in the field of machine learning [13]. SVM is a supervised learning approach, and it can be used to solve regression and classification problem. SVM is used for the leakage detection purpose because of its special characteristics which enhance the geometric margin and meanwhile reduces the classification error.

In this experiment, tabulated dataset is used for the classification and prediction of leak size. A total of 100–1000 cases are needed in SVMs [9]. This dataset consists

of 120 observations and 3 variables. In this, two variables, namely Pressure_diff and
Flow_diff are numeric and Leak_size is a factor variable with three levels: small
(S), medium (M), and large (L). Using this numeric variable, a classification model
is built which will correctly predict the leak condition. Scatter plot for leak size is
shown in Fig. 3.

In this, SVM type is c-classification and the response variable is a factor variable,
it is not continuous in nature, so the classification is used. In this, radial basis function
is used as kernel. For the above model, a total of 29 support vectors are required. For
large leak size 9, for medium leak size 12, and for small leak size 8, support vectors
are required as shown in Fig. 4.

In Fig. 4, visualization of data, support vector, and decision boundary have been
shown. Leak size definition used is as per Wu et al. [12]. For the scenario as per
Table 3, there is no misclassification error and the model is correctly predicting the
leak size type.

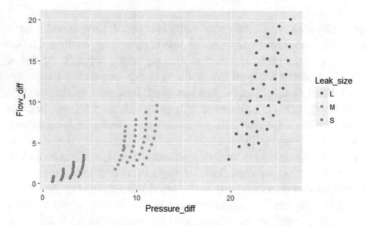

Fig. 3 Scatter plot for the leak size

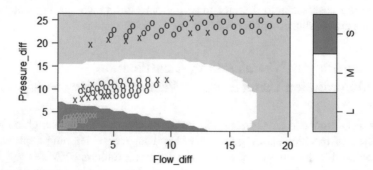

Fig. 4 SVM classification plot

4 Conclusion

Water is the most valuable resource available in the earth. It is a basic necessity of the live beings. Research in this direction can give new dimensions for solving the water crisis problem. This paper incorporates the idea to deal with the hydraulic parameter pressure as well as flow for the leak size detection and modeling. In this paper, radial basis function of SVM is used for the modeling purpose and it is working well for the given scenario. Although there are various researches already done in this area, then also there is a need of fast, efficient, integrated, intelligent, and automatic system for the detection of leakage in pipes. Technological advancement offers lot of research scope for the researchers in this area.

References

1. http://timesofindia.indiatimes.com/city/mumbai/Mumbai-loses-25-water-to-leakage-theft/articleshow/51401216.cms
2. http://timesofindia.indiatimes.com/home/environment/the-good-earth/World-Water-Day-UN-report-predicts-grim-scenario-for-India-experts-pitch-for-making-water-conservation-a-national-obsession/articleshow/32507693.cms
3. http://timesofindia.indiatimes.com/edit-page/India-Water-challenges-and-the-way-forward/articleshow/32488030.cms?
4. Hunaidi, O., Chu, W.T.: Acoustical characteristics of leak signals in plastic water distribution pipes. Appl. Acoust. **58**, 235–254 (1999)
5. Martin, L., Angus, S., John, V., Xiao-Jian, W., Pedro, L.: A Review of Leading-edge Leak Detection Techniques for Water Distribution Systems (2003)
6. Casillas, M.V., Puig, V., Garza-Castanon, L.E., Rosich, A.: Optimal sensor placement for leak location in water distribution networks using genetic algorithms. In: Sensors 2013, pp. 14984–15005 (2013)
7. Caputo, A.C., Pelagagge P.M.: Using neural networks to monitor piping systems. Process Saf. Progr. **22**(2), 119–127(2003)
8. Feng, J., Zhang, H.: Algorithm of pipeline leak detection based on discrete incremental clustering method. In: Huang, D.S., Li, K., Irwin, G.W. (eds.) Computational Intelligence. ICIC 2006. Lecture Notes in Computer Science, vol. 4114. Springer, Berlin, Heidelberg (2006)
9. Mashford, J., De Silva, D., Marney, D., Burn, S.: An Approach Leak Detection in Pipe Networks Using Analysis of Monitored Pressure Values by Support Vector Machine, pp. 534–539. IEEE (2009)
10. Kwok, T.J.: Support vector mixture for classification and regression problems. In: Proceedings of the International Conference on Pattern Recognition (ICPR), Brisbane, Australia, pp. 255–258 (1998)
11. Kumar, A., Kumar, K., Bharanidharan, B., Matial, N., Dey, E., Singh, M., Thakur, V., Sharma, S., Malhotra, N.: Design of water distribution system using EPANET. Int. J. Adv. Res. **3**(9), 789–812 (2015)
12. Wu, Z.Y., Sage, P.: Water loss detection via genetic algorithm optimization-based model calibration. In: ASCE 8th International Symposium on Water Distribution System Analysis, Cincinnati (2006)
13. Vapnik, V.: The Nature of Statistical Learning Theory. Springer, NY (1995)

Context-Sensitive Thresholding Technique Using ABC for Aerial Images

Kirti and Anshu Singla

Abstract Image anatomization is a remarkable notch in image processing which entails the scrutinization of the number of non-overlapping and homogeneous regions that exist in the input image. Thresholding is the most popular algorithm of image segmentation. In this article, the authors have utilized energy curve to incorporate spatial contextual information to inspect the regions where most favourable threshold(s) exist. The thresholding technique automatically computes the count of objects present in input image. To determine the optimal thresholds present in the image, artificial bee colony algorithm has been deployed. The results achieved have been compared with GA-based technique to ensure the efficacy of the proposed technique.

Keywords Artificial bee colony (ABC) · Image segmentation · Optimization
Thresholding

1 Introduction

Image segmentation segregates the input image into different non-overlapping and homogeneous regions to identify the various objects exist in the input image. Generally, image segmentation can be classified as (1) texture, (2) thresholding, (3) clustering and (4) region [1]. Thresholding technique extracts non-overlapping and homogeneous regions from the background. The three major issues thresholding technique need to address are (i) inclusion of context-sensitive information of the input image, (ii) determination of optimal number of non-overlapping regions that exist in the image and (iii) optimal thresholds to segment the image. Thresholding techniques can further be classified based on two criteria [2]: (i) number of thresholds determined and (ii) spatial information. Generally, the histogram-based threshold-

Kirti (✉) · A. Singla
Department of CSE, CUIET, Chitkara University, Chandigarh, India
e-mail: kirti@chitkara.edu.in

A. Singla
e-mail: asheesingla@gmail.com

© Springer Nature Singapore Pte Ltd. 2019
J. Wang et al. (eds.), *Soft Computing and Signal Processing* ,
Advances in Intelligent Systems and Computing 898,
https://doi.org/10.1007/978-981-13-3393-4_10

85

ing techniques [3–7] are context-insensitive. The complexity of the thresholding techniques is directly proportional to the number of thresholds required to segment image. To consider spatial contextual information, thresholding techniques [8, 9] were developed where the computational burden increased significantly.

In literature, multilevel thresholding techniques employed nature-inspired algorithms to mitigate the complexity of algorithm with increase in number of thresholds required to segment the image. Different nature-inspired algorithms such as GA, PSO, CS and WDO have been utilized to develop thresholding techniques [10–16]. The aforementioned state-of-the-art thresholding techniques are efficient but cannot ascertain the count of objects that exist in the input image automatically. The authors in [17] developed a context-sensitive thresholding technique which automatically establishes the count of objects that exist in the input image using GA.

In this paper, an automatic context-sensitive thresholding technique is designed to determine optimal thresholds for aerial images using an algorithm on the basis of foraging behaviour named artificial bee colony (ABC) optimization algorithm. The rest of the paper is collocated as follows: a concise overview of ABC and genetic algorithm (GA) is presented in Sects. 2 and 3, respectively. The proposed ABC-based thresholding technique is elucidated in Sect. 4. Section 5 elucidates the comprehensive explanation of the analytical settings and the anatomization of results attained on the contemplated dataset of aerial images. Finally, Sect. 6 depicts the epilogue of the proposed technique and the diverse opportunities hereafter in upcoming time.

2 Artificial Bee Colony (ABC)

The authors in [18] propounded a survey of different model on the basis of foraging behaviour of a foragers (bees) colony called ABC algorithm. In ABC algorithm, the forager's population comprises of three phases: (a) employed foragers, (b) onlooker foragers and (c) scout foragers. Employed foragers hunt for the food sources surrounding the colony and apportion the collected information about the food origins with onlooker foragers. Onlooker foragers decide which food sources are to be kept. Scout foragers are those employed foragers that abandon their food origins and hunt for new foragers. Each food origin is associated with one employed forager. ABC induces a randomly dispensed inceptive set of nominee solution P of SN solutions for cycle 1. Here, solutions symbolize the position of each food origin in the population. SN symbolizes count of employed or onlooker foragers. $x \in \{1, 2, ..., SN\}$ is a D-dimensional vector that symbolizes the position of each food source. D symbolizes the count of optimization variables. The position of each solution once initialized changes after each cycle $C \in \{1, 2, ..., MCN\}$. MCN represents the maximum cycle number. Onlooker bees choose which foods are to be kept using Eq. 1:

$$p_i = \frac{fit_i}{\sum_{n=1}^{SN} fit_n}. \tag{1}$$

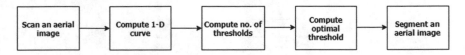

Fig. 1 Block diagram of the proposed technique

where fit_i symbolizes richness of solution of the food origin at position i, which is equivalent to nectar quantity of food origins at position i. The new position of the food origins can be enumerated using Eq. 2:

$$v_{ij} = x_j + \phi_{ij}(x_{ij} - x_{kj}), k \in \{1 \ldots SN\}, \; j \in \{1 \; \ldots \; D\} \tag{2}$$

3 Genetic Algorithm (GA)

Genetic algorithm was propounded by John Holland and his students at the University of Michigan [19]. GA is encouraged by the genetic process of natural selection from set of solutions known as population. The strategy of genetic algorithm mimics the Darwinism approach of population of structures subject to the fierce forces narrated in "survival of the fittest" theory. It is based on the notion that rich chromosomes have proportionately more chance to survive. GA takes set of solutions as input and produces the suitable solution as output by using the fitness function. To generate new offsprings, GA practises three different operators: selection operator, crossover operator and mutation operators.

4 Proposed Thresholding Technique

In this paper, the authors concentrated on context-sensitive thresholding technique based on meta-heuristic ABC algorithm. The various existing non-overlapping and homogeneous regions that exist in the input aerial image are detected automatically. The major steps are explicated in Fig. 1.

1. Scan the input aerial image say, AI of dimension $M * N$ where $AI = \{l_{ij}\}$, $1 \leq i \leq M$, $1 \leq j \leq N\} \cdot l_{ij}$ is the gray value of input aerial image AI at picture element (i, j) and can procure value 0 and L. L delineates the extreme gray value of aerial input image AI.
2. Compute the energy for each gray level of input aerial image AI employing energy function propounded by [18]. The energy of input aerial image AI at gray value l $(0 \leq l \leq L)$ is calculated using Eq. 3.

$$E_l = -\sum_{i=1}^{M} \sum_{j=1}^{N} \sum_{pq \in S_{ij}^2} b_{ij} \cdot b_{pq} + \sum_{pq \in S_{ij}^2} C_{ij} \cdot C_{pq}. \qquad (3)$$

where $b_{ij} = 1$ if $l_{ij} > l$; else $bij = -1$ and $cij = 1 \; \forall(i,j)$. S_{ij}^2 symbolizes the second-order neighbour for the picture element at position (i,j) where $1 \le i \le M$ and $1 \le j \le N$.

3. The energy curve procured comprises of apexes and valleys. In each region of two successive apexes, there may exist a potential threshold. The centre-most loci of line uniting the successive apexes are contemplated as inceptive contender thresholds. Suppose there exist n initial probable thresholds $\{th_1, th_2, ..., th_n\}$, then there exist preliminary $n+1$ overlapping regions of the input aerial image AI.

4. Compute the DB-index [20] of the initial clumps of the input aerial image AI. DB-index is interpreted in terms of inter- and intra-scatterance of objects which exist in image. Suppose $\omega_1, \omega_2 ... \omega_k$ are the k objects explicated using thresholds $t_1 < t_2 < t_3 < \cdots < t_{k-1}$. Then, the DB-index is interpreted as Eq. 4:

$$R_{ij} = \frac{\sigma_i^2 + \sigma_j^2}{d_{ij}^2}$$

$$R_i = \max_{\substack{j=1...k \\ i \ne j}} \{Rij\}. \qquad (4)$$

$$DB = \frac{1}{k} \sum_{i=1}^{k} R_i$$

where σ_i^2 and σ_j^2 symbolize variances of ω_i and ω_j objects, respectively, d_{ij}^2 symbolizes distance of the centres of objects ω_i and ω_j. If DB-index value is low, better segmentation results will be achieved because less dispersion and large distance between the object leads to compact values of R_{ij}.

Inceptively, the count of regions procured may be more than the optimal count of regions that exist in the input aerial image. Repetitively, clumps are merged one after another separately till the count of clumps are minimized to 2 by abolishing $t_1, t_2, ..., t_n$ thresholds one by one. Corresponding to each amalgamation of clumps, DB-index value is computed. This abolishment of threshold is a two-phase sequential procedure and thus very fruitful. In the first phase, each threshold is deleted one after another from the set of threshold which exists. The step of deletion is iterated till only one threshold is procured, i.e. TH_1. In the second phase, the set for which the DB-index is least gets selected.

5. Let the preliminary thresholds exist in set TH_k as preliminary acquired in step 3 is preliminary best position to be supplied to the ABC algorithm to spot the nearmost favourable threshold values for segmentation of image. The modes of energy curve in which preliminary threshold(s) exist are considered as the region to procreate inceptive positions of all foragers in the colony.

| (a) Image1 | (b) Image2 | (c) Image3 | (d) Image4 |

Fig. 2 Original aerial image dataset

6. Segment the input aerial image to analogize the results on the basis of accuracy and quality. Let $\{t_1, t_2, \ldots t_k\}$ be the most favourable or optimal thresholds procured in step 4. Let the original aerial image be $AI(i, j)$, and the segmented aerial image $Seg\_AI(i, j)$ for k thresholds will be calculated using Eq. 5:

$$Seg\_AI(i, j) = \begin{cases} 0, & 0 \leq AI(i, j) \leq t_1 \\ (l-1) \times \left(\frac{256}{k} - 1\right), & t_{l-1} \leq AI(i, j) \leq t_l \forall 2 \leq l \leq k-1. \\ 255, & t_k \leq AI(i, j) \leq 255 \end{cases} \quad (5)$$

5 Implementation

5.1 Experimental Elucidation

The dataset of four aerial images as shown in Fig. 2 has been contemplated from http://sipi.usc.edu/database/database.php?volume=aerials and https://pxhere.com/en/tag/97650. To evaluate the efficacy of the proposed technique, the images with different number of thresholds have been included in experimental results.

Both the ABC- and GA-based thresholding techniques have been deployed in MATLAB (R2017a). The parameters used for both the proposed ABC-based thresholding technique as well as GA-based thresholding technique are shown in Table 1.

5.2 Experimental Results

The energy curve of the respective images is shown in Fig. 3. It can be concluded from the energy curves of all the images that it consists of apexes and valleys which distinguish the different non-overlapping consistent regions that exist in the input aerial image.

Table 1 Parameters for both ABC- and GA-based thresholding technique

Algorithm	Parameter	Value
ABC	Population	100
	Abandonment limit	0.6
	Acceleration coefficient	1
GA	Population	20
	Crossover rate	0.8
	Mutation probability	0.01

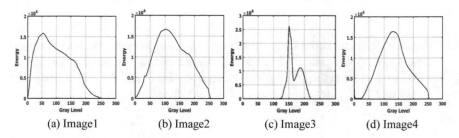

(a) Image1   (b) Image2   (c) Image3   (d) Image4

Fig. 3 Energy curves of the aerial image dataset

Table 2 Optimal thresholds, DB-index and standard deviation obtained using ABC and genetic algorithms for different aerial images

Image	Using ABC			Using GA		
	Threshold	DB-Index	Standard deviation	Threshold	DB-Index	Standard deviation
Image 1	125	0.1079	1.9849e−05	125	0.1079	2.8141e−05
Image 2	150	0.1664	2.2959e−05	150	0.1664	1.9527e−16
Image 3	112,184	0.0785	2.8362e−04	109,185	0.0769	1.7088e−04
Image 4	55,221	0.1170	4.1093e−04	58,220	0.1169	1.2877e−04

Along with the DB-index, the standard deviation has been computed to ensure the consistency of results. Table 2 depicts the most favourable threshold(s) and respective DB-index and standard deviation procured using the proposed ABC-based and existing GA-based thresholding techniques.

It has been contemplated that the DB-index cluster validity measure is very less for the considered image dataset for both the proposed and GA-based techniques. To evaluate the efficacy of propounded technique, comparative analysis of standard deviation has been performed upon the achieved results (DB-index of input aerial image) accompanying the GA-based context-sensitive thresholding-based technique. For qualitative analysis, Figs. 4 and 5 depict the segmented images for different aerial images using ABC algorithm and GA, respectively.

To extenuate the performance of the technique, simulations are accomplished on large dataset of aerial images. Figure 6 showcases the convergence graph of aerial

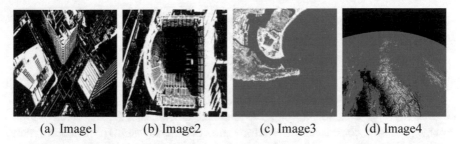

(a) Image1   (b) Image2   (c) Image3   (d) Image4

Fig. 4 Segmented aerial images using ABC algorithm

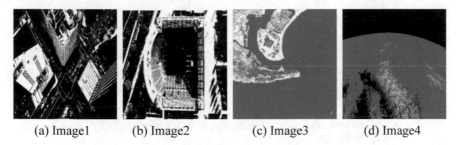

(a) Image1   (b) Image2   (c) Image3   (d) Image4

Fig. 5 Segmented aerial images using GA algorithm

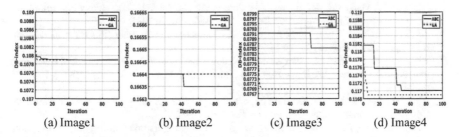

(a) Image1   (b) Image2   (c) Image3   (d) Image4

Fig. 6 Convergence graphs of aerial images using ABC and GA algorithms

images present in the dataset using ABC and GA, respectively. It can be monitored from the plots that fitness value (DB-index) converges and proves consistent results in both the proposed and existing GA-based techniques.

6 Conclusion

ABC has been a prominent strategy employed to rectify optimization problems because of its outrageous performance. ABC relies on foraging behaviour of foragers in lieu of mutation or crossover. This paper practises a novel approach collaborating context-sensitive thresholding-based technique with ABC to encounter most favourable thresholds to segment input aerial image. The chief aspects of this paper are outlined below:

(i) The technique includes spatial contextual information.
(ii) Determines the count of objects present in the input aerial image automatically.

However, in future the researcher may explore how: (i) to scrutinize the modification of ABC optimization algorithm to refine the outcomes of segmentation of input image procuring context-sensitive information of the image and (ii) to achieve more optimal thresholds, one can examine varieties of different nature-inspired algorithms.

References

1. Pal, N.R., Pal, S.K.: A review on image segmentation techniques. Pattern Recognit. **26**(9), 1277–1294 (1993)
2. Sezgin, M., Sankur, B.: Survey over image thresholding techniques and quantitative performance evaluation. J. Electron. Imaging **13**(1), 146–165 (2004)
3. Otsu, N.: A threshold selection method from gray level histograms. IEEE Trans. Syst. Man Cybern. **9**, 62–66 (1979)
4. Chang, C.C., Wang, L.L.: A fast multilevel thresholding method based on lowpass and highpass filter. Pattern Recognit. Lett. 1469–1478 (1997)
5. Akay, B.: A study on particle swarm optimization and artificial bee colony algorithms for multilevel thresholding. Appl. Soft Comput. **13**, 3066–3091 (2013)
6. Ali, M., Ahn, C.W., Pant, M.: Multi-level image thresholding by synergetic differential evolution. Appl. Soft Comput. **17**, 1–11 (2014)
7. Hammouche, K., Diaf, M., Siarry, P.: A multilevel automatic thresholding method based on a genetic algorithm for a fast image segmentation. Comput. Vis. Image Underst. **109**(2), 163–175 (2008)
8. Abutaleb, A.S.: Automatic thresholding of gray-level pictures using two-dimensional entropy, Comput. Vis. Graph. Image Process. **47**(1), 22–32 (1989)
9. Xiao, Y.Y., Cao, Z., Zhong, S.: New entropic thresholding approach using gray-level spatial correlation histogram. Opt. Eng. **49**(12), 127007 (2010)
10. Ghamisi, P., Couceiro, M.S., Benediktsson, J.N.A., Ferreira, N.M.: An efficient method for segmentation of images based on fractional calculus and natural selection. Expert Syst. Appl. **39**(16), 12407–12417 (2012)
11. Bhandari, A.K., Singh, V.K., Kumar, A., Singh, G.K.: Cuckoo search algorithm and wind driven optimization based study of satellite image segmentation for multilevel thresholding using kapurs entropy. Expert Syst. Appl. **41**(7), 3538–3560 (2014)
12. Bhandari, A.K., Kumar, A., Singh, G.K.: Modified artificial bee colony based computationally efficient multilevel thresholding for satellite image segmentation using kapurs, otsu and tsallis functions. Expert Syst. Appl. **42**(3), 1573–1601 (2015)
13. Zhong, F., Li, H., Zhong, S.: A modified abc algorithm based on improved-global-best-guided approach and adaptive-limit strategy for global optimization. Appl. Soft Comput. **46**, 469–486 (2016)
14. Sun, H., Wang, K., Zhao, J., Yu, X.: Artificial bee colony algorithm with improved special centre. Int. J. Comput. Sci. Math. **7**(6), 548–553 (2016)
15. Karaboga, D., Kaya, E.: An adaptive and hybrid artificial bee colony algorithm (aabc) for an s training. Appl. Soft Comput. **49**, 423–436 (2016)
16. Sahoo, G., et al.: A two-step arti cial bee colony algorithm for clustering. Neural Comput. Appl. **28**(3), 537–551 (2017)

17. Singla, A., Patra, S.: A fast automatic optimal threshold selection technique for image segmentation. Signal Image Video Process. **11**(2), 243–250 (2017)
18. Karaboga, D., Gorkemli, B., Ozturk, C., Karaboga, N.: A comprehensive survey: artificial bee colony (abc) algorithm and applications. Artif. Intell. Rev. **42**(1), 21–57 (2014)
19. Goldberg, D., Holland, J.H.: Genetic Algorithms in Search, Optimization, and Machine Learning (1989)
20. Davis, D.L., Bouldin, D.W.: A cluster separation measure. IEEE Trans. Pattern Anal. Mach. Intell. PAMI-**1**(2), 224–227 (1979)

Vision-Based Inceptive Integration for Robotic Control

Asif Khan, Jian-Ping Li, Asad Malik and M. Yusuf Khan

Abstract Robots are used frequently nowadays to reduce the human effort due to its efficient capability and performance. However, interaction with human-friendly environment is needed to integrate the robot accurately. Recently, robotic vision plays an important role in the real-world phenomenon to achieve this goal. This is because robotic vision has the capability to identify and determine the accurate positions of all related objects within the working area of the robot. In robotic vision, images are used as input, which analyzes the content of images. These have been used for the output of the system based on the criteria of image analysis, image transformation, and image understanding. Hence, the system may be able to capture the related information using the motion of the objects. And it updated this information for verification, tracking, acquisition, and extractions of images to adopt in the database. The main objective of this work is to enable the algorithm to understand the visual world. In order to achieve the goal, we have proposed an algorithm which is based on computer vision theories. Experiments have been performed on real-world image which shows our algorithm has better performance.

Keywords Computer vision · Objects grasping
Manipulation of human–robot interaction

A. Khan (✉) · J.-P. Li
UESTC, Chengdu, China
e-mail: asifkhan@uestc.edu.cn

J.-P. Li
e-mail: jpli2222@uestc.edu.cn

A. Malik · M. Yusuf Khan
IITR, Roorkee, India
e-mail: yusufkhan.iitr2010@gmail.com

© Springer Nature Singapore Pte Ltd. 2019
J. Wang et al. (eds.), *Soft Computing and Signal Processing* ,
Advances in Intelligent Systems and Computing 898,
https://doi.org/10.1007/978-981-13-3393-4_11

1 Introduction

In the era of robotic technology, vision-based information system has been applied in the last couple of years. The application of robotics can be found from our home to big industries, e.g., vacuum cleaner [1, 2], self-driving vehicles, and industrial robots. Similar to human processing system, for eyes cameras are used and for brain processors are used.

Nowadays, some of the ideas can be underlined such as visual perception, exception handling, video monitoring, smart yearly warning alert, efficient image storage, and many other fields as shown in Fig. 1. By this perceive the complex situation is easily detected by human and accordingly can take the action to get the location, similarly type of the target object identify correctly. But visual understanding is a complex task for the machine.

The delivery robot works in the hospital and operates in the presence of people. Also, people travel on robotic car and some robots can operate over people by robotic surgical device. However, camera is a static device that catches the patterns of energy, reflected from the visual scene [3]. On the basis of self-experience, eyes are a very much effective sensor for navigation, recognition, preventing obstacles, and manipulation. Therefore, researchers of robotics are keenly interested in vision. One of the important limitations of a camera is that to infer 3D structures from the 2D images. Stereo vision is an alternate approach, where one may use more than one camera to calculate the 3D structure of the world. Suppose you drive a car where front window is covered over, just look at the GPS system [4, 5]. If there were only roads, you can probably drive successfully although slowly from A to B. But if

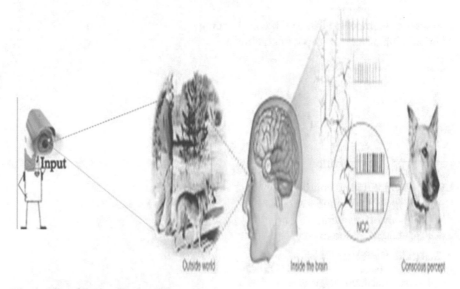

Fig. 1 Idea of the proposed model

there were other objects like cars, foot walkers, traffic signals, then you would be in trouble. To deal with this situation, you need to look outside to sense the real-world objects and take your actions accordingly. Humans can deal this situation very easily, but it is very difficult for machines to do the same. To represent the orientation and position of objects in real environment is a basic requirement in robotics and computer vision. Basic feature-based alignment techniques are used in camera calibration and geometric alignment. This problem can be overcome either using linear or nonlinear least square method. It depends upon the motion, robust regression, and uncertainty weighting. These are necessary to make real-world systems work. As a building block feature-based alignment is used for three-dimensional pose estimation and camera calibration techniques to photograph alignment. The field of computer graphics is still not perfect in limited domains. For example, rendering a motionless scene comprises everyday objects and/or applies animation on destroyed creatures, e.g., dinosaurs.

2 Problem Formulation

Robotics approach to perform the task at particular object in real world is required to understand the world environment, like human being system used to learn and perform as per previous knowledge. To use a robot like "crane" to perform task involuntary mean make robot capable to perform the task by particular environment is a challenging issue. The perceiving the idea from the surrounding by using onboard camera developed computer vision system to detect the autonomous object/target. Above vision-based control requires the inceptive integration of equipment and algorithms' best support [6].

In this paper, we add

1. Essential, generally, measured capacity skill of computer vision.
2. Range obtained and sensor-based control of mechanical robotics.
3. Tightly incorporated, synchronized, and controlled in the administration of some vehicle/eye control.

At last, however, all segments with eye cycle must interface in an exact way to accomplish powerful robot eye coordination as in Fig. 2.

A. System Configuration

Following, motion and control essential for a settled verbalized robot like crane with a portable device like mobile sensor to arrange the objects accordingly with actuator arms and onboard camera, with visible device that able to position controllable world stage So these genuine exploratory methodologies consolidate with an algorithmic coding of MATLAB, and that signal activity executed as control and process.

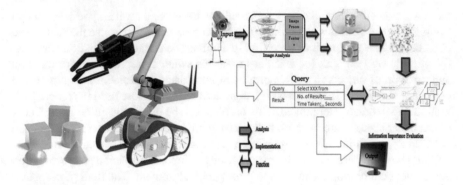

Fig. 2 Environmental learning by crane with camera

3 Visual Tracking

Visual tracking for a robot has been an active research work of robotics vision with respect to computer vision where different interfaces are used to interact with the outside environment. The experimental study which is based on the design of a robotic operation is shown in Fig. 3 to track object by using visual information. Our basic experiment has been done, by assuming the 2D plane objects which is shown in Fig. 4. A typical approach form—input/output relationship—is as follows

$$f_0(x) = \begin{cases} A, \, for \, f_i(x, y) \le t \\ B, \, for \, f_i(x, y) > t \end{cases} \tag{1}$$

where t is the threshold value, and $f1$ and $f0$ are input and output image functions, respectively (1). It is simple to implement apparent contrast target and its background. Control design is a process of manipulating robot toward the tracked target using the position of target in camera input process. The Webcam included in the system makes an image captured to be processed. During this process, the camera feedback is constantly checked to see the template matching for each frame of object extraction. When the sum squared error between two objects is less than the captured image and Canny results in predetermined threshold, then we can say that the object is found—the one we were looking for. The described process can find multiple objects in a single image illumination [7, 8].

A. Edge Detection in Color Image

Applying the edge detection algorithm to extract the edge of the image effectively, the input color image can be transformed into a grayscale image, then combining independent component analysis de noise smoothly for the image Figs. 3 and 4. As image with color are bring J when

$$J = R, G, B$$

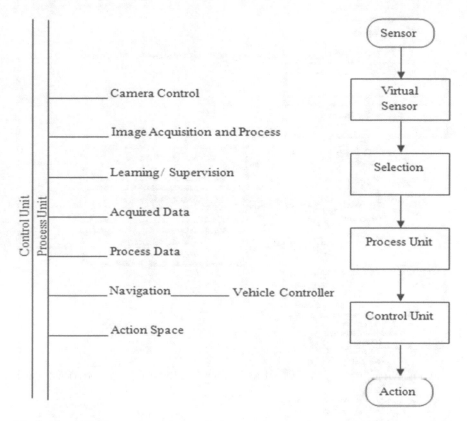

Fig. 3 Visual tracking process and control units

So for each pixel a, b, c, and d are with intensity value over related pixel by

$$\begin{bmatrix} a\,b \\ c\,d \end{bmatrix}$$

Are as

$$E = \sum (|a - d| + |b - c|); \qquad (2)$$

So at level T, binary threshold E for each pixel where if

$$E \geq T \text{ then } E = 1$$
$$\text{Else } E = 0$$

So for each pixel over array calculation

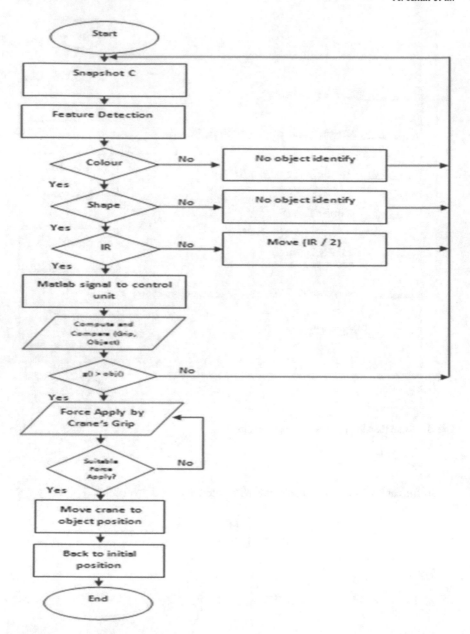

Fig. 4 Geometric calculation and comparison

$$E_R = |a - d| + |b - c|$$
$$E_G = |a - d| + |b - c|$$
$$E_B = |a - d| + |b - c|$$

So binary threshold for—each pixel according to

$$\text{If } (E_R \geq T) \text{ OR } (E_G \geq T) \text{ OR } (E_B \geq T)$$
$$\text{Then } E = 1 \text{ ELSE } E = 0. \tag{3}$$

or more identification of object in as image 3 using a transformation so for each pixel, re-continuing then (2), (3) Image for original value T, for exact substitute are

$$\sqrt{(I_R)^2 + (I_G)^2 + (I_B)^2}$$

For I; when

$$J = R, G, B.$$

B. Range-Based Scene Segmentation

Image segmentation is a technique by which individual pixel is labeled to recognize its association in a connected set of pixels that define a visually distinct image portion as in Fig. 3. When object as in range image, G, R, out of range, background diction object, all of scene. So connected component of binary image, when image array (objects are 0's and background 1's) be given by

$$\begin{bmatrix} a_{11} & & a_{1m} \\ & a_{ij} & \\ a_{N1} & & a_{NM} \end{bmatrix} \text{ where } a_{ij} \leftarrow \begin{Bmatrix} 0 \\ 1 \end{Bmatrix}$$

So proving operation takes a_{ij} into a'_{ij} in forward from left to right end top to bottom, where

$$\Rightarrow \text{ if } a_{ij} = 1 \text{ then } a'_{ij} = 1$$
$$\Rightarrow \text{ if } a_{ij} = 0 \text{ then } a_{i-1j-1} = a'_{1-N}$$
$$\Rightarrow \text{ if } a'_{1,j-1} = 1 \text{ then } a'_{ij} = U_k$$

where U_k are set of encounter red labels. If $a_{ij} = 0$ and each a'_{ij} as 1 or –same U_k smallest with labeled ij. So when more than one U_k is involved, then object encountered situation is more efficient to mark the connection point.

C. Perception, Sensing, and Learning

The robot cannot conclude in advance of interaction from camera, where the object in sign is real or just a visual on the image. So vision object perception is the most interesting for live image matrix with a simple snapshot without image acquisition process. Moving robots camera is need to enough geometrically space.

4 Vision-Based Control

To consider for non celibate situation using pinhole camera taking snapshot condition and after object detection compute to robot ideal range is existing or not. Viewing the some physical points from two different viewpoints allows dispatch triangulation as in Fig. 8.

A. Tracking

So goal to achieve depth map of a word scene as $z = f(x, y)$ where x, y co-ordinates one of the range plan and z is the height above the respective image plane. Now here consider four intrinsic parameters convert from pixel matrix value are S_x, S_y, C_x and C_y So processing technique of images (4) cover all as sequences to recognition the object where the camera operate in RGB format with default RGB is 24-bit color depth and 320×240 pixels. In the current study, the algorithm is such that the robots are controlled, if and only if the target is smaller than the griper of the arm.

B. Motion

Here, motion goal is to navigate the target control to a design location in the image as in Figs. 5 and 6. When TL(), TF() TM(), TBL, TB, TBR all arm base of geometric location of object.

$$IR = \begin{cases} x, if\ x \in IR \cong |MR - (MR - IR)| \\ 0, otherwise \end{cases} \qquad (4)$$

of that robustness achieve more, when the target is hidden like Eq. 4 from the camera for a small movement of tiny.

C. Control

After motion when crane robot achieves the "ideal range" with target object. New image a snapshot by camera captured to extracts size (edge of color based image) object captured and now use algorithm to compare the size of object with gripper range of crane robot. So computing position of both crane robot's gripper (maximum range) and target object T_{obj} as per current environment condition

$$e = T_{obj} - T_{robg}$$

But observation on behalf of computing error based on image of crane robot grips and object where consider as error based image are *obj*

$$e \cong T_{obj} - T_{obg}$$

So as per crane gripper mapping and gripper control unit following MATLAB signal as toward,

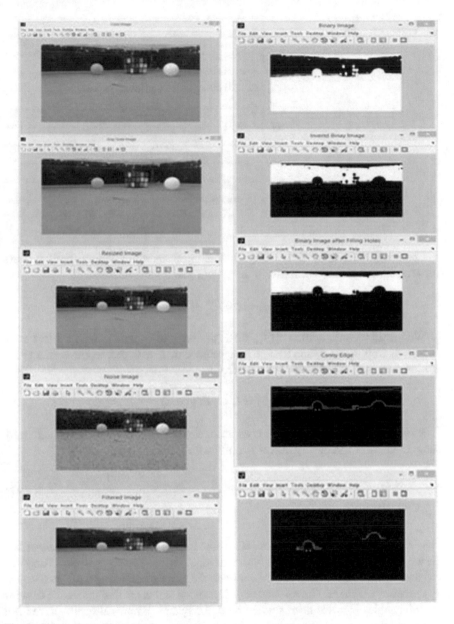

Fig. 5 Object observation output

$$e = \begin{cases} e(obj, obg) = 0 & if\, T(f_1, f_2) \to 0 \\ 0, otherwise \end{cases}$$

Fig. 6 PDF for matching
ratio (xy object motion)

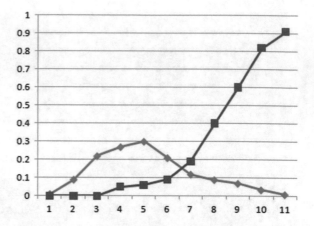

Considering above all equations resultant approach with all algorithmic process allow a suitable position to perform the task (T) so if suitable then

$$e \to 0 \Rightarrow T \to 0$$

With feature configuration $f(f_1, f_2)$ as per setup of camera C with actual observation where, $Y = C(f)$ and though image encoding E with new features after every motion forward as final task done effectively.

$$T(f) = 0 \leftrightarrow E(y) = 0$$

So specific task at target object is experimentally performed (as Figs. 5 and 6) by crane robot more effectively and efficiently with vision-based control [8, 9].

5   Conclusion

The process of extraction, characterization, and interpretation of the information from the world of images is the exact definition of robot vision. The present paper proposes a vision-based control system to manipulate the robot for tracking a targeted object by the use of a camera. The relation between vision and control of the image data for the joint angles has also been studied here. Acceptable empirical results have been obtained using the proposed strategy. This technique provides a simple and efficient approach for vision-based control in comparison with the conventional or traditional one. However, the performance may influence the limitations of the hardware such as computer architecture and image processing required. Speedy image processing and motion estimation may be taken into consideration for the future work.

References

1. Cabre, T.P., Cairol, M.T., Calafell, D.F., Ribes, M.T., Roca, J.P.: Project-Based learning example: controlling an educational robotic arm with computer vision. IEEE RevistaIberoamericana De Technologias Del Apppendiaje **8**(3), 135–141 (2013)
2. Lichtigstein, A., Or-El, R., Nakhmani, A.: Autonomous robot control with dsp and video camera using matlab state flow chart. In: 4th European Education and Research Conference, pp. 170–174 (2010)
3. Rai, N., Rai, B., Rai, P.: Computer vision approach for controlling educational robotic arm based on object properties. In: IEEE 2nd International Conference on Emerging Technology Trends in Electronics, Communication and Networking, pp. 1–9 (2014)
4. Mashali, M., Alqasemi, R., Sarkar, S., Dubey, R.: Design implementation and evaluation of a motion control scheme for mobile platforms with high uncertainties. In: 5th IEEE RAS & EMBS International Conference on Biomedical Robotics and Biomechatronics, pp. 1091–1097 (2014)
5. Ali, H., Seng, T., C., Hoi, L., H., Elshaikh, M.: Development of vision based sensor of smart gripper for industrial applications. In: IEEE 8th International Colloquium on Signal Processing and its Applications, pp. 300–304 (2012)
6. Song, K., T., Chnag, J., M.: Experimental study on robot visual tracking using a neural controller. In: IEEE IECON 22nd International Conference on Industrial Electronics, Control, and Instrumentation, vol. 3, pp. 1850–1855 (1996)
7. Crowley, J.L.: Integration and control of reactive visual processes for visual navigation. In: Proceedings of the Intelligent Vehicles Symposium, pp. 32–38 (1994)
8. Zhang, B., Wang, J., Rossano, G., Martinez, C.: Vision-guided robotic assembly using uncelebrated vision. In: IEEE International Conference on Mmechatronics Automation, pp. 1384–1389 (2011)
9. Saegusa, R., Natale, L., Metta, G., Sandini, G.: Cognitive robotics—active perception of the self and others. In: 4th International Conference on Human System Interactions, pp. 419–426 (2011)

A Study on Convolutional Neural Networks with Active Video Tubelets for Object Detection and Classification

R. Rajkumar and J. Arunnehru

Abstract Convolutional neural networks are a powerful learning model inspired from biological concept of neurons. This deep learning model allows us to replicate the complex neural structure seen in living beings to be applied on data sets and to structure convulsions consisting of several layers. A study on convolutional neural networks have been proven to be an effective class of models for object recognition, taking those results into consideration we intend to apply convolutional neural networks for video classification in two different ways. Generalization of the results obtained by the application of convolutional neural networks on existing data sets for videos, namely Sports 1-M and YouTube object data set (YTO) and their implementation of two distinct CNNs.

Keywords Convolutional neural networks · Deep learning · Object detection
Video Tubelets · Video classification

1 Introduction

Convolutional neural networks [1] as mentioned earlier are proven to be very effective on image recognition this is owed to the following reasons such as how CNN allows the mapping of millions of parameters and being able to accommodate massive data sets. And also the development in image recognition is very high compared to video classification. The benchmark set-up by image classification is very high due to the existing data set sets, namely NORB, MNIST [2], Caltech-101/256 and CIFAR-10/100. Videos are extensively harder to annotate and store. While modelling the

R. Rajkumar · J. Arunnehru (✉)
Department of Computer Science and Engineering, SRM Institute of Science and Technology,
Vadapalani Campus, Chennai, Tamil Nadu, India
e-mail: arunnehru.aucse@gmail.com

© Springer Nature Singapore Pte Ltd. 2019
J. Wang et al. (eds.), *Soft Computing and Signal Processing*,
Advances in Intelligent Systems and Computing 898,
https://doi.org/10.1007/978-981-13-3393-4_12

CNN for video classification from a computational perspective requires extensively long period of training time to effectively optimize the millions of parameters that parameterize the model. This process is for a single image when this is to be done for a video which is nothing but several frames of images, the process gets much more complex. Unlike images the temporal and spatial aspects in video are hard to keep track of and incorporate to obtain high benchmarks, CNN has been applied to various data sets and the ability to be able to implement such complex structures has proven useful especially with object recognition. The detection of objects in an image is easy compared to detection of objects in a video with respect to the change in time. The main aspects to be considered for video classification are (i) existence of large data sets to work with, (ii) object detection and ability to differentiate various objects, (iii) object tracking over a specific time frame.

The standard approach to video classification [3–6] involves three major stages: First, local visual features that describe a region of the video are extracted either densely [7] or at a sparse set of interest points [8, 9]. Next, the features get combined into a fixed-sized video level description. One popular approach is to quantize all features using a learned k-means dictionary and accumulate the visual words over the duration of the video into histograms of varying spatio-temporal positions and extents [10]. The video classification domain is lagging behind compared to image domain. This can be accounted to the fact that there is very less data set to train videos compared to images. CNN has been successful in image domain, and this can be traced back to the fact that image domain has high benchmarks. It is very essential to explore the CNN limitations in video domain. In [11] and [12] uses the CNN by adding the space and time dimensions of the input video and consider these extensions for possible generalization. Convolutional gated restricted Boltzmann machines (CGRBM) [13] and independent subspace analysis (ISA) [14] use unsupervised learning approach for training the spatio-temporal features. In contrast, our models are trained end to end fully supervised [15, 16].

In this work, we propose two distinct way in which CNN is applied on two separate data sets (Sports 1-M and YTO) for video classification. (a) Large-scale video classification with CNN on the data set Sports 1-M and (b) object detection from video Tubelets with CNN on the data sets ImageNet VID and YTO.

2 Large-Scale Video Classification with CNN

Large-scale classification as the name implies is the classification of videos on a very large data set. Unlike images, videos do not have data sets on the scale of millions. CNN methods implemented on image or object classifications have shown some insane increase in performance. So this model will take a video and split it in to individual frames and to these frames the CNN methods applied on images will be used. But obviously we will come across hurdles to overcome since videos keep changing with respect to time. The amount of processing required for a video is far more higher compared to images. The GPU requirements and run-time performances are quite high. In order to speed up the run-time process, we have to create CNN architecture with two processing streams, namely a context stream and fovea stream. This proposed method increases the performance on video classification.

To summarize as follows:

1. We do a study on application of CNN on large-scale video classification.
2. We draw attention to an architecture involving input at two spatial resolutions.
3. We discuss the CNN architecture applied on the video data sets, namely Sports 1-M.

2.1 Time Information Fusion in CNN

Video classification involves three major stages: (1) local visual feature extraction, (2) quantization of the extracted features into a fixed point using means like k-means dictionary and (3) a classifier is applied on the obtained points and is further trained. But when we are to apply CNN and its class of deep learning models, it can replace the above-mentioned stages with a single neural network that can train end to end from raw pixel values to classifier outputs. This neural network consists of layers, convolutions and max pooling.

- Layers—local filters.
- Convolutions—sharing of parameters/attributes.
- Max pooling—internal spatial structure and its connectivity.

Great success has been obtained on applying CNN to image classification, and this is partly due to the availability of large training set and also the lack of large-scale video classification benchmarks limits the development of video classification. Here we are going to discuss the various approaches for fusing information over the dimensions through the neural networks. Different fusion techniques are shown in Fig. 1.

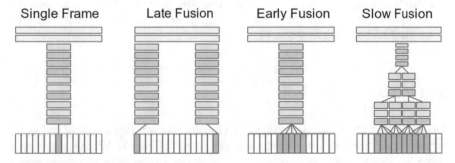

Fig. 1 Different fusion techniques in CNN

Single Frame This indicates a convolutional layer with d filters of spatial size $f \times f$, with input of strides, a fully connected layer with n nodes. So to this layer static input is fed, which is here nothing but a single frame. But this input is of size $170 \times 170 \times 3$ pixels instead of the original size $224 \times 224 \times 3$ pixels.

Early Fusion The single frame in the convolutional layer is further reduced in size of $11 \times 11 \times 3 \times T$ pixels where T is some temporal extent. This reduction in input leads to direct and better connectivity to the pixel data and produces precise results.

Late Fusion This involves two separate single frame networks which are placed 15 frames apart from each other and share parameters which further goes on to merge and form a fully connected layer which is capable of global motion characteristics.

Slow Fusion This fusion model is a slow and balanced approach which is basically the extending the connectivity of all convolutional layers and includes the application of both temporal and spatial convolutions.

2.2 Multi-resolution CNNs

The process of applying CNN on large data sets normally required order of weeks even on the fastest GPUs. The run-time performance is influenced by architecture and parameters. So now there arises a need to improve the models but at the same time retain the performance. The basic way to go about reducing run-time is to reduce the size of the neural network and neurons (parameters) in each layer. But this leads to lack of performance. Hence to overcome such limitations, the multi-resolution architecture was introduced.

Fovea and Context Streams This multi-resolution architecture aims at processing two separate spatial streams, namely context stream and fovea stream. The input to the context stream is give by, half of the original resolution and whereas the input for the fovea stream is given by the original resolution focusing on the centre region of the video, this is done to take advantage of the camera bias usually the object of interest. So basically a context stream models low-resolution frames, whereas as fovea stream models high-resolution centre cropped frames. This reduction in dimensionality in input allows for a much fast processing of frames. After both the streams are processed separately, they are concatenated and the output is fed to the first convolutional layer is shown Fig. 2.

3 Limitations

This method of CNN on large-scale video classification is focused mainly on feature extraction and being able to discriminate objects in a video which is being achieved with great success using this method but to obtain more precise results the temporal aspect of the object in the video must also be incorporated without losing the ability to discriminate objects. The next method we are going to discuss will try to overcome

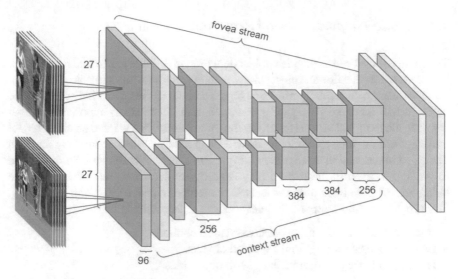

Fig. 2 Multi-resolution CNN architecture with fovea and context stream

this limitation and provide with a better method of video classification. That being said this is by far the largest scale at which video classification has been done with such fine feature extraction using the time information fusion and multi-resolution methods.

4 Object Detection and Tracking Methods

A convolutional neural network is very effective on object detection. They are able to recognize objects in still images thanks to powerful deep networks. Here we try to detect objects in a video domain and the challenges faced in that. This would involve object detection and object tracking in a video. Similar to how object detection in still images has led to image classification, object segmentation and image captioning object detection in videos can lead to an increase in performance in video classification, video captioning and surveillance-related application. The reason for implementation of object tracking is due to the fact that unlike a still image in video the object's position keeps varying so in order to extract information about the object at that particular instance object tracking is necessary. So in this implementation of CNN for video classification is primarily based on two factors: (i) object detection and (ii) object tracking.

The CNN framework proposed here consists of two major modules

1. Tubelet object proposal module.
2. Tubelet classification and re-scoring module.

4.1 Spatio-temporal Tubelet Proposal

The main issues faced with video object detection is that one has to find the fine line between object detection and object tracking with respect to time and space (location). If we were to go about straightforward object detection, then every individual frame detected will be based purely on appearance, and this leads to importance of that frame in the temporal aspect (time). On the other hand, if we were to go about object tracking method, then the entire focus will be in regard to the temporal aspect of an identified object and thus lacks the ability to be able to differentiate. So our aim is to implement the discriminative ability from object detection and at the same time being able to keep track of the objects in the temporal space using object tracking. The proposed spatio-temporal Tubelet consists of three major steps:

- Image object proposal using selective search method.
- Object proposal scoring from extracted objects.
- High-confidence proposal tracking with object classes.

Image Object Proposal To obtain the potential objects from the video, a selective search (SS) algorithm is used. The SS method returns outputs in the form of objects extracted from a single frame in a video. Majority of these proposed objects are removed if it does not cross the minimum threshold.

Object Proposal Scoring The classification of the extracted objects as background or one of the object classes by cross referencing it with pre-existing SVM models. The higher the SVM score, the higher the possibility that is an object corresponding to a particular class.

High-Confidence Proposal Tracking We are done with the object detection and classification of those objects into respective object classes based on their confidence scores. Now we move on to object tracking that is to information about the change in object pose and scalar changes. First we start off by selecting the object classes with the highest confidence scores. The early detectors (object classes with very high-confidence scores from object proposal scoring) used for tracking are referred to as anchors. Taking this anchor as the starting point, we start tracking forwards and backwards that is from the first frame to the last frame and vice versa. Thus, the two separate tracklets obtained and then concatenated.

The tracking must be done such that the tracker does not drift away from its initial or chosen object class during the process; in order to avoid such drifting, a minimum threshold has been set. After the tracking of the first object, we move on to selecting the next object class to be tracked. The second object class should not necessarily be the object class with the next highest confidence score, and this is the case because most of the time object classes of higher confidence scores tend to cluster both spatially and temporally so in order to avoid such overlaps, we apply a suppression method while deciding the next object class selection. This suppression method basically lies upon a certain condition; the next object class to be chosen will be rejected if it overlaps with the existing tracks beyond a certain threshold and will not be chosen as new anchor and complete framework for object detection and tracking. At the end of the above discussed three steps, we now have tracks of high-confidence anchors for each class. These tracks are Tubelet proposals for Tubelet classification.

4.2 Tubelet Classification and Re-scoring

If we were to classify these Tubelets similar to the method followed to classify object class models, the performance is at most modest. This is because of the following reasons

- The number of available Tubelets is lesser than the number of available object classes.
- Tubelet models are less sensitive compared to object classes.
- Suppression on Tubelets leads to false negatives.
- Temporal information must be incorporated before classification of Tubelets.

In order to overcome such problems in Tubelet classification, we implement the following two methods:

- Tubelet box perturbation and max pooling.
- Temporal convolution and re-scoring.

Tubelet Box Perturbation and Max Pooling There are two kinds of perturbation discussed here: The first method is to obtain a uniform distribution by taking the boundaries of a Tubelet box.

$$\Delta x \sim U(-r.w, r.w) \tag{1}$$

$$\Delta y \sim U(-r.h, r.h) \tag{2}$$

where U is the uniform distribution, w and h are the width and height of the Tubelet box r is the sampling ratio hyper parameter.

The second method is to replace the Tubelet box with original object detections with a threshold overlap value. Greater the overlap value, greater the confidence on the Tubelet box. Tubelet box with maximum detection score is alone replaced. The max pooling processing is to improve the spatial robustness of detector and utilize the actual object detections around the Tubelet.

Temporal Convolution and Re-scoring The way in which this CNN framework is different from still image object detection is because of the incorporation of temporal convolutional network (TCN) consisting of the following features, namely detection scores, tracking scores, anchor offsets, in order to generate temporally dense prediction on every Tubelet box. This TCN is a four-layer 1-D fully convolutional network. This 1-D convolutional network is of better supervision than single Tubelet-level labels.

5 Conclusion

In this study, we first discuss CNN and its highlighting features. Then we took upon the task of video classification using two distinct CNN proposed architecture. The first one being on a very large-scale data set Sports 1-M using time information fusion and multi-resolution CNN architecture, which resulted in a more features and object distinction-based video classification. Then we tried to overcome the limitations observed in the previous proposed system which was to incorporate temporal aspect of an object into video classification. This is where the concept of object detection from video Tubelet with CNN comes into play. This framework is based on the space and temporal aspect of an object detected in a video and tracking it with change in time using Tubelet classification. This allows the video classification to be able to get better at classifying the videos.

References

1. LeCun, Y., Bottou, L., Bengio, Y., Haffner, P.: Gradient-based learning applied to document recognition. Proc. IEEE **86**(11), 2278–2324 (1998)
2. Zeiler, M.D., Fergus, R.: Visualizing and understanding convolutional networks. In: European Conference on Computer Vision, pp. 818–833. Springer, Cham (2014, September)
3. Wang, H., Ullah, M.M., Klaser, A., Laptev, I., Schmid, C.: Evaluation of local spatio-temporal features for action recognition. In: BMVC 2009-British Machine Vision Conference, p. 124-1. BMVA Press (2009, September)
4. Liu, J., Luo, J., Shah, M.: Recognizing realistic actions from videos "in the wild". In: IEEE Conference on Computer Vision and Pattern Recognition, 2009. CVPR 2009, pp. 1996–2003. IEEE (2009, June)
5. Sivic, J., Zisserman, A.: Video Google: a text retrieval approach to object matching in videos. In: Null, p. 1470. IEEE (2003, October)
6. Niebles, J.C., Chen, C.W., Fei-Fei, L.: Modeling temporal structure of decomposable motion segments for activity classification. In: European Conference on Computer Vision, pp. 392–405. Springer, Berlin, Heidelberg (2010, September)
7. Wang, H., KlÃd'ser, A., Schmid, C., Liu, C.L.: Action recognition by dense trajectories. In: 2011 IEEE Conference on Computer Vision and Pattern Recognition (CVPR), pp. 3169–3176. IEEE (2011, June)
8. Laptev, I.: On space-time interest points. Int. J. Comput. Vis. **64**(2–3), 107–123 (2005)
9. DollÃar, P., Rabaud, V., Cottrell, G., Belongie, S.: Behavior recognition via sparse spatio-temporal features. In: 2nd Joint IEEE International Workshop on Visual Surveillance and Performance Evaluation of Tracking and Surveillance, 2005, pp. 65–72. IEEE (2005, October)
10. Laptev, I., Marszalek, M., Schmid, C., Rozenfeld, B.: Learning realistic human actions from movies. In: IEEE Conference on Computer Vision and Pattern Recognition, 2008. CVPR 2008, pp. 1–8. IEEE (2008, June)
11. Baccouche, M., Mamalet, F., Wolf, C., Garcia, C., Baskurt, A.: Sequential deep learning for human action recognition. In: International Workshop on Human Behavior Understanding, pp. 29–39. Springer, Berlin, Heidelberg (2011, November)
12. Ji, S., Xu, W., Yang, M., Yu, K.: 3D convolutional neural networks for human action recognition. IEEE Trans. Pattern Anal. Mach. Intell. **35**(1), 221–231 (2013)
13. Taylor, G.W., Fergus, R., LeCun, Y., Bregler, C.: Convolutional learning of spatio-temporal features. In: European Conference on Computer Vision, pp. 140–153. Springer, Berlin, Heidelberg (2010, September)

14. Le, Q.V., Zou, W.Y., Yeung, S.Y., Ng, A.Y.: Learning hierarchical invariant spatio-temporal features for action recognition with independent subspace analysis. In: 2011 IEEE Conference on Computer Vision and Pattern Recognition (CVPR), pp. 3361–3368. IEEE (2011, June)
15. Karpathy, A., Toderici, G., Shetty, S., Leung, T., Sukthankar, R., Fei-Fei, L.: Large-scale video classification with convolutional neural networks. In: Proceedings of the IEEE Conference on Computer Vision and Pattern Recognition, pp. 1725–1732 (2014)
16. Kang, K., Ouyang, W., Li, H., Wang, X.: Object detection from video tubelets with convolutional neural networks. In: Proceedings of the IEEE Conference on Computer Vision and Pattern Recognition, pp. 817–825 (2016)

Cloud-Enabled WBANs for Ubiquitous Healthcare Applications: Data Collection and Privacy Preserving Approach

K. Praveen Kumar, Makula Vani and G. Shankar Lingam

Abstract With the snappy progression of wireless body area network (WBAN), some continuous social protection watching and unavoidable e-prosperity organizations are traded from the human body to investigate gatherings. Regardless, different specific issues and troubles are developed with the blend of WBAN and cloud enlisting. A critical test for social assurance is the route by which to give updated and secure associations to a party of individuals with an obliged measure of cloud assets. Another major test for WBAN is that quality-of-service (QoS) prerequisites, including throughput, postponement, and package affliction rate. This paper demonstrated a blueprint of cloud drew in WBAN which encourages body a territory structures with flowed enrolling. In this paper, the two hindrances of QoS fundamentals and the characteristics of the cloud are considered. We feature the techniques for transmitting indispensable sign sensor information to the cloud by utilizing centrality valuable get-together, handover, information security structures, and scattered amassing. Exploratory outcomes are shown for our proposed work with past techniques in updating handover achievement rate, stop up, information precision, engineer ability and lifetime of system, distribution, and reversal/assign rate.

Keywords Wireless body area network · Cloud computing
Quality of service and security

K. Praveen Kumar (✉) · M. Vani · G. Shankar Lingam
Computer Science & Engineering, Chaitanya Institute of Technology & Science,
Hanamkonda 506001, Telangana, India
e-mail: praveen.katukojwla99@gmail.com

M. Vani
e-mail: makula.vani@gmail.com

G. Shankar Lingam
e-mail: shankar_lingam@hotmail.com

K. Praveen Kumar · M. Vani · G. Shankar Lingam
Department of Computer Science & Engineering, KU College
of Engineering & Technology, Warangal, India

© Springer Nature Singapore Pte Ltd. 2019
J. Wang et al. (eds.), *Soft Computing and Signal Processing* ,
Advances in Intelligent Systems and Computing 898,
https://doi.org/10.1007/978-981-13-3393-4_13

1   Introduction

Change of the total masses is broadening very much arranged, and it faces three key issues: 1. ascend in social security costs; 2. exceedingly increase of people considered after WW2; 3. demographically growth of future. Especially in Australia, any longing everlastingly has stretched out in 1960 from 70.8 years to 78.2 years. In the USA, desire forever common enlargements to 13.5% from 1960 to 2010 (ages from 69.8 years to 78.2 years) [1]. Unpreventable restorative administrations are a present creating development which ensures increases in enduring quality, precision, openness, and adequacy of remedial treatment in view of the enormous advances in various new thoughts, for instance, wearable therapeutic contraptions, unavoidable remote broadband trades, appropriated processing, and body an area frameworks. In this circumstance, wireless body area networks as a dynamic research field which is by and by a work in advance to offers some progressing organizations [2].

With the developing request from clients and the ponder of creating individuals, redress recognizing information is required for positive treatment and the commitment from research social affairs such as therapeutic exhortation, physicians, and specialists are beside engaging. In addition to WBANs provides a variety of various services in diverse fields including consumer electronics, sports or fitness, entertainment and military applications [3]. Remote sensors and actuators are the important constituents of WBANs. They remain as the structure between the physical world and electronic gadgets. If all else fails, sensors and actuators are wearing with patients by especially or even embedded in tolerant. Different parameters in WBANs for continuous monitoring of human vital signs such as blood pressure, ECG (Electrocardiography), temperature, EMG, EEG, blood glucose, carbon dioxide gas sensor (CO_2), pulse oximetry and gyroscope and also some sensors in sensor networks [4]. Scattered enrolling is a stage that can be utilized for anchoring, isolating, recovering, dealing with, appropriating, and securing any sort of crucial and superfluous information. This field manages the preferred standpoint necessities issue. Particularly, cloud-based associations utilized for bioinformatics including data as a service (DaaS), software as a service (SaaS), infrastructure as an association (IaaS), and Platform as a Service (PaaS). Patient's physiological status is anchored, organized, overseen, and isolated finished a gigantic piece of time. This human data is anchored in the cloud orchestrate, and to help, it can be appropriated to focus, remedial assistants, experts, and relatives, etc. Body zone framework circuits are all the more great since it included low-power and sharp sensor center points that objective is to grow the framework execution to the extent accuracy, speed, and resolute nature of the data. So the cloud helped WBANs will extraordinarily redesign the e-prosperity.

Figure 1 addresses the general consolidated outline of cloud-enabled WBAN. Cloud-enabled WBANs is versatile, shrewd, and data-driven human administration system. Body zone frameworks have a couple of results including confined storage space and memory, low taking care of intensity, interface quality, and imperativeness use. These obstacles and resource constraints are defeats by fusing WBAN and appropriated registering.

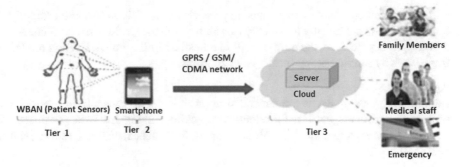

Fig. 1 Integrated architecture

2 Wireless Body Area Network

A body region organize (BAN), also implied as a remote body zone compose (WBAN) or a body sensor arrange (BSN) or a versatile body region organize (MBAN), is a remote course of action of wearable figuring gadgets. Boycott gadgets might be implanted inside the body, installed, surface-mounted on the body in a settled position Wearable advancement might be kept running with contraptions which people can pass on various positions, in bits of attire pockets, by hand or in different packs. While there is a case toward the cutting back of gadgets, especially, systems containing several scaled back body sensor clients (BSUs) together with a solitary body focal unit (BCU). Greater decimeter (tab and cushion) assessed sharp contraptions, kept running with gadgets, still acknowledged an essential part likewise as going about as an information center point, information passage and giving a UI to see and oversee BAN applications. The progress of WBAN improvement began around 1995 around utilizing a remote individual zone deal with (WPAN) advances to execute trades on, close, and around the human body. Around six years thereafter, the verbalization "Boycott" came to intimate structures where the correspondence is completely inside, on, and in the energetic closeness of a human body.

Concept:
The rapid headway in physiological sensors, low-control encouraged circuits, and remote correspondence has connected with some other time of remote sensor systems, now utilized for purposes, e.g., checking activity, things, structure, and flourishing. The body of a territory dealing with field is an interdisciplinary zone which could permit unassuming and steady thriving checking with advancing updates of remedial records through the Internet. Distinctive careful physiological sensors can be encouraged into a wearable remote body district plan, which can be utilized for PC helped revamping or early zone of healing conditions. This zone depends upon the trustworthiness of embedding little biosensors inside the human body that are charming and that do not incapacitate standard exercises. The introduced sensors in the human body will gather different physiological changes with a specific extreme goal to screen the patient's success status paying little regard to their zone. The data

will be transmitted remotely to an outside preparing unit. This gadget will promptly transmit all data continually to the geniuses all through the world. On the off chance that a crisis is perceived, the masters will in a flash edify the patient through the PC framework by sending fitting messages or cautions.

Applications:
Beginning usages of BANs are depended upon to show up basically in the social protection territory, especially for relentless checking and logging basic parameters of patients encountering interminable sicknesses, for instance, diabetes, asthma, and heart ambushes.

- A BAN's arrangement set up on a patient can alert the specialist's office, even before they appear no less than a touch of graciousness attack, through assessing changes in their basic signs.
- A BAN's sort out on a diabetic patient could automatically mix insulin through a pump, when their insulin level declines.

Diverse employments of this advancement consolidate recreations, military, or security. Extending the development to new regions could in like manner help correspondence through steady exchanges of information between individuals, or among individual and machines.

3 Cloud Computing

Dispersed figuring is a data progression (IT) point of view that connects with unpreventable access to shared pools of configurable framework assets and greater entirety benefits that can be promptly provisioned with irrelevant association exertion, as regularly as conceivable finished the Internet. Passed on enlisting depends after sharing of points of interest for accomplish soundness and economies of scale, like an open utility.

Outsider mists connect with relationship to spin around their center relationship as opposed to weakening assets on PC framework and maintenance [2]. Advocates watch that appropriated handling engages relationship to keep away from or restrict early IT foundation costs. Promoters in like way ensure that passed on preparing enables attempts to get their applications up and running speedier, with enhanced sensibility and less upkeep, and that it draws in IT social occasions to all the more quickly change focal points for meet fluctuating and flighty demand [2–4]. Cloud suppliers generally utilize a "pay-as-you-go" appear, which can incite sudden working costs if boss is not balanced with cloud-assessing models [1].

4 Qualities of Service and Security

Nature of framework (QoS) is the depiction or estimation of the general execution of an association, e.g., a correspondence or PC arrange or a passed on handling association, especially the execution seen by the clients of the structure. To quantitatively gauge nature of association, two or three related parts of the system advantage are much of the time seen as distribute, bit rate, throughput, transmission delay, accessibility, jitter, and whatnot.

In the field of PC dealing with and other bundle exchanged media transmission structures, the nature of association recommends development prioritization and asset reservation control instruments instead of the master association quality. The nature of association is the capacity to give distinctive need to various applications, clients, or information streams, or to ensure a specific level of execution to an information stream.

4.1 Applications

A portrayed nature of organization may be needed or required for particular sorts of framework action, e.g.:

- Streaming media especially;
- Internet tradition TV (IPTV);
- Audio over Ethernet;
- Audio over IP;
- IP correspondence generally brought Voice over IP (VoIP);
- Videoconferencing;
- Telepresence.

4.2 Simulation Comes About for Our Proposed Work

4.2.1 Execution Measurements

This subsection passes on us about the generation parameters that are incorporating into improving QoS of our proposed structure. The execution parameters that are considered are handover accomplishment rate, blockage, data precision, sort out viability and lifetime of framework, securing, and inversion/package botch rate. Each one of these results shows our work with better displays.

i. **Accuracy**:

Accuracy is the most critical factor in WBANs. The aggregated sensor information is in all likelihood of low reliability and quality due to sensors remaining essentialness.

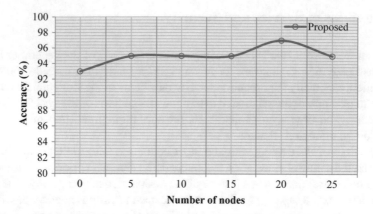

Fig. 2 Results of accuracy

The exactness of human administrations structure relies upon data. Along these lines, watching out for precision at both distinguishing and transmission level is basic for remedial applications. In our system, we used 12 bio medical sensors for sensing data and coordinator used for transmitting data to traffic classifier.

Figure 2 addresses the results of precision. In our proposed work, that is, cloud-empowered WBAN-based social insurance application displays the high-precision rate since our proposed WBAN unequivocally distinguishing essential signs from all body sensors that are embedded on human body. Information improvement classifier sees the right information and keeps up a vital separation from flawed qualities with a specific genuine target to check imperfect examination.

ii. **Congestion rate**:

In a WBAN, the blockage can pass on high importance utilization and package episode and its immediate impacts in the QoS. The blockage will incite drop out the package on the structure.

Figure 3 exhibits the obstruct rate similar to number of rounds. Reenactment comes to fruition which shows that our proposed system gives better blockage rate since obstruct is lessened and improve orchestrate execution by data traffic classifier. To affirm the execution, in this work we portrayed framework into various traffics (emergency, on-demand, and common).

iii. **Handover success rate**:

The handover advance rate (HOSR) is gained through movement estimation. HOSR is essential for human administrations applications since HOSR regard impacts the customer experience. HOSR is addressed as takes after:

$$HOSR = \text{Successful handovers/hand overrequest} \quad (9)$$

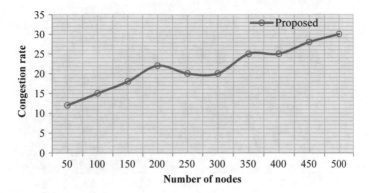

Fig. 3 Results of congestion rate

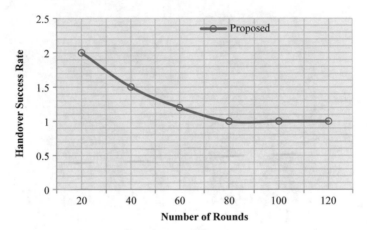

Fig. 4 Results of handover success rate

Figure 4 implies that the performance of handover success rate. The proposed IEEE 802.15.6 MAC protocol provides better results.

iv. **Inversion/packet error rate**:

Reversal/packet blunder rate is a vital parameter in QoS. It is characterized as the way toward investigating blunders in systems amid the parcel transmission. In light of this examination, the execution of system is assessed. A consequence of the parameter is shown in Fig. 5.

v. **Preemption**:

Acquisition is required for body territory systems. In any case, the seizure is accommodating crisis information which devours a high measure of vitality. This outcome to corrupt the system execution as far as high postponement for critics information, to defeat this issue we proposed non-preemptive line for putting away information in need way. Figure 6 demonstrated the aftereffects of seizure.

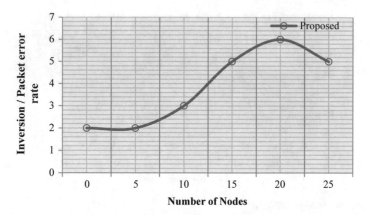

Fig. 5 Results of inversion/packet error rate

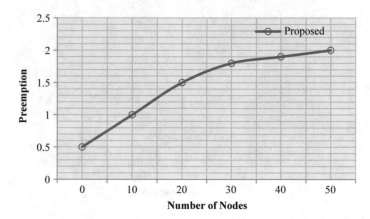

Fig. 6 Results for preemption

vi. **Network efficiency**:

Figure 7 depicts the framework capability which unmistakably insinuates lifetime of the framework. It is portrayed as the time taken for the sensor center point to complete generation. Exactly when a center point misses the mark on essentialness, the present time of the framework is lessened. In our work, MAC tradition and gathering are considered for extending the center point essentialness and lifetime which improve the framework execution. Figure 7 shows the delayed consequences of framework capability.

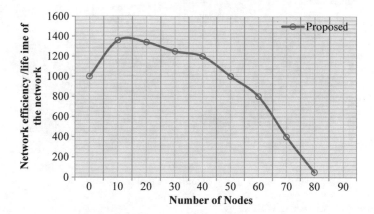

Fig. 7 Results of network efficiency/lifetime of the network

4.3 Comparative Analysis

In the midst of the relationship amusement, we grasped the going with estimations to check the execution of cloud-enabled WBAN and each one of these estimations shows our work with happens better than past displays.

4.3.1 Packet Delivery Ratio (PDR)

PDR is portrayed as the total number of viably received data divides for the total number of produces packs in the midst of proliferating. Change of this metric should reliably be extended with respect to the amount of center points. Subsequently, using the condition we evaluate the level of data packages that are got between end customers.

$$PDR(\%) = \frac{\sum number\ of\ packets\ delivered}{\sum number\ of\ packets\ sent} \times 100$$

Figure 8 exhibits that the graphical outcomes of PDR. The outline shows the packet delivery ratio of proposed and ENSA-BAN. As it to be seen from the outline, the PDR of proposed system is greater than the ENSA-BAN. IEEE 802.15.6 is considering the speedy data transmission and clustering prompts collect the data with same order. The higher data transmission rate and the lower allocate could provoke the higher package transport extent. This change infers that our proposed structure performs better data transmission.

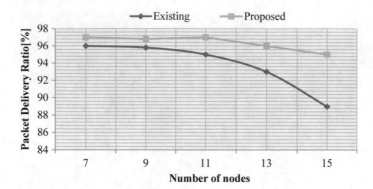

Fig. 8 Comparison result of PDR

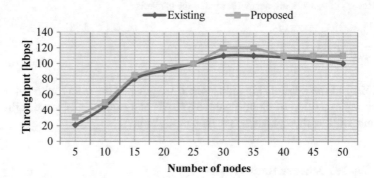

Fig. 9 Comparison result of network throughput

Network Throughput:

Throughput is a great degree tremendous and basic factor in human administration applications. It is registered with respect to the total number of received data allocated with the total number of transmitted data.

Throughput is figured by utilizing Eq

$$\text{Throughput} = \frac{\text{Total number of received datapackets}}{\text{Total number of transmitted datapackets}}$$

In Fig. 9, the execution of the framework throughput is outlined. The proposed structure gives better execution when stood out from the present PA-MAC. The present PA-MAC has the high package crash rate and high question multifaceted nature. In any case, the data action prioritization and the handover for constant data transmission can fabricate the framework throughput. We in like manner used IEEE 802.15.6 MAC tradition which grows the framework capability.

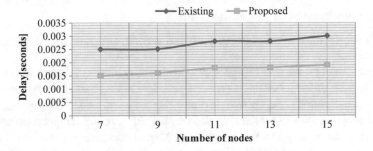

Fig. 10 Comparison result of delay

Average End-to-End Delay:

It is the typical time required by the data groups to accomplish the therapeutic staff/understanding/others. End-to-end defer is portrayed as the typical time taken for the data packs to be transmitted to objective.

$$\text{Average end-to-end delay} = \frac{\sum time\ spent\ to\ deliver\ packets}{Number\ of\ packets\ received\ by\ destination}$$

(Or)

$$\text{Delay} = \{\text{Packet Arrival Time} - \text{Packet Start Time}\}$$

Figure 10 shows the result of end-to-end delay. From the graph surely, the conclusion to-end put off of proposed achieved less when stood out from the present structure ENSA-BAN. As determined already, PDR has the effect on slightest deferment in the proposed structure.

5 Conclusion

Various responses for avoid correspondence overhead and to guarantee quality-of-service necessities have been proposed for cloud-helped WBANs in the earlier years. In any case, these courses of action are not suitable for helpful applications for therapeutic administrations watching, where physiological parameters should be sent to the examination organizes by inside a foreordained time delay. In this investigation work, we proposed, made, and evaluated a novel in cloud-based framework respond in due order regarding compose the transmission of patient crucial signs for end customers using cloud-engaged body zone frameworks. Exploratory results exhibit that our proposed work offers ideal execution over standard strategies.

Future work can be dense as takes after:

i.   We survey the execution of our proposed work in authentic restorative circum-
     stances.
ii. Pushed hail taking care of strategies will similarly be asked about and associated
     with recover the lost data or even the principal basic signs at the authority.
iii. We perceive and pick appropriate preventive measures and restorative treatment.
iv. We propose dynamic channel errand way to deal with avoid impedance issue and
     besides upgrading security by realizing pushed customer approval frameworks
     (e.g., by methods for voice affirmation and face affirmation) on mobile phone
     or some other contraption.

References

1. Movassaghi, S., Abolhasan, M., Lipman, J., Smith, D., Jamalipour, A.: Wireless Body Area
   Networks: A Survey. IEEE Communications Surveys & Tutorials (2013)
2. Negraa, R., Jemilia, I., Belghitha, A.: Wireless body area networks: applications and technologies. In: The Second International Workshop on Recent Advances on Machine-to-Machine
   Communications, Procedia Computer Science, vol. 83, pp. 1274–1281 (2016)
3. Zhang, K., Liang, X., Barua, M., Lu, R., Shen, X.: PHDA: A Priority Based Health Data Aggregation with Privacy Preservation for Cloud Assisted WBANs. Elsevier (2016)
4. Chen, M., Gonzalez, S., Vasilakos, A., Cao, H., Leung, V.C.M.: Body area networks: a survey.
   Mobile Netw. Appl. **16**, 171–193 (2011). https://doi.org/10.1007/s11036-010-0260-8

Reconstruction of Gene Regulatory Network Using Recurrent Neural Network Model: A Harmony Search Approach

Biswajit Jana, Suman Mitra and Sriyankar Acharyaa

Abstract Gene regulatory network (GRN) is an artificial network comprising a gene set and their mutual interactions. One gene interacts with other through protein production which is called 'gene expression.' The computational reconstruction of GRN is made using gene expression data where gene expression values are provided in a matrix. Here, a recurrent neural network (RNN) model is used to represent the GRN. In this paper, a meta-heuristic algorithm, namely harmony search (HS), has been modified and applied to train the network parameters in reconstructing GRN. The proposed HS-GRN algorithm is applied to a standard artificial gene expression dataset, and the experimental results are compared with several other methods. It shows that HS-GRN performs better than the literature methods on this dataset. Furthermore, HS-GRN has also been applied to two real gene datasets, namely IRMA-ON and IRMA-OFF. The experimental results state that HS-GRN performs satisfactorily well in real datasets also.

Keywords Gene regulatory network · Recurrent neural network
Gene expression · Harmony search · Meta-heuristics

B. Jana · S. Mitra · S. Acharyaa (✉)
Maulana Abul Kalam Azad University of Technology, Kolkata, West Bengal, India
e-mail: srikalpa8@gmail.com

B. Jana
e-mail: biswajit.cseng2012@gmail.com

S. Mitra
e-mail: smsumanmitra@gmail.com

© Springer Nature Singapore Pte Ltd. 2019
J. Wang et al. (eds.), *Soft Computing and Signal Processing* ,
Advances in Intelligent Systems and Computing 898,
https://doi.org/10.1007/978-981-13-3393-4_14

1   Introduction

Gene is a molecular functional unit of heredity that carries the genetic information from one generation to the next generation. Every living organism has a genomic structure at their cellular level. Gene produces protein (gene expression) to regulate the protein production of other genes. Therefore, multiple interactions appear among genes and form an artificial regulatory network, called gene regulatory network (GRN) [1, 2]. In a GRN, nodes stand for genes and edges stand for interactions. An interaction between the genes is categorized into two types: influence and inhibition. The computational reconstruction of GRN from gene expression data (available from DNA microarray) [3] is a great challenge for the researchers in gene expression analysis domain. The objectives of the reconstruction of GRN are associated with disease analysis, drug design, etc.

There are various modeling schemes to reconstruct the GRN. A type of neural network, namely recurrent neural network (RNN), has been widely used to model GRN reconstruction. In RNN modeling, a neuron is represented as gene i and an input to gene i is considered as a regulation from some other gene j to gene i. There are three parameters associated with GRN: $W = [w_{ij}]_{NN}, B = [b_i]_{NX1}$, and $\mathsf{T} = [\mathsf{T}_i]_{NX1}$. The weight parameter w_{ij} is associated with the connection (edge) between the genes i and j. The value of weight parameter w_{ij} is either positive (the expression of target gene i is influenced by the regulator gene j) or negative (the expression of target gene i is inhibited by the regulator gene j). A basal expression b_i which is associated with gene i is the default expression level of gene i and T_i is the time constant [5]. To reconstruct GRN, several meta-heuristic algorithms have been used so far to train the network parameters [4].

In this paper, a modified variant of harmony search, namely HS-GRN, has been proposed. Three operations, namely *mute*1, *mute*2, *and mute*3, have been incorporated into HS to enhance its performance in terms of better parameter tuning for reconstructing GRN. *mute*1 used the Gaussian mutation, *mute*2 used the Cauchy mutation, and *mute*3 used the opposition-based mutation. A new harmonic is created by one of the three alternative operations. An operation is selected by the roulette wheel selection strategy using uniform probability. The proposed variant HS-GRN has been applied to artificial gene dataset (gold standard GRN) and compared its performance with other methods. Furthermore, HS-GRN has been applied to two other real gene datasets, namely IRMA-ON and IRMA-OFF, and compared its performance with the literature methods. The results obtained by HS-GRN outperform the state-of-the-art methods in case of gold standard GRN.

The rest of the paper is organized as follows: RNN modeling technique has been described in Sect. 2. HS and its proposed variant HS-GRN have been described in Sect. 3. Results and discussions have been presented in Sect. 4. Finally, Sect. 5 concludes the paper.

2 RNN Modeling

The computational reconstruction of GRN is a reverse process to find which genes are connected to form a GRN. An RNN [1, 2, 4] has feed forward and feedback loop, and here it is applied to model the GRN. Gene expression dataset $G = [g_i(t)]_{NXT}(g_i(t) =$ expression level of gene i at time point t) contains the expression value of expressed genes (N) at time points (T). The expression level of each gene $(g_i(t))$ is measured at small time intervals $(t + \Delta t)$ using Eq. 1.

$$g_i(t + \Delta t) = \frac{\Delta t}{T_i} f \left(\sum_{i=1}^{N} w_{ij} g_j(t) + b_i \right) + \left(1 - \frac{\Delta t}{T_i} \right) g_j(t) \qquad (1)$$

where Δt is a small time interval and T_i are the weight parameter, basal expression value, and time instance, respectively. A candidate solution (X) of three genes and its corresponding GRN is depicted in Fig. 1a and Fig. 1b, respectively. Using X, a simulated gene expression matrix can be generated by Eq. 1. The summation of differences between the original time series $G = [g_i(t)]_{NXT}$ and the simulated time series $G' = g_i'(t)]_{NXT}$ has been considered as the objective function (Eq. 2) which measures the mean square error. The objective function $C(X)$ is to be minimized.

$$C(X) = \frac{1}{TN} \sum_{t=0}^{T} \sum_{i=1}^{N} \left(g(t) - g_i'(t) \right)^2 \qquad (2)$$

In GRN reconstruction process, the total numbers of parameters are $N \times (N + 2)$ ($N =$ number of genes). The curse of dimensionality problem [6] can arise with the dataset G, as the number of genes is much larger than the number of time points. When the number of genes is increased, the value of $(N \times (N + 2))$ becomes very large and it makes the parameter learning problem more complex. Therefore, meta-

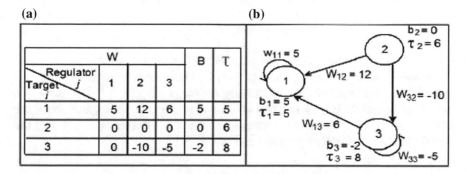

(a)   (b)

	W			B	T
Regulator j Target i	1	2	3		
1	5	12	6	5	5
2	0	0	0	0	6
3	0	-10	-5	-2	8

Fig. 1 Candidate solution and the corresponding three-gene GRN. **a** A set of GRN parameters (candidate solution). **b** The internal connection between three genes [5]

heuristic algorithms can be used for parameter learning of GRN to provide near-optimal solution in a reasonable amount of time.

3 Methodologies

Here, in this paper, the meta-heuristic method, harmony search, is used as a training algorithm for learning of GRN parameters.

3.1 Harmony Search

Harmony search is a meta-heuristic algorithm proposed by Geem et al. [7], based on extemporization process of music players. The goal of harmony search is to improvise a pleasing music from an aesthetic point of view by adjusting pitches (frequencies), timbers (sound qualities), and amplitudes (loudness) of the instruments. The performance of a newly improvised music is determined by the aesthetic estimation of it which can be viewed as an optimization problem to optimize the objective function associated with it. Aesthetic estimation is determined by the set and order of the notes jointly played on different instruments, and it can be improved through repeated practice which is treated as the iteration process. In HS, there are three important components: usage of harmony memory (HM), pitch adjustment, and randomization. Harmony memory (HM) contains the good quality of harmonies which can be used to create new harmonies in turn. A parameter, named harmony memory consideration rate ($r_{accept} \in [0, 1]$) (also known as harmony accepting rate), has been introduced to utilize harmony memory effectively. Usually, the range of r_{accept} is $0.7 \sim 0.95$. Pitch adjustment normally depends on the frequencies of the music, and it is determined by the pitch adjustment rate (r_{pa}) and the bandwidth of pitch (b_{range}). Pitches can be adjusted linearly by Eq. 3.

$$x_{new} = x_{old} + b_{range} * \epsilon \tag{3}$$

where x_{new} is the adjusted pitch after linear pitch adjustment. x_{old} is the old pitch, and ϵ is a random number with the range of $[-1, 1]$. The pitch adjustment is controlled by the pitch adjustment rate (r_{pa}). Normally, the range of $r_{pa}(= 0.1-0.5)$ is used in most applications. In HS, the randomization is used to extend the exploration of the search process. The probability of randomization is determined by Eq. 4, and the actual probability of pitch adjustment is determined by Eq. 5

$$P_{rand} = 1 - r_{accept} \tag{4}$$

$$P_{pitch} = r_{accept} * r_{pa} \tag{5}$$

The steps of HS algorithm are described as follows:

1. Initialize HM, r_{accept}, r_{pa}.
2. Improvise a new harmonic by randomly selecting a harmonic from the existing harmonics in HM with probability r_{accept}.
   Otherwise, improvise a new harmonic randomly and go to Step 3.

   2.1. With probability r_{pa}, adjust pitch using Eq. 3 to generate a new harmonic.

3. If the new harmony is better than the worst harmony in HM, then add the new harmony to HM and the worst harmony will be discarded from HM.
4. If stopping conditions are not reached, then go to Step 2.

3.2 Proposed Method

Here, in this paper, a new variant of HS has been proposed to reconstruct the GRN using RNN model (HS-GRN). Meta-heuristic algorithms can suffer from premature convergence when the search process does not explore enough in solution space. Another problem is that if a solution is trapped in local optima there is no exact method for jumping out of the local optima. Randomization is one of the main characteristics of meta-heuristic algorithms that explores the solution space and prevents it from premature convergence. Every meta-heuristic algorithm has its own randomization strategy to perturb the solution. In HS-GRN, an element (harmonic) of its candidate solution (harmony) is mutated by a small amount randomly. Here, three mutation (perturbation) strategies, i.e., Gaussian (**mute1**), Cauchy (**mute2**), and opposition-based (**mute3**), have been incorporated in HS to enhance its performance in terms of better parameter tuning to reconstruct the GRN. The algorithm is randomly selecting a harmonic from the existing harmonics in HM with probability r_{accept}; otherwise, it improvises a new harmonic randomly by choosing a randomization strategy out of the three strategies using roulette wheel selection. One of three alternative mutation strategies has been selected with equal probability ($r_1, r_2, r_3 = \frac{1}{3}$) and applied to an element of a current candidate solution to alter by a random small amount.

Algorithm.1 HS-GRN

Set HS parameters: r_{accept}, r_{pa}, b_{range}, $\in$
Initialize HM
Set probabilities r_1=r_2 = r_3 = $\frac{1}{3}$
while $t = 1$ *to* t_{max} *do*   /* t_{max} is amaximum number of iteration*/
   if rand $\le r_{accept}$ /* rand is a uniform random number in [0,1]*/
      Randomly select an harmonic from existing harmonics in HM
      if rand $\le r_{pa}$
         Adjust pitch randomly depending on b_{range} and $\in$
      end if
   else
      Choose a strategy(*mute1/mute2/ mute3*) according to roulette wheel
         selection with probability $r_1/r_2/r_3$,
      mute1: begin
         Generate new harmonics using Gaussian mutation
      end /*end *mute1*/
      mute2: begin
         Generate new harmonics using Cauchy mutation
      end /*end *mute2* */
      mute3: begin
         Generate new harmonics using Opposition Based mutation
      end /*end *mute3* */
   end if
   Accept the new harmony *if* it is better than the worst harmony in HM
end while

---

4 Experimental Results

In this section, HS has been applied to reconstruct the GRN, using RNN model. The proposed method has been applied to a gene expression time series data representing a four-gene artificial network [1, 2, 8]. The experimental results have been compared with the state-of-the-art results. The machine and software specifications are: Intel(R) Core(TM) i7-3770 Processor, speed 3.4 GHz, 8 GB RAM, 64-bit Windows 7 (Professional), MATLAB 2015a.

The quality of simulated GRN has been compared with the original GRN (gold standard GRN) by checking the mutual existence of each individual connection (edge) in the two networks. On this aspect, edges can be categorized into four types, namely *True Positive (TP)*, *False Positive (FP)*, *True Negative (TN)*, *False Negative (FN)*. A *True Positive (TP)* edge is an edge between a particular node pair which exists in both simulated GRN and original GRN. When an edge exists in simulated GRN but does not exist between the same node pair in original GRN is called *False Positive (FP)*. A *True Negative (TN)* edge is an edge between a particular node pair which does not exist in both simulated GRN and original GRN. When an edge does not exist in simulated GRN but exists between the same node pair in original GRN is called *False Negative (FN)*. On the basis of these edge categories, six performance metrics, namely

Positive Predictive Value (*PPV*), *Specificity* (*SPc*), *Accuracy* (*Acc*), and f_1 score (f_1), have been computed for evaluating the quality of simulated GRN in the respective ways, $TPR = \frac{TP}{TP+FN}$, $FPR = \frac{FP}{FP+TN}$, $PPV = \frac{TP}{TP+FP}$, $SPc = \frac{TN}{TN+FP}$, $ACC = \frac{TP+TN}{TP+FP+TN+FN}$ and $f_1 = 2 * \frac{TPR*PPV}{(TPR+PPV)}$.

4.1 Parameter Settings

GRN parameters have been taken from the previous studies [1, 2]. The range of the strength magnitude of w_{ij}, b_i, and T_i is $-30 \leq w_{ij} \leq 30$, $-10 \leq b_i \leq 10$, and $1 \leq T_i \leq 15$, respectively. The gene expression dataset consists of 4 genes ($N = 4$) with 11 time points ($T = 11$). For HS, parameter values of r_{accept}, r_{pa} have been set to 0.8 and 0.2, respectively. Maximum number of iterations is set to 100, and the harmony memory (HM) size is set to 20.

4.2 Results and Discussion

The proposed HS-GRN has been applied to a synthetic gene dataset with four genes reported in previous studies [1, 2, 8, 9]. HS-GRN has been executed 10 times independently. Therefore, ten GRNs are produced corresponding to ten independent runs. The final GRN is obtained from ensemble averaging [5] of those ten GRNs. Here, the ensemble averaging is performed by checking the simultaneous existence of each individual edge in GRNs. For a node pair, the existence of an edge contributes 1; otherwise, it is 0. A voting score for a particular node pair has been calculated by averaging the total contribution on ten GRNs. If the voting score > 0.7 (threshold), then there exists an edge between the nodes in final GRN. The final GRN topology has been represented in Fig. 2.

Figure 2 shows that among eight actual edges of the gold standard GRN, seven edges have been identified in simulated GRN. Two extra edges are absent in gold standard GRN but present in the simulated GRN, and only one edge appears in gold standard GRN but absent in simulated GRN. Therefore, the final values of *TP*, *FP*, *TN*, and *FN* are 7, 2, 6, and 1, respectively. The values calculated for *TPR*, *FPR*, *PPV*, *SPc*, *ACC*, and f_1 are 0.88, 0.25, 0.78, 0.75, 0.81, and 0.83, respectively. Based on the performance metrics, the quality of the simulated GRN has been compared with that obtained literature methods [1, 8, 9, 10–15] and comparative results have been presented in Table 1. It signifies that the proposed HS-GRN performs better than the other methods in respect of TPR and f_1.

HS-GRN is also applied to two real gene datasets: IRMA-ON and IRMA-OFF [16]. IRMA-ON contains 5 genes with 16 time points, and IRMA-OFF contains 5 genes with 21 time points. The computationally obtained simulated networks for IRMA-ON and IRMA-OFF are describe in Fig. 3 and Fig. 4, respectively.

Fig. 2 Final simulated GRN topology of artificial gene dataset

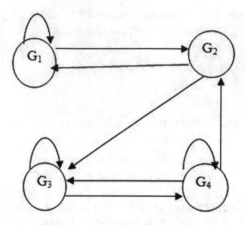

Table 1 Comparison of HS-GRN with the literature method based on artificial gene dataset

Method	TPR	PPV	f_1
HS-GRN	**0.875**	0.777	**0.823**
Kordmahalleh [9]	0.750	0.660	0.700
Kentzoglanakis [8]	0.625	**0.833**	0.714
Morshed [10]	0.75	0.75	0.75
BITGRN [11]	0.63	0.83	0.71
TDARACNE [12]	0.63	0.71	0.67
NIR and TSNI [13]	0.50	0.80	0.62
Xu et al. [1]	0.500	0.444	0.470
ARACNE [14]	0.60	0.50	0.54
BANJO [15]	0.25	0.33	0.27

Fig. 3 Final simulated GRN topology of IRMA-ON

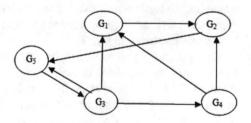

In Fig. 3 for IRMA-ON, it shows that six edges have been identified in simulated GRN among eight actual edges existing in original GRN. Four extra edges are absent in original GRN but present in the simulated GRN, and only two edges exist in original GRN but absent in simulated GRN. Therefore, the value of *TP*, *FP*, *TN*, and *FN* is 6, 4, 13, and 2, respectively. In Fig. 4 for IRMA-OFF, it shows that five edges have been identified in simulated GRN among eight actual edges in original GRN. Three extra edges are present in the simulated GRN but absent in original GRN, and three edges exist in original GRN but does not exist in simulated GRN.

Fig. 4 Final simulated GRN
topology of IRMA-OFF

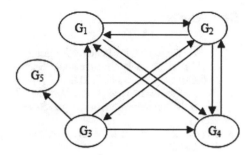

Table 2 Comparison of HS-GRN with the literature methods based on IRMA-ON and IRMA-OFF gene dataset

Method	IRMA-ON			IRMA-OFF		
	TPR	PPV	f_1	TPR	PPV	f_1
HS-GRN	**0.75**	0.60	0.67	0.625	0.625	0.625
Morshed and Chetty [10]	**0.75**	0.75	**0.75**	**0.75**	**0.67**	**0.71**
TDARACNE [12]	0.63	0.71	0.67	0.60	0.37	0.46
TSNI [13]	0.50	**0.80**	0.62	0.38	0.60	0.47
BANJO [15]	0.25	0.33	0.27	0.38	0.60	0.46
ARACNE [14]	–	0.50	0.54	0.33	0.25	0.28

Therefore, the value of TP, FP, TN, and FN is 5, 3, 14, and 3, respectively. Based on the performance metrics, the quality of the simulated GRN has been compared with that of the literature methods [10, 12–15] and comparative results for IRMA-ON and IRMA-OFF have been presented in Table 2. It signifies that the proposed HS-GRN performs better than most of the literature methods.

5 Conclusion

In this paper, a modified variant of HS, named HS-GRN, has been proposed and applied to reconstruct the GRN using RNN modeling. The proposed HS-GRN has been applied to a four-gene artificial dataset available in the past literature. The experimental results have been compared with the several literature methods based on gold standard GRN. The experimental results confirm that proposed HS-GRN performs better than the state-of-the-art method. Furthermore, HS-GRN has been applied to two real gene datasets, namely IRMA-ON and IRMA-OFF. The comparative results ensure that HS-GRN performs better than most of the literature methods on IRMA-ON and IRMA-OFF.

Acknowledgements We are thankful to TEQIP-III Maulana Abul Kalam Azad University of Technology (MAKAUT), West Bengal, India, for supporting our research.

References

1. Xu, R., Wunsch II, D., Frank, R.L.: Inference of genetic regulatory networks with recurrent neural network models using particle swarm optimization. IEEE/ACM Trans. Comput. Biol. Bioinform. **4**(4), 681–692 (2007)
2. Xu, R., Venayagmoorthy, G.K., Wunsch, D.C.: Modelling of gene regulatory networks with hybrid differential evolution and particle swarm optimization. Neural Netw. **20**(8), 917–927 (2007)
3. Eisen, M., Brown, P.: DNA arrays for analysis of gene expression. Methods Enzymol. **303**, 179–205 (1999)
4. Biswas, S., Acharyya, S.: Neural model of gene regulatory network: a survey on supportive meta-heuristics. Theory Biosci. **135**(1–2), 1–19 (2016)
5. Biswas, S., Acharyya, S.: A Bi-objective RNN model to reconstruct gene regulatory network: a modified multi-objective simulated annealing approach. IEEE/ACM Trans. Comput. Biol. Bioinform. (99), 1–8 (2017)
6. Donoho, D.L.: High-dimensional data analysis: the curses and blessings of dimensionality. AMS Math. Chall. Lect. **1**, 1–32 (2000)
7. Geem, Z.W., Kim, J.H., Loganathan, G.V.: A new heuristic optimization algorithm: harmony search. Simulation **76**(2), 60–68 (2001)
8. Kentzoglanakis, K., Poole, M.: A swarm intelligence framework for reconstructing gene networks: searching for biologically plausible architectures. IEEE/ACM Trans. Comput. Biol. Bioinform. **9**(2), 358–371 (2012)
9. Kordmahalleh, M.M., Sefidmazgi, M.G., Harrison, S.H., Homaifar, A.: Identifying time-delayed gene regulatory networks via an evolvable hierarchical recurrent neural network. Bio-Data Min. **10**(1), 29 (2017)
10. Morshed, N., Chetty, M.: Reconstructing genetic networks with concurrent representation of instantaneous and time-delayed interactions. In: Proceedings of the 2011 IEEE Congress on Evolutionary Computation (CEC), pp. 1840–1847 (2011)
11. Morshed, N., Chett, M.: Information theoretic dynamic Bayesian network approach for reconstructing genetic networks. In: Proceedings of the 11th IASTED International Conference on Artificial Intelligence and Applications, pp. 236–43 (2011)
12. Zoppoli, P., Morganella, S., Ceccarelli, M.: TimeDelay-ARACNE: reverse engineering of gene networks from time-course data by an information theoretic approach. BMC Bioinform. **11**(1), 154 (2010)
13. Gatta, G.D., Bansalm, M., Ambesi-Impiombato, A., Antonini, D., Missero, C., di Bernardo, D.: Direct targets of the TRP63 transcription factor revealed by a combination of gene expression profiling and reverse engineering. Genome Res. **18**(6), 939–948 (2008)
14. Margolin, A., Nemenman, I., Basso, K., Wiggins, C., Stolovitzky, G., Favera, R.D.: ARACNE: an algorithm for the reconstruction of gene regulatory networks in a mammalian cellular context. BMC Bioinform. **7**(1), S7 (2006)
15. Yu, J., Smith, J.V., Wang, P., Hartemink, A., Jarvis, E.: Advances to Bayesian network inference for generating causal networks from observational biological data. Bioinformatics **20**(18), 3594–3603 (2004)
16. Cantone, I., Marucci, L., Iorio, F., Ricci, M.A., Belcastro, V., Bansal, M., Cosma, M.P.: A yeast synthetic network for in vivo assessment of reverse-engineering and modelling approaches. Cell **137**(1), 172–181 (2009)

New Gossiping Protocol for Routing Data in Sensor Networks for Precision Agriculture

K. Sneha, Radhika Kamath, Mamatha Balachandra and Srikanth Prabhu

Abstract Automation in the field of agriculture is essential for its development. In this digital era, we can make use of various technologies to enhance the agricultural process. One such technology is network technology. Now, precision agriculture being the emerging concept in this field, a network of sensors can be deployed in order to study the environmental parameters such as temperature, soil moisture content, nitrogen level, type of crop. Therefore, the internetworking of wireless devices is very important in order to manage and modernize agricultural practices. The restriction faced in this process is the efficient use of energy. Since the wireless sensors are battery-powered devices and it is not feasible in terms of cost and automation for a human being to keep changing batteries every now and then while deploying several sensors in a field. In order to tackle this situation, our goal is to come up with a routing protocol that tries to maintain network lifetime. Many protocols have been introduced to solve this problem but not very efficient when it comes to reducing the energy consumption of sensors. Therefore, there is a need for an algorithm which can be a modification made to the existing protocols and achieve improvement in terms of network lifetime, energy consumption, and time taken to route the data through the sensor nodes.

Keywords Gossiping · Routing protocol · Precision agriculture
Energy efficiency · Wireless sensor networks

K. Sneha · R. Kamath (✉) · M. Balachandra · S. Prabhu
Computer Science and Engineering, Manipal Institute of Technology, Manipal, India
e-mail: radhika.kamath@manipal.edu

K. Sneha
e-mail: naiksneha95@gmail.com

M. Balachandra
e-mail: mamtha.bc@manipal.edu

S. Prabhu
e-mail: srikanth.prabhu@manipal.edu

© Springer Nature Singapore Pte Ltd. 2019   139
J. Wang et al. (eds.), *Soft Computing and Signal Processing* ,
Advances in Intelligent Systems and Computing 898,
https://doi.org/10.1007/978-981-13-3393-4_15

1 Introduction

Agriculture plays a very important part in every nation's economic progress and survival of its people. All the rural areas in India solely depend on agriculture for existence in terms of food and earning a living. There have been a lot of improvements in farming since modernization in agriculture was introduced. But as the population is increasing, the need for basic needs is also increasing rapidly. As a result, one of the major problems that will be faced by any country is to manage food supply. The demand increases and the supply needs to be well managed in order to avoid problems like lack of food supply and rise in the cost. Therefore, our main intention here is to introduce a cost- and energy-efficient system in the field of agriculture [1].

As the population is increasing and there are limited resources, irrigation becomes a necessary tool to increase the supply of food in the future. The entire process depends on the amount of water supply, soil moisture content, temperature, type of crop, location, and area. This requires proper understanding of the crops and the amount of water or moisture content that is essential for the crop. Therefore, lack of resources required for cultivation may lead to low yield. Introducing automation techniques can help improve the state of agriculture in India. This will ensure an increase in the production of crops and also result in inexpensive and better quality agricultural products [2].

Wireless sensor network (WSN) is very efficient and a popular technology that can help in providing automated services to the farmers. This involves a couple of sensors that are deployed in fields and can help in monitoring environmental factors like moisture and temperature. These sensors send real-time data that is captured from the fields to the nearby transmitters. This data that is received will be analyzed, and the required measures will be taken. In the near future, wireless sensor networks will become the most cost-effective techniques for automating the agricultural process as it does not require manpower and is wireless in nature. Since battery-powered sensors are used, we must also consider the fact that sensors do consume a lot of energy for data transmission, and lack of energy in sensors might lead to loss of data [3].

The main reason for energy consumption is the time taken by the sensors to collect and send data. If the energy consumption of the sensors increases, then the lifetime of sensors decreases. Therefore, we need to overcome the scarcity of energy, ability to process, and memory space limitations of the sensors. To tackle these limitations, certain protocols and algorithms have been introduced that control the path taken by the data packets during transmission and also improve the energy consumption of sensors and its lifetime.

2 Literature Review

A lot of work has been done in the field of wireless sensor networks with the main intention being to reduce energy consumption and increase the network lifetime, finally making it cost-effective. Several protocols have been introduced, still being unable to get maximum efficiency hence providing a platform to perform more research to improvise the existing protocols. Techniques like flooding and gossiping have been introduced. In flooding, each sensor sends the data packet to all its neighboring nodes, thus resulting in duplicate packets and excess energy consumption. Whereas in gossiping, each sensor node that receives a packet will transmit it to any random neighbor. This technique has its own drawbacks like packet loss or congestion. Both the flooding and gossiping techniques were combined to give rise to flossiping method where a threshold value was set and a random neighbor was chosen for packet transmission [3–6]. Other nodes in this network would listen to the transmitting packet and generate random numbers. Such that, the neighbor with the lowest random number would transmit the packet.

A new protocol called as Flooding Energy Location (FEL) was proposed based on both flooding and gossiping technique. It had three phases:

- Initialize;
- Gather information; and
- Routing.

Initializing involved setting up a gradient, the second phase requires gathering information regarding the energy left in each node, and third was routing the packets based on the information gathered. It will also check which path forms the shortest path to the sink node. In case of a situation where more than one node had the same distance to the sink node at that point, the node with more energy left was considered for further transmission [3, 5, 6].

Another routing protocol was called as the Single Gossiping Directional Flooding (SGDF) which had two phases. The first phase is a generation of the gradient by each sensor node. Here gradient is nothing but the number of hops to the sink node. In the second phase, the SGDF protocol was applied. This protocol did overcome limitations like packet delay but failed due to redundant packets. Next, the location-based algorithm was also applied which reduced the delay in packet transmission because the gossiping technique was used and packets were not flooded to all the nodes. Still, this protocol needed improvisation hence the protocol was called the Energy Location Gossiping (ELG) protocol.

In this protocol, there was no random selection of neighboring nodes like in typical gossiping technique. Here the packet was sent to all the neighboring nodes that existed within each node's transmission radius. The shortest path with minimum hops to the sink node was considered, and also, the amount of energy left in each node was checked before transmitting the packets. The only advantage of this protocol was it solved the problem of latency and delay [7].

Talking about energy-efficient routing techniques, many routing methods like clustering, Low Energy Adaptive Clustering Hierarchy (LEACH), Power-Efficient

Gathering in Sensor Information System (PEGASIS), and Threshold-Sensitive Energy-Efficient Network (TEEN) exist. But each of them has its own advantages as well as disadvantages. LEACH method did reduce the energy consumption but it supported single-hop transmission; therefore, it was difficult to deploy this technique in larger areas such as agricultural fields. PEGASIS approach was just an extension of LEACH technique, and it used the greedy method to form data chain which sometimes resulted in long data chains and the sensors would die due to lack of energy. TEEN was another efficient routing algorithm for smaller areas but consumed a lot of energy in long-distance transmission [8, 9].

Another routing approach called as the Dynamic gossiping protocol involves gossiping and diffusion techniques, where the path along which the packet has to be routed is decided based on packet header information. The best path is chosen and the routing tables are updated. All this includes the concept of the energetic unit which contains the mean energy that is required to forward the packet. Here the performance is measured using parameters like power consumption, node lifetime, throughput, and delay. This approach is compared to existing flooding and static approaches and is proved to be efficient, but not as efficient in terms of power consumption because the improvement is just 50% with less delay and efficient transmission. Since this approach requires updating of routing tables, it still does not satisfy the power consumption problem [10].

Several routing protocols in wireless sensor networks are energy-efficient since the sensors are static, and in case of a process, automation does not involve human intervention. These routing protocols now include sleep mechanism in the sensors in order to further decrease energy consumption and improve network lifetime. When few nodes in the field are active, the idle nodes are put to sleep. Some protocols make use of probabilistic techniques to incorporate sleep mode in sensors [11].

Some protocols considered the position of sensors deployed in a large area. Only a few selective sensors were active, and remaining sensors most of the time were in sleep mode. Based on the decision made, few sensors aimed to forward the packets to sink node and few sensors were used only when required [12].

Since every node has a unique identification number, each sensor was assigned a clock that was in sync with the other nodes in the network. So, even if the nodes were in sleep mode, based on the clock of neighboring nodes, these would get active at the required time. Their aim was that more the number of nodes was in sleep mode, lesser would be the energy consumed. The amount of energy consumed by the nodes is balanced, and the time required to successfully forward the packets is reduced. Considering the pros and cons of the existing methods, we realize that there is a lot of scopes to improvise these methods for efficient automation of certain fields as it will be energy-efficient and cost-effective.

3 Methodology

In order to overcome the restrictions faced in wireless sensor networks, we have come up with a new protocol which is a modification to the existing gossiping protocol. As we know, there are several challenges faced in the area of wireless sensor network like topology, robustness, scalability, communication, computation, delay, locating neighboring nodes, and energy consumption. Most of the challenges are resolved by several protocols, but the main challenge here is to reduce the energy consumption and to improve the system efficiency.

3.1 Proposed System Architecture

The system architecture is a wireless network of several sensors deployed in an agricultural field in Fig. 1. Each sensor collects information from the field and forwards the packet to its appropriate neighboring node. It consists of a base station that receives the information collected by these sensors and forwards the data to the server. The base station not only forwards packets but also provides information about active sensors based on its energy level to the server.

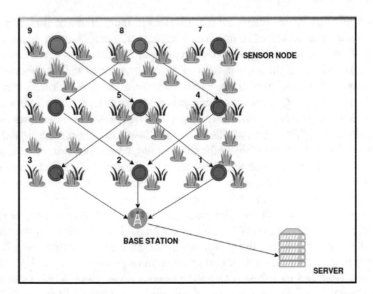

Fig. 1 System architecture with nine sensor nodes and a sink node

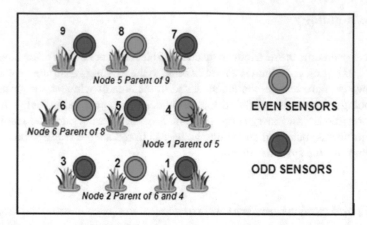

Fig. 2 Odd- and even-numbered sensors along with parent–child relationship

3.2   New Gossiping Algorithm

The new algorithm consists of six phases; they are initial phase, send probe message phase, check battery level phase, forward data packet phase, no response or discard packet phase, and sleep phase.

(1) Initial Phase: In this phase, all the sensors are numbered and further grouped as odd-numbered and even-numbered nodes, and a parent–child relationship is established between the nodes of each group. Each time a node tries to forward a packet to its neighboring node, it will make sure that the neighboring node belongs to the same group as the current node with the data packet. For example, referring to Fig. 2, suppose the sensor node 9 wants to forward a packet since node 9 belongs to odd-numbered group, it will further forward the data to its odd-numbered neighbor such as node 5. The parent–child relationship is established among the nodes that belong to the same group, where the node with the greater number is the child node and the node with the smaller number becomes the parent node. Therefore, every node that collects the data will always forward the packet to its parent node in its group.

(2) Send Probe Message Phase: This phase involves communication between sensor nodes, where the node which wants to send a packet will send a probe message to the desired nodes in the group at a specified interval of time until it gets a reply from any of the desired parent nodes in its group.

(3) Check Battery Level Phase: Once the current node which wants to send data receives a response to the probe message, the current node will check the battery level of its neighboring node. If the battery level is equal to some specified threshold value, the current node will forward the data packet to the desired node. If the battery level is low, then the current node will again send a probe message to check if there is any other alternative node in the same group. If it

finds any alternate node that is active, it will check for its battery level with the threshold value and perform the required action.

(4) Forward Data Packet Phase: When a neighboring node responds to the probe message and the battery level matches the threshold value, the current node will forward the packets to the desired node. This node on receiving the packets performs the same actions depending on its distance from the base station. If the node has another level of nodes, then it will perform the probe message and battery level checking phase before it further forwards the data. If it does not contain further nodes in its group which may act as the parent node, then it will directly forward the packet to the base station.

(5) No Response or Discard Packet Phase: What happens to the data in the current node? After sending many probe packets, if the child node does not get any response from its parent node, then it drops all the data which it wanted to forward.

(6) Sleep Phase: Sleep mode phase controls when a sensor can collect data and forward it to the appropriate node. When a node wants to send a packet, it will check if the neighboring node is active or not. If active, it will forward the packet else it will check if any other node from the same group is active and forwards the packet. Here assuming that the data is forwarded only twice a day when the sensor is not performing any action, the sensor is put to sleep for a certain amount of time and it wakes up after completion of the sleep period. The main intention here is to prevent unnecessary wastage of battery power in sensors when it is idle.

(7) Role of Base Station: The basic role of the base station is to wait for the data packets from the sensors and forward the field condition related data to the server. Based on the response from the nodes which are closest to the base station, it makes required decisions. Suppose a particular node that is expected to send data packets to the base station is not responding for quite a long time, the base station assumes that the battery of that particular sensor is completely drained out or may be due to some hardware issues the sensor has failed to respond. In that situation, the base station sends a message to the server in order to notify it about the inactive sensor node.

4   Comparative Analysis

After observing the techniques used in various routing protocols, the major issue faced in each one of them is the level of energy consumption. It shows that none of the protocols is 100% efficient in terms of power consumption. The comparison of how much power is consumed by the existing protocols for wireless sensor networks is shown in Fig. 3. These routing protocols were compared and it was found that power usage in PEGASIS protocol was maximum, whereas in case of LEACH and TEEN, it was relatively low.

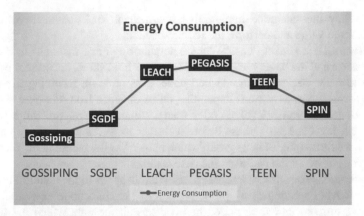

Fig. 3 Energy consumption in the existing protocols

PROTOCOL	FEATURES
Gossip	Low energy consumption, Increased latency
SGDF	Limited energy consumption, Reliable
PEGASIS	Maximum energy consumption, Greater Performance
Dynamic Gossip	Limited energy consumption, Better network lifetime
LEACH	Limited energy consumption, Cluster head ambiguity
TEEN	Limited energy consumption, Event based delivery
Flossip	Low energy consumption, Reliable
SPIN	Request Packet Collision, Limited energy consumption

Fig. 4 Feature comparison among various protocols

It was observed that gossiping protocol consumed very less power than SGDF which was introduced using the concept of gossiping [11]. Therefore, we assume that by making changes in the way the packets are forwarded and reducing unnecessary transmission of redundant data, a lot of improvement can be expected in automation process using wireless sensor networks. The feature comparison among the various protocols that are discussed in this paper is shown in Fig. 4. As we can see, the table shows that energy consumption is a drawback in all the protocols. We can try to come up with an algorithm that transmits the packets with minimum energy usage.

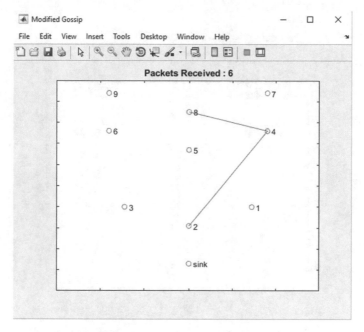

Fig. 5 Modified gossiping protocol showing communication between even-numbered sensor group

5 Results

The performance analysis of gossiping, new gossiping protocol and new gossiping protocol with sleep mechanism was done using MATLAB2017b. The simulation first displays the working of modified gossiping protocol as shown in Fig. 5, where each node selects the appropriate sensor node from the desired group either odd-numbered group or even-numbered group and forwards the packet to the sink node. Another window displays gossiping protocol using flooding mechanism (Fig. 6). The modified gossiping protocol is further enhanced by introducing sleep mechanism in the sensors as shown in Fig. 7, with an intention to reduce power consumption when sensors are idle and improve the network lifetime. The simulation displays the packets received by the sink node and also displays battery information once the sensor's power gets completely discharged (Figs. 8 and 9).

The simulation displays three scenarios: gossiping protocol, New gossiping protocol, and New gossiping protocol with sleep mechanism. Finally, it displays performance analysis of all the three protocols in the form of a graph where the graph provides information about the residual energy in each sensor at the end of the simulation as shown in Fig. 10. A line plot shows the average residual power in the sensors (Fig. 11), for different sleep time set in the gossiping protocol with sleep mechanism (Fig. 12). The results show that the New gossiping protocol with sleep mechanism performs better than the earlier gossiping protocols.

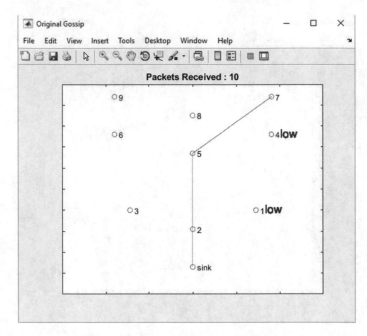

Fig. 6 Gossiping protocol using flooding mechanism

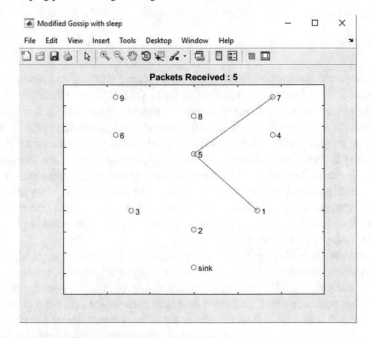

Fig. 7 Modified gossiping protocol along with sleep mechanism

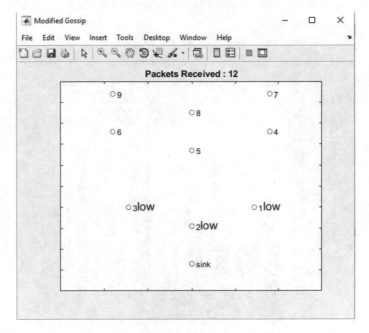

Fig. 8 Modified gossiping protocol displaying low when sensor battery gets completely discharged

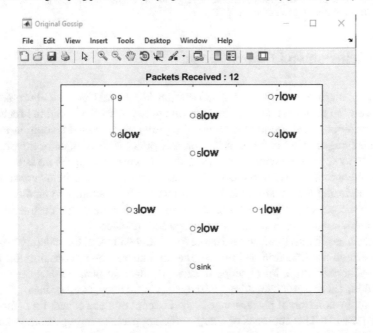

Fig. 9 Gossiping protocol displaying low when sensor battery gets completely discharged

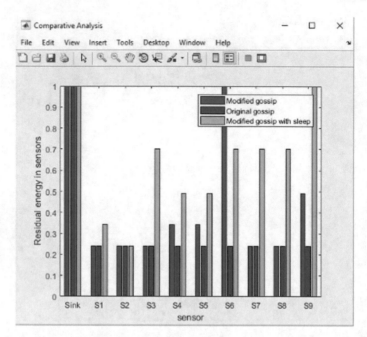

Fig. 10 Performance comparison of all the three protocols displaying the residual energy in each sensor after completion of simulation

6 Conclusion and Future Scope

This paper highlights the need for automation in the field of agriculture and also focuses on the problems that are faced in this process. We reviewed the features of various protocols and found that even if the protocols provided a solution to some of the challenges faced in wireless sensor networks (WSNs), none of the protocols were able to solve the major issue faced, that is, power usage. There is a need for improvement in the existing protocols because a lot of energy is being consumed by the sensor during the transmission process. Therefore, an attempt to modify the way in which the packets are transmitted from the source node to the destination node without redundancy and excess energy usage can be made.

After observing all the facts, we came to a conclusion that the existing protocols in WSN need modification to improve the amount of power consumption by the sensors. There is still a lot of scope because till date no protocol has been proved to be 100% energy-efficient. Along with energy efficiency, another factor that is an add-on to the solution of power usage is the concept of sleep scheduling among the sensors [13–15]. This approach of energy-efficient packet routing incorporated with sensor sleep mode can help in achieving maximum efficiency and also increase the network lifetime.

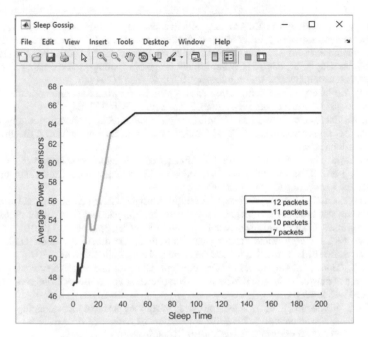

Fig. 11 Average residual power and packet transmission with sleep time (Sec)

Sleep Time (Sec)	Packets Received	Avg. Residual Power	Sleep Time (sec)	Packets Received	Avg. Residual Power
0	12	47%	12	10	54.43%
1	12	47.33%	13	10	54.43%
2	12	47.33%	14	10	52.90%
3	12	47.33%	15	10	52.90%
4	12	49.44%	16	10	52.90%
5	12	47.90%	17	10	52.90%
6	11	48.83%	30	7	63%
7	11	48.93%	50	7	65.10%
8	11	49.96%	100	7	65.10%
9	10	51.43%	200	7	65.10%
10	10	51.43%	500	7	65.10%
11	10	53.80%	1000	7	65.10%

Fig. 12 Average residual energy and packet delivery information in modified gossiping protocol for different sleep time

References

1. Stefanos, A., Kandris, D., Dimitrios, D., Douligeris, C.: Energy efficient automated control of irrigation in agriculture by using wireless sensor networks. J. Comput. Electron. Agric. **113**, 154–163 (2015)
2. Diaz, S., Perez, J., Mateos, A., Guerra, B.: A novel methodology for the monitoring of the agricultural production process based on wireless sensor networks. J. Comput. Electron. Agric. **76**, 252–265 (2011)
3. Pishyar, S., Ghiasian, A., Khayyambashi, M.R.: Gossip based energy aware routing algorithm for wireless sensor network. J. Comput. Netw. Commun. Secur. **3**, 164–172 (2015)

4. Bandyopadhyay, S., Coyle, E.: An energy efficient hierarchical clustering algorithm for wireless sensor networks. In: IEEE Proceedings of the 22nd Annual Joint Conference of the Computer and Communications Societies (INFOCOM 2003), San Francisco, pp. 1713–1723 (2003)
5. Dutta, R., Gupta, S., Paul, D.: Improvement on gossip routing protocol using TOSSIM in wireless sensor networks. J. Comput. Appl. **97**, 0975–8887 (2014)
6. Zhang, Y., Cheng, L.: Flossiping: a new routing protocol for wireless sensor networks. In: IEEE International Conference on Networking, Sensing, and Control (2004)
7. Yen, W., Chen, C.W., Yang, C.H.: Single gossiping with directional flooding routing protocol in wireless sensor networks. In: IEEE International Conference on Industrial Electronics and Application (2008)
8. Karthikeyan, A., Jagadeep, V., Rakesh, A.: Energy efficient multihop selection with PEGASIS routing protocol for wireless sensor networks. In: IEEE International Conference on Intelligence and Computing Research (2014)
9. Wu, B., Zhen, C., Linping, W., Jeng, W.Z.: Improved algorithm of pegasis protocol introducing double cluster heads in wireless sensor network. In: International Conference on Computer Mechatronics, Control and Electronic Engineering (CMCE), pp. 148–153 (2010)
10. Francesco, C.: Energy efficient routing algorithms for application to agro-food wireless sensor networks. In: IEEE International Conference on Communication (2014)
11. Hao, X., Tipper, D.: Gossip-based sleep protocol (GSP) for energy efficient routing in wireless ad hoc networks. In: IEEE Wireless Communication and Network Conference (WCNC) pp. 1305–1310 (2004)
12. Bulut, E., Korpeoglu, I.: Dssp: A dynamic sleep scheduling protocol for prolonging the lifetime of wireless sensor networks. In: IEEE International Conference on Advanced Information Networking and Applications Workshop pp. 725–730 (2007)
13. Swain, A.R., Hansdah, R.C., Chouhan, V.K.: An emergency aware routing protocol with sleep scheduling for wireless sensor networks. In: IEEE International Conference on Advanced Information Networking and Applications Workshop pp. 933–940 (2010)
14. Jain, S., Gupta, S.: Maximizing system lifetime of WSN by scheduling of wireless sensor nodes. J. Comput. Sci. Module. Comput. **3**, 639–650 (2014)
15. Sundaran, K., Ganapathy, V.: Energy efficient wireless sensor networks using dual cluster head with Sleep/Active mechanism. J. Sci. Technol. **9**, (2016)

Factual Instance Tweet Summarization and Opinion Analysis of Sport Competition

N. Vijay Kumar and M. Janga Reddy

Abstract Spilling information study now factual instance remains fetching the best ever then the majority well-organized method to obtain useful knowledge from what is happening now, letting group to respond rapidly once problematic originate hooked on opinion or to classify newest tendencies portion to recuperate their performance. The problem we try to solve is that cutting-edge absence of existence living cutting-edge obverse of TV usual, shape information dispensation scheme that make informative information concerning contest competitions then relate view of admirers toward competition production. By means of tweet information, we discover sub-events cutting-edge willing and then view of admirers position Twitter associated toward game. We endorse a scheme aimed at factual instance summarization of arranged sub-events aimed at sporting race by means of tweet information. We too suggest a method that examines spirits of persons placement Twitter. We focused on summarizing sporting events, specifically FIFA World Cup 2017 and IPL 2017. For a system using social media like twitter toward retain path of belongings trendy about, we appearance on behalf of next qualities: (I) gratitude of bursty subject by way of rapidly as the situation arises; (II) summarization of linked bursty theme; (III) examining viewpoint of followers then relating view toward ready.

Keywords Summarization · Opinion examination · Large information

N. Vijay Kumar (✉)
JJT University, Jhunjhunu, India
e-mail: vijaya59@gmail.com

M. Janga Reddy
CSE Department, CMR Institute of Technology, Hyderabad, India
e-mail: principalcmrit@gmail.com

© Springer Nature Singapore Pte Ltd. 2019   153
J. Wang et al. (eds.), *Soft Computing and Signal Processing* ,
Advances in Intelligent Systems and Computing 898,
https://doi.org/10.1007/978-981-13-3393-4_16

1 Introduction

Micro-blogging nowadays needs to develop an actual well-liked message instrument among Internet operators. Millions of messages are found on social networking sites like twitter1, facebook2, tumblr3 which might remain recycled publicity then communal educations. Education of the operator information of communal nets remains unique to the new tendencies of the periods. Large information container remains cast off aimed at the meeting of enormous then formless information which remains firm though consuming conservative record. The tweets that are done may be related to different topics then the commercial gentleman needs toward distinguish additional around the situation. Nevertheless the immensity of information ended tweet limits to them after experiencing impression. Consequently the essential of summarization remains nearby which resolve deliver an improved explanation.

We focus on summarizing sporting events, specially World Cup soccer competitions, meanwhile respectively occasion takings residence ended a small definite old-fashioned of period, here remains a substantial capacity of twitters around apiece occasion, then here remains media journalism of apiece occasion toward help by way of a golden customary. Intended for a scheme by communal broadcasting similar tweet toward retain path of belongings trendy everywhere, unique would remain observing on behalf of the subsequent qualities: (I) detection of a busty topic as soon as it emerges; (II) summarization of associated busty subjects; (III) examining sensation of soccer followers then relate sentiment toward competition performance.

2 Literature Survey

2.1 Literature Survey of Factual Instance Summarization

Primary, discovery substantial instants throughout competition similar sub-events articulating objective, permits, shots, etc., throughout competition. Next, decision insufficient significant twitters which best express the documented sub-event cluster. So inactivity fashionable dataset is summary. Chakrabarti et al. [1] recycled adaptive hidden Markov chain to study preparation of actions then recycled football match twitters detained America. Then it prepared not to identify actual sub-events. As it used unsupervised approach, it remained not likely to identify contributors and sub-events except competition were played many stints among similar binary players.

Arkaitz et al. [2] projected two-step procedure for the factual instance summarization of events sub-event recognition then tweet collection and inspected and evaluated dissimilar methods for individually of these two steps. Corney et al. [3] planned a transfer to aim at follower discovery then sub-event finding by classifying word or phrase that demonstration rapidly rise in incidence.

2.2 Literature Survey of Opinion Analysis

Opinion tokenization problems similar control HTML and XML markup, Twitter markup (names, Hashtags), Capitalization, Figures remain difficult to grip, though methods similar using n-grams, part of speech classification must remained working efficiently in [4, 5] for discovery the twitter sentiment using the machine learning techniques and other methodologies.

Dissimilar sentiment lexicon as lengthy as dissimilar lessons of positive, negative, strong, pronoun, quantifier, and various additional has been used to create WordNet of positive and negative sentiment lexicons [6–8]. There are two methods for it. Primary contain of expressions and rubrics anywhere we treasure recurrent incidence expressions, i.e., fish, paneer formerly sieve these by rubrics similar happen correct afterward opinion word, i.e., inordinate fish. Second find aspect in advance and treasure dataset linked to it must remained deliberate by Jiang et al. [6], Bollen et al. [4].

3 Methodology

we fetch the tweets using Twitter API v 1.1. In order to remove stop words then gross available geographies; we perform information spring cleaning then standardization. We search using twitter API v 1.1 toward wrinkle info by a diversity of hashtags like #BRAZIL, #USA, #FIFAWORLD CUP aimed at gathering Twitter linked toward competition. The collected tweets have to texture numerous phases near distinguish sub-events besides opinion of followers through competition production by way of shown in Fig. 1. In following subsection we will describe each stage broadly.

This work also presents identifying opinion of followers through competition by admiration toward group. We use a hybrid approach of removing opinion consuming straight then unintended facemask arrival of tweet information founded happening support vector machines (SVM), naive Bayes, logistic regression, and AFINN counting. Figure 1 displays entire planning of construction.

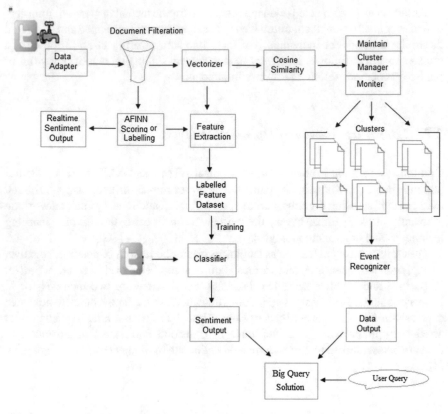

Fig. 1 Classification planning

3.1 Preprocessing

3.1.1 Data Adapter

The data adapter unit remains as the connection bordered through the communal stream plus NED. This unit remained envisioned toward becoming spinal unique text at a time after the waterway in consecutive mandate. This document is then passed on to the document filtration component.

3.1.2 Document Filtration

The flashy countryside of communal watercourses requires a percolation coating. The manuscript percolation constituent was envisioned to stretch such a coating. Leaflets which are not event related are cleanly available earlier to as extended as it for sub-event appreciation. This project removes additional calculation on irrelevant documents. The document filtration part receives the leaflets since the data adapter besides permits them on to the vectorizer if they permit the clean trial. This idea also eliminates hyperlinks, username, etc., before providing it to vectorizer. The application of the clean exam will differ in contingent on the social waterway.

3.1.3 Document Match and Clustering

Clustering algorithms are additional efficient methods to the NED duty. These procedures variety groups of identifications by the area that the identifications in a group altogether declaration on the alike occasion. Once categorizing a novel document, these procedures associate the centroids of the groups to the novel document. Uncertainty the detachment is inside a sure brink formerly that text is auxiliary to that group. Then it is off the record as NOVEL plus a novel group is shaped. Algorithm 1 shows a basic clustering approach.

For sample, Allan et al. recycled a changed form of tf.idf allowance that factored in file age [3]. Brants et al. used an incremental tf.idf model where they updated the idf values periodically [9].

Algorithm 1 Clustering algorithm for the NED task

Require: t $\longleftarrow$ input threshold
1: for all d in documents do
2:   dis_{min},c $\longleftarrow$ min_c { distance(d, clusters) }
3:     if dis_{min}> t then
4:           d $\longleftarrow$ NEW
5:           create new cluster(d)
6:     else
7:           d $\longleftarrow$ OLD
8:           c $\longleftarrow$ d

3.2 Event Detection

Watercourse processor chomps single document at a period since the communal watercourse. This process remained envisioned to stand for apiece document in course interplanetary and calculates its adjacent national. The tuple of text besides adjacent national is formerly directed to the group executive procedure.

3.2.1 Cluster Manager

Cluster Manger maintains and monitors clusters of similar identifications. This process obtains apiece original text besides its adjacent national since the waterway workplace procedure. Movements remain noticed by checking the amplification beat of the groups. Any period an occasion obligates remained noticed; the corresponding group is directed to the occasion recognizer process. The group executive process is intended to do binary purposes: to last groups besides to checked groups. The essential constituent to this process is the group.

3.2.2 Event Recognizer

The part of this process is to be acquainted by these proceedings assumed information of the planned proceedings the construction is annoying to sign. By this information, the arrangement efforts to classify apiece group addicted to a probable occasion. If charming, this occasion is production after the construction. Or other, the group is rejected. These strategy assistances comfort approximately untrue positives after the preceding procedures. These procedure consequences are advanced to occasion classifier.

3.2.3 Event Classifier

The project undertakes an angle of predefined proceedings that is obtainable, plus a set of keywords or expressions for apiece occasion. For sample, probable occasion strength be 'goal' besides the slope of keywords related by that occasion might be ['score,' 'scores,' 'scored,' 'goal,' 'goals']. The algorithm for categorizing an occasion is shown in Algorithm 2.

Algorithm 2 Event Classification

```
1: for all doc in cluster do
2:        for all event in predefined events do
3:               lexicons  get_lexicons(event)
4:               if any lexicon in doc then
5:                      votes[event]++
6: event ←——— max(votes)
7: if event.count > (cluster.count=2) then
8:        output event
9: else
10:       output unclassified
```

3.3 Sentiment Analysis

Followers fluent their feeling then after lateral to lateral opinion analysis we classify the opinion battered to any play actor or group with tall opinion to slightly feature similar transitory, goal, kick, etc. We employed a cross originate near to of removing opinion using straight and secondary facemask arrival of Twitter information then categorize opinion using support vector machines (SVM), naive Bayes, logistic regression, and SentiWordNet.

3.3.1 Opinion Classification Using SentiWordNet

The term's glosses are then used to train a machine learning classifiers. SentiWordNet is typically second hand in factual instance opinion organization as here there is no preparation usual. It is too recycled for organization dataset that is to be used by machine learning classifier.

3.3.2 Opinion Classification Using Machine Learning Techniques

Traditional is built up using exercise dataset. Advanced than emerging perfect, it is jumble sale to extra categorize the challenging dataset. We comprise using naive Bayes, SVM, and logistic regression classifiers with topographies removed from Twitter information consuming feature abstraction approaches for sentiment analysis.

4 Dataset and Results

4.1 Dataset

The construction is verified on twitters since five dissimilar competitions. IPL 2017 challenger 1, eliminator, challenger 2, and last competition twitters recycled. The assembly was likewise verified on FIFA 2017 last competition. The tweets were raised consuming #ipl and #UCLfinal hashtags.

4.2 Event Detection

Generous proceedings contain a sequence of instants, respectively, of which strength comprises movements by companies, the arbitrator, the followers, etc. Figure 2 shows spears in twitters through spirited.

The construction is cheerful to be acquainted by sub-events in sports through closely 79%. In cricket, it remained ingenious to know sub-events like wicket, six, four, halftime, etc. In football, it documented sub-events like goal, penalty, halftime, red card, yellow card, and corner. Table 1 displays proceedings then performers rank given to tweets for challenger 1 in IPL 2017 match.

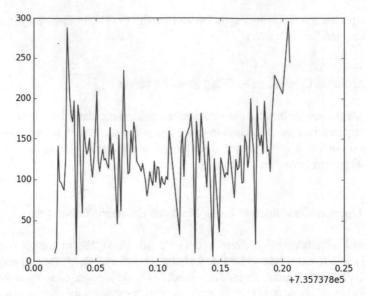

Fig. 2 Numeral of Twitter versus period

Table 1 Qualifier1 rank

Rank	Demonstration wise			
	Performer		Result	
	Term	Tweet total	Label	Tweet total
1	Virat	838	Run	790
2	Joe Root	974	Ball	890
3	Williamson	676	Out	971
4	David	523	Over	656
5	Rohit	487	Wicket	456

4.3   Opinion Analysis

The construction charity SVM, naive Bayes, and logistic regression classifier for categorizing opinion of followers. Future than likening the significances, we finish that SVM is crafty to categorize twitters opinion by resources of advanced precision additional than needs extra instance for teaching and challenging. Figure 3 demonstrates precision of unlike classifier for unlike competitions.

Table 2 demonstrates the period essential for exercise besides challenging dissimilar classifier in part.

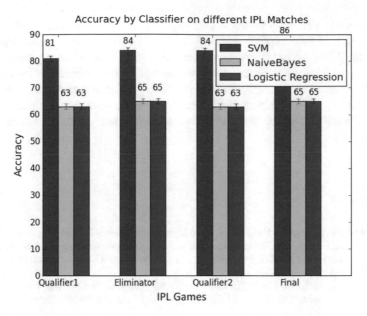

Fig. 3 Emotion classifier accuracy

Table 2 Sentiment classifier time analysis

IPL game	Dataset number of tweet	Machine learning classifier (Time in s)		
		SVM	Naive Bayes	Logistic regression
Challenger 1	Train (10,000)	7.082	0.43	0.261
	Test	5.117	0.89	0.213
Eliminator	Train (10,000)	5.73	0.32	0.145
	Test	4.28	0.56	0.04
Challenger 2	Train (10,000)	5.7	0.31	0.19
	Test	5.55	0.89	0.020
Final	Train (10,000)	5.69	0.34	0.27
	Test	6.76	2.18	0.025

5 Conclusion and Summary

The construction needs classifying sub-events in competition then usages burst in tweet amount to classify it. The construction examines opinion consuming supervised change to similar support vector machine, naive Bayes, and logistic regression. The schemes review tweets to trace greatest informative tweet around occasion, which needs starting clusters besides recording tweets. The construction examines a variation of requests like discovery group followers, opinion of follower with respect to group, greatest communicated sub-event besides. This paper concludes with a conversation of the likely future work. Originally, use of summaries would get well the detection of little capacity events which do not have a real vocabulary related with them. Then, the use of out-of-vocabulary (OOV) indulgence is probable to get healthier production of a scheme due to the meager syntax and terminology recycled by Twitter operators throughout alive proceedings. Solitary additional part of the future work would be in machine learning to mark a topic-conditioned classifier. Any additional part of the future work would be to classify opinion with respect to proceedings.

References

1. Chakrabarti, D., Punera, K.: Event summarization using tweets. ICWSM **11**, 66–73 (2011)
2. Arkaitz, Z., Damiano, S., Enrique, A., Julio, G.: Towards real-time summarization of scheduled events from twitter streams. In: Proceedings of the 23rd ACM Conference on Hypertext and Social Media, pp. 319–320 (2012)
3. Corney, D., Martin, C., Göker, A.: Two sides to every story: subjective event summarization of sports events using Twitter. In: ICMR2014 Workshop on Social Multimedia and Storytelling, pp. 662–672 (2014)
4. Bollen, J., Pepe, A., Mao, H.: Modeling public mood and emotion: Twitter sentiment and socio-economic phenomena (2009). arXiv:0911.1583
5. Seol, Y.-S., Kim, H.-W., Kim, D.-J.: Emotion recognition from textual modality using a situational personalized emotion model. Int. J. Hybrid Inf. Technol. **5**(2), 169–174 (2012)
6. Jiang, L., Yu, M., Zhou, M., Liu, X., Zhao, T.: Target-dependent twitter sentiment classification. In: Proceedings of the 49th Annual Meeting of the Association for Computational Linguistics: Human Language Technologies, vol. 1, pp. 151–160 (2011)
7. Sharifi, B., Hutton, M.-A., Kalita, J., Automatic summarization of twitter topics. In: National Workshop on Design and Analysis of Algorithm, Tezpur, India (2010)
8. Sakaki, T., Okazaki, M., Matsuo, Y.: Earthquake shakes twitter users: real-time event detection by social sensors. In: Proceedings of the 19th International Conference on World Wide Web, pp. 851–860 (2010)
9. Turney, P.D.: Thumbs up or thumbs down?: semantic orientation applied to unsupervised classification of reviews. In: Proceedings of the 40th Annual Meeting on Association for Computational Linguistics, pp. 417–424 (2002)
10. Lin, C., Lin, C.: Generating event storylines from microblogs. In: Proceedings of the 21st ACM International Conference on Information and Knowledge Management, pp. 210–216 (2012)
11. O'Connor, B., Balasubramanyan, R.: From tweets to polls: linking text sentiment to public opinion time series. In: Proceedings of the International AAAI Conference on Weblogs and Social Media, Washington, DC, pp. 511–519, May 2010
12. Stone, P.: Sentiment Lexicon General Inquirer: A Competitive Approach to Content Analysis. The MIT Press (1966)
13. Hole, V., Takalikar, M.: A survey on sentiment analysis and summarization for prediction. Int. J. Eng. Comput. Sci. (IJECS) **3**(12), 9503–9506 (2014). ISSN 2319-7242
14. Hatzivassiloglou, V., McKeown, K.: Predicting the semantic orientation of adjectives. In: Proceedings of the 35th Annual Meeting of the Association for Computational Linguistics and Eighth Conference of the European Chapter of the Association for Computational Linguistics: Association for Computational Linguistics, pp. 174–181 (1997)

A New Bat Algorithm with Distance Computation Capability and Its Applicability in Routing for WSN

Shabnam Sharma, Sahil Verma and Kiran Jyoti

Abstract Bat algorithm (BA) is developed by Xin She Yang in 2010 and gaining popularity due to its astonishing feature of echolocation. It has drawn the attention of many researchers, to contribute in the performance enhancement of the algorithm. The proposed variant of Bat algorithm computes 'distance' by calculating the similarity among the pulse emitted by artificial bats and the received echo. This work also focuses on the applicability of the proposed variant of BA for finding optimal route in wireless sensor network, while reducing the delay, which may occur due to heavy traffic on the optimal path. The results of the proposed algorithm are evaluated, in terms of best, mean, worst, median and standard deviation, for the time required to obtain optimal results on the basis of distance (as fitness value) between the sensing nodes and outperforms standard BA.

Keywords Bat algorithm · Routing · Swarm intelligence
Wireless sensor network

1 Introduction

Combinatorial problems and optimization are those fields, which are intertwined inextricably. Most of the combinatorial problems are treated as optimization problems. Combinatorial problems prefer those optimization techniques which can provide the solution in reasonable computational time and can select finite solutions among an infinite set of solutions. Many papers blend different categories of

S. Sharma (✉) · S. Verma
Lovely Professional University, Jalandhar, Punjab, India
e-mail: shabnam09sharma@hotmail.com

S. Verma
e-mail: sahil.21915@lpu.co.in

K. Jyoti
Guru Nanak Dev Engineering College, Ludhiana, Punjab, India
e-mail: kiranjyotibains@yahoo.co.in

© Springer Nature Singapore Pte Ltd. 2019   163
J. Wang et al. (eds.), *Soft Computing and Signal Processing* ,
Advances in Intelligent Systems and Computing 898,
https://doi.org/10.1007/978-981-13-3393-4_17

optimization techniques. In [1], author has discussed broad categories of optimization techniques, which include trajectory techniques, simulated annealing, tabu search, meta-heuristic approaches, explorative local search techniques, population-based techniques, evolutionary computational techniques and hybridization of these techniques. Traveling salesman problem, routing problem, job scheduling problem, training sports session problem and many more fall in the category of combinatorial optimization problems. Based on the nature of the problem, selection of appropriate optimization technique is done, as the implementation of the same optimization technique for different problems or the same problem in different scenarios, would not yield optimal results.

Due to scalability, mobility of sensing nodes, heterogeneity of nodes, on the fly creation of network, wireless sensor network has gained popularity since last two decades. Sender sensing node needs to communicate with other intermediate sensing nodes, lying in its proximity, in order to transfer the data to destination sensing node. To find the optimal path between the sensing nodes, require the implementation of best suited optimization technique. This problem is named as 'Routing' in WSN and is the main focus of this research. For the selection of optimal path, swarm intelligence technique, namely, bat algorithm is reviewed and selected to carry out the proposed work. The selection of bat algorithm is done, due to its astonishing feature of echolocation. In the next section, review of various traffic avoidance routing protocols is carried out, followed by the study of applicability of bat algorithm.

2 Literature Survey

Author of [2] reviewed the application areas of WSN, their QoS requirements and the routing protocols which fulfill these QoS requirements. Depending upon the network structure, routing protocols are broadly classified as flat-based routing, hierarchy-based routing (also known as, sometimes, cluster-based routing) and location dependent routing [3]. A review of challenges and security concerns associated with wireless sensor network is done by the author in [4]. Various researchers have contributed in the field of swarm intelligence, either by developing a new nature-inspired algorithm or by enhancing the performance of existing algorithms. In [5], author has reviewed variants of BA and also suggested key areas that can be explored to further improve its performance. LEACH protocol, one of the famous routing protocols, is described in [6]. In [7], swarm intelligence-based routing protocols are categorized as ant colony-based, artificial bee colony-based and SLIM. Opportunistic based routing protocol makes the use of 'broadcast' feature of WSN, in [8]. Author has proposed a novel technique to achieve better energy efficiency in [9]. Using BA, author has suggested congestion free path in vehicular ad hoc network [10]. The author of [11] has described how genetic algorithms can be used for routing in WSN environment. One can classify WSN, as mentioned in [12], on the basis of network requirements, which include, energy level of sensing nodes, memory requirement of sensing nodes and their processing power. In [13], author has proposed a variant of BA and named as guidable BA and validated the results using mathematical functions. The concept of binary bat algorithm relies on the mapping of continuous search space to binary search space, proposed in [14] and validated on engineering problem. In [15], author has proposed MBO (multi-objective bat algorithm) by enhancing the performance of BA and result validation is carried out on uni-modal and multi-modal functions. In [16], based on the upper bound of queue, author selects the optimal path. The drawback of this approach of reducing the traffic over the channel is overhead and delay in exchanging control information related to the queue size. In the next section, variant of bat algorithm is proposed which computes distance based on the emitted pulse and received echo. The proposed algorithm is then applied to solve the routing problem of WSN. At last, the algorithm is proposed for suggesting traffic-free optimal path, followed by result validation section.

3 Bat Algorithm

Bat algorithm is inspired from echolocation behavior of real bats. Real bats work in three phases: approach phase, target phase and terminal phase, while capturing their targets [17]. To execute all the three phases of capturing process, real bats have fraction of seconds. Xin She Yang got inspired from this way of accomplishing the task and proposed bat algorithm for obtaining the optimal solution in lesser time period. Real bats have the ability to select their food source in the presence of many food sources. They made the decision based on the quality of food source, and the quality is measured in terms of amount of energy, real bats will gain, after eating that food [18]. Real bats can also decide which prey to target depending upon the distance between them and targets. In [19], author has proposed BA, with the assumption of computing distance in 'magical' way. Many researchers have contributed in the way of computing distance, either by using hamming code concept [20] or by using two-opt and three-opt concept. Artificial bat algorithm adjusts its parameters, as represented using Eqs. 1–3, to explore more number of solutions and selection of local optima is carried out. This phase is known as diversification or exploration phase.

$$f_i = f_{min} + (f_{max} - f_{min}) * \beta \tag{1}$$

$$v_i^t = v_i^t - 1 + (x_i - x_{best}) * f_i \tag{2}$$

$$x_i^t = x_i^t - 1 + v_i^t \tag{3}$$

Result Analysis		On the basis of Execution Time						On the basis of Cost					
		Standard Bat Algorithm			Proposed Algorithm (BA-DC)			Standard Bat Algorithm			Proposed Algorithm (BA-DC)		
Number of Transmissions	Performance Evaluation	Bat Population						Bat Population					
		10	15	20	10	15	20	10	15	20	10	15	20
10	Best	3.976	4.250	4.633	4.006	4.280	4.663	155.263	165.977	180.914	152.263	162.977	177.914
	Median	4.058	4.337	4.728	4.088	4.367	4.758	156.746	167.111	182.150	153.746	164.111	179.150
	Worst	4.123	4.407	4.804	4.153	4.437	4.834	158.147	168.097	183.225	155.147	165.097	180.225
	Mean	4.057	4.337	4.727	4.087	4.367	4.757	156.757	167.093	182.131	153.757	164.093	179.131
	SD	0.041	0.044	0.048	0.039	0.037	0.038	0.100	0.082	0.070	0.092	0.089	0.077
15	Best	4.187	4.552	4.684	4.217	4.582	4.714	163.492	177.761	182.905	160.502	174.771	179.915
	Median	4.273	4.645	4.780	4.303	4.675	4.810	165.053	178.975	184.154	162.063	175.985	181.164
	Worst	4.342	4.720	4.857	4.372	4.750	4.887	166.528	180.031	185.241	163.538	177.041	182.251
	Mean	4.272	4.644	4.779	4.302	4.674	4.809	165.065	178.956	184.135	162.075	175.966	181.145
	SD	0.043	0.047	0.048	0.041	0.046	0.046	1.048	0.068	0.070	0.099	0.067	0.070
20	Best	4.234	4.889	5.142	4.264	4.919	5.172	165.356	190.915	200.815	161.516	187.075	196.975
	Median	4.321	4.989	5.248	4.351	5.019	5.278	166.934	192.220	202.187	163.094	188.380	198.347
	Worst	4.391	5.070	5.333	4.421	5.100	5.363	168.426	193.354	203.380	164.586	189.514	199.540
	Mean	4.320	4.988	5.247	4.350	5.018	5.277	166.946	192.199	202.166	163.106	188.359	198.326
	SD	0.044	0.050	0.053	0.042	0.046	0.047	1.060	0.074	0.077	1.060	0.071	0.067

Fig. 1 Performance evaluation of BA-DC on the basis of execution time and cost

To obtain global optimal result, relevant parameters are required to update, as mentioned in Eqs. 1–3. Searching optimal result near local optima and obtaining global optimal result, is known as intensification or exploitation phase. During exploitation phase, as artificial bats reach closer to the target, the pulse emission rate increases, while, the loudness decreases, as written in Eqs. 3 and 4.

$$A_i^{t+1} = 0.9 * A_i^t; \tag{4}$$

$$r_i^{t+1} = r_i^t * (1 - \exp(-1 * 0.9 * i)); \tag{5}$$

Inspired from real bats, problem-solving strategy adopted by artificial bats is depicted in Fig. 1.

4 Proposed Algorithm

The author of [19] has considered the computation of distance in 'some magical way' and, hence, no technique is adopted for the same. Since then, authors have proposed different ways to calculate the distance between artificial bats and the solutions available. In this work, a way of computing distance, using cross correlation, is proposed and based on that variant of BA is proposed. Further, to avoid the traffic over selected optimal path, a technique is proposed in subsequent sections.

Algorithm 1: BAT Algorithm
Data: Assign initial value to position x_i, velocity v_i, pulse rate r_i, loudness a_i
and frequency f_i of bat population
Result: Optimized Solution
Begin
 Initialize max as iteration's count
    while (curr_iter<max)
            Explore new solutions by updating x_i, v_i, f_i.
            if (random > r_i)
            Select optimal solution among existing solutions
            Compute the new local solution by exploring the neighborhood
            End if
             if ((random < a_i) AND (f(x^*) > f(x_i))
            Explore other solutions by increasing r_i and decreasing the value of a_i.
            End if
             Rank the solutions and find the best solution, x_* among all.
     end
    Post-process the solutions.
End

4.1 Bat Algorithm for Distance Computation (BA-DC)

This algorithm focuses on computing the distance between artificial bats and feasible solutions by finding the similarity among the emitted pulse and echo received. Moreover, the distance computed is also used as fitness value, rather than relying on mathematical function, namely, Rosenbrock. Higher the fitness values of any solution, lesser the chances of its selection as optimal result. The entire concept of bat algorithm will remain the same, except the way of computing fitness value. The concept of the proposed algorithm is the same as that of standard bat algorithm, with a different way of computing fitness value, using distance as a parameter. The steps to compute the distance between the artificial bat and target are given in Algorithm 2. Thus, Algorithm 2 is named as 'Bat Algorithm for Distance Computation (BA-DC)'.

4.2 Bat Algorithm for Traffic Avoidance over Optimal Path (BA-TAOP)

To provide optimal solution to routing problem in WSN, researchers have proposed numerous optimization techniques. Based on the nature of problem and expected outcome, different optimization techniques are applied in different scenarios. To check the performance of the proposed variant of bat algorithm and its applicability in solving the routing problem, Algorithm 4 illustrates the traffic-free optimal path in WSN. In WSN, many sensing nodes may communicate with other sensing nodes, where there could be the chances of assigning same optimal path to multiple source–destination nodes pair. It may result into the traffic over the optimal path for specific interval of time. To avoid such situation, another algorithm is proposed in this research work and named as 'Bat Algorithm for Traffic Avoidance over Optimal Path (BA-TAOP)'.

Algorithm 2: Bat Algorithm for Distance_Computation()
Data: Input Solution.
Result: Fitness Value of each solution, i.e. distance
Begin
  if (Obstacle)
  If present, delay, β, and attenuation, α, will affect the solution.
    $RS_i = (SS_i * \alpha) + \beta$
    else
    $RS_i = SS_i + \beta$
  end if
  Compute the similarity among sound produced SS_i and Echo received RS_i.
  [Corr,Lags]= xcorr(SS_i, RS_i) and compute the delay samples w.r.t. SS_i and RS_i
  Delay_Samples= Lags(find(Corr==max(Corr)))
  Compute the distance, using Delay_Samples and Time_Samples.
  Select the 'best solution', having minimum fitness value of 'distance'.
End

Algorithm 3: Traffic Avoidance over Optimal Path
Data: Input number of bats, N and number of Transmissions T.
Result: Traffic free optimal path
Begin
  for i = 1 to T
    Select the starting' sensing node and ending' sensing node.
    Compute the optimal path using BA-DC proposed algorithm.
    Maintain the record of starting' sensing node, represented as visited node.
    if starting' sensing node == visited node (i, :)
        Increment counter and check for other starting' sensing node
    end if
  end
   for i=1 to T
       if count(i,:)>threshold
           Compute the optimal path, again, using BA-DC proposed algorithm
       end if
   end
  Optimized route by calling Standard BA, using 'distance' as fitness value
  Display Cost for each route assigned to each packet.
End

In this section, an algorithm is proposed to offer traffic-free optimal path to the nodes. The algorithm is written above. Offering traffic-free path will also reduce the delay. Here bats, N, represent the number of nodes present in WSN and T represents number for transmissions, ongoing at a particular interval of time. For each transmission, number of bats deployed for selecting the optimal route may vary.

5 Results and Validation

To implement the proposed algorithms, MATLAB is used as simulation tool. To evaluate the performance of the proposed algorithm (BA-DC) while finding optimal route, comparison is carried out with respect to standard bat algorithm. For number of transmissions [11, 16, 19], the bat population is varied over [11, 16, 19] for 10 rounds. Firstly, the performance evaluation of proposed algorithm (BA-DC) is done with respect to standard bat algorithm, without considering traffic over the optimal path, without and using traffic-free optimal path algorithm and results are evaluated using best, worst, median, mean and standard deviation. Performance evaluation of BA-DC and BA-TAOP on the basis of execution time and cost are depicted in Figs. 1 and 2.

5.1 Performance Evaluation of Proposed Algorithm

From the results, it has been concluded that the proposed algorithm BA-DC, using Algorithm 2, produce more optimal results in comparison with standard bat algorithm, in terms of cost. Execution times of proposed algorithms are more, but, it gives more prominent results.

Result Analysis		On the basis of Execution Time						On the basis of Cost					
		Standard Bat Algorithm			Proposed Algorithm (BA-TAOP)			Standard Bat Algorithm			Proposed Algorithm (BA-TAOP)		
Number of Transmissions	Performance Evaluation Parameters	Bat Population						Bat Population					
		10	15	20	10	15	20	10	15	20	10	15	20
10	Best	4.030	4.503	4.823	4.045	4.351	4.723	151.273	162.987	176.924	149.293	159.157	174.944
	Median	4.070	4.379	4.832	4.078	4.391	4.794	152.756	164.121	178.160	150.776	160.291	176.180
	Worst	4.223	4.982	4.901	4.732	4.412	4.801	154.157	165.107	179.235	152.177	161.277	177.255
	Mean	4.079	4.632	4.862	4.067	4.311	4.872	152.767	164.103	178.141	150.787	160.273	176.161
	SD	0.040	0.043	0.047	0.038	0.038	0.037	0.985	0.885	0.647	0.954	0.815	0.647
15	Best	4.151	4.343	4.386	4.173	4.442	4.785	159.502	174.641	179.915	156.672	170.941	177.495
	Median	4.241	4.432	4.993	4.363	4.691	4.673	161.063	175.855	181.164	158.233	172.155	178.744
	Worst	4.349	4.423	4.999	4.413	4.812	4.866	162.538	176.911	182.251	159.708	173.211	179.831
	Mean	4.213	4.446	4.343	4.176	4.773	4.671	161.075	175.836	181.145	158.245	172.136	178.725
	SD	0.041	0.040	0.048	0.043	0.043	0.043	1.044	0.635	0.664	1.011	0.585	0.564
20	Best	4.233	4.777	4.983	4.332	4.777	4.992	162.366	187.795	197.825	160.386	186.925	195.405
	Median	4.124	4.982	4.783	4.213	4.671	4.675	163.944	189.100	199.197	161.964	188.230	196.777
	Worst	4.381	5.012	5.012	4.399	5.016	5.031	165.436	190.234	200.390	163.456	189.364	197.970
	Mean	4.230	4.992	4.784	4.312	4.773	4.673	163.956	189.079	199.176	161.976	188.209	196.756
	SD	0.042	0.049	0.049	0.048	0.046	0.047	1.049	0.984	0.677	1.049	0.674	0.673

Fig. 2 Performance evaluation of BA-TAOP on the basis of execution time and cost

5.2 Avoiding Traffic over Optimal Path in WSN

Further with the inclusion of BA-TAOP, using Algorithm 3, the cost of obtaining traffic-free optimal path is obtained in lesser cost, in comparison with standard BA and BA-DC.

6 Conclusion and Future Work

The main goal of this research is to review the routing techniques and versions of BA, to propose a new technique of routing in WSN which can help in avoiding traffic over optimal path. Two variants of BA were developed in this work and results are compared with BA, which proves to be beneficial to use while selecting optimal path. In the future, applicability of the proposed algorithm will be evaluated in other fields. Many new meta-heuristic techniques have been proposed, so hybridization of proposed algorithm with those techniques will result in producing more optimal solutions. Real bats possess the capability of jamming the pulse emitted or echo received. This feature can also be explored to develop another variant of bat algorithm.

References

1. Blum, C., Roli, A.: Metaheuristics in combinatorial optimization: overview and conceptual comparison. ACM Comput. Surv. (CSUR) **35**(3), 268–308 (2003)
2. Prathap, U., Shenoy, P.D., Venugopal, K.R., Patnaik, L.M.: Wireless sensor networks applications and routing protocols: survey and research challenges. In: 2012 International Symposium on Cloud and Services Computing (ISCOS), pp. 49–56, Dec 2012
3. Singh, S.P., Sharma, S.C.: A survey on cluster based routing protocols in wireless sensor networks. Procedia Comput. Sci. **45**, 687–695 (2015)
4. Sharma, M.: Wireless sensor networks: routing protocols and security issues. In: 2014 International Conference on Computing, Communication and Networking Technologies (ICCCNT), pp. 1–5. IEEE
5. Sharma, S., Luhach, A.K., Jyoti, K.: Research and analysis of advancements in BAT algorithm. In: 2016 3rd International Conference on Computing for Sustainable Global Development (INDIACom), pp. 2391–2396. IEEE (2016)
6. Tyagi, S., Kumar, N.: A systematic review on clustering and routing techniques based upon LEACH protocol for wireless sensor networks. J. Netw. Comput. Appl. **36**(2), 623–645 (2013)
7. Zengin, A., Tuncel, S.: A survey on swarm intelligence based routing protocols in wireless sensor networks. Int. J. Phys. Sci. **5**(14), 2118–2126 (2010)
8. Jadhav, P., Satao, R.: A survey on opportunistic routing protocols for wireless sensor networks. Procedia Comput. Sci. **79**, 603–609 (2016)
9. Jung, S.G., Kang, B., Yeoum, S., Choo, H.: Trail-using ant behavior based energy-efficient routing protocol in wireless sensor networks. Int. J. Distrib. Sens. Netw. (2016)
10. Bhatt, M., Sharma, S., Prakash, A., Pandey, U. S., Jyoti, K.: Traffic collision avoidance in VANET using computational intelligence. Int. J. Eng. Technol. (2016)
11. Shahi, B., Dahal, S., Mishra, A., Kumar, S.V., Kumar, C.P.: A review over genetic algorithm and application of wireless network systems. Procedia Comput. Sci. **78**, 431–438 (2016)

12. Camilo, T., Carreto, C., Silva, J. S., Boavida, F.: An energy-efficient ant-based routing algorithm for wireless sensor networks. In: International Workshop on Ant Colony Optimization and Swarm Intelligence, pp. 49–59. Springer, Berlin, Heidelberg Sep 2006
13. Chen, Y.-T., Shieh, C.-S., Horng, M.-F., Liao, B.-Y., Pan, J.-S., Tsai, M.-T.: A guidable bat algorithm based on Doppler effect to improve solving efficiency for optimization problems. Comput. Collect. Intell. Technol. Appl. **8733**, 373–383 (2014)
14. Mirjalili, S.M., Yang, X.-S., Mirjalili, S.: Binary bat algorithm. Neural Comput. Appl. 663–681 (2014)
15. Zhou, Y., Li, L.: A novel complex-valued bat algorithm. Neural Comput. Appl. **25**(6), 1369–1381 (2014)
16. Manshahia, M.S., Dave, M., Singh, S.B.: Improved bat algorithm based energy efficient congestion control scheme for wireless sensor networks. Wirel. Sens. Netw. **8**(11), 229 (2016)
17. Kalko, E.K.: Insect pursuit, prey capture and echolocation in pipestirelle bats (Microchiroptera). Anim. Behav. **50**(4), 861–880 (1995)
18. Simmons, J.A.: A view of the world through the bat's ear: the formation of acoustic images in echolocation. Cognition **33**(1–2), 155–199 (1989)
19. Yang, X.S.: A new metaheuristic bat-inspired algorithm. In: Nature Inspired Cooperative Strategies for Optimization, pp. 65–74. Springer, Berlin, Heidelberg (2010)
20. Osaba, E., Yang, X.S., Diaz, F., Lopez-Garcia, P., Carballedo, R.: An improved discrete bat algorithm for symmetric and asymmetric traveling salesman problems. Eng. Appl. Artif. Intell. **48**, 59–71 (2016)

Optimizing Network QoS Using Multichannel Lifetime Aware Aggregation-Based Routing Protocol

Uma K. Thakur and C. G. Dethe

Abstract Improvement in quality of service (QoS) of wireless networks has always been a study subject for wireless network designers worldwide. Optimization of end-to-end communication delay, reduction in energy consumption, improvement in network throughput and reduction in end-to-end communication delay jitter are some of the parameter optimizations which are used to improve the QoS of the wireless networks. In this paper, we propose a QoS aware routing protocol which uses a combination of delay and energy aware routing with data aggregation and multichannel communication in order to reduce the energy consumption, reduce the end-to-end delay and improve the network throughput. The simulation results show that there is a more than 20% improvement in network communication speed, and at least 15% improvement in the network lifetime after using the proposed QoS aware routing protocol.

Keywords QoS · Delay · Throughput · Aggregation · Multichannel
Energy aware

1 Introduction

QoS optimization in wireless networks has been a topic of study for more than a decade now. While researchers have claimed to improve the QoS by using various optimization techniques like route optimization, selected node optimization, multichannel usage, and several others, but the study is still ongoing and is a NP-hard problem to solve. Thus, researchers can only provide a finite number of solutions to the problem, but the best solution is solely dependent on the technology in place and will always keep on changing based on the advances in network optimizations.

One such solution is proposed in this paper, which uses a QoS aware routing algorithm, a layer of compression and data aggregation and multichannel routing with

U. K. Thakur (✉) · C. G. Dethe
Sant Gadge Baba Amravati University, Amravati, India
e-mail: umapatel21@gmail.com

© Springer Nature Singapore Pte Ltd. 2019
J. Wang et al. (eds.), *Soft Computing and Signal Processing* ,
Advances in Intelligent Systems and Computing 898,
https://doi.org/10.1007/978-981-13-3393-4_18

the help of distance and energy measures. QoS aware routing usually refers to the field of routing where the routing protocol is developed in such a manner that the QoS parameters like end-to-end communication delay, throughput, energy and others are optimized. This is done by incorporating these parameters while selecting the route for communication. For example, if we need to optimize the delay and energy while routing, then the node selection process uses a metric like distance/energy and then minimizes it for each of the selected nodes. Thus, the nodes with minimum distance to energy ratio are selected for routing, this ensures that the nodes selected are having lowest distance of communication, and have the highest energy of communication, thus if a node has high energy for communication, then it will have higher lifetime when compared to a node with lower energy, thereby improving the delay and the energy consumption of the network, thus optimizing the overall QoS of the network, and such a protocol is called as a QoS aware routing protocol [1].

1.1 Overview

In this paper, we have designed a novel protocol for QoS optimizing in terms of end-to-end communication delay, network throughput, network lifetime and network jitter, which is a measure of the delay consistency in the network. This is achieved by using a QoS aware protocol and combining it with a multichannel distance to energy-based data aggregation routing protocol. The details about this proposed protocol are described in Sect. 3 of this paper. The next section describes some of the standard protocols which are used while QoS optimizations in wireless networks, followed by the proposed protocol, which is followed by a result analysis and comparison section, where we have performed an in-depth analysis of the proposed protocol and compared it's performance with the existing protocols in order to evaluate the superiority of our proposed protocol over other standard techniques. Finally, we have laid out some finer points which describe how the work can be taken further by other researchers in order to further optimize the QoS of the overall wireless network.

2   Literature Review

Researchers have worked rigorously on improving the overall network performance of wireless networks using different optimization techniques. In the research done by Zhao [2] and team on low power and lossy networks, they have used a region-based AODV routing protocol which is based on the P2P protocol of communication. The node discovery is restricted to a particular region based on the selected source and destination, thereby saving energy which is needed for node discovery in networks. They have obtained an improvement in both network delay and network lifetime by using the region-based protocol. In contrast to this research, the work done by Barcelo [3] and their team focuses on addressing the mobility in AODV by using position assisted metrics, wherein they apply the Kalman filter for finding the nodes which will be used for routing. The paper introduces Kalman position-based AODV protocol, which reduces the network load and thereby reduces the network cost and increases the overall network lifetime.

Another interesting protocol MoMoRo [4] is proposed by Jeong Gil Ko and Marcus Chang. In this protocol, the researchers have provided mobility assistance to low power wireless devices. Basically, it is a separate layer in the communication stack which collects network data and converts it into fuzzy decision values in order to estimate the quality of link for the routing. This allows the protocol to be both flexible and highly run-time reconfigurable. The MoMoRo technique uses the widely accepted AODV protocol for routing and achieves a high packet delivery ratio of 96%.

Some researchers have also proposed how computational intelligence can be applied to wireless sensor networks. A detailed description of this is given by Kulkarni and his team [5], wherein they have compared various algorithms for computational intelligence like neural networks, fuzzy logic, evolutionary algorithms like genetic algorithm, swarm intelligence techniques like particle swarm optimization, artificial immune systems and reinforcement learning. Using this research, we evaluated that reinforcement learning which is a part of machine learning is one of the best choices for routing in wireless networks and can be exploited further for increased network performance. Clique, Q-Routing, DRQ-Routing, Q-RC, RL-Flooding, TPOT-RL, SAMPLE, AdaR, Q-PR, Q-Fusion, RLGR are some of the examples of algorithms which are developed by various researchers over the globe to demonstrate how reinforcement learning can be used to optimize the network performance.

Another example of reinforcement learning is given by Forster and Murphy [6] in their paper on FORMS, wherein they are using reinforcement learning to optimize multiple sinks in the WSN networks [7–9]. They have used Q-Routing technique to the multiple sink problems and obtained a low network overhead while maintaining an acceptable network QoS in practical WSNs. Their algorithm natively supports node failure and sinks mobility and reduces the routing cost [10–12]. Our work is inspired by these algorithms, and we thus developed a novel technique which is based on machine learning, for routing the data in the given wireless network environment. The next section describes our algorithm in details.

3 QoS Optimization with Multichannel Data Aggregation

Our routing algorithm can be described as follows,

Deploy a network of N nodes placed randomly in an area of $X \times Y$ m^2.
Select any source (S) and destination (D) from the network for routing process.
Let the Euclidean distance between node S and D be dref.
Select all nodes from the network, where the following conditions are satisfied,

$$dsn + dnd > dref$$
$$dsn < dref$$
$$dnd < dref$$

where

dsn   Distance between source to selected node
dnd   Distance between selected node to destination.

This filters in only those nodes which are in the routing path and removes all other nodes.

For each node in the path, evaluate the following metric,

$$Metric = di/Ei$$

where

di   Distance between the nodes
Ei   Energy of the source node.

Start the node selection from the source till the destination node is reached.
Once reached, send the data on the selected path.
Before sending the data, apply data aggregation at the source node.
Split the aggregated data into k parts, where k is the number of channels available for routing.

Send the data on all the k channels from the source node to the destination.
Repeat this process for all communications.

The above algorithm makes sure that the data is sent from the source to destination with minimum delay, and minimum energy due to data aggregation, multichannel communication and incorporation of d/E factor in the routing process. The throughput is optimized as well due to improvement in delay and reduced packet loss due to multichannel communication. This makes sure that the packet is transmitted in the almost same timing interval as the previous packets, thereby reducing the jitter of the network. The detailed result analysis is mentioned in the next section.

4   Results and Analysis

We simulated our routing protocol in the network simulator version 2.34 environment, under the following network conditions in Table 1, the configuration for network is as follows.

For comparison purpose, we compared our proposed protocol with the AODV routing for the wireless network, and the following parameters are obtained in Table 2.

Table 1 Network configuration

Network parameter	Value
Network type	Wireless
Number of nodes	30–100
Network area	300 × 300 m
Routing protocols	QoS aware
Packet size	1000 bits @ 0.001 packets per second
Number of communications	2–20
Initial node energies	Randomized, with maximum energy of 1000 mJ per node
Energy model	2 mJ per transmission 1 mJ per reception 0.1 mJ idle energy

Table 2 AODV comparison

Nodes	Comms.	Delay AODV (ms)	Delay proposed (ms)	% Improv.
20	2	0.31	0.24	21.74
20	3	0.35	0.27	22.78
20	4	0.38	0.26	31.10
20	5	0.41	0.33	18.45
20	6	0.44	0.32	27.39
20	7	0.45	0.35	22.17
20	8	0.48	0.36	24.85
20	9	0.49	0.33	31.65
20	10	0.56	0.45	19.47
50	5	0.37	0.32	12.49
50	6	0.39	0.28	27.31
50	8	0.52	0.41	20.32
50	12	0.63	0.49	23.00
50	15	0.65	0.50	22.83
50	20	0.73	0.52	28.61
75	5	0.66	0.46	30.32
75	10	0.69	0.54	22.64
75	15	0.72	0.53	26.16
75	20	0.79	0.55	29.92
100	5	0.46	0.32	30.39
100	10	0.70	0.47	32.43
100	15	0.79	0.59	26.18
100	20	0.82	0.60	27.20
Mean improvement		0.556	0.412	26%

Table 3 Performance comparison

Parameter	AODV	Proposed	% Improvement
Avg. delay (ms)	0.556	0.412	26
Avg. energy (mJ)	3.126	2.198	29
Avg. PDR (%)	99.5	98.6	0
Avg. throughput (kbps)	137.8	134.9	−0.5
Avg. jitter (ms)	0.0062	0.0058	7

Similar comparisons were made for energy, packet delivery ratio, throughput and jitter. The following table shows the performance comparison for all the 5 parameters,

From the above Table 3, we can observe that the network delay has been minimized, the energy consumption has been reduced by maintaining a constant average packet delivery ratio and average throughput. The overall packet delivery jitter has also been slightly improved using the machine learning approach. The delay is reduced due to selection of minimum distance nodes for routing, while energy is reduced because of it's inclusion in the routing metric as an inversely proportional parameter. Due to reduction in delay, the jitter is also reduced, and thus, it makes the network more reliable and consistent in terms of packet delivery times at the receiver. The PDR and throughput of AODV are already optimized, and thus there is minimal scope of improvement in that area. We recommend researchers to further evaluate this machine learning routing technique in order to check it's viability for the applications for which they would be designing the communication network.

5 Conclusion

The proposed approach when applied to the wireless network gives a significant improvement in network performance, when compared with the recent de facto AODV routing algorithm. The performance improvement to network lifetime is more than 25%, while the delay minimization is more than 20% for a wide variety of network simulation parameters. This causes the network throughput to reduce by an infinitesimal percentage which is admissible by the wireless networks, due to the fact that our algorithm increases the energy consumption efficiency for the network, that can be used effectively by low power devices.

6 Future Work

As a future work, we plan to realize the protocol using hardware implementation in a real-time wireless-based network, due to the low-cost nature of Arduino-based wireless nodes, the hardware realization can be done in a closed laboratory environment. We also intend to research more into the QoS improvement of the wireless networks by incorporating more parameters into our machine learning protocol and also adding Q-Learning and deep nets into the routing algorithm, which can adapt to the network patterns and select the most optimum route intelligently and in real time, with minimum on-the-fly complexity.

References

1. Bizagwira, H., Toussaint, J., Misson, M.: Multi-channel routing protocol for dynamic WSN. In: Wireless Days (WD). IEEE (2016)
2. Zhao, M., Ho, I.W.-H., Chong, P.H.J.: An energy-efficient region-based AODV routing protocol for low-power and lossy networks. IEEE Int. Things J. (2012)
3. Barcelo, M., Correa, A., Vicario, J.L., Morell, A., Vilajosana, X.: Addressing mobility in AODV with position assisted metrics. IEEE Sens. J. 16(7) (2016)
4. Ko, J., Chang, M.: MoMoRo: providing mobility support for low-power wireless applications. IEEE Syst. J. 9(2) (2015)
5. Kulkarni, R.V., Förster, A., Venayagamoorthy, G.K.: Computational intelligence in wireless sensor networks: a survey. IEEE Commun. Surv. Tutor. 13(1) (2011)
6. Forster, A., Murphy, A.L.: FROMS: feedback routing for optimizing multiple sinks in WSN with reinforcement learning. National Competence Center in Research on Mobile Information and Communication Systems (NCCR-MICS) (2010)
7. Kaur, R., Rai, M.K.: A novel review on routing protocols in MANETs. Undergrad. Acad. Res. J. (UARJ) 1(1) (2012). ISSN 2278-1129
8. Usha, M., Jayabharathi, S., Banu R.W.: REAODV: an enhanced routing algorithm for QoS support in wireless ad-hoc sensor networks. In: IEEE International Conference on Recent Trends in Information Technology (2011)
9. Gupta, P., Kumar, P.R.: The capacity of wireless networks. IEEE Trans. Inf. Theory 46(2), 388–404 (2007)
10. Khabbazian, M., Bhargava V.K.: Localized broadcasting with guaranteed delivery and bounded transmission redundancy. IEEE Trans. Comput. 57(8), 1072–1086 (2008)
11. Shen, Z., Thomas, J.P.: Security and QoS self-optimization in mobile ad hoc networks. IEEE Trans. Mob. Comput. 7, 1138–1151 (2008)
12. Asutkar, G.M., Rangaree, P.: Design of self-powered wireless sensors network using hybrid PV-wind system. In: IEEE 10th International Conference on Intelligent System and Control (ISCO) (2016)

Soft-Computing Techniques for Voltage Regulation of Grid-Tied Novel PV Inverter at Different Case Scenarios

T. Lova Lakshmi and M. Gopichand Naik

Abstract In this paper, the voltage regulation of large-scale grid-tied photovoltaic power plant (GTPVPP) operating during nonlinear PV generation has been discussed. This research proposes the comparative voltage regulation of a novel multilevel inverter with soft-computing techniques such as fuzzy and adaptive neuro-fuzzy inference system (ANFIS)-based control for regulating the voltage of GTPVPP. Due to the interruptible PV generation and at worst-case scenarios, the proposed control scheme is useful to satisfy the load demand by grid integration. In this comparison, the ANFIS-based control scheme improves the dynamic performance, reduces the THD, and improves the efficiency. The fuzzy and proposed ANFIS-based control schemes are developed in MATLAB/Simulink environment and are compared at worst-case solar generation, rapid change of loads, and grid faults.

Keywords Solar PV generation · DC–DC converter
Resonant switched-capacitor converter (RSCC) · Multilevel inverter
Soft-computing techniques

1 Introduction

Power distribution systems experience a substantial change due to the innovative advents, technologies, and control schemes in the grid-tied PV inverter [1]. Nowadays, a renewable energy source plays a significant role in electrical power generation [2]. In fact, the use of these PV DGs with the proposed control schemes in grid integration presents some significant advantages, such as reduction of voltage stresses and switching losses of devices, environmental-friendly, uninterruptible service, and improved power quality [3]. In addition to this, an increased amount of renewable

T. Lova Lakshmi (✉) · M. Gopichand Naik
Andhra University, Visakhapatnam, India
e-mail: tlakshmi.sch@gmail.com

M. Gopichand Naik
e-mail: gopi_525@yahoo.co.in

© Springer Nature Singapore Pte Ltd. 2019
J. Wang et al. (eds.), *Soft Computing and Signal Processing* ,
Advances in Intelligent Systems and Computing 898,
https://doi.org/10.1007/978-981-13-3393-4_19

DGs initiates a novel dynamics which has significant adverse impacts on the voltage profile at the point of common coupling (PCC) [4]. Conventional voltage regulators such as proportional–integral (PI), proportional-resonant (PR), and fuzzy are not effectively reducing the THD in test system. In those cases, the efficiency of grid-tied multilevel converter-based solar PV system is increased as they can reduce the harmonic percentage, stresses on power electronic devices are increased, power conversion capability is increased, power factor is increased, losses are reduced, and filtering components required for harmonic compensation are reduced. However, the PI-based control scheme cannot obtain the proper parameters, and power oscillations are more [5]; with PR controller, harmonic distortion is far high as per IEEE standards; the fuzzy controller is not robust to the nonlinear systems and in the real-time application.

There are huge varieties of intelligent control techniques that are being used nowadays, which are fuzzy controllers, artificial neural networks (ANNs), neuro-fuzzy systems, etc. The exact system model is not required to operate these intelligent and active control techniques, which are secure to system dynamics. Hence, these techniques are advantageous in operating to a great extent the nonlinear systems [6]. Along with the above control techniques, ANFIS has the essential characteristics such as easier to implement, more accurate, providing a damping control to the oscillated power system, fastest learning system, muscular in generalization skills, and easier in understanding [7, 8]. In controlling the multilevel converter-based GTPVPP, an ANFIS has enormously high capability in handling uncertainties during nonlinear operating conditions.

To overcome the problems faced by fast nonlinear dynamics grid integration, active and automatic control techniques are necessary.

The main objectives to develop these novelties in this paper are:

- Dynamic performance improvement of the GTPVPP with multilevel inverter during interruptible solar PV generation.
- Addressing fault-ride-through capability and a grid support for the inverter.

This manuscript is prepared as follows: The mathematical representation of the system is detailed in Sect. 2. Section 3 establishes the proposed multilevel inverter working principle. The proposed active control technique ANFIS is detailed in Sect. 4. The case studies and simulation results are detailed in Sect. 5. The conclusion derived from the work is established in Sect. 6.

2 Mathematical Representation of the System

The grid-integrated solar PV-based proposed inverter ac side in the synchronous reference frame (SRF: dq^p) can be expressed as (Fig. 1):

$$e_{dq}^p = L\frac{d}{dt}i_{dq}^p + U_{dq}^p + (j\omega_s L + R)i_{dq}^p \tag{1}$$

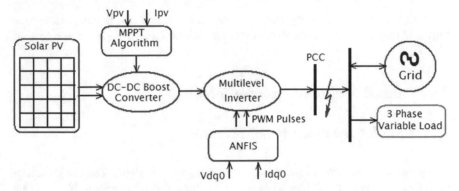

Fig. 1 GTPVPP controlled by ANFIS controller

where R and L are the grid filter resistance and inductance, respectively, and e_{dq} are the inverter ac-side transformed voltages. U_{dq} are transformed grid voltages to SRF. Superscript P represents positive sequence components, and superscript N represents negative sequence components. Detailed mathematical modeling of this system under unbalanced conditions was presented in [9–11]; here, only a brief description is presented, but this is only for the balanced condition. Under the unbalanced network condition, the three-phase system equation may be decomposed into positive and negative sequence voltage and current components. The sequentially stationary reference frame ($\alpha\beta$) is transformed into dq reference frame rotating at angular speeds of ω_s and $-\omega_s$, respectively.

The positive and negative sequence components' representation in dq frame of grid-tied solar PV-based inverter is as follows:

$$\frac{d}{dt}\begin{bmatrix} i_d^P \\ i_q^P \end{bmatrix} = \begin{bmatrix} R/L & -\omega_s \\ \omega_s & R/L \end{bmatrix}\begin{bmatrix} i_d^P \\ i_q^P \end{bmatrix} + \frac{1}{L}\begin{bmatrix} e_d^P - U_d^P \\ e_q^P - U_q^P \end{bmatrix} \tag{2a}$$

$$\frac{d}{dt}\begin{bmatrix} i_d^N \\ i_q^N \end{bmatrix} = \begin{bmatrix} R/L & \omega_s \\ -\omega_s & R/L \end{bmatrix}\begin{bmatrix} i_d^N \\ i_q^N \end{bmatrix} + \frac{1}{L}\begin{bmatrix} e_d^N - U_d^N \\ e_q^N - U_q^N \end{bmatrix} \tag{2b}$$

The injected current is defined as follows:

$$\begin{bmatrix} i_d^P \\ i_q^P \\ i_d^N \\ i_q^N \end{bmatrix} = \begin{bmatrix} e_d^P & e_q^P & e_d^N & e_q^N \\ e_d^N & e_q^N & e_d^P & e_q^P \\ e_q^N & -e_d^N & -e_q^P & e_d^P \\ e_q^P & -e_d^P & e_q^N & -e_d^N \end{bmatrix}^{-1}\begin{bmatrix} P_0^* \\ P_{c2}^* = 0 \\ P_{s2}^* = 0 \\ Q_0^* \end{bmatrix} \tag{3}$$

Here, the terms P_{c2} and P_{s2} are varied with unbalanced sags. To compensate these terms, the injection is in such a way that to maintain P_{c2} and P_{s2} are closer to zero.

3 Proposed Multilevel Inverter Topology

3.1 Circuit Arrangement and Features

The proposed novel seven-level dc–ac inverter is shown in Fig. 2. The source voltage is acting as a voltage divider by connecting three series capacitors (C_1, C_2, and C_3) across it. The voltage which is from the voltage divider is broadcasted to H-bridge developed by power electronic devices and maintained at output terminals. The generated output voltage of the multilevel inverter is seven-level output voltage. The mixed advantages of semiconductor devices (IGBT and MOSFET) are high power rating, able to withstand at high voltage stress, and low switching frequency. Reduction of power electronic switches is the main aspect of this seven-level inverter; with this, the switching losses are reduced, which results in improved efficiency.

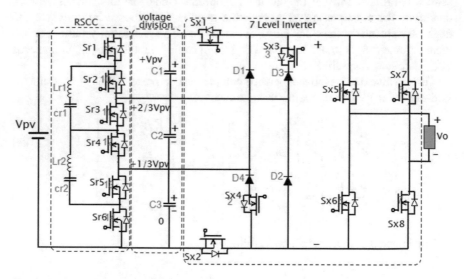

Fig. 2 Seven-level inverter with RSCC

Table 1 Switching combinations with their respective output voltage

Output voltages v_o	Switching sequences							
	S_{x1}	S_{x2}	S_{x3}	S_{x4}	S_{x5}	S_{x6}	S_{x7}	S_{x8}
$1/3v_{dc}$	ON	OFF	OFF	OFF	ON	OFF	OFF	ON
$2/3v_{dc}$	ON	OFF	OFF	ON	ON	OFF	OFF	ON
v_{dc}	ON	ON	OFF	OFF	ON	OFF	OFF	ON
$-1/3v_{dc}$	OFF	ON	OFF	OFF	OFF	ON	ON	OFF
$-2/3v_{dc}$	OFF	ON	ON	OFF	OFF	ON	ON	OFF
$-v_{dc}$	ON	ON	OFF	OFF	OFF	ON	ON	OFF
0	OFF	OFF	OFF	OFF	ON	OFF	ON	OFF

3.2 Working Principle

The proposed seven-level inverter topology comprises eight switches ($S_{x1}-S_{x8}$) and four diodes ($D_{x1}-D_{x4}$) as shown in Fig. 2. The generated output voltages in the seven levels are: $\pm 1/3\, v_{pv}, \pm 2/3\, v_{pv}, \pm v_{pv}, zero$.

Table 1 gives the switching combinations to generate different voltage levels at the output.

3.3 Seven-Level Inverter with RSCC

The origin of the larger harmonics in the output voltage is the voltage deviation. The voltage balancing circuits play a vital role for the capacitors in the proposed seven-level inverter [12, 13]. To reduce the harmonic distortions, the voltage balance of source-side capacitors is necessary, which can be implemented by using RSCC. The circuit configuration of RSCC is shown as a part in Fig. 2. To apply RSCC at seven-level inverter configuration, two switches S_{r5} and S_{r6}, resonant inductor L_r, and resonant capacitor C_r are added. The functioning of the switches in this application is S_{r1}, S_{r3}, and S_{r5} are turned on at the same time. S_{r2}, S_{r4}, and S_{r6} are turned on at the same time. The duty of each switch is equal to 50%.

4 Proposed Active Control Technique: ANFIS

The frequently used technique in control systems is the fuzzy logic controller (FLC), because to use this exact model of the system is not required. FLC is not sensitive to the parameter variations taking place in the test system. It is completely based on

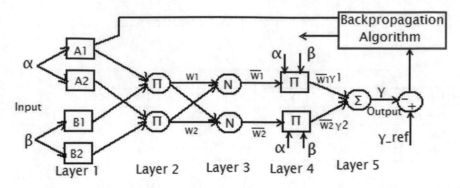

Fig. 3 General five-layered adaptive neuro-fuzzy controller structure

fuzzy rules and membership functions, which are commonly obtained by trial-and-error method and are like a human mind (time-consuming and may cause an error). FLC and ANNs are complementary technologies in the design of intelligent control systems [14]. Neuro-fuzzy systems combine the inference ability of fuzzy logic like a human and the learning and parallel data processing abilities of ANNs. With these systems, the development time is reduced and the accuracy of the fuzzy model is improved.

The system with an adaptive neuro-fuzzy controller (can be called an adaptive neuro-fuzzy inference system (ANFIS)) is an intelligent control technique [15–17]. This is based on learning and parallel data processing of artificial neural networks (ANNs). A general architecture of an ANFIS is shown in Fig. 3, which is of five-layered, and for simplicity, we assumed two inputs (α and β) and one output (γ). In this architecture, circle node indicates a fixed node and a square node is used to reflect different adaptive capabilities. Here, it presents a five-layered feed-forward network, in which each one has a significant function.

In Layer 1, the nodes (A_1, A_2, B_1, and B_2) contain the membership function assigned to each input (α and β). It is a fuzzification layer. In Layer 2, nodes are represented as circles. Here, the incoming signals are multiplied and forwarded to the next layer. In this, the system identifies two rules, such as rule 1—"$\alpha = A_2$ and $\beta = B_1$"; rule 2—"$\alpha = A_2$ and $\beta = B_2$". In Layer 3 also, nodes are represented as circles. It calculates the normalized firing strength of each rule and forwards it to the next layer $\overline{W}_n$. Layer 4 includes the linear functions, which are represented as "$\gamma_n = p_n \alpha + q_n \beta + r_n$". Finally, Layer 5 computes the output by summing all incoming signals as "$\gamma = (\overline{W}_1 \gamma_1 + \overline{W}_2 \gamma_2)(W_1 + W_2)$". As compared to fuzzy controllers, with this the implementation is easy, the combination of fuzzy and ANN makes it fast learning and more accurate, and numeric knowledge and linguistic can be easily incorporated for problem-solving.

5 Case Studies and Discussion

The performance of the proposed ANFIS- and fuzzy-based GTPVPP control scheme under load variations and its performance with an energy management system (EMS) for voltage regulation at no PV cases have been simulated. Initially, high PV generation with different load scenarios (RC, R, and RL) has been considered, and sudden voltage fluctuations due to voltage sags were presented. The proposed ANFIS and fuzzy logic controllers under variable load and unbalanced voltage sags are developed. As compared to FLC, the proposed ANFIS has high fault-ride-through capability for the test system, and the following results are only with ANFIS.

A. Case 1: High Solar PV Generation with Increased Loads and Under Unbalanced Voltage Sags

In this case, the power generated from PV is enough to satisfy load demand according to load variations and under unbalanced voltage sags. Here, the voltage at PCC is controlled by using soft-computing techniques. The ANFIS reduces the THD effectively than fuzzy controller. By applying ANFIS, the voltage profile has been improved. In both the soft-computing techniques, load variations are taking place at 0.5 and 1 s, and voltage fluctuations are 2.3–2.4 s. The THD is observed at three different time instants, which are at low dynamic load point (0.2 s); low dynamic load point after load variation (0.8 s); and peak load point (1.4 s). The THD analysis of PCC current in ANFIS-based control of GTPVPP is shown in Fig. 4. In this THD analysis shown in Table 2, it is observed that ANFIS improves the performance to a greater extent.

B Case 2: No Solar PV Generation with Increased Loads and Under Unbalanced Voltage Sags

In this case, no solar PV is generated but to satisfy load demand according to load variation energy storage system (ESS) discharges and delivered power to load. In addition to this, at peak load condition and insufficient ESS, grid plays an active role to improve the voltage profile. Here, the voltage at PCC is controlled by using soft-computing techniques (fuzzy controller and ANFIS). In this case also, the ANFIS reduces THD effectively than fuzzy controller. The load variations and voltage sags are considered and implemented as Case 1, but no PV case scenario is simulated.

Table 2 THD analysis of fuzzy and ANFIS techniques

Case scenarios	High PV generation case THD (%)			No PV case THD (%)		
Dynamic load (s)	0.2	0.8	1.4	0.2	0.8	1.4
Fuzzy controller	2.13	3.03	3.19	0.39	6.06	1.4
ANFIS	1.82	3.00	3.02	0.37	0.48	1.16

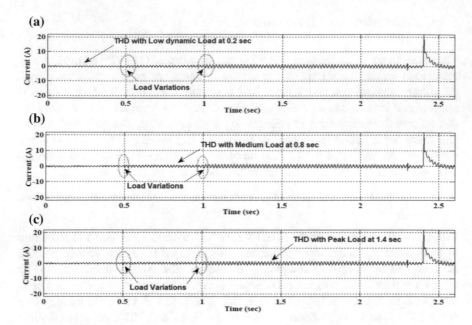

Fig. 4 PCC current at different load points with ANFIS under unbalanced voltage sags: **a** PCC current at low dynamic load, **b** PCC current at medium load, and **c** PCC current at peak load

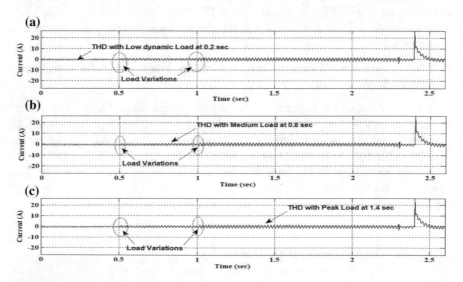

Fig. 5 PCC current at different load points with ANFIS under unbalanced voltage sags: **a** PCC current at low dynamic load, **b** PCC current at medium load, and **c** PCC current at peak load

The THD analysis of PCC current in ANFIS-based control of GTPVPP is shown in Fig. 5.

6 Conclusion

Voltage regulation plays a most significant role to improve the system performance. The main idea of the proposed seven-level inverter is to reduce the number of power electronic devices. Soft-computing techniques regulate PCC voltage of proposed two-stage seven-level inverter. Case studies showed that in the ANFIS-based GTPVPP scheme by minimizing PCC voltage fluctuations, THD is improved in both high PV generation and no PV with EMS. Finally, it is observed that the ANFIS-based control scheme damps the oscillations greatly and provides robust response at different case scenarios. The efficiency of the proposed system increases with seven-level inverter. The effective control of the proposed two-stage seven-level inverter with soft-computing techniques has been implemented in MATLAB/Simulink environment.

References

1. Sun, Y., Ma, X., Xu, J., Bao, Y., Liao, S.: Efficient Utilization of Wind Power: Long-Distance Transmission or Local Consumption. Higher Education Press and Springer-Verlag, Berlin, Heidelberg (2017)
2. Pertl, M., Weckesser, T., Rezkalla, M., Marinelli, M.: Transient Stability Improvement: A Review and Comparison of Conventional and Renewable-Based Techniques for Preventive and Emergency Control. Springer-Verlag, GmbH Germany (2017)
3. Etemadi, A.H., Davison, E.J., Iravani, R.: A decentralized robust control strategy for multi-DER microgrids—part II: performance evaluation. IEEE Trans. Power Del. 27(4) (2012)
4. Rogers, B., Taylor, J., Mimnagh, T., Tsay, C.: Studies on the time and locational value of DER. CIRED Open Access Proc. J. 2017(1), 2015–2018 (2017)
5. Mirhosseini, M., Pou, J., Karanayil, B., Agelidis, V.G.: Resonant versus conventional controllers in grid-connected photovoltaic power plants under unbalanced grid voltages. IEEE Trans. Sustain. Energy 7(3) (2016)
6. Li, H. Shi, K.L., McLaren, P.G.: Neural-network-based sensorless maximum wind energy capture with compensated power coefficient. IEEE Trans. Ind. Appl. 41(6), 1548–1556 (2005)
7. García, P., García, C.A., Fernández, L.M., Llorens, F., Jurado, F.: ANFIS-based control of a grid-connected hybrid system integrating renewable energies, hydrogen and batteries. IEEE Trans. Ind. Informa. 10(2) (2014)
8. Jang, J.-S., Sun, C.T., Mizutani, E.: Neuro-Fuzzy and Soft Computing. Prentice-Hall, Englewood Cliffs, NJ, USA (1997)
9. Hu, J., He, Y.: Multi-frequency proportional-resonant (MFPR) current controller for PWM VSC under unbalanced supply conditions. J. Zhejiang Univ. Sci. A 8(10), 1527–1531 (2007)
10. Hu, J., Hr, Y.: Modeling and control of grid-connected voltage-sourced converters under generalized unbalanced operation conditions. IEEE Trans. Energy Convers. 23(3), 903–913 (2008)
11. Mirhosseini, M., Agelidis, V.G.: Performance of Large-Scale Grid-Connected Photovoltaic System under Various Fault Conditions. IEEE (2013)
12. Sano, K., Fujita, H.: Voltage-balancing circuit based on a resonant switched-capacitor converter for multilevel inverters. IEEE Trans. Ind. Appl. 44(6), 1768–1776 (2008)
13. Suroso, S., Noguchi, T.: New generalized multilevel current-source PWM inverter with no-isolated switching devices. In: Proceedings of the IEEE International Conference Power Electronics Drives Systems (PEDS), pp. 314–319 (2009)

14. Altin, N., Sefa, I.: DSPACE based adaptive neuro-fuzzy controller of grid interactive inverter. Energ. Convers. Manag. **56**, 130–139 (2012)
15. Basarir, H., Elchalakani, M., Karrech, A.: The Prediction of Ultimate Pure Bending Moment of Concrete-Filled Steel Tubes by Adaptive Neuro-Fuzzy Inference System (ANFIS). Springer, Australia (2017)
16. Heddam, S.: Modeling Hourly Dissolved Oxygen Concentration (DO) Using Two Different Adaptive Neuro-Fuzzy Inference Systems (ANFIS): A Comparative Study. Springer Science, Business Media Dordrecht (2013)
17. Dubey, S.K., Jasra1, B.: Reliability Assessment of Component Based Software Systems Using Fuzzy and ANFIS Techniques. Springer, India (2017)

Deep Sentiments Extraction for Consumer Products Using NLP-Based Technique

Mandhula Trupthi, Suresh Pabboju and Narsimha Gugulotu

Abstract The growth in the field of e-commerce and product availability over the Internet is the higher availability of the consumable items is making the customers seek for higher quality and comparative price points. The primary reason for this ambiguity is the lack of in hand experience for the customers before the purchase. The customers mostly tend to rely on the feedbacks of the other buyers. The feedbacks on the products are often made in thousands in numbers, and it is difficult for the potential buyers to decide by looking into these feedbacks or reviews. Thus the demand of the modern research is to automate the process for extracting the true feedback matching their needs based on usage or price or location constraints. The feedback or the review system can be easily manipulated by the incorrect feedbacks. Hence it is important to reduce the influence of those feedbacks during extracting the overall sentiment of any product. Also, yet another challenge is that most of the feedbacks are not in formal English, thus making it difficult to extract the accurate feedback. This work proposes a novel-automated frame for extracting the deep sentiments from the reviews or the feedbacks on e-commerce websites. Another major outcome of this work is to detect the false reviews and making the sentiment true for any decision making. The research work generates a trustable sentiment extraction process to justify the need of true feedbacks for customer decision making.

Keywords Deep sentiment extraction · False review detection
Weighted sentiment analysis · Semantic orientation analysis
Pointwise mutual information validation

M. Trupthi (✉)
Jawaharlal Nehru Technological University Hyderabad, Hyderabad, India
e-mail: trupthijan@gmail.com

S. Pabboju
Chaitanya Bharathi Institute of Technology, Hyderabad, India

N. Gugulotu
Jawaharlal Nehru Technological University Sultanpur, Sultanpur, India

© Springer Nature Singapore Pte Ltd. 2019
J. Wang et al. (eds.), *Soft Computing and Signal Processing* ,
Advances in Intelligent Systems and Computing 898,
https://doi.org/10.1007/978-981-13-3393-4_20

1   Introduction

Sentiment analysis has become the predominant field in data mining and natural language processing (NLP) for determining the subjective information, i.e., sentiments extraction from the text [1]. Sentiment extraction is the primary and specific task in the field of opinion mining. Sentiment extraction mainly focuses on sentiment classification or topic identification on the particular domain. Sentiment classification aims to categorize the sentiment into positive or negative classes.

Bag-of-words model is very popular for representing the text data. It treats text data as a vector of independent words. These words will be analyzed by using statistical approaches like support vector machines (SVM) and Naïve Bayes classifiers (NB). Even this model is simple and efficient for text classification, but it is not appropriate for sentiment extraction because it could not understand the underlying syntactic structures and semantics between the words.

Subsequently, a group of novel researches in sentiment analysis are intended to find the research gap and trying to enhance the capabilities of bag-of-words model by integrating linguistic knowledge to it [2]. However, most of these efforts are unimplemented because of its difficulties and accuracy concerns. Polarity shift opposite words are considered to be very similar; is the primary difficulty to discourse in this model.

Several approaches were proposed to address the polarity shift problem, but most of them are manual in providing annotations and required deep knowledge on the language semantics [3, 4]. Some other approaches defined the same with the absence of extra annotations and human knowledge, but the outcome is poor and unsatisfied.

2   Outcome of the Parallel Research

Sentiment analysis tasks are majorly categorized as sentence, phrase, aspect and document based on their granularities.

Wilson et al. [5] defined the effects of complex polarity shift. Initially, they estimated the polarities and developed a lexicon of words. Next, identify the "contextual polarity" of phrases based on supervised annotations.

Nakagawa et al. [6] established a dependency graph and semi-supervised model to detect the sentiments at different phrases. The model predicts the polarities based on the graph node interactions and motivates toward negation. In aspect level, the polarity shift problem was measured in both text corpus and lexicon-based methods [7, 8].

Data extension is one of the major techniques in hand-written recognition systems. The performance of such system is measured by providing training data set consisting of set of hand-written samples [9].

Agirre and Martinez et al. proposed a model for word-sense disambiguation. This model collects large amounts of labeled data through Web search using WordNet by identifying synonyms and word expressions [2].

Fujita and Fujino et al. proposed reliable training method that provides data using sample sentences from an external thesaurus [10].

The outcomes from the parallel researches formulate the research problem.

3 Problem Formulation

To form an initial set of positive words, the corpus is scanned through two passes.

First phase ending with negative words is extracted from the corpus with a variable length.

$$[S_p] = \frac{d\langle Corp \rangle}{dx} \tag{1}$$

Here,

S_p denotes the positive list of phrases
Corp denotes the total corpus for analysis
X   length of the phrase.

Further, the phrase list is pruned based on the frequency of use. The list contains the phases which are appearing more frequently without the negative notations.

$$[S'_p] = \prod_{Non-Negative} S_p \tag{2}$$

Next, the corpus is classified based on the positive sentiments extracted in the previous phase. The classification boundaries for each phase are the sentences between the punctuations. Thus for each sentence part, the score must be calculated based on the number of positive sentiments are extracted.

$$S = \sum \left(\prod_{positive} S'_p \right) \Big/ \Delta y \tag{3}$$

Here,

S   denotes the collection of sentence parts between the punctuations
Y   denotes the variable length of the parts between the punctuations

$$Senti\_Score = \sum \frac{S_i \cdot N_i}{\Delta y} \tag{4}$$

Senti_Score   denotes the collective score of the review corpus for each product
S_i denotes the sentence part of the feedback or review corpus a specific
                product
N_i denotes the number of positive sentiments for those sentence parts.

Nevertheless, the primary reasons for trust factors on the online reviews or the feedbacks are, firstly, the numeric feedback and corpus review feedbacks can differ based
on the algorithm used for sentiment extraction. Thus the final validation of the reviews
or the feedbacks needed to be calculated as weighted average of both the components.

$$ES = Senti\_Score \cdot w_i + Rating \cdot w_j \qquad (5)$$

Here,

ES   denotes the total sentiment score
Senti_Score   denotes the score obtained from the review text
Rating   denotes the score from the numerical feedback
w_i and w_j denotes the weight adjustments for both the feedbacks.

It is natural to understand that the numerical ratings are more specific than the
text review feedbacks. Hence,

$$w_i < w_j \qquad (6)$$

Nonetheless, the perfect adjusted values are to be identified to calculate the measures. Secondly, it has been observed that the feedbacks of the reviews are often
posted by anonymous reviewers with the hidden intension of damaging the brands.
Thus, the reduction of the false feedbacks also needs to be achieved through the
automate sentiment extraction process.

4   Proposed Framework

The major objective of this work is to automate the sentiment analysis for the consumer products. The components' description is formulated in Fig. 1.
    The components 1–4 are used to deploy crawler to extract the Amazon products.

4.1   Sentiment Extraction Module (Layer-3)

In this module, the proposed work behind this component is to identify the positive
sentiments from the feedbacks or the reviews. The English language does not always
follow this principle. Thus the enhanced method used by this work is to deploy
the method of the initial phrase extraction and reduction of the phrases by multiple

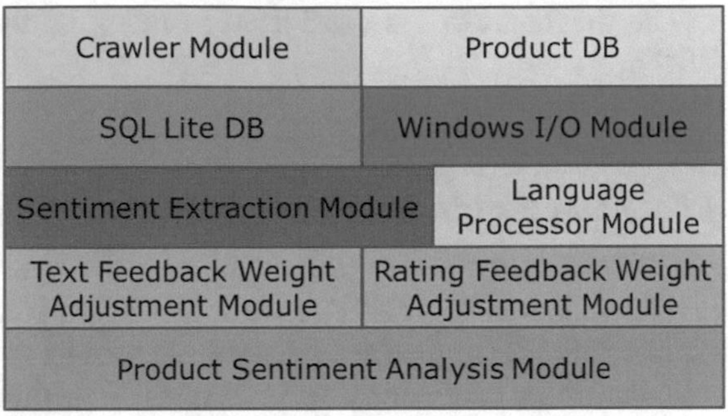

Fig. 1 Proposed novel framework

Crawler Module	Product DB
SQL Lite DB	Windows I/O Module
Sentiment Extraction Module	Language Processor Module
Text Feedback Weight Adjustment Module	Rating Feedback Weight Adjustment Module
Product Sentiment Analysis Module	

passes to understand the positive use of the language or the vocabulary makes a strong impact on positive phrases as there are many strongly positive words are available in the English language, but the use of those phrases are highly uncommon.

So, this work makes the final enhancement to the sentiment extraction strategy by adopting a language-independent method for extracting the positive phrases. The main idea behind this improvement is to build a correlation metric, which defines the relation and differentiation of positive and negative phrases.

This result into a semantic orientation of the word or the phrase using the pointwise mutual information:

$$SO(S) = PMI(S, Excellent) - PMI(S, poor) \tag{7}$$

Here,

SO(S)—semantic orientation of the phrase.
PMI-Excellent is positive sentiment and PMI-Poor is negative sentiment.

4.2 Language Processing Module (Layer-3)

The ReText is built on unitext API framework. Unified is an interface for processing text using syntax trees. retext, and rehype, allows for processing between multiple syntaxes.

4.3 Weight Adjustment Modules for Text and Ratings (Layer-4)

In this module, weights are assigned to the text and the rating feedbacks of the reviews. The weight for the rating feedback will be more compared to the text feedback as the rating system is more prominent in order to understand the emotion.

The weight matrix is considered as rating feedback (70%) and text feedback (30%).

4.4 Final Sentiment Analysis Module (Layer-5)

The final module for this framework is the final sentiment analysis module. This module presents the final sentiment of any product to the consumers.

In the next section, this work elaborates the method and algorithm used for powering this framework.

5 Proposed Algorithm for Sentiment Extraction

The sets P and N denote the positive and negative reference words. Next, the correlation measures between the phrases in the feedback or reviews and the positive and negative sets are calculated.

$$SO(S) = PMI(S, [P]) - PMI(S, [N]) \tag{8}$$

Further, only the phrases or the words are considered with the highest correlation as $SO(S) > 1$.

Henceforth, the mathematical model is converted into the workable algorithm and presented below.

Step -1. Build the positive and negative word sets
Step -2. Accept the list of products and analysis for ID
      A. For each product
      B. Break the sentences into zones
          i. For each zone
              a. Calculate the correlation score
              b. if the score is higher than 1
                  • Then consider the score
              c. Else
                  • Reject the word from the set
              d. Repeat for minimal set
      C. Consider the rating score
      D. Calculate the final score = rating *0.7 + text score * 0.3
      E. Leverage the final sentiment

---

6 Results and Discussion

The results are classified in terms of the review extraction, deep sentiment extraction, final sentiment analysis, fake review detection, and time complexity analysis.

6.1 Experimental Setup and Review Extraction Results

The experiment setup and extraction results are shown in Tables 1 and 2.

Further, the number of attempts and the review extraction success rates are measured (Table 3), and number of feedbacks available and number of feedbacks extracted are compared in Table 4. The extractions of the reviews are 100%.

6.2 Deep Sentiment Extraction

Secondly, this work extracts the deep sentiments from the reviews (Table 5). And the rating is matched with the extracted sentiments are matched (Tables 6 and 7).

6.3 Final Sentiment Analysis

Final sentiment analysis is shown in Table 8.

6.4 Fake Review Detection and Time Complexity Analysis

The proposed work analyses the time complexity to build every product feedbacks and ratings based sentiments the average time to extract the deep sentiment is 1.60 s in Table 9.

Table 1 Setup for the experiment

Computing environment	Amazon web services (Basic starter plan)
Review source	Amazon.com
Number of nodes (AWS)	5
Crawler module API	Amazon product-based crawler (Java 4 Amazon)
Sentiment extraction base	Application is built after the customization of stanford API

Table 2 Review extraction result

Record ID	Date added	Review date	Verified buyer	Total votes	Up votes	Rating	Review ID	Item ID
4580	13-02-2018 13:27	17-08-2017	TRUE	−1	0	4	R39Y4UJ31VXA44	B077Y8DN87
4740	13-02-2018 13:27	18-05-2017	TRUE	−1	0	5	R3JPQL0OCYV2QX	B077Y8DN87
4741	13-02-2018 13:27	16-07-2017	TRUE	−1	0	2	R2A3HREQHVF27P	B077Y8DN87
11339	13-02-2018 19:58	13-12-2015	TRUE	−1	0	3	R2UDTU9968O419	B00OLT7QSU
11340	13-02-2018 19:58	27-12-2016	TRUE	−1	0	5	R1BM78DPDCCDOF	B00OLT7QSU
11341	13-02-2018 19:58	02-04-2015	TRUE	−1	0	3	RVU4CEH330F4E	B00OLT7QSU
11196	13-02-2018 17:59	12-01-2018	TRUE	−1	0	5	R210QI1EHLFQTH	B06XD1K5S6

Table 3 Availability analysis

No. of attempts	No. of reviews	Success rate
5701	5660	99.28

Table 4 Products and review analysis

Product ID	No. of reviews available	No. of reviews extracted
B00OLT7QSU	31	31
B00WF988BW	198	198
B00YD545CC	3439	3439
B06XCM9LJ4	41	41
B06XD1K5S6	30	30
B06XDMD7WK	50	50
B077Y79RBY	211	211
B077Y8DN87	1439	1439

Table 5 Sentiment extraction results

Sentiment score	Sentiment type	Very pos	Pos	Neu	Neg	Very neg
3	Positive	24	66	6	3	2
0	Very neg	1	1	12	39	47
2	Neutral	1	9	64	24	2
1	Negative	2	8	41	43	6
2	Neutral	2	40	55	3	1
1	Negative	1	5	22	57	14
1	Negative	1	3	26	60	10
1	Negative	1	1	11	58	30
4	Very pos	61	32	5	1	1
2	Neutral	1	8	46	39	5

Table 6 Ratings and sentiments comparison

Extracted sentiment	Rating given by the user
Positive	4
Very neg	1
Neutral	3
Negative	1
Neutral	3
Negative	1
Very pos	5
Neutral	3

Table 7 Accuracy analysis

Record ID	User rating (Given) (A)	Extracted sentiment type	Extracted sentiment (Numeric Rep.) (B)	Accuracy (%) (B/A)*100
11362	3	Neutral	3	100
11363	5	Very pos	5	100
11341	3	Neutral	3	100
11342	3	Neutral	3	100
5299	3	Neutral	3	100
11343	5	Very pos	5	100
11345	5	Very pos	5	100
11346	5	Very pos	5	100

Table 8 Ratings and sentiments comparison

Sentiment score * 0.3	Rating given by the user * 0.7	Final sentiment
0.9	2.8	3.7
0	0.7	0.7
0.6	2.1	2.7
0.3	0.7	1
0.6	2.1	2.7
0.3	0.7	1
0.3	0.7	1
0.3	0.7	1
1.2	3.5	4.7
0.6	2.1	2.7

Table 9 Certified buyers and potential fake reviews

Product ID	No. of certified buyers reviews	No. of potential fake reviews
B00OLT7QSU	28	3
B00WF988BW	191	7
B00YD545CC	3262	177
B06XCM9LJ4	41	0
B06XD1K5S6	30	0
B077Y79RBY	203	8
B077Y8DN87	1363	76

7 Conclusion

The growth in the e-commerce and the support from the technology has made the e-shopping highly popular across the globe. The online shopping is challenged by various factors such as the customer experience can only be achieved after the purchase. Thus, majority of the customers tends to depend on other consumer feedback on the same products. Nonetheless, the feedback or the review or the rating system is not perfect and for a single customer, it is highly impossible to identify the overall sentiment about the product and make decision. Thus this work builds a novel framework for automatically extracting the reviews and extracts deep sentiments for the products. The work demonstrated a significant accuracy of 100% true sentiment extraction. Further, this work also proposes a novel algorithm to identify the false reviews in order to make the consumer and e-commerce world more likely to be safe and decision compatible.

References

1. Abbasi, A., France, S., Zhang, Z., Chen, H.: Selecting attributes for sentiment classification using feature relation networks. IEEE Trans. Knowl. Data Eng. (TKDE) **23**(3), 447–462 (2011)
2. Dave, K., Lawrence, S., Pen-nock, D.: Mining the peanut gallery: opinion extraction and semantic classification of product reviews. In: Proceedings of the International World Wide Web Conference (WWW), pp. 519–528 (2003)
3. Cano, J., Perez-Cortes, J., Arlandis, J., Llobet, R.: Training set expansion in handwritten character recognition. In: Structural, Syntactic, and Statistical Pattern Recognition, pp. 548–556 (2002)
4. Wilson, T., Wiebe, J., Hoffmann, P.: Recognizing contextual polarity: An exploration of features for phrase-level sentiment analysis. Comput. Linguist. **35**(3), 399–433 (2009)
5. Nakagawa, T., Inui, K., Kurohashi, S.: Dependency tree-based sentiment classification using CRFs with hidden variables. In: Proceedings of the Annual Conference of the North American Chapter of the Association for Computational Linguistics (NAACL), pp. 786–794 (2010)
6. Ding, X., Liu, B.: The utility of linguistic rules in opinion mining. In: Proceedings of the 30th ACM SIGIR Conference on Research and Development in Information Retrieval (SIGIR) (2007)
7. Ding, X., Liu, B., Yu, P.S.: A holistic lexicon-based approach to opinion mining. In: Proceedings of the International Conference on Web Search and Data Mining (WSDM) (2008)
8. Varga, T., Bunke, H.: Generation of synthetic training data for an HMM-based handwriting recognition system. In: Proceedings of the IEEE International Conference on Document Analysis and Recognition (ICDAR) (2003)
9. Fujita, S., Fujino, A.: Word sense disambiguation by combining labeled data expansion and semi-supervised learning method. In: Proceedings of the International Joint Conference on Natural Language Processing (IJCNLP), pp. 676–685 (2011)
10. Nallapati, R., Zhou, B., Gulcehre, C., Xiang, B., et al.: Abstractive text summarization using sequence-to-sequence RNNs and beyond (2016). arXiv:1602.06023

Music Generation Using Deep Learning

Aishwarya Bhave, Mayank Sharma and Rekh Ram Janghel

Abstract Deep learning has recently been used for many art-related activities such as automatic generation of music and pictures. This paper deals with music generation by using raw audio files in the frequency domain using Restricted Boltzmann Machine and Long Short- Term Memory architectures. The work does not use any information about musical structure to aid the learning, instead, it learns from a previous permutation of notes and generates an optimal and pleasant permutation. It also serves as a comparative study for music generation using Long Short-Term Memory and Restricted Boltzmann Machine.

Keywords Deep Learning · Long short-term memory network
Music Generation · Restricted Boltzmann Machine

1 Introduction to Music Generation

Music composition [1] is an art, even the task of playing composed music takes considerable effort by humans. Given this level of complexity and abstractness, designing an algorithm to perform the task is not obvious and would be a fruitless effort. It is thus easier to model this as a learning problem [2] where composed music is used as training data to extract useful musical patterns. Our goal is to generate music that is pleasant to hear but not necessarily one that resembles how humans play music. We are hoping for the learning algorithm to find spaces where music sounds pleasant without enforcing any restrictions on whether it adheres to guidelines of musical theory. For this reason, we do not use any features such as notes, notation,

A. Bhave (✉) · M. Sharma · R. R. Janghel
National Institute of Technology, Raipur, India
e-mail: aishwaryabhave54@gmail.com

A. Bhave
e-mail: mayanksharma.nitrr@gmail.com

R. R. Janghel
e-mail: rrjanghel.it@nitrr.ac.in

© Springer Nature Singapore Pte Ltd. 2019   203
J. Wang et al. (eds.), *Soft Computing and Signal Processing*,
Advances in Intelligent Systems and Computing 898,
https://doi.org/10.1007/978-981-13-3393-4_21

or chords to aid the learning or generation process, instead, we directly deal with the end result which are audio waveforms.

2 Related Work

Long Short-Term Memory Network have been previously used by Nayebi and Vitelli [3] for music generation on waveform of the music files. They also used Gated Recurrent Unit Network for the same. Van den Oord [4] used variational recurrent network along with autoencoder to associate advantages of language model in music generation using raw files. Yu-Siang [5] uses similarity embedding network to generate music via a game puzzle. Greg Bickerman and others have used Deep Belief Network(a stacked RBM Network) to generate Jazz Melody [6]. We take on the wave as our initial input of music files instead of midi files. The model requires no knowledge of music for training, and we work on numpy matrices derived after applying Fast Fourier Transform [7] on the wave input.

3 Methodology

A playlist of piano covers of the latest pop songs [5] were collected. The playlist consisted of 87 covers. The audio files were converted in WAV format (Fig. 1).

3.1 Data Preprocessing

The preprocessing was done in four stages: Sampling, Numpy audio to sample blocks, Fast Fourier transform, populating training data.

Sampling Sampling [8] is used to convert continuous time signal into discrete time signal. Sampling period is the uniform interval of time taken to two consecutive amplitude values for the discrete signal. This sampling period is denoted by Timeperiod. The sampling frequency or sampling rate, fs, is the average number of samples obtained in one second (samples per second); thus, fs = 1/T. 44,100 Hz sampling

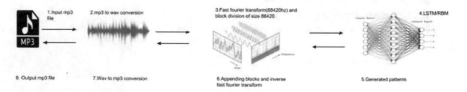

Fig. 1 Methodology Flowchart

rate is used for Audio CD. Then the sampled function is given by the sequence: F(n(Timeperiod)), for integer values of n.

Numpy audio to sample blocks A block [9] size of 44,100 data units was used. The audio data that was obtained in the form a numpy array was divided into small blocks of the above stated size. The last block if any smaller than the blocksize is padded with zeros.

Fast Fourier transform A Fast Fourier transform [10] is an algorithm that samples a signal over a period of time (or space) and divides it into its frequency components.

Populating training data The data blocks are appended to form a sequence unit called chunks. Each chunk consists of 10 data blocks. These chunks are accumulated in a list X which forms the input data for the deep learning models. The first data block is skipped and the rest of the blocks are added to the list Y. In order to populate the data as time series data, we assign Input of Block(Xi-1) has Block(Yi) as output.

3.2 Model Used

Long short-term memory network LSTM were introduced by Hochreiter and Schmidhuber [11] and were refined and popularized by many people in following work. An LSTM block is composed of four main components: a cell, an input gate, an output gate, and a forget gate.

$$\tilde{C} = \tanh(W_{cx} X_T + W_{ch} h_{t-1} + b_c) \tag{1}$$

$$C_t = gateforget \cdot C_{t-1} + gateinput \cdot \tilde{C} \tag{2}$$

$$h_t = gateout \cdot \tanh(C_t) \tag{3}$$

The architecture involves passing the input at each timestep through a fully connected layer and feeding the output of this layer to the LSTM. The LSTM is then used to make predictions. Rectified Linear Units(RELUs) is used as the nonlinearity in the fully connected layer. Experimentally we found that setting the size of the fully connected layer equal to the size of the LSTM hidden state resulted in the best predictions.

Restricted Boltzmann Machine A Restricted Boltzmann machine (RBM) is a type of neural network introduced by Smolensky [7] and further developed by Hinton [6, 12–16]. It is a bipartite undirected graph and contains m visible units V i.e. V_{1-m}, and n hidden units H i.e. H_{1-n}. In binary RBM, Gibbs distribution gives the joint probability distribution as specified:

$$p(v, h) = \frac{1}{Z} * e^{-E(v,h)} \tag{4}$$

where energy function E(v, h) is specified as:

$$E(v, h) = -\sum_{i=1}^{n}\sum_{j=1}^{m} w_{ij}h_i v_j - \sum_{j=1}^{m} b_j v_j - \sum_{i=1}^{n} c_i h_i \tag{5}$$

4 Simulation

All of the approaches deal with the frequency representation of the audio samples. The audio used is sampled at sampling rate of 44,100, and we only consider a single channel of audio for the sake of simplicity. Consider the raw audio samples to be represented as <a0, a1, a2, ...an>, the Fourier transform is obtained by considering n of these samples at a time to obtain a n-dimensional vector. Each of the values in the n-dimensional frequency vector consists of real and imaginary components, to simplify calculations, the vector is unrolled as 2n-dimensional vector (D) where the imaginary values are appended after the n real values. The LSTM used in all the approaches are the vanilla LSTMs.

As part of dataset collection, piano tunes were scraped from the youtube as mp3 files. These files were then converted to mono channel wav files sampled at 16KHz using wav module. The dataset consists of 84 piano tunes of duration varying between 2 and 3 min. 70 songs in the collected set are randomly picked for the training phase and remaining 14 songs are used as random starting seeds for prediction. In order to construct the dataset appropriate for the models, each of the songs is sampled at 0.25 s and converted to frequency domain using Fourier transform to obtain 4000 frequency components. Each of the frequency components consists of both real and the imaginary parts which when unrolled give us an 8000- dimensional vector where the imaginary values are appended to the real values. The 8000-dimensional vector forms the input to the model at each timestep. The model has been trained to utilize 40 timesteps worth of information for each iteration allowing 10 s worth of information to be processed by the model. Each of the audio files is segmented into 10 s intervals giving nearly 300 samples for training the model.

The models were implemented and executed using the Keras [17] wrappers with Theano framework as the backend to run the models on the Nvidia GTX960M GPU. Each of the models was trained for 500, 1000, 1500, and 2000 epochs using the RMSProp optimizer and a learning rate of 0.0001. The mean square error function used for designing the loss function for all the models. For the fully connected layers in the models, combination of L2 regularization and dropout was used. The L2 values were tested with 0.0001 along with dropout values of 0.2. For the LSTM models, dropouts of 0.5 were used to provide regularization for the hidden state weight matrix and the input weight matrix.

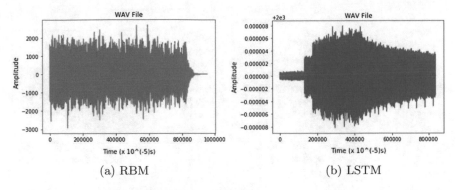

(a) RBM   (b) LSTM

Fig. 2 Output wave graph of both the Networks for 20 Epochs

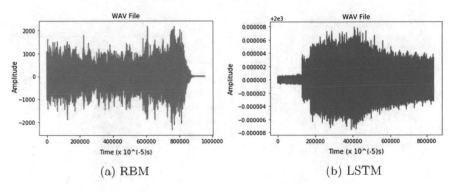

(a) RBM   (b) LSTM

Fig. 3 Output wave graph of both the Networks for 50 Epochs

5   Result

Figures 2, 3, 4, 5, 6, 7, and 8 show the resultant output for variable number of epochs in an amplitude v/s time graph. Figure 9 shows comparison graphs for RBM and LSTM model by depicting the loss values for epochs 20, 50, 100, 200, 500, 1000 and 2000.

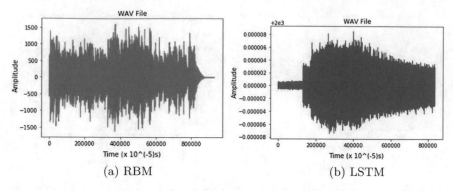

Fig. 4 Output wave graph of both the Networks for 100 Epochs

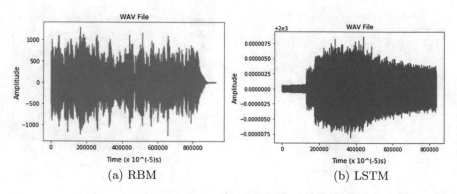

Fig. 5 Output wave graph of both the Networks for 200 Epochs

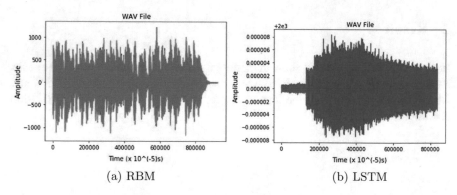

Fig. 6 Output wave graph of both the Networks for 500 Epochs

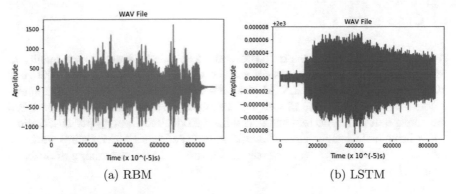

(a) RBM   (b) LSTM

Fig. 7 Output wave graph of both the Networks for 1000 Epochs

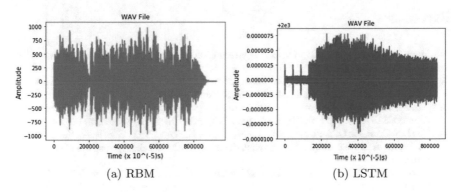

(a) RBM   (b) LSTM

Fig. 8 Output wave graph of both the Networks for 2000 Epochs

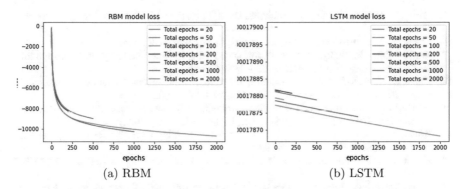

(a) RBM   (b) LSTM

Fig. 9 Loss Function Values of both the Networks corresponding to each Epoch

6 Conclusion

A comparative study is done on RBM and LSTM model based on the quality of music generated and loss values for each of the 20, 50, 100, 200, 500, 1000, and 2000 number of iterations. LSTM produces music with less noise. It also shows a considerably lower value of loss and a sharper decrease in the same than RBM indicating a more regularized and generalized model. However, as per observation, RBM being a stochastic neural network trains faster than LSTM. Future work can include usage of Generative Adversarial Networks (GANs) for mixing of different genres of music.

References

1. Horner, A., Goldberg, D.E.: Genetic algorithms and computer-assisted music composition. Urbana **51**(61801), 437–441 (1991)
2. Sutskever, I., Vinyals, O., Le, Q.V.: Sequence to sequence learning with neural networks. In: Advances in Neural Information Processing Systems (2014)
3. Nayebi, A., Vitelli, M.: Gruv: algorithmic music generation using recurrent neural networks. Course CS224D: Deep Learning for Natural Language Processing (Stanford) (2015)
4. Van Den Oord, A., Dieleman, S., Zen, H., Simonyan, K., Vinyals. O., Graves, A., Kalchbrenner, N., Senior, A., Kavukcuoglu, K.: Wavenet: A Generative Model for Raw Audio. arXiv preprint (2016)
5. Hochreiter, S., Schmidhuber, J.: Long short-term memory. Neural Comput. **9**(8), 1735–1780 (1997)
6. Bickerman, G., et al.: Learning to Create Jazz Melodies Using Deep Belief Nets. In: ICCC (2010)
7. Earley, S., Obama, T., Note Identification Using FFT.: Note Identification Using Fast Fourier Transform
8. Fliege, N.J.: Multirate Digital Signal Processing, vol. 994. Wiley, New York (1994)
9. Yu, G., Mallat, S., Bacry, Emmanuel: Audio denoising by time-frequency block thresholding. IEEE Trans. Signal Process. **56**(5), 1830–1839 (2008)
10. Katoh, K., et al.: MAFFT: a novel method for rapid multiple sequence alignment based on fast Fourier transform. Nucl. Acids Res. **30**(14), 3059–3066 (2002)
11. Hinton, G.E.: A practical guide to training restricted Boltzmann machines. In: Neural Networks: Tricks of the Trade, pp. 599–619. Springer, Berlin, Heidelberg (2012)
12. Hinton, G.E., Osindero, S., Teh, Yee-Whye: A fast learning algorithm for deep belief nets. Neural Comput. **18**(7), 1527–1554 (2006)
13. Hinton, G.E.: To recognize shapes, first learn to generate images. Prog. Brain Res. **165**, 535-547 (2007)
14. Salakhutdinov, R., Mnih, A., Hinton, G.: Restricted Boltzmann machines for collaborative filtering. In: Proceedings of the 24th International Conference on Machine Learning. ACM (2007)
15. Susskind, J.M., et al.: Generating facial expressions with deep belief nets. Affective Computing. InTech (008)
16. Smolensky, P.: Foundations of harmony theory: cognitive dynamical systems and the subsymbolic theory of information processing. Parallel Distrib. Process. Explor. Microstruct. Cogn. **1**, 191–281 (1986)
17. Chollet, F.: "Keras (2015)" (2017)
18. Huang, A., Wu, R.: Deep learning for music. arXiv preprint arXiv:1606.04930 (2016)

19. Eck, D., Schmidhuber, J.: Learning the long-term structure of the blues. In: International Conference on Artificial Neural Networks. Springer, Berlin, Heidelberg (2002)
20. Huang, Y.-S., Chou, S.-Y., Yang, Y.-H.: Generating Music Medleys via Playing Music Puzzle Games
21. Hinton, Geoffrey E.: Training products of experts by minimizing contrastive divergence. Neural Comput. **14**(8), 1771–1800 (2002)

An Implementation of Malaria Detection Using Regional Descriptor and PSO-SVM Classifier

Dhanshree Dawale and Trupti Baraskar

Abstract Malaria is a serious worldwide health issue which causes an expected 13,444 individuals in danger of malaria in 2017. The estimated cost of detection of malaria in India is 11,640 crores per year. So there is an urgent need for a new tool to diagnose malaria. Malaria is completely preventable and treatable disease. In this project, we make a new tool to diagnose malaria using regional descriptor and PSO-SVM classifier. The proposed work used various image processing techniques like image acquisition, image pre-processing, image segmentation, feature extraction and classification. The implementation work is mainly focusing on detection accuracy, computational time, less estimation time for parasite detection. In this way, the new tool for the detection of malaria parasites gives faster and accurate results, and by using this proposed methods pathologists can easily detect malaria parasites, and they can achieve 98% accuracy. This new tool is useful to reduce deaths.

Keywords Image acquisition · Segmentation · Image pre-processing
Feature extraction · Classification

1 Introduction

Malaria distributes widespread in tropical and subtropical climatic region. Transmission of protozoan parasites from one person to another by anopheles mosquito causes malaria. The protozoa parasite of the genus Plasmodium infects human liver cells first, then the red cells, and then the insect hosts alternatively. Plasmodium parasite has four species, namely Plasmodium falciparum, Plasmodium vivax, Plasmodium ovale and Plasmodium malariae, and in comparison with Plasmodium parasite,

D. Dawale (✉) · T. Baraskar
Department of Information Technology, Maharashtra Institute
of Technology, Pune, Maharashtra, India
e-mail: dhanshreedawale1000@gmail.com

T. Baraskar
e-mail: trupti.baraskar@mitpune.edu.in

© Springer Nature Singapore Pte Ltd. 2019
J. Wang et al. (eds.), *Soft Computing and Signal Processing* ,
Advances in Intelligent Systems and Computing 898,
https://doi.org/10.1007/978-981-13-3393-4_22

we understood that Plasmodium falciparum is the most deadly type of parasite. In human's host, each species has three stages, i.e. trophozoite, schizont and gameto-cyte.

1.1 Blood Tests to Detect Malaria

The blood tests are two types for detecting malaria:

1. **Slide test**:

In the glass slide, spread the drop of blood which is taken from a finger. The glass slide is examined by a trained laboratory technician under a microscope. The glass slide test is reported positive; technician sees Plasmodium organisms in the smears. The slide is made by health worker at home in a village, sent to the laboratory and then gets back to the patient at several days. The main advantage of this method is that it can detect both types of malaria.

2. **Rapid Diagnosis Test (RDT)**:

A drop of blood is taken from a finger and immediately placed on a test strip. A few drops of a solution are added, and a few minutes later a red line appears on the strip. The test is positive when two lines appear for falciparum malaria. At present, this test can detect only falciparum malaria, and its the dangerous form. Then less dangerous form of parasite vivax will be similar test present. The main advantage of this method is that it is easy to learn, and there is no need for a laboratory; it takes only 15 min to get the result. The dangerous form of malaria is detected by rapid diagnosis test (RDT) faster and saving lives. It is expensive, but it is supplied by the Government of India for free of cost.

2 Proposed Methodology

In the proposed methodology, system architecture encompasses the modules about the organization of a software system including the selection of the structural elements and their interfaces by which the system is composed (Fig. 1).

The system architecture illustrates five main sections:

(1) Image acquisition,
(2) Image pre-processing,
(3) Image segmentation,
(4) Feature extraction using regional descriptor,
(5) Classification using PSO-SVM classifier.

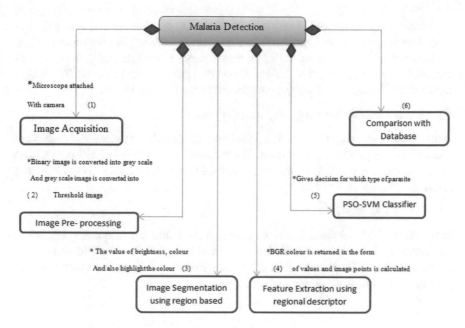

Fig. 1 System architecture

(1) Image acquisition

The project aims to collect database of stained blood smear images which are to be taken from a laboratory by a microscope attached with camera.

(2) Image pre-processing

In this stage, the blood image is enhanced, and the database which is taken is made noise-free. They are filtered, smoothed and normalized which are for better quality image. In the image pre-processing, the sample image is converted into greyscale image and greyscale image is converted into threshold image. There are six types of threshold images: histogram shape, clustering, entropy, object to attribute, spatial and local threshold. In the implementation, we used the binary threshold image and simple local image.

(3) Image segmentation using region based

In image analysis, identifying and separating, according to certain criteria of homogeneity and separation, the regions of the image are the important step of segmentation. Divide the image of parts that have a strong correlation between objects or areas of the real world contained in the image which is the main objective of segmentation. In the image segmentation using region based, the binary blood image of the selected portion of parasites is bright. The commonly used segmentation methods operate essentially based on characteristics such as the value of brightness, colour and reflection of the individual pixels, identifying groups of pixels that correspond to spatially connect regions. The blood image of the parasite portion is bright. The blood image gives the contour, and they highlight the parasite with blue colour. Then the blood image is converted into greyscale image. Greyscale image is converted into threshold image. The threshold image is the type of Otsu threshold. The image captured BGR colour, and we get the parasite in the form of black dot.

(4) Feature extraction using regional descriptor

In the feature extraction using regional descriptor, the full binary image parasite point is calculated by geometric descriptor. The point of the image is in BGR colour, and the colour range is 95–255; the colour is returned to the form of values. Image feature colour is blue, green, red and alpha.

(5) PSO-SVM classifier

In the PSO-SVM classifier, the binary images are converted into greyscale image and greyscale image is converted into threshold image. This threshold image gets the parasite in the black dot, and PSO-SVM gives the decision for which types of parasite. In this way, we detect malaria.

3 Input

(1) Image acquisition

This is a sample image of Plasmodium falciparum. In the image acquisition, microscope is attached to camera (Figs. 2 and 3).
This is binary blood image, and selected portion is highlighted with the help of image segmentation (Fig. 4).

(2) Image pre-processing

In the first image, it is sample image of Plasmodium falciparum. In the pre-processing, the sample image is converted into greyscale image and greyscale image are converted into threshold image. Type of threshold image is the binary image (Fig. 5).

Fig. 2 Sample image of
Plasmodium falciparum

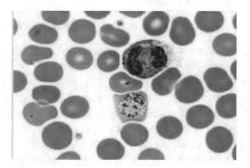

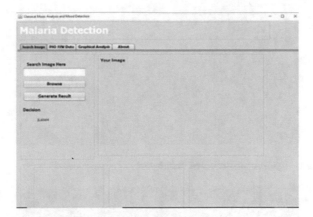

Fig. 3 GUI of malaria detection

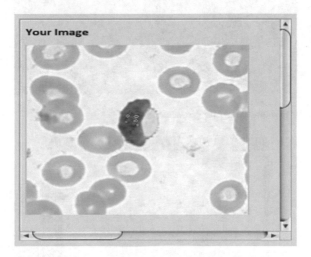

Fig. 4 Binary blood image

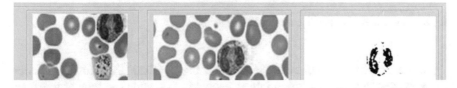

Fig. 5 Greyscale image converted into threshold image

(3) **Image segmentation using region based**

In the blood image, it gives the contour and they highlight the parasite with blue colour (Figs. 6 and 7).

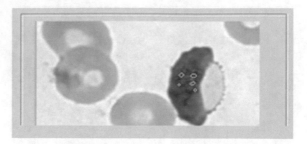

Fig. 6 Highlight the parasite in the blood image

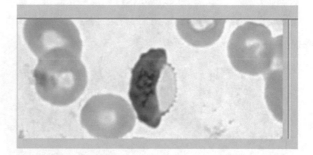

Fig. 7 Binary image is converted into greyscale image

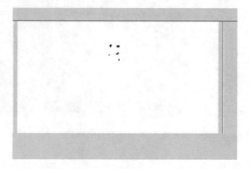

Fig. 8 Greyscale is converted into threshold image

The image captured BGR colour, and we get the parasite in the form of black dot. The greyscale image is converted into threshold image, and Otsu type of threshold is used in the image (Fig. 8).

(4) **Feature extraction using regional descriptor**

In the feature extraction, the full binary image parasite point is calculated and the point is in BGR colour, and the colour is returned in the form of values (Fig. 9).

Fig. 9 BGR colour returns
in the form of values

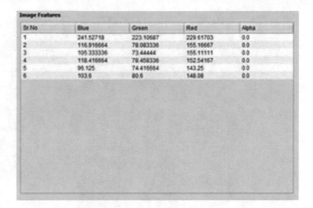

(5) **PSO-SVM classifier**

In the PSO-SVM classifier, it gives the decision of which type of parasites is in the blood image (Figs. 10 , 11, 12, 13 and Table 1).

Fig. 10 Which type of
parasite?

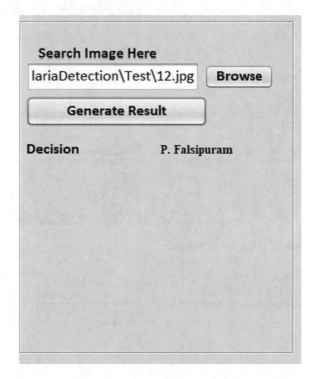

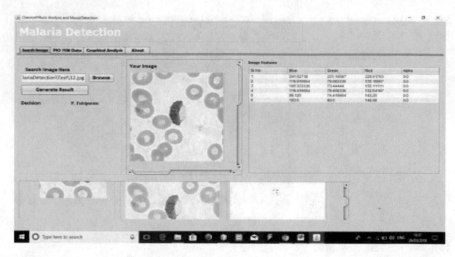

Fig. 11 Malaria detection process

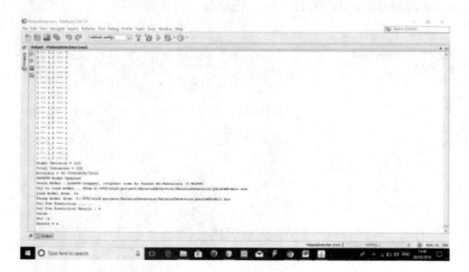

Fig. 12 Decision on parasites

Table 1 Accuracy of algorithm for different methods of classification	Method	Neural network	SVM	PSO-SVM
	Accuracy	78.53%	98.25%	98.70%

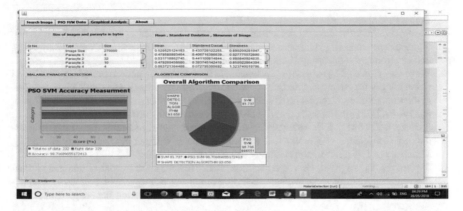

Fig. 13 Comparison between existing systems, and we generate the automatic tool for malaria detection

4 Conclusion

Malaria is serious infectious disease. This system can detect malaria quickly by using regional descriptor and PSO-SVM classifier. In the image acquisition, microscope is attached to camera and in the pre-processing, the GUI of malaria detection searches an image of Plasmodium parasite and this GUI generates result of image pre-processing. The region is captured and get BGR colour and also get the parasite with the points in the image segmentation, and the system is calculated by the points of the image of the form of values using geometric descriptor. The system has calculated the points in the form of BGR colour, and PSO-SVM gives the decision of the parasite. The parasite is detected and then starts the treatment as per the type of parasites. The system gives 98.70% accuracy. This process is estimated detection accuracy, computational time and less estimation time for parasite detection is reduces as compare to early process and it helps to more accurate detection of malaria. In this way, the detection of malaria parasites gave faster and accurate results.

References

1. Suwalkar, S., Sanadhya, A., Mathur, A., Chouhan, M.S.: Identify malaria parasite using pattern recognition technique. In: 2012 International Conference on Computing, Communication and Applications, Dindigul, Tamilnadu (2012)
2. Patel, M.N., Tandel, P.: A survey on feature extraction techniques for shape based object recognition. Int. J. Comput. Appl. (0975–8887) 137(6) 2016
3. Zheng, H., Zhang, S., Sun, X.: Classification recognition of anchor rod based on PSO-SVM. In: 2017 29th Chinese Control and Decision Conference (CCDC), Chongqing, pp. 2207–2212 (2017). https://doi.org/10.1109/ccdc.2017.7978881
4. Mohammed, H.A., Abdelrahman, I.A.M.: Detection and classification of malaria in thin blood slide images. In: International Conference on Communication, Control, Computing and Elec-

tronics Engineering, Khartoum, Sudan (2017)

5. Bashir, A., Mustafa, Z.A., Abdelhameid, I., Ibrahem, R.: Detection of malaria parasite using digital image processing. In: International Conference on Communication, Control, Computing and Electronics Engineering Khartoum, Sudan (2017)
6. Widiawati, C.R.A., Nugroho, H.A., Ardiyanto, I.: Plasmodium detection methods in thick blood smear images for diagnosing malaria: a review. In: 2016 1st International Conference on Information Technology, Information Systems and Electrical Engineering (ICITISEE), Yogyakarta, pp. 142–147 (2016). https://doi.org/10.1109/icitisee.2016.7803063
7. Savkare, S.S., Narote, S.P.: Automated system for malaria parasite identification. In: 2015 International Conference on Communication, Information & Computing Technology (ICCICT), Mumbai, pp. 1–4 (2015). https://doi.org/10.1109/iccict.2015.7045660
8. Saputra, W.A., Nugroho, H.A., Permanasari, A.E.: Toward development of automated plasmodium detection for malaria diagnosis in thin blood smear image: an overview. In: International Conference on information Technology Systems and Innovation Bandung—Bali, October 24–27 (2016)
9. Savkare, S.S., Narote, S.P.: Blood cell segmentation from microscopic blood images. In: International Conference on Information Processing, December 16–19 (2015)
10. Vikhar, P., Karde, P.: Improved CBIR system using edge histogram descriptor (EHD) and support vector machine (SVM). In: 2016 International Conference on ICT in Business Industry & Government (ICTBIG), Indore, pp. 1–5 (2016). https://doi.org/10.1109/ictbig.2016.78926784

Interactive Print Media Using Augmented Reality

Winston Fernandes, Thelma Gomes, Ashley Fernandes, Sweedle Mascarnes and Dakshata Panchal

Abstract Augmented Reality is one of the revolutionary technologies gaining fast pace in enhancing user interaction. Education is evolving rapidly and needs to keep accomplishing its goal of providing knowledge seamlessly. The use of mere images in print media like textbooks limits learner's understandability. A much more interactive approach is required to boost the overall learning experience. The existing systems are quite capable of delivering useful content such as 3D models, by using special markers for tracking done in images. This limits their potential as it is not feasible to create markers for every image and is time-consuming. This research work focusses on developing an Android application using Marker-less AR. The proposed system encompasses image recognition using a mobile camera, rendering relevant YouTube videos pertaining to the image, and their superimposition in AR. The aim of this system is to enhance the productivity of education to enliven images with interactive videos.

Keywords Augmented Reality (AR) · Marker-less · YouTube videos
Normalization · YouTube videos ranking

W. Fernandes · T. Gomes (✉) · A. Fernandes · S. Mascarnes · D. Panchal
Department of Computer Engineering, St. Francis Institute of Technology, Mumbai, India
e-mail: thelmagomes.11@gmail.com

W. Fernandes
e-mail: winston23fernandes.wf@gmail.com

A. Fernandes
e-mail: ashleefernandes@gmail.com

S. Mascarnes
e-mail: swedle.mascarenhas@gmail.com

D. Panchal
e-mail: dakshatapanchal@sfitengg.org

© Springer Nature Singapore Pte Ltd. 2019
J. Wang et al. (eds.), *Soft Computing and Signal Processing* ,
Advances in Intelligent Systems and Computing 898,
https://doi.org/10.1007/978-981-13-3393-4_23

223

1 Introduction

In today's age, the level of education is elevating rapidly to new heights by combining with technology. But a vast number of institutions still rely on traditional teaching methods like books and face-to-face interactions. Though it is an effective way of providing better results, it somewhat lacks in enhancing the caliber of the learner. Predominantly, the diagrams or images in books do not render all adequate information that a user entails and further depend on external sources such as the internet. Generally, YouTube videos impart good information and are quite preferred by learners. Thus, this formed the basis of the proposed study which aims to provide dynamic content directly to users, while learning through books, thereby saving time and boosting their performance. A novel approach to this is by using Augmented Reality (AR). By using AR, the proposed system focusses on rendering dynamic and useful content directly to the user through the image at hand.

Augmented Reality (AR) is a technology that superimposes an image or a video on a user's view of the real world, thus providing a composite view. AR consists of two types- Marker-based and Marker-less. Marker-based AR uses fiducially identifying markers to provide data. Marker-less AR uses natural features of the image and does not need user-specified markers.

In the proposed system, using Marker-Less technology, the range of images on which the application can be implemented is enlarged. Thus, it detects and recognizes an image from books and gets a keyword that describes the image and all the relevant YouTube videos with respect to the keyword are fetched and displayed in an augmented view. The user then selects the required video and it is played using Augmented Reality in the mobile phone. Thus, the proposed system uses a Marker-less AR approach by presenting relevant YouTube videos in augmented view.

2 Review of Literature

With the boom of Virtual Reality in various technologies, Augmented Reality is fast gaining momentum. In the research work stated in [1], AR works on the three aspects: Combination of the real and virtual environment, Real-time interaction and Registering in 3 Dimension. They are very important for AR technology as combining them will give us an appealing visual technology.

In the research work done in [2], the potential of AR in enhancing education is studied upon. An interesting application of advanced teaching methods is the use of augmented reality textbooks. Though the books are the usual printed ones, what makes it stand out, is that pointing a camera at it, brings out visualizations and interactions. The use of AR can create dynamic sources of information out of mere pages of books.

In another research work stated in [3] towards using AR in education, the primary school students were provided an app which helped in understanding history better.

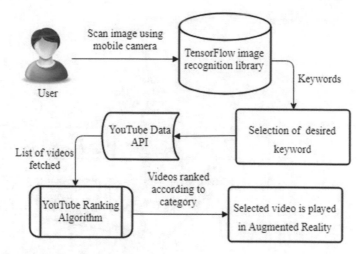

Fig. 1 Architectural diagram of the system

This was done by providing real-time content in an augmented view using a mobile application. It used images (uploaded by users as markers) as a trigger to display content and allows the user to link content to be viewed in augmented view.

To overcome the limitation of manually uploading images for markers, the research work presented here introduces a novel method to provide interactive YouTube videos for images that can be taken from all education-related books and not any custom-made one, in particular, thereby boosting availability of the system. Its underlying motive is to save the user's time and efforts.

3 Proposed System

In this research work, an Android application is built which allows the user to scan the image using the device camera. The scanned image is further recognized and assigned keywords using Google TensorFlow object detection library [4]. The user then selects the appropriate keyword matching the image. Furthermore, the YouTube videos are fetched based on the selected keyword. The user chooses any one video from the list to be played. The selected video is played in augmented view.

The workflow of the overall system (as shown in Fig. 1) is presented below.

3.1 Image Recognition

(1) Convolutional neural network

TensorFlow uses a convolutional neural network architecture codenamed "Inception" which is responsible for image recognition [4]. A convolution helps in finding out whether a particular feature is present in an image or not [5, 6].

Fig. 2 A full neural network

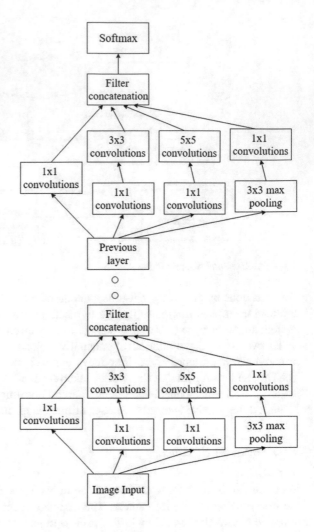

A bigger convolution helps in finding the higher level details of an image like shapes and smaller ones help in finding the lower details like edges and corners. As these "Inception modules" are stacked on top of each other (as shown in the Fig. 2 below), the output of one layer is concatenated and then given to the next layer as the input.

(2) **Label assignment and calculation of similarity index**

The final layer of the convolutional neural network is the softmax layer. The similarity index of the keyword is calculated on the basis of the softmax regression. After this layer, the list of suitable keywords and their similarity indexes is displayed.

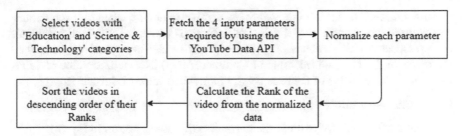

Fig. 3 Flowchart of the Ranking Algorithm

3.2 Relevancy of YouTube Videos (Ranking Algorithm)

Once the user-desired keyword is selected from the list provided, all the relevant YouTube videos associated with it are fetched using a ranking algorithm. The values of the parameters considered are normalized to calculate the video rank. The YouTube Data API [7] helps in acquiring the videos. The algorithm works as shown in Fig. 3.

(1) **Fetch the videos using the keyword**

Once the keyword is obtained and given to the YouTube Data API [7] in string format, the API provides all the video data matching that keyword in JSON format. With the count of the number of videos set to a number 20, the category IDs of videos with the categories 'Education' and 'Science & Technology' will be considered while they are being fetched.

(2) **Obtain the values of parameters from the Video ID and Channel Id of the video**

According to the proposed ranking algorithm, 4 parameters are considered in order to rank the videos for gaining its relevancy. The parameters are Number of Comments (C), Number of Likes (L), Number of Dislikes (D), all of which are fetched from Video ID and Number of Subscribers (S), fetched from Channel ID.

(3) **Normalization by scaling between 0 and 1**

Row normalization on each parameter considered for each video is performed using the method approached in the research work [8]. Since multiple videos are be fetched, multiple rows each having these 4 parameters are created.
For example, the normalized value of e_i for variable E in the ith row is calculated as:

$$Normalized(e_i) = \frac{e_i - E_{min}}{E_{max} - E_{min}} \tag{1}$$

where E_{min} = the minimum value for variable E
E_{max} = the maximum value for variable E
Similarly, normalized values for the above-stated parameters are calculated.

For the minimum and maximum values, respective values are taken for each parameter from the rows containing the data of the videos fetched. Thus, each parameter of the respective video will have a normalized value scaled between 0 and 1 and they are respectively: Normalized value of Comments (N_C), Normalized value of Likes (N_L), Normalized value of Dislikes (N_D), Normalized value of Subscribers (N_S).

(4) **Rank Calculation**

To calculate the rank, we will use the normalized values as shown in Eq. (2),

$$Rank = (N_C * N_L * (1/N_D) * N_S)^{1/N} \tag{2}$$

where $N = 4$ since 4 parameters are considered.

(5) **Selection of relevant videos**

After all the rows have been assigned a rank, they will be sorted in the decreasing order of their ranks.

3.3 Fetching and Displaying Video in Augmented View

(1) **Streaming the YouTube video in AR:**

The video that is played in AR requires an RTSP (Real Time Streaming Protocol) link so as to stream it in the media player view. The Video ID acquired from the chosen video to be played is used in the link that looks like below:

https://m.youtube.com/watch?v=<YouTubeID>&app=m

The address of the YouTube video in the above-mentioned webpage is in RTSP which looks like the following.

rtsp://r5---sn-5mlrnez.googlevideo.com/Cj0LENy73B4732/yt6/1/video.3gp

As the proposed system is an Android application, the above link is fetched using JSoup library [9] and provides video in 3gp format.

(2) **Layers in Augmented Reality**

The following are the steps involved in displaying the video in Marker-less AR (Fig. 4).

(1) The image which was taken by the camera for recognition is taken as the reference image for displaying the video in AR. While placing the camera on it, its corners are detected from the frame. The OpenCV library [10] is used for this and the other steps.
(2) The aspect ratio of this image is further calculated. This helps to convert the size of the above image to fit the size of the screen.
(3) The features are extracted from the scene and the descriptors are computed using the ORB (Oriented FAST Rotated BRIEF) algorithm [11]. Feature descriptors are computed from the pixels around each interest point.

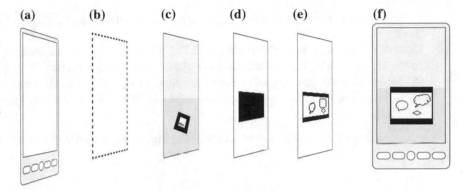

Fig. 4 Layers in augmented reality: **a** Phone. **b** Frame layout. **c** Camera view. **d** GL surface view. **e** Texture view. **f** Final view

focalLengthXInPixels	0	centerXInPixels
0	focalLengthYInPixels	centerYInPixels
0	0	1

Fig. 5 Camera calibration matrix

(4) The calibration matrix calculates the camera matrix using extrinsic and intrinsic parameters. Extrinsic parameters represent a rigid transformation from 3-D world coordinate system to the 3-D camera's coordinate system. Intrinsic parameters represent a projective transformation from the 3-D camera's coordinates into the 2-D image coordinates in (Fig. 5).

The above matrix is 3×3 one and is stated as: The focal length is the distance between the camera's sensor and the lens system's optical center when the lens is focused at an infinite distance.

(5) The frame containing the image is checked for enough feature points. Since the reference image is captured the first time by clicking on the screen, its feature points are checked (Fig. 4a). For that, it waits till the camera focusses on the image to provide clarity and hence extract features [12].

(6) Homography Estimation: The homography between a reference image and the image in the scene is found. By applying the homography to a rectangle, we can get an outline of the tracked object (Fig. 4b).

(7) Check for the perspective of the image in the Camera view (Fig. 4c).

(8) 8If the camera is moved from the reference image, the image corners of this frame are also calculated in order to check if it matches the reference image. Steps 2–7 are repeated to obtain data of this new frame. At step 5 for this frame, the features are checked and if enough, they are compared with the reference image to see if they match.

(9) Using the GL SurfaceView (Fig. 4d) in OpenGL, the surface (in rectangular form) on which the video will be placed is created. This is done using OpenGL

shader language. In the proposed system, this shader is used to keep the surface black.

(10)   Once the video frame is available after the RTSP link is generated, it is placed in the Texture View (Fig. 4e) which is positioned over the GL SurfaceView. A Texture View can just as easily be used to embed an OpenGL scene in your application. The video is played using the media player which is embedded in this view.

The proposed application will appear as shown in Fig. 4f on completing all the steps.

4   Implementation and Experimental Results

For validating the present system, experiments were conducted on a set of images of various animals such as tiger, elephant, anemone fish, etc. As shown in Fig. 6a, the image of the tiger was scanned using the mobile camera. The appropriate keyword detected was 'tiger' as shown in Fig. 6a. Upon clicking on the keywords section in the app, a dialog box appears. It contains the list of keywords stated at the time of clicking on that area (Fig. 6b). The user selects the keyword 'tiger' and a list of all the YouTube videos related to this keyword are fetched as per the ranking algorithm (Fig. 6c). The desired video is then selected and its corresponding RTSP link is generated. From the Fig. 6e, the desired video is played in augmented view. The user can also change or edit their keyword directly using the edit option (Fig. 6d). This aids in optimizing the user's search which may not always be possible using the image at hand.

This system is created for educative content at present. Therefore, only 'Education' and 'Science and Technology' videos are fetched. The ranked videos with these categories help provide easier access, which aim at sharing informative knowledge.

The video search results delivered by the system (refer Fig. 6c) when compared by the YouTube search results (refer Fig. 6f) shows that the former efficiently provides helpful and educative videos only, on entering the same keyword. The latter provides videos related to movies, music as well which are not needed in context to the image. The user would also require more time to search for required videos in the latter case.

The proposed system is also suitable for images whose keywords may portray different mediums. For example, the keyword 'python' will provide highly ranked videos on both the snake python as well as the programming language python. This will help the user to easily decide which video they'd opt for using a single image.

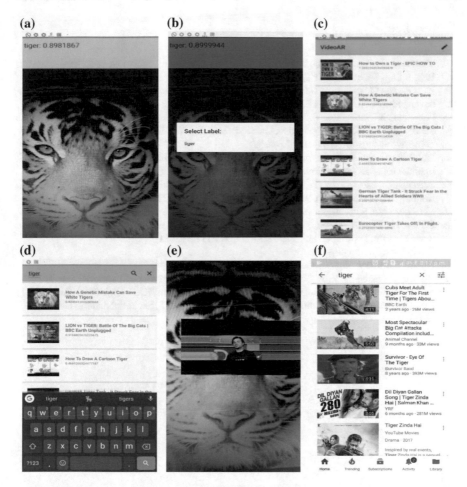

Fig. 6 Snapshots of the system (**a–e**) and 'tiger' keyword search on YouTube (**f**)

5 Conclusion

The proposed system provided a platform which combined AR and provision of YouTube videos in a way which would enhance the usual way of learning. This system incorporated the use of a mobile camera for image recognition, by developing an Android application. To provide relevant YouTube videos, the system uses an efficient ranking algorithm. Relevant videos related to the image are displayed for user selection, after which user can play the desired video in augmented view. The validation of the system was carried out on animal images. The future scope of the proposed system could be in medicine, newspapers, and entertainment.

References

1. Tiwari, V., Tiwari, V.P., Chudasama, D., Bala, K.: Augmented reality and its technologies. Int. Res. J. Eng. Technol. (IRJET) **3**(4) (2016)
2. Kesima, M., Ozarslan, Y.: Augmented reality in education: current technologies and the potential for education. Procedia Soc. Behav. Sci. **47**, 297–302 (2012) (Elsevier)
3. Persefoni, K., Tsinakos, A.: A mobile augmented reality application for primary school's history. IOSR J. Res. Method Educ. (IOSR-JRME) **6**(6 Ver. III) (2016)
4. TensorFlow. https://www.tensorflow.org/
5. Dalal, N., Triggs, B.: Histograms of Oriented Gradients for Human Detection. French National Institute for Research in Computer Science and Automation (INRIA)
6. Szegedy, C., Liu, W., Jia, Y., Sermanet, P., Reed, S., Anguelov, D., Erhan, D., Vanhoucke, V., Rabinovich, A.: Going Deeper with Convolutions. Cornell University Library (2014)
7. YouTube Data API. https://developers.google.com/youtube/v3/
8. Dubey, H., Roy, B.N.: An improved page rank algorithm based on optimized normalization technique. Int. J. Comput. Sci. Inf. Technol. (IJCSIT) **2** (2011)
9. JSoup Library. https://jsoup.org/
10. OpenCV Library. https://opencv.org/
11. Rublee, E., Rabaud, V., Konolige, K., Bradski, G.: ORB: an efficient alternative to SIFT or SURF. In: IEEE International Conference on Computer Vision (ICCV), pp. 2564–2571 (2011)
12. Howse, J.: Android Application Programming with OpenCV 3. Packt Publishing, Birmingham (2015)

Effects of Different Queueing Models on Migration of Virtual Machines

Surabhi Sachdeva and Neeraj Gupta

Abstract Virtualization is an act of creating virtual resources that are made accessible by the application of cloud computing. In case of failure of a virtual machine, it is imperative to shift the processes running on this machine to another. This activity is known as live virtual machine migration and is classified under the major issues of research. Xia (2015) proposed a mathematical model to analyze this problem in which the work has been done in two phases. It is showing a state transition model system in which the M/M/1/K model has been utilized in phase1 to find the rejection probability of the jobs in the phase2. Each virtual machine is considered to have a buffer to store the incoming buffer. The major issue being discussed in this paper is the relation of the rejection probability of jobs with the changing size of the buffer. It actually provides an exhaustive analysis of three different queueing models, i.e., $M/M/1/\infty$, $M/M/\infty$, and $M/M/1/K$. The simulations are carried out in MATLAB, and the results are analyzed based on the rejection probability of the jobs. It is observed that with an increase in the request arrival rate, the rejection probability of jobs increases. However, with an increase in execution rate, the rejection probability of jobs decreases. If we change the model to $M/M/\infty$, actually, the formulas of request rejection probability and job rejection probability got changed that resulted in a continuous decrease in values of rejection rate lines as compared to the values of the author. Hence, we can say that changing the queueing model is beneficial.

Keywords Load balancing · Queueing models · Virtual machine migration
Buffer size · Rejection probability

S. Sachdeva (✉) · N. Gupta
K. R. Mangalam University, Gurugram, Haryana, India
e-mail: surabhisachdeva@gmail.com

N. Gupta
e-mail: neerajgupta3729@gmail.com

© Springer Nature Singapore Pte Ltd. 2019   233
J. Wang et al. (eds.), *Soft Computing and Signal Processing* ,
Advances in Intelligent Systems and Computing 898,
https://doi.org/10.1007/978-981-13-3393-4_24

1 Introduction

Delays and queueing issues are more normal in everyday circumstances as well as in technical environments like telecommunication. A branch of mathematics that deals with the study and modeling the act of waiting in lines is named as queueing theory. In a system, assuming that the jobs arrive with the rate λ jobs/minute, service rate being μ jobs/minute, and K refers to the finite capacity of the system. Each virtual machine is considered to have buffer to store the incoming buffer. This paper provides an exhaustive analysis of different queueing models. The queueing models that we have considered are M/M/1/∞, M/M/∞, and M/M/1/K. By taking into consideration the two stages, one being M/M/1/∞ and the other being M/M/1/K, both of them have different buffer sizes. M/M/1/∞ on the one hand has an infinite buffer, whereas M/M/1/K model on the other hand has a finite capacity buffer. This simply means that in the first stage all the jobs can be served, whereas in the second stage only the K number of jobs can be served and the (K + 1)th job gets rejected. The primary fundamental point of this paper is to relate the rejection probability of jobs to different queueing models. So, by taking the same assumed parameter values as that of [1], the simulations are carried out in MATLAB and fluctuations of graph lines of the rejection rate formula of different models are compared in Table 1. This is a scientific paper featuring the points of interest of the rejection rate formula and utilization of the other two models on the same.

The prime motive of this paper is to explore the details of different queueing models for migration of virtual machines and draw a contrast among them so as to arrive at a conclusion about the best queueing model for migration. The remaining sections of the paper are systematically categorized as follows: In Section 2, we reviewed studies of migration and queueing theory in the literature survey. The comparison of various queueing models in live migration scenario is there in Sect. 3. Section 4 gives an insight into the MATLAB implementations and results, and Sect. 5, the last section consists of details of conclusion, recommendations, and future work.

Table 1 Definitions of symbols used

$\lambda = 10$ jobs/min	Request arrival rate
$\mu = 20$ jobs/min	Service rate of jobs
N_c, N_p	Finite queue capacity
$\rho = 10/20 = 1/2$	Utilization factor
P_i	Steady-state probability of ith request
RJC	Request rejection probability in phase1
RJQ	Job rejection due to insufficient queue capacity
RJH	Job rejection due to non-existing hot PM
RJP	Job rejection probability
RJ	Overall rejection probability (rejection rate)

2 Literature Survey

There has been a lot of work done in the area of the live migration. Various research challenges have been addressed in [2–4]. There are different types of ways to tackle the problem of migration in [5–7]. In [8], M/G/1/K * PS queueing model for a Web server has been introduced, and expressions for average response time, throughput, and blocking probability were made. Keeping in mind the end goal of anticipating the execution of administration uncovered by the cloud as in [9], it proposed a queueing hypothesis based model and analyzed expected request completion time. Also, a percentile of response time used as a QoS metric, introduced M/M/1 model taking arrival and service rates as inputs but failed to include failure/repair rates. In [10], a similar study using sequential queueing network was carried out ignoring failure/repair rates. In [11], a theory-based probability distribution of service time considering failure/repair rate was derived. A multivariate probabilistic model was given in [12] to study the effect of VM relocation plans. A comprehensive QoS model taking into consideration all the aspects was given in [13]. Keeping in mind the main aim to spare vitality, [14] proposed a packing algorithm-based strategy. The issue of demonstrating PC benefit execution as for reaction time, throughput, usage of system has been examined in [9, 11, 14, 15]. Proposal and validation of a Web application execution toll are in [4]. Authors in [10, 16] proposed a queue-construct display for execution administration in light of the cloud where the web applications are modeled as service centers. Authors in [1] utilized the M/M/1/K queueing model for the migration of virtual machines. It is showing a state transition model system where job arrival rate is represented by λ jobs/minute and service rate is represented by μ jobs/minute. In numerous bits of research, the term 'dirty pages' or 'dirty blocks' has been utilized to indicate the assets like memory or capacity which really gets adjusted in the relocation process [17]. Chanchio and Thaenkaew [18] utilized M/M/1 model to ascertain the limit with a specific end goal to control the relocation. Zheng et al. [19] reenacted the movement procedure and assessed utilizing expectation edge technique. Quantitative examination of live relocation inside a data center has been given in [20–22]. Numerical model used to create arbitrary factors for fake occasions has been discussed. Also, the analysis of process migration using genetic algorithm has been discussed in [23, 24].

3 Comparison of Different Queueing Models in Live Migration Scenario

According to Fig. 1, we are interested in finding the behavior of the queues to discover the rejection probability of jobs in light of M/M/1/K model [1]. In this paper, we are assuming the M/M/1/∞, M/M/1/K, M/M/∞ queueing models for the two buffers. The above migration model actually works in two phases. One is CMU handling phase, and the other is job execution phase. M/M/1/K queueing model was utilized

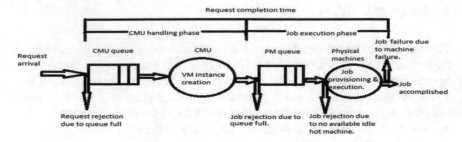

Fig. 1 Migration model [1]

by the authors in phase1. The main objective under consideration is to relate the rejection probability of jobs to the changing size of the buffer and further analyze the changes. We know that the nature of the output will change according to the buffer size. Upon considering all the models one by one, and keeping in mind the end goal to discover the rejection rate of that model.

We know that utilization factor,

$$\rho = \lambda/\mu_c \tag{1}$$

$$P_i = \rho^i P_0 \tag{2}$$

$$\text{RJC} = P_{Nc} \tag{3}$$

After doing the analysis of the formulas of [1] and carrying out calculations on them, the generalized formula for request rejection probability in phase1 can be framed as:

$$\text{RJC} = 1 - \left(1 - \rho^i P_0\right) \tag{4}$$

If the change is 0 at steady state, then

$$\mu P_1 = \lambda P_0.$$

$$\text{Or } P_1 = \left(\frac{\lambda}{\mu}\right)P_0, \quad P_2 = \left(\frac{\lambda}{\mu}\right)P_1, \quad P_3 = \left(\frac{\lambda}{\mu}\right)P_2 \text{ and so on} \tag{5}$$

Hence,

$$\left.\begin{array}{l} P_1 = \rho P_0 \\ P_2 = \rho^2 P_0 \\ P_3 = \rho^3 P_0 \\ \quad \text{or} \\ P_n = \rho^n P_0 \end{array}\right\} \tag{6}$$

Table 2 Comparison of assumed queueing models with [1]

S. No.	Queueing model	No. of servers	Buffer size	Jobs served	Jobs rejected
1.	M/M/1/∞	1	Infinite	No limit	Never
2.	M/M/1/K [1]	1	K	Only K jobs	(K + 1)th job
3.	M/M/∞	Infinite	Infinite	No limit	Never

Table 3 Parameter values for drawing graphs

Parameters	M/M/1/K	M/M/1/∞	M/M/∞
RJC	1/63	1/64	0.0001
RJQ	0000000	0000000	0000000
RJH	1000101	1000101	1000101
RJP	1/63	1/64	0.0001
RJ	$1 - \left(1 - \rho^i \times \left(\frac{1-\rho}{1-\rho^{K+1}}\right)\right)$	$1 - (1 - \rho^i \times (1 - \rho))$	$1 - (1 - \rho^i \times \exp^{(-\rho)}/i\,!)$

Hence,

$$P_0 = 1 - \rho \tag{7}$$

$$P_n = \rho^n(1 - \rho) \tag{8}$$

The value of P_0 is different for the different queueing models. RJ formula is having P_0 as a variable. Hence, the RJC formula (4) gets manipulated accordingly. So, by putting the values of P_0 in (4) according to the different queueing model, we can derive the expressions for request rejection probability in phase1 as:

$$M/M/1/\infty = 1 - \left(1 - \rho^i \times (1 - \rho)\right) \tag{9}$$

Similarly, on comparative grounds, the request rejection probability of jobs of other queueing models can be found at:

$$M/M/1/K \text{ model} = 1 - \left(1 - \rho^i \times \left(\frac{1 - \rho}{1 - \rho^{K+1}}\right)\right) \tag{10}$$

$$M/M/\infty \text{ model} = 1 - \left(1 - \rho^i \times \exp^{(-\rho)}/i!\right) \tag{11}$$

The net effect of the change in the queueing model affected the values of the RJC and RJP as given in Table 2 [1]. So, further on solving each model, firstly, we get the values of the different parameters and then the graphs are drawn to check the fluctuations of RJ lines in Table 3.

So, general formula for

$$RJ = 1 - \left(1 - \rho^i P_0\right) \times (1 - RJC)^{\wedge}2 \tag{12}$$

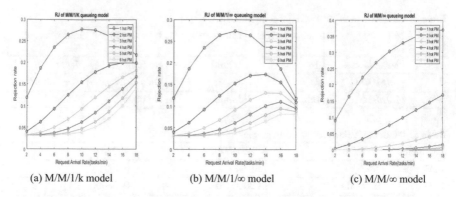

(a) M/M/1/k model   (b) M/M/1/∞ model   (c) M/M/∞ model

Fig. 2 RJ of different models versus arrival rate with different number of initial hot PMs

$$M/M/1/\infty \quad 1 - (1 - \rho^i \times (1 - \rho)) * (63/64).^2) \tag{13}$$

$$M/M/1/K \quad 1 - (1 - \rho^i \times (1 - \rho 1 - \rho^{K+1})) * (62/63).^2 \tag{14}$$

$$M/M/\infty \text{ model} \quad 1 - (1 - \rho i \times \exp(-\rho)/i!) * ((\rho)5/5!)^2 \tag{15}$$

4 Results

In this section, we carried out the implementation of the work with MATLAB 9.3 platform. The implementations carried out are capable of calculating RJ (the overall rejection probability), provided input request arrival rate, queue capacity, service rate of jobs, number of initial hot PMs. We considered $\lambda = 10$ and $\mu = 20$. Figure 2a, b, and c shows RJ versus the arrival rate of [1], M/M/1/∞, and M/M/1/∞ models, respectively. For a specific number of hot PMs, the request arrival rate is showing a direct relation to the rejection probability. In [1], the initial number of hot PMs is showing an inverse relation to the rejection probability of jobs.

Figure 2b shows an inverse relation between the number of hot PMs and the job rejection probability. In Fig. 2c, by taking the number of hot PMs into consideration, it shows that with an increase in the initial number of hot PM, rejection probability decreases. However, if we take into consideration an individual instance of 1 hot PM, it is visible that with an increase in the request arrival rate, the rejection probability of jobs additionally continued increasing. Figure 3a, b, and c represents RJ versus the execution rate of M/M/1/K [1], M/M/1/∞, and M/M/∞ models, respectively. From these figures, it is clearly visible that execution rate has an inverse relation with the rejection probability of jobs. Also, for a specific number of hot PMs, with an increase in the execution rate, the rejection probability of jobs decreases. The initial value of RJ lines is least in the case of M/M/∞ model, i.e., Fig. 3c.

Figure 4a and b shows comparison of the rejection rate versus request arrival rate and execution rate of all the models, respectively. As can be seen in this that with an

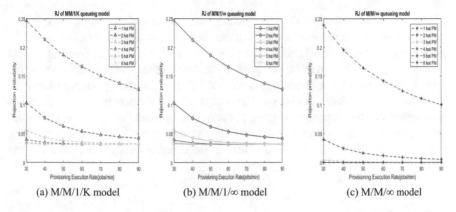

(a) M/M/1/K model　　　　(b) M/M/1/∞ model　　　　(c) M/M/∞ model

Fig. 3 RJ of different models versus execution rate with different number of initial hot PMs

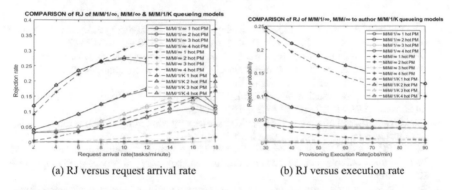

(a) RJ versus request arrival rate　　　　(b) RJ versus execution rate

Fig. 4 Comparison of RJ of different queueing models versus arrival and service rate

increment in the number of hot PMs, there is a higher increase in the job rejection probability of M/M/1/K [1] when compared to M/M/∞ model. It can be seen in the graph that on changing the model to M/M/∞, i.e., by taking an infinite buffer, there is a decrease in the overall rejection probability. Hence, M/M/∞ is the best possible choice of queueing model for migration.

5　Conclusion

This is a mathematical study highlighting the details of different queueing models relating to virtual machine migration. [1] is based on the application of the M/M/1/K model to compute the RJ (overall rejection probability of jobs). We further consider applying M/M/1/∞ and M/M/∞ models for comparison with work already done. After implementing, the basic conclusion of the paper is that on considering a node with a single server, finite buffer size (K) = 5 jobs, the graphs denote a direct relation of the request arrival rate with the values of the RJ. In considering a node with a single server or infinite servers and infinite buffers, it is clearly readable from the graph that the values of RJ decrease. And with an increase in execution rate, the RJ is decreasing. With a specific end goal of displaying the changing conduct of the system, we changed the queueing model and by changing the queueing model the formulas for RJC, RJP got changed that graphically outcomes in a decrease in RJ. Consequently, the proposed changing of the queueing model in the system turns out to be gainful. In the future, we need to execute some general distribution queueing models like $M/G/1/\infty$, $M/G/1/K$ and look at the comparisons of estimations of the RJ of these with the $M/M/1/K$ system.

References

1. Xia, Y., Zhou, M., Luo, X., Zhu, Q., Li, J., Huang, Y.: Stochastic modeling and quality evaluation of infrastructure-as-a-service clouds. IEEE Trans. Autom. Sci. Eng. 162–170 (2015)
2. Sahoo, J., Mohapatra, S., Lath, R.: Virtualization: a survey on concepts, taxonomy and associated security issues. In: 2010 2nd International Conference on Computer and Network Technology (ICCNT), pp. 222–226. IEEE (2010)
3. Strunk, A.: Costs of virtual machine live migration: a survey. In: 2012 IEEE 8th World Congress on Services (SERVICES), pp. 323–329. IEEE (2012)
4. Loganayagi, B., Sujatha, S.: Enhanced cloud security by combining virtualization and policy monitoring techniques. Procedia Eng. **30**, 654–661 (2012)
5. Chen, H.P., Li, S.C.: A queueing based model for performance management on cloud. In: International Conference on Advanced Information Management and Service (IMS), pp. 83–88. IEEE (2011)
6. Anala, M.R., Shetty, J., Shobha, G.: A framework for secure live migration of virtual machines. In: 2013 International Conference on Advances in Computing, Communications and Informatics (ICACCI), pp. 243–248. IEEE (2013)
7. Baghshahi, S.S., Jabbehdari, S., Adabi, S.: Virtual machine migration based on Greedy algorithm in cloud computing. Int. J. Comput. Appl. **96**(12) (2014)
8. Cao, J., Andersson, M., Nyberg, C., Kihl, M.: Web server performance modeling using an M/G/1/K*PS queue. In: 10th International Conference on Telecommunications, 2003. ICT 2003, vol. 2, pp. 1501–1506. IEEE (2003)
9. Xiong, K., Perros, H.: Service performance and analysis in cloud computing. In: Services-I, 2009 World Conference, pp. 693–700. IEEE (2009)
10. Dai, Y.S., Yang, B., Dongarra, J., Zhang, G.: Cloud service reliability: modeling and analysis. In: 15th IEEE Pacific Rim International Symposium on Dependable Computing, pp. 1–17. IEEE (2009)
11. Yang, B., Tan, F., Dai, Y. S., Guo, S.: Performance evaluation of cloud service considering fault recovery. In: IEEE International Conference on Cloud Computing, pp. 571–576. Springer, Berlin, Heidelberg (2009)
12. He, S., Guo, L., Ghanem, M., Guo, Y.: Improving resource utilisation in the cloud environment using multivariate probabilistic models. In: 2012 IEEE 5th International Conference Cloud Computing(CLOUD), pp. 574–581. IEEE (2012)
13. Ghosh, R., Trivedi, K.S., Naik, V.K., Kim, D.S.: End-to-end performability analysis for infrastructure-as-a-service cloud: an interacting stochastic models approach. In: 2010 IEEE 16th Pacific Rim International Symposium on Dependable Computing (PRDC), pp. 125–132. IEEE (2010)

14. Li, B., Li, J., Huai, J., Wo, T., Li, Q., Zhong, L.: Ena cloud: an energy saving application live placement approach for cloud computing environments. IEEE (2009)
15. Karlapudi, H.: Web application performance prediction. In: Proceedings of International Conference on Communication and Computer Networks, IASTED, pp. 281–286 (2004)
16. Mastelic, T., Brandic, I.: Recent trends in energy efficient cloud computing. J. Latex **11**(4) (2012)
17. Sarker, T.K., Tang, M.: Performance-driven live migration of multiple virtual machines in datacenters. In: International Conference on Granular Computing (GrC), pp. 253–258. IEEE (2013)
18. Chanchio, K., Thaenkaew, P.: Time-bound, thread-based live migration of virtual machines. In: 14th IEEE/ACM International Symposium on Cluster, Cloud and Grid Computing (CCGrid), 2014, pp. 364–373. IEEE (2014)
19. Zheng, J., Ng, T.E., Sripanidkulchai, K., Liu, Z.: Pacer: a progress management system for live virtual machine migration. IEEE Trans. Cloud Comput. **10**(4), 369–382 (2013)
20. Vilaplana, J., Solsona, F., Teixidó, I., Mateo, J., Abella, F., Rius, J.: A queueing theory model for cloud computing. J. Supercomput. 492–507 (2014)
21. Pham, C., Hong, C.S.: Using queueing model to analyse the live migration process in data centers, pp. 1136–1138. IEEE (2014)
22. Yu, L., Chen, L., Cai, Z., Shen, H., Liang, Y., Pan, Y.: Stochastic load balancing for virtual resource management in datacenters. IEEE Trans. Cloud Comput. IEEE (2014)
23. Kumar, N., Saxena, S.: Migration performance of cloud applications—a quantitative analysis. Procedia Comput. Sci. **45**, 823–831 (2015)
24. Sandhya, S., Revathi, S., NK, C.: Performance analysis and comparative analysis of process migration using genetic algorithm. Int. J. Sci. Eng. Technol. Res. **5**(11) (2016)

A Deep Learning-Inspired Method for Social Media Satire Detection

Sayandip Dutta and Anit Chakraborty

Abstract In this paper, we put forward an effective approach of segmentation of sentiment in social media texts that may include informal language or pop culture texts. We introduce a method to churn out vector representations from phrase-level sentences. We train a recurrent neural network combining quantitative and qualitative methods with lexical features stored in gold standard array of lexicons. In this work, we extract opinion expression using deep RNNs in the form of a token-level sequence-labeling sentiment from variable length of text corpuses. Furthermore, in this paper, we have introduced a novel approach to determine whether the article is satirical or not via the combination of computational linguistics and machine learning tools. We have compared the performance of our algorithm with respect to the benchmark methods, on satire detection as well, on benchmark datasets, news articles, and social media platforms for better reflection of the experiment, and we yielded competitive and satisfactory results.

Keywords Recurrent deep neural network · Word2vec
Sentiment analysis in social media · Deep learning · Machine learning

1 Introduction

Owing to an unforeseen growth of the internet, we are seeing a flood of online communication. A terrific effort has been put into research in recent times for fast and accurate sentiment classification.

A different approach to learning based on latent feature model is constructed on hidden layers of distributed dense vectors. A recurrent neural network (RNN) can

S. Dutta (✉)
MCKV Institute of Engineering, Howrah, India
e-mail: sayandip199309@gmail.com

A. Chakraborty
RCC Institute of Information Technology, Kolkata, India
e-mail: ianitchakraborty@gmail.com

© Springer Nature Singapore Pte Ltd. 2019
J. Wang et al. (eds.), *Soft Computing and Signal Processing* ,
Advances in Intelligent Systems and Computing 898,
https://doi.org/10.1007/978-981-13-3393-4_25

also be applied as a sequence labeler as this is one such learner which can operate on sequential stream of data of variable length.

Our approach aims to concoct a sentiment analysis engine which gives birth to a novel approach to determine whether the article is satirical or not via the combination of computational linguistics and machine learning tools.

We have organized the paper in following way: a brief review of related works in Sect. 2. Subsequently, in our proposed method of deep bidirectional neural network, architecture for social media sentiment analysis and satire detection is explained in Sect. 3. Section 4 depicts the experimental results and analysis part. Section 5 concludes the paper with the plan for further improvement and future possibilities.

2 Related Works

Previously, sentiment classification problems have been solved through logistic regression and support vector machines (SVM) [1]. Naïve-based classifiers and maximum entropy model have also been studied. In [1], Maas et al. tried to tackle this issue using an unsupervised model by learning word vectors, in order to encompass layered sentimental information.

Most of the techniques for sentiment classification heavily look up to intermediate sentiment lexicons. A few popular lexicons are LIWC [2], GI [3], Hu-Liu [4], etc., where every word is divided into different binary classes, and another kind of lexicons, like ANEW [5], SentiWordNet [6], and SenticNet [7] associate valence-based values to words for sentence-level sentiment intensity.

In numerous sequential prediction tasks, as briefly explained by Elman [8], one significant class of naturally deep neural architecture has been explored. In the context of natural language processing, the deep RNN architecture takes sentences as a sequence of tokens at every time-steps and has been applied to various tasks (Mikolov et al. [9]) which involve language modeling, understanding of language and expression (Mesnil et al. [10]), etc.

3 Proposed Method

Here, we introduce the text modeling algorithm in order to calculate compact and efficient vector representation in line with the corresponding position of the word in vocabulary. Thereafter, we introduce deep neural sentiment classification method that has three layers via recurrent neural network.

3.1 Text Modeling

In this method, we approach negative sampling of word embedding. Conditional probabilities $p(c|w)$ is considered, where w and c represent corpus of words and their contexts, respectively, and given a corpus *Text*, we aim to maximize the corpus probability by setting the parameters θ of $p(c|w;\theta)$.

$$\arg\max_{\theta} \prod_{w \in Text} \left[\prod_{c \in C(w)} p(c|w;\theta) \right],$$

here, set of contexts of word corpus w is represented by $C(w)$. Alternatively:

$$\arg\max_{\theta} \prod_{(w,c) \in D} p(c|w;\theta),$$

here, we extract a set of all words and their corresponding context pairs from the text, which is represented by D.

3.2 Sentiment Classification with Deep Neural Networks

We parse an n-gram into a simple binary tree with the help of compositional models where every corresponding leaf node of a word gets a vector symbol. Each and every word is taken as d-dimensional vector. We initialize all of the word vectors after random sampling of their subsequent score following uniform distribution curve: U $(-r, r)$, where $r = 0.0001$.

As recurrent networks are well suited to work with any arbitrary spatiotemporal dimensions, these kinds of structures are ideal for many NLP tasks. Here we consider Elman-type network; in this type of network, a hidden layer, denoted by h_t at time step t, is derived with the principle of a nonlinear transformation of x_t and h_{t-1}, the then current input layer and the immediately previous hidden layer, respectively. Final output y_t is calculated using h_t. An Elman-type RNN computes the following output and memory sequences, when given a set of vectors $\{x_t\}_{t=1...T}$:

$$h_t = f(Wx_t + Vh_{t-1} + b), \tag{1}$$

$$y_t = g(Uh_t + c), \tag{4}$$

where f and g represent a nonlinear function (sigmoid function) and the output nonlinearity, respectively (softmax function). W is the weight matrix between the input and hidden layer, and V is the weight matrix among the hidden units themselves. Output weight matrix is U. b is bias vector connected to hidden unit, and c is the bias vector connected to output unit. Here h_0 is assumed to be 0.

Bidirectionality. As mentioned earlier, we rely on past information while calculating x_t. That is the restraining factor for the most NLP methods. But this problem could be simply resolved by including the future context of fixed size around a token (i.e., single input vector).

Depth in Space. We stack Elman-type RNNs on top of each other [11] to construct deep RNNs. Intuitively, every layer of the deep RNN takes its memory sequence of the layer before it as the input sequence, and subsequently, own memory representation is calculated.

3.3   Satire Detection

The satire corpus that we prepared contains a total of 233 satire news articles, 4000 newswire documents, and 3500 NY Times articles split into two sets, one for training and the other for test, as detailed in Table 3. We randomly sampled the newswire documents from the English Gigaword Corpus. All of the material was post-edited to remove any textual metadata that may refer to the source.

Standard Text Classification Approach. In our text classification approach, we reinforce bag-of-words model along with feature weighting, with the help of the following methods:

Binary feature weights: In this architecture, all features are tagged with the exact same weight, regardless of their frequency of appearance in each article. It has been found for most of the examples in the topic of sentiment classification that binary feature weightage performs better than the other available substitutes.

Bi-normal separation feature scaling: The bi-normal separation feature scaling [12] scheme has been beyond any doubts proven to outshine other standard feature representation schemes of text classification tasks on a wide. This advantage is remarkably staggering for collections which have a low proportion of positive class instances. Unlike in binary feature weighting, in BNS, every feature is given a score in accordance with the formula:

$$\left| F^{-1}(tpr) - F^{-1}(fpr) \right|,$$

where F^{-1} symbolizes inverse normal cumulative distribution function, and *tpr* and *fpr* stand for the true positive rate ($P(feature|positive\,class)$) and false positive rate ($P(feature|negative\,class)$), respectively. In BNS, highest weights are produced against features that are in strong correlation with either the positive or negative class. Lowest weights are given to the features that are uniform across the training instances.

Targeted Lexical Features. Here, three kinds of methods are described for identifying characters of satirical documents.

Headline features: The first line in most of the documents is their headlines. For a human reader, a headline alone would be sufficient to determine vast number of satirical documents, indicating that our classifiers may work a lot better and faster if the headline contents can be explicitly determined in the feature vector.

Profanity: It is evident that in true news articles, there is scarcely any profanity involved but upon a review of the corpus, it can be seen that satirical articles mostly use profanity as humorous element.

Slang: It is intuitively understandable that slangs are mostly avoided in true news articles. A rigorous review of articles suggests, satirical articles tend to use more informal language, which often contains slangs. We measure the article's informality as:

$$i \overset{\text{def}}{=} \frac{1}{|T|} \sum_{t \in T} s(t),$$

where T corresponds to set of unigram tokens. The function s becomes equal to 1 if the token corresponds to slang in dictionary entry and 0 otherwise. Discrete features, high informality, and low informality, i.e., *highi* and *lowi*, are set as:

$$highi \overset{\text{def}}{=} \begin{cases} 1 & v > \bar{\imath} + 2\sigma; \\ 0 \end{cases}$$

$$lowi \overset{\text{def}}{=} \begin{cases} 1 & v < \bar{\imath} - 2\sigma; \\ 0 \end{cases}$$

where σ and $\bar{\imath}$ symbolize the standard deviation and mean of i throughout all the articles.

Semantic Validity. News articles that are satirical in nature have a tendency to imitate genuine news articles in terms of style of organization, content, and tone; lexical approaches are not adequate for this task.

We started with an assumption that true events must are bound to be reported in different news communities and various media, and hence, the documents with similar arrangement or combination of material will be much greater than with satire documents. Stanford Named Entity Recognizer [13] is used for identifying a set of organization and person, E, from all the articles present in the corpus. The validity of the arrangement of entities in a given article is estimated as:

$$v(E) \overset{\text{def}}{=} |g(E)|$$

where the set of matching documents is denoted as g (the matching documents are found using Google). Here v will be lower if E includes made-up entity names or contains unusual combination of entities.

4 Experimental Results

For comparative performance measure, we cross-check the codependent relation of generated value of sentiment intensity with the sentiment rating mean value obtained from human raters (10 prescreened persons), in addition to the six-class classification (i.e., very positive, very negative, positive, negative, neutral, compound) with parameters like F1 score, precision, and recall.

Experimental Analysis. We have made a comparative analysis with respect to eight benchmark sentiment analysis lexicons, namely: Linguistic Inquiry Word Count, Valence Aware Dictionary for sEntiment Reasoning [14], Word Sense Disambiguation using WordNet, General Inquirer, Hu-Liu04 opinion lexicon, Affective Norms for English Words, SenticNet, and the SentiWordNet.

As seen in Tables 1, 2, and 3, our approach is superior in all the cases. The social media posts I can be seen that this architecture provides an altogether better F1 score, precision, and recall as compared to the human raters. The general accuracy of this algorithm analyzed against different benchmark approaches in classifying sentiments is graphically depicted in Figs. 1, 2, 3, and 4.

Figure 3 shows the decreasing *training error* with the increasing number of leaf nodes. The *validation error* is increasing when the number of *leaf nodes* is greater than 14.

Table 1 RNN model training data

Steps	Twitter posts	Positives	Neutrals	Negatives
Word embedding	100 M	–	–	–
Pre-training	30 M	12 M	8 M	10 M
Training	20104	10914	3551	5549
Validation	3000	1417	622	961
Test	21132	8816	7851	5165

Table 2 Classification performance comparison of Twitter data

Correlation to ground truth (mean of 10 human raters)	Metrics of classification accuracy (overall)		
	Recall	Precision	F1 score
Twitter Data (5,000 Tweets)			
Human Raters	**0.908**	0.80	0.87	0.82
Our Approach	0.812	**0.92**	**0.98**	**0.95**
Hu-Liu '04	0.712	0.66	0.88	0.73
VADER	0.798	0.91	**0.96**	**0.95**
GI	0.511	0.50	0.78	0.66
SCN	0.541	0.69	0.78	0.71
LIWC	0.605	0.48	0.90	0.59
SWN	0.440	0.59	0.73	0.60
WSD	0.400	0.43	0.68	0.51
ANEW	0.450	0.45	0.74	0.56

Table 3 Satire classification performance comparison of NY Times Editorial

Methods	Metrics classification accuracy (overall)		
	Precision	Recall	F1 Score
Sarcasm-labeled long texts (NY Times 3500 article review)			
Ind. Human	0.98	0.95	0.92
Ours	0.79	0.65	0.61
Joshi et al. [15]	0.42	0.51	0.55
Bouazizi and Ohtsuki [16]	0.54	0.49	0.50
Ghosh and Veale [17]	0.30	0.50	0.42
Amir et al. [18]	0.44	0.48	0.57
Abercrombie and Hovy [19]	0.39	0.51	0.51

Fig. 1 Comparison of performance: analysis on tweets

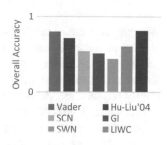

Fig. 2 Comparison of performance: analysis on newspaper editorials

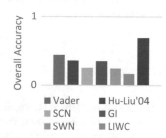

Fig. 3 Validation and training error versus number of leaf nodes

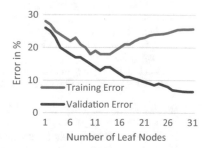

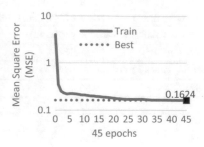

Fig. 4 Demonstration of the asymptotic decline in the improvement in error rate that occurs after approximately 40 *epochs*. The halting of improvement held true for the three networks that were trained once the epochs reached over 40 *epochs*

5 Conclusion

We present a novel technique in sentiment classification domain via deep recurrent neural network based on sentence-level vector representation (e.g., Word2vec). Empirically, we have evaluated our deep RNNs against the traditional shallow wide RNNs having only a single hidden layer. Our novel approach of satire detection, as briefly demonstrated in our comparative analysis, has shown immense potentiality when compared with other benchmark algorithms. In the near future, we are planning to incorporate more robust features by the virtue of pre-training the deep RNNs for various use cases, like patter finding in social media text, analyzing personality over social media texts.

References

1. Maas, A.L., Daly, R.E., Pham, P.T., Huang, D., Ng, A.Y., Potts, C.: Learning word vectors for sentiment analysis. In: Proceedings of the 49th Annual Meeting of the Association for Computational Linguistics: Human Language Technologies, Vol. 1, pp. 142–150. Association for Computational Linguistics (June, 2011)
2. Pennebaker, J.W., Chung, C.K., Ireland, M., Gonzales, A., Booth, R.J.: The Development and Psychometric Properties of LIWC2007. LIWC.net, Austin, TX (2007)
3. Stone, P.J., Dunphy, D.C., Smith, M.S.: The General Inquirer: A Computer Approach to Content Analysis (1966)
4. Hu, M., Liu, B.: Mining and summarizing customer reviews. In: Proceedings of the Tenth ACM SIGKDD International Conference on Knowledge Discovery and Data Mining, pp. 168–177. ACM
5. Bradley, M.M., Lang, P.J.: Affective Norms for English Words (ANEW): Instruction Manual and Affective Ratings (1999)
6. Fellbaum, C.: WordNet: An Electronic Lexical Database. MIT Press, Cambridge, MA (1998)
7. Cambria, E., Havasi, C., Hussain, A.: SenticNet 2. In: Proceeding of AAAI IFAI RSC-12 (2012)
8. Elman, J.L.: Finding structure in time. Cogn. Sci. **14**(2), 179–211 (1990)

9. Mikolov, T., Kombrink, S., Burget, L., Cernocky, J.H., Khudanpur, S.: Extensions of recurrent neural network language model. In: 2011 IEEE International Conference on Acoustics, Speech and Signal Processing (ICASSP), pp. 5528–5531. IEEE (2011)
10. Mesnil, G., He, X., Deng, L., Bengio, Y.: Investigation of recurrent-neural network architectures and learning methods for spoken language understanding. In: Interspeech (2013)
11. Hermans, M., Schrauwen, B.L.: Training and analysing deep recurrent neural networks. In: Advances in Neural Information Processing Systems, pp. 190–198 (2013)
12. Forman, G.: BNS scaling: an improved representation over TF-IDF for SVM text classification. In: Proceedings of the 17th International Conference on Information and Knowledge Management, pp. 263–270, Napa Valley, USA (2008)
13. Finkel, J.R., Grenager, T., Manning, C.: Incorporating non-local information into information extraction systems by gibbs sampling. In: Proceedings of the 43rd Annual Meeting on Association for Computational Linguistics, pp. 363–370. Association for Computational Linguistics (2005, June)
14. Hutto, C.J., Gilbert, E.: Vader: a parsimonious rule-based model for sentiment analysis of social media text. In: Eighth International AAAI Conference on Weblogs and Social Media (May, 2014)
15. Joshi, A., Tripathi, V., Patel, K., Bhattacharyya, P., Carman, M.: Are word embedding-based features for sarcasm detection? In: EMNLP 2016 (2016)
16. Bouazizi, M., Ohtsuki, T.: Sarcasm detection in Twitter: "All Your Products are Incredibly Amazing!!!"-are they really?. In: 2015 IEEE Global Communications Conference (GLOBE-COM), pp. 1–6. IEEE (2015)
17. Ghosh, A., Veale, T.: Fracking sarcasm using neural network. In: WASSA NAACL 2016 (2016)
18. Amir, S., Wallace, B.C., Lyu, H., Silva, P.C.M.J.: Modelling context with user embeddings for sarcasm detection in social media. In: CoNLL 2016, p. 167 (2016)
19. Abercrombie, G., Hovy, D.: Putting Sarcasm detection into context: the effects of class imbalance and manual labelling on supervised machine classification of Twitter conversations ACL 2016(2016), 107 (2016)

Medical Diagnosis of Ailments Through Supervised Learning Techniques on Sounds of the Heart and Lungs

Shantanu Patil, Abha Saxena, Tarun Talreja and Vidushi Bhatti

Abstract Auscultation is a medical technique to decipher ailments of human body through careful observation of heart and lung sounds. This project is a sincere effort to develop a low-cost stethoscope solution which can record and store the internal body sounds of patients' heart and lungs which will assist physicians in taking careful observations and tagging illnesses to the sounds, and training a machine learning model on recorded and tagged sounds to autosuggest the illness. There exist much matured medical diagnostic solutions in the country, but all of these solutions require a specific amount of time as well as incur a great cost to the patient. Machine learning in computation has evolved as a key driving force in the industry nowadays serving in a range of applications. Hence, we have extensively used various classifiers to train and test the audio data of patients and generate a viable diagnostic output which will serve as a primary guide to a medical practitioner. We hope this project serves its cause and has an impact on society through proper channelization of resources.

Keywords Telemedicine · Auscultation · Supervised learning
Sound classification

1 Introduction

In a developing country like India, getting access to the quality time of physicians is becoming more and more difficult for patients and unproductive for doctors.

S. Patil (✉) · A. Saxena · T. Talreja · V. Bhatti
National Institute of Technology Delhi, New Delhi, India
e-mail: 141100011@nitdelhi.ac.in

A. Saxena
e-mail: 141100058@nitdelhi.ac.in

T. Talreja
e-mail: 141100026@nitdelhi.ac.in

V. Bhatti
e-mail: vidushibhatti@nitdelhi.ac.in

© Springer Nature Singapore Pte Ltd. 2019
J. Wang et al. (eds.), *Soft Computing and Signal Processing*,
Advances in Intelligent Systems and Computing 898,
https://doi.org/10.1007/978-981-13-3393-4_26

253

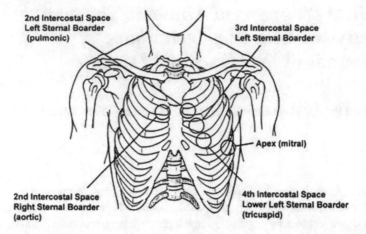

Fig. 1 Points of cardiac auscultation (*Source* National Center for Biotechnology Information)

In suburban areas, patients change their physicians very often and there is no consistent way to store and share their observations as fundamental as stethoscope sounds. Some of the digital alternatives, costing from Rs. 10,000 to Rs. 50,000 [1], are available to very wealthy patients and rich doctors in Fig. 1. However, even they do not provide a consistent interface to store and share. It is believed that many of the common lung problems along with some critical illnesses (like pneumonia and rheumatic heart disease) could be diagnosed at very early stages by careful stethoscope observations [2–4].

To overcome these difficulties, we have come up with innovative solutions as a part of our project. As a primary object, we would be developing a low-cost stethoscope solution which can record and store the internal body sounds of patients' heart and lungs which will assist physicians in taking careful observations and tagging illnesses to the sounds, and training a machine learning model on recorded and tagged sounds to autosuggest the illness.

2 Proposed Methodology

In machine learning and statistics, classification is the problem of identifying to which of a set of categories also called as a sub-population, a new observation belongs, on the basis of a training set of data that contains observations whose category membership is known. Classification belongs to the study of pattern recognition.

In the context of machine learning, classification is considered an object of supervised learning, i.e., learning where a training set of data which is already identified is available. The corresponding unsupervised equivalent is known as clustering and involves grouping data into classes based on some measure of fundamental similarity.

Audio classification is a sub-category in classification which deals with the study and classification of sound signals. We will be strictly meaning classification as audio classification from hereon. Audio classification involves the following procedures.

- Feature extraction
- Audio preprocessing
- Boundary-based classification
- Repetitive training.

3 Model Architecture

3.1 Apparatus

Auscultation requires collection of sounds of heart and lungs which need to be recorded and stored. There are no feasible means currently in the market that can easily meet this requirement. Hence, we have indigenously developed a low-cost digital stethoscope to record the sounds into a mobile application.

Below is description of each of the parts of the stethoscope and its role in Fig. 2.

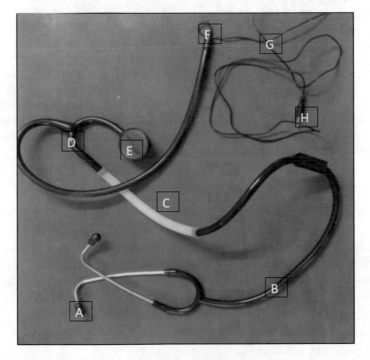

Fig. 2 Devised apparatus—digital stethoscope to record data

A. Earpiece—Normal stethoscopic earpieces to listen to the recorded sound in real time.
B. Conducting Tube—Medical grade tube to conduct the audio signal to earpieces.
C. Connection Tube—Latex tube to join the stethoscope to its digital part.
D. Y-connector—Two-way connector that combines the diaphragm with digital and analog parts of stethoscope.
E. Diaphragm—A sensitive vibrating sheet of conducting material to produce and amplify auscultatory sounds.
F. Microphone—High-sensitivity microphone to digitally record the sound.
G. Conducting Wire—Takes audio signal to a digital channel like phone.
H. Audio Jack—Connects to the phone to store sound as a file in the system.

3.2 Classifiers Used During Testing

We tested our audio classification model on an initial test data of heart sounds [5] that had four types of traits, namely murmur, normal heart sound, extra heart sound, and artifacts in the signal. The classifiers used were as follows:

- Support vector machine (SVM)
- Neural network [6]
- Convolutional neural network [7]
- K-nearest neighbor (kNN)
- Random forest classifier [8]
- Extra trees classifier [9]
- Recurrent neural network
- Gradient boosting classifier.

The same classifiers were later implemented upon actual data collected from Government Medical College, Chiplun, Maharashtra, India, with the help of Ms. Rucha Aurangabadkar, a medical student at the institute.

3.3 Apparatus Refining

We tried a lot of different orientation while designing the stethoscope before coming to the final design. As a result, we were able to remove artifacts and noise from the signal to a great extent. The following are the time series visualisations of the recordings in the initial and final phase of design as shown in Fig. 3.

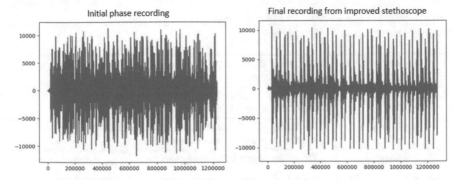

Fig. 3 Comparison

3.4 Feature Extraction

The features extracted from each of the sound recordings are used to develop the model on the dataset. Two stages are followed in the audio feature extraction—short-term feature extraction and mid-term feature extraction.

The short-term feature extraction divides the audio into short-term windows or frames and computes a number of features for each frame to generate short-term feature vectors for the entire audio. The features calculated here include zero-crossing rate and MFCCs. The mel-frequency cepstral coefficients (MFCCs) represent the envelope of an audio signal in the form of features using spectral entropy, spectral spread, spectral energy, etc. These features are the most important in classifying audio data.

The mid-term feature extraction extracts a number of statistical features like mean and standard deviation from the short-term feature vectors.

The total number of short-term features implemented is 34. In the following Table 1, the complete list of the implemented features is presented.

4 Devised Input Dataset

Data related to auscultation is limited. And hence, with the help of our digital stethoscope, we are already collecting it. For each patient, the below fields were recorded along with the sound samples.

- Name
- Age
- Date of visit to hospital
- Gender
- Provisional diagnosis
- Confirmed diagnosis.

Table 1 List of implemented features on audio data

Feature ID	Feature name	Description
1	Zero-crossing rate	The rate of sign changes of the signal during the duration of a particular frame
2	Energy	The sum of squares of the signal values, normalized by the respective frame length
3	Entropy of energy	The entropy of sub-frames' normalized energies. It can be interpreted as a measure of abrupt changes
4	Spectral centroid	The center of gravity of the spectrum
5	Spectral spread	The second central moment of the spectrum
6	Spectral entropy	Entropy of the normalized spectral energies for a set of sub-frames
7	Spectral flux	The squared difference between the normalized magnitudes of the spectra of the two successive frames
8	Spectral roll-off	The frequency below which 90% of the magnitude distribution of the spectrum is concentrated
9–21	MFCCs	Mel-frequency cepstral coefficients form a cepstral representation where the frequency bands are not linear but distributed according to the mel-scale
22–33	Chroma vector	A 12-element representation of the spectral energy where the bins represent the 12 equal-tempered pitch classes of western-type music (semitone spacing)
34	Chroma deviation	The standard deviation of the 12 chroma coefficients

Along with these fields, the four points where sounds were recorded from body were:

- Right lung top
- Left lung top
- Right lung bottom
- Left lung bottom.

Due to shortage of diversity and number of patients, we are currently focusing on only four types of test subjects, namely

- Mild bronchospasm patients
- Mild wheez patients
- Portal hypertension patients
- Normal test subjects.

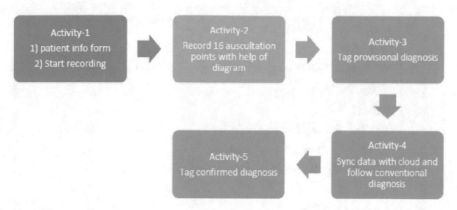

Fig. 4 Activity flow of the system

5 Classification, Analysis, and Results on Test Data

5.1 Comparative Study

Table 2 shows the comparative performance of all the classification models implemented on the test data. Below are the results in a summary.

6 Activity Flow During Supervised Learning

6.1 Current Process

We have prototyped a system that will automate the process of data collection and
training in Fig. 4.

Table 2 Comparison of all classifiers implemented

Classifier	F1 score	Accuracy
NN	77.8	77.8
Extra trees	74.1	76.4
Random forest	74.1	75.4
SVM	71.8	74.7
Gradient boosting	73.2	73.7
kNN	67	70.3
CNN	–	65.0
RNN	–	61.9

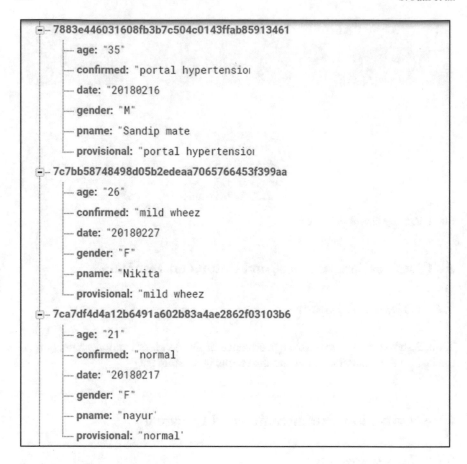

Fig. 5 Data collected from patients

1. The digital stethoscope records sounds via a mobile application.
2. The other data inputs are manually fed into and are sent to a cloud server.
3. The classification model is trained on the fly with the proxy script, and results are evaluated.
4. The result is sent back as a response to the mobile application.

6.2 Server JSON Data Sample

The data is stored in JSON format and the audio files in MP4 format on a firebase server. The following is the screenshot in Fig. 5.

```
tt@tt-HP-15-Notebook:~/Desktop/Multiple$ python knn.py
(0, array([ 1.,  0.,  0.,  0.]), ['Mild_Bronchospasm', 'Mild_Wheez', 'Normal', 'Portal_Hypertension'])
(1, array([ 0.,  1.,  0.,  0.]), ['Mild_Bronchospasm', 'Mild_Wheez', 'Normal', 'Portal_Hypertension'])
(0, array([ 1.,  0.,  0.,  0.]), ['Mild_Bronchospasm', 'Mild_Wheez', 'Normal', 'Portal_Hypertension'])
(0, array([ 1.,  0.,  0.,  0.]), ['Mild_Bronchospasm', 'Mild_Wheez', 'Normal', 'Portal_Hypertension'])

Resultant array
[ 0.75 0.25 0.   0. ]

Diagnosis of patient :
Mild_Bronchospasm
```

Fig. 6 Output of the multi-point classifier to predict diagnosis

6.3   Diagnostic Tests Carried Out

Once we established a suitably accurate classifier, we tested it for actual prediction and diagnosis of the ailment of real data. Since the initial classification model was built only for single audio classification, adjustments and modifications were made to allow classification based on a group of audio files. The below code output snippet in Fig. 6 shows the results of the model developed and the results it yielded.

As it shows, the classifier is able to predict the diagnosis correctly. A sample containing four point sounds of the class mild bronchospasm is classified using the multiple-point classifier. The results are the four arrays denoting probabilities of each point belonging to the different classes and the resultant array denoting the average probability. The sample is correctly classified as mild bronchospasm.

7   Conclusion

In summary, we proposed an approach to record the auscultation sounds of patients through our low-cost digital stethoscope using an android app which will record and store the audio files of the patient on firebase storage. We have started the data collection with our apparatus and the accompanying android application.

We are now successfully able to diagnose and classify three major respiratory/pulmonary disorders, namely

1. Bronchospasm
2. Wheezing
3. Portal hypertension
4. Along with ascertaining whether a patient is normal or not.

This project has been a revelation also with respect to our knowledge and understanding of telemedicine. The following are the key outcomes and conclusions of the project.

- Instead of type of classifier, the nature and size of training data decided the accuracy.
- Multiple-point classification is better and more accurate than single-point classification.

Declaration We declare that we have taken the required permissions for use of patients' involvement in the study and taken responsibility if any issues arise later.

References

1. Thinklabs One Digital Stethoscope. www.thinklabs.com
2. Wu, M., Der-Khachadourian, G.: Digital Stethoscope—A Portable, Electronic Auscultation Device. Cornell University (2012)
3. Leng, S., Tan, R., Chai, K., Wang, C., Ghista, D., Zhong, L.: The electronic stethoscope. Biomed. Eng. Online (2015)
4. Moriarty, M.: Heart and breath sounds: listening with skill. RN, BSN, Modern Medicine Network online (May 01, 2002)
5. PhysioNet/CinC Challenge 2016: Training Sets. www.physionet.org/physiobank/database/challenge/2016
6. Lippmann, R.P.: An introduction to computing with neural nets. In: IEEE ASSP Magazine, pp. 4–22 (1987)
7. Salamon, J., Bello, J.: Deep convolutional neural networks and data augmentation for environmental sound classification. In: IEEE Signal Processing Letters (Accepted November 2016)
8. Geurts, P., Ernst, D., Wehenkel, L.: Extremely Randomized Trees. Springer (2 March 2006)
9. Fernando, C.: Mobile diagnostics for rural patients in Columbia. Telemed. E-Health **23**(11), 934–937 (2017)

Fleet Management and Vehicle Routing in Real Time Using Parallel Computing Algorithms

A. Mummoorthy, R. Mohanasundaram, Shubham Saraff and R. Arun

Abstract Algorithms which take uncertainty into account make better systems for fleet management and lead to greater efficiency. These algorithms are faster and can manage real-time traffic and even compute the location and status of vehicles with miscellaneous requests from users. Parallel computing technologies enable us to implement fuzzy-based algorithms which route the traffic in a much more efficient mechanism. This also improves the overall system of user request management by using meta-heuristics.

Keywords Parallel computing · Fleet management · Fuzzy logic · Meta-heuristics
Routing

1 Introduction

Both sectors of the economy, i.e. public and private, are impacted by the logistics and fleet management systems. VRPs or vehicle routing problems are a crucial part in handling these systems [1–3]. These problems determine the routes that the vehicle should opt for which is set by certain constraints and bounds. There are multiple problems in these systems which need real-time attention. Some of the important problems are:

A. Mummoorthy (✉)
Malla Reddy College of Engineering and Technology, Hyderabad, India
e-mail: amummoorthy@gmail.com

R. Mohanasundaram · S. Saraff
School of Computer Science and Engineering, VIT University, Vellore, India
e-mail: mohan.sundhar@gmail.com

S. Saraff
e-mail: saraffshubham@gmail.com

R. Arun
Builders Engineering College, Kangayam, India
e-mail: ra.cse@builderscollege.edu.in

© Springer Nature Singapore Pte Ltd. 2019
J. Wang et al. (eds.), *Soft Computing and Signal Processing* ,
Advances in Intelligent Systems and Computing 898,
https://doi.org/10.1007/978-981-13-3393-4_27

1. Fleet Management: Dynamic management is needed for a large-scale logistic operation involving real-time dispatching of trucks and vehicles for shipment purposes.
2. Distribution Systems managed by Vendors: Replenishing the inventory before running out of stocks is an important function of the distribution system.
3. Couriers: Local distribution of outbound parcels.
4. Repair Services: Critical services such as rescue- or repair-related services.
5. Point-to-Point Ride: Fixed rides between source and destination at fixed times; Needs advanced booking.

GPS-enabled vehicles for dynamic customers are also an important management problem which is enabled due to parallel computing the big data generated by the vehicles on the street [4–6]. This engulfs meta-heuristics and uses fuzzy-based algorithms to enable a more efficient and dynamic system.

2 Problem Classification

There are two classifications of VRPs, namely static or dynamic and deterministic or stochastic.

When the input data does not depend explicitly on time, it is called as static; otherwise if any dependency of data occurs with time, it is referred to as dynamic.

VRPs are deterministic if the input data can be inferred before the vehicle is designed; else, it is then referred to as stochastic.

(i) **Static Vehicle Routing Problems**:

A VRP, which is static, can be classified as stochastic or deterministic.
Total information is known ahead of time, and also, time is not considered expressly in deterministic and static VRPs. For VRPs that are static and stochastic, designing of vehicle paths is done at initial phase of horizon, which is done as such before dubious data is known [7, 8]. Various factors are influenced by the uncertainty such as which service requests are available, client (user) requests, client benefit times, or travel times. In the event that information is unverifiable, it is typically difficult to fulfil the requirements for all realizations of the arbitrary factors. Considering a broad scenario, before any dubious data is available, the initial phase arrangement is completed. The main idea behind this is to reduce the cost of normal response along with reducing cost of initial phase.

(ii) **Dynamic Vehicle Routing Problems**:

When all the data are known prior and depend on time, it is referred as deterministic and dynamic problem.
In real-time routing or problems that are dispatching—or in more general terms, VRPs that are dynamic and stochastic—the processes that are stochastic represent

random or uncertain information. For example, the VRP with time windows investigated has a place with this class of issues. More importantly, the TSP, i.e. a representative of travelling sales, with travel times that are time-dependent is dynamic and deterministic [9–11]. In this issue, a heading out sales representative needs to locate the most limited shut visit among a few urban areas going through all urban communities precisely once, and travel times may change for the duration of the day. At long last, in stochastic and dynamic issues (otherwise called ongoing defeating and dispatching issues) unverifiable information is spoken to by stochastic procedures. For example, client solicitations can carry on as a Poisson procedure. Since indeterminate information is step by step revealed amid the operational interim, courses are not developed. Rather, as the new information keeps arriving, demands of clients are dispatched to vehicles in an ongoing fashion.

A number of events that occur lead to the needs for the modification of plan:
The three steps are:

- When a new request from the user arrives;
- When a vehicle reaches the destination; and
- Travel time updates.

As the vehicle operator sets the strategies, every vehicle is prepared accordingly. On reception of a new demand, we should choose if the demand could be set or adjusted around the same time, or if it should be rejected or postponed. As soon as the demand is acknowledged, it is allotted to a position where the vehicle is on the path. The request is adequately overhauled as arranged if no other occasion happens meanwhile. Else, the demand is relegated to a position alternately, in the same path of the vehicle, or can be relegated to an alternate vehicle. Henceforth, every time a vehicle achieves a goal it must be allotted another goal. Reassignment of the position of the vehicle that is moving is difficult. For instance, assume that a startling congested driving conditions happens; some client administrations can be conceded. It is significant that when the request rate is low, it is helpful to move sit without moving vehicles keeping in mind the end goal to anticipate future requests or to get away from a gauge activity blockage.

3 Particular Features

A VRP has various features, some of which have recently been portrayed.

(i) **Quick response**: Dynamic and stochastic calculations are meant to give fast response, with the goal that path alterations can be conveniently transmitted to fleet. Two methodologies can be utilized: simple strategies (FCFS) or tabu search (TS). As will be explained, the decision between them depends for the most part on the goal, the level of dynamism as well as the rate of request.

(ii) **Deferred or Denied service**: There are applications in which it is substantial to refuse assistance to a few clients or lead them to another contender, so as to maintain a strategic distance from excessive delays or unsatisfactory expenses. For example, if the demands cannot be fulfilled in a given time frame, they are rejected. At the point when no time frames are forced, a few client solicitations are put off due to their troublesome area.

(iii) **Congestion**: If the request rate surpasses a given limit, the framework winds up noticeably soaked; i.e. the normal holding up time of one demand develops to infinity.

4   Degree of Dynamism of a Problem

A dynamic and stochastic algorithm depends highly on the dynamism of the problem.

Consider the planning horizon to be interim [0, T], perhaps separated into limited number of littler interims.

Given ns and nd as the quantities of static and dynamic request, respectively.

Service request, time of occurrences: $ti \in [0, T]$

Static requests: $ti = 0$

Dynamic requests: $ti \in [0, T]$.

So,

$$\delta = \frac{nd}{ns + nd}$$

whose value may lie between [0, 1].

It was observed by Lund et al. that for a particular value of δ, the problem will be highly dynamic if requests occur in [0, T].

So, they came up with a new measure of dynamism:

$$\delta' = \frac{\sum_{i=1}^{ns+nd} ti/T}{ns + nd}$$

This ranges between 0 and 1.

$\delta' = 0$ if the user requests are known beforehand.

$\delta' = 1$ if they occur at time T.

Lastly, Larsen's definition of δ'' for considering time for servicing the user

$ai = $ ready time;

$bi = $ deadline time of client i;

for $(ti <= ai <= bi)$.

Do,

$$\delta'' = \frac{\sum_{i=1}^{ns+nd} [T - (bi - ti)]/T}{ns + nd}$$

δ'' belongs to range [0, 1]. Also, if time windows are not imposed (i.e. $ai = ti$ and $bi = T$), then $\delta'' = \delta'$.

5 Parallel Algorithms

Meta-heuristics are being used to solve dispatching and real-time problems on routing, and route re-optimization computation time is made acceptable using parallel implementation. This parallelization using meta-heuristics could be achieved in numerous ways depending on the framework of the problem and the hardware support under consideration. We observe TS-based procedures in the following section.

A. Tabu search-based parallelization strategies

The TS-based parallel procedures are classified according to the given three parameters: search control cardinality, differentiation and type. The first criterion depicts whether the parallel control search is performed by one processor or mapped across multiple processors. The second criterion depicts the manner in which communication is performed in multiple processors. Synchronous communication maybe independent of the synchronized computation status or maybe not as in the case of knowledge-based communication. The processor that finds the best-case solution in asynchronous communication broadcasts the message across other various processors. In the simplest case, a single solution is broadcast, whereas information is transmitted in addition to receivers in knowledge-based collegial communication. The final third parameter gives a description of the way in which the different searches are performed. The most commonly used configurations are:

(i) **Master–Slave strategy (RS/SPSS/1C)**:
  A single master processor performs one search distributing time-consuming task over other slave processors. The slave processors do not communicate with each other.

  (i) **1C/KS/SPSS strategy**:
    In comparison to the master–slave strategy, the slaves may be used by the master to stop computation. The slaves in this strategy do not communicate with each other but only with the master. This is the base case used when slaves are assigned primary search cases.

  (ii) **p-C/RS/MPMS, p-C/RS/MPSS and p-C/RS/SPMS strategies**:
    Many independent search tasks are processed simultaneously in parallel. The processors carry no communication among themselves until at the end to establish a best solution.

  (iii) **p-C/C/SPMS, p-C/C/MPSS and p-C/C/MPMS strategies**:
    All processes perform searches independent to each other. Once a processor finds a solution that is the best, it sends the same to other processors to reinitialize the search.

B. Strategies for DVRPs to Parallelize

Three main factors should be taken into consideration when we choose a parallelization strategy for stochastic and dynamic VRP: (i) computational effort to obtain an optimal solution, (ii) total available computation power and (iii) arrival time of requests. These factors determine the performance of re-optimization. An important factor affecting design is the architecture of parallel machines on which implementation takes place. One of the motivations behind this is guided by the fact that parallel computing has recently been trending as to moving away from platforms which specialize in traditional supercomputing to cost-effective and general-purpose systems based on shared memory multiprocessing rather than message passing multiprocessing, or workstations or single PCs that run over networks with high speed.

Development in applications of clustered computers corresponds to cost efficiency and high-performance ratios of clustered systems when they are compared with other parallel processors. Distributed supercomputers are more feasible due to provision of easy configuration of communication capacity and additional computing into the system. The master–slave model is most commonly used for distributed computing. Problems requiring only a small computation effort using execution and evaluation in large neighbourhoods use the 1C/RS/SPSS parallelization strategy presenting feasible ways to generate an optimal solution whenever a request occurs. After every iteration, master process generates information from the slave operations and executes and evaluates either the best move or through a sequence of continuous moves. After which every slave is reinitialized with a new space of neighbour solutions that are to be executed in parallel. In case of a look-ahead implementation, a fixed sequence of iterations is performed by slaves before the execution of the synchronization phase. It is most preferred as it is characterized by low interprocess communication and few synchronization events, thus making implementation more suitable on clustered computers. The solutions obtained are further merged to attain optimal solutions.

Strategies on multithread parallelization present alternative approaches towards optimization problems on receipt of a new request. This parallelization is characterized by executing several threads or processes that determine the same space solution in parallel, beginning from all possible different solutions and using the TS strategies with all possible parameter settings [1, 2, 12]. These search processes can be cooperative or independent. There is no exchange of information between the threads in the first attempt to determine search of the solution space using multiple trajectories. Synchronization and selection of best-case solution takes place after the termination of each thread and the initial point for the new loop of independent search initialization. Hence, multistart sequential heuristics are used for implementation. The shared information can be represented in the master's local memory as a state space of solutions, in a network of computers. Communication only occurs between the master and the slaves and not between any other processes. The threads are also responsible for controlling access to shared knowledge. In the strategies discussed above, one of the loopholes rests in the determination of the total iterations required for the execution of the algorithm. This number is usually defined based on the temporal distribution of the incoming service requests.

6 Conclusions

Although many papers have been analysed for real-time VRPs, a few major issues have not been addressed yet, and the vague literature usage calls for future redirections.

The issues in this system can be:

(i) It is advisable to choose heuristics with look-ahead capability for route re-optimization problems.
(ii) The dynamicity factor should be considered for developing new meta-heuristics to obtain best solutions.

Certain other issues which specifically need to be addressed:

(i) The advancement in intelligent transportation might pose time variations in routing problems. The same has been addressed by Daskin and Malandraki [13] by providing heuristics for static cases.
(ii) There is a need to study booking problems in more depth.
(iii) Ambulance allocation and re-allocation problems of dynamic nature require more focus.
(iv) Automated guided vehicle (AGV) optimization problems require more attention.

References

1. Ichoua, S., Gendreau, M., Potvin, J.-Y.: Diversion issues in real-time vehicle dispatching. Transp. Sci. **34**, 426–435 (2000)
2. Caricato, P., Ghiani, G., Grieco, A., Guerriero, E.: Parallel tabu search for a vehicle rounting problem under track contention. Parallel Comput. (forthcoming)
3. Miller-Hooks, Mahmassani, H.: Least possible time paths in stochastic, time-varying transportation networks. Comput. Oper. Res. **25**, 1107–1125 (1998)
4. Gao, Chabini, I.: Best routing policy problem in stochastic time-dependent networks. Transp. Res. Rec. **1783**, 188–196 (2002)
5. Marianov, V., ReVelle, C.S.: Siting emergency services. In: Drezner, Z. (ed.) Facility location, pp. 199–223. Springer, New York (1995)
6. Mohanasundaram, R., Periasamy, P.S.: A meta heuristic algorithm for optimal data storage position in wireless sensor networks. Pak. J. Biotechnol. 463–468 (2016)
7. Pallottino, S., Scutella, M.G.: Shortest path algorithms in transportation models: classical and innovative aspects. In: Marcotte, P., Nguyen, S. (eds.) Equilibrium and Advanced Transportation Modelling, pp. 245–281. Kluwer, Boston (1998)
8. Mohanasundaram, R., Periasamy, P.S.: Swarm based optimal data storage position using enhanced bat algorithm in wireless sensor networks. Int. J. Appl. Eng. Res. **10**(2), 4311–4328 (2015). ISSN 0973-4562
9. Repede, J.F., Bernardo, J.J.: Developing and validating a decision support system for locating emergency medical vehicles in Louisville, Kentucky. Eur. J. Oper. Res. **75**, 567–581 (1994)
10. ReVelle, C.S.: Review, extension and prediction in emergency services siting models. Eur. J. Oper. Res. **40**, 58–69 (1989)

11. Mohanasundaram, R., Periasamy, P.S.: Clustering based optimal data storage strategy using hybrid swarm intelligence in WSN. Wirel. Pers. Commun. (2015) (Springer)
12. Brotcorne, L., Laporte, G., Semet, F.: Ambulance location and relocation models. Eur. J. Oper. Res. 147(3), 451–463 (2003)
13. Malandraki, C., Daskin, M.S.: Time dependent vehicle routing problems: formulations, properties, and heuristics algorithms. Transp. Sci. 26, 185–200 (1992)
14. Mohanasundaram, R., Periasamy, P.S.: Hybrid swarm intelligence optimization approach for optimal data storage position identification in wireless sensor networks. Sci. World J. (2015)

Automated Bug Reporting System with Keyword-Driven Framework

Palvika, Shatakshi, Yashika Sharma, Arvind Dagur and Rahul Chaturvedi

Abstract In this paper, a keyword-driven framework approach has been investigated which is used for the automation testing. In this approach, we make a separate java file of each and every object, i.e., actions, test setup, and test scripts. It generates the report according to their status of execution (e.g., pass and fail). The report is an HTML format such as an excel sheet having columns, named as test cases name, keyword, description, execute, and result. In the proposed methodology, we have a keyword function library in which we define all the keywords belonging to the Web applications. Here, keywords are the different Web elements present in the Web application, and actions are performed on it. These actions are the functions which are a call from execution engine. After performing the entire test, it will write the status of the test cases in the report and then send it to the concern team. The implementation results show that the proposed approach has generated better results as compared to the existing approaches.

Keywords Bug report · Test case · Test script · Test suits · Execution engine Framework

Palvika · Shatakshi · Y. Sharma · A. Dagur (✉) · R. Chaturvedi
Department of Computer Engineering, Krishna Engineering College, Ghaziabad, UP, India
e-mail: arvinddagur@gmail.com

Palvika
e-mail: palvikasinha@gmail.com

Shatakshi
e-mail: guptashatakshi235@gmail.com

Y. Sharma
e-mail: yashika.1707@gmail.com

R. Chaturvedi
e-mail: shreyansh12@gmail.com

© Springer Nature Singapore Pte Ltd. 2019   271
J. Wang et al. (eds.), *Soft Computing and Signal Processing* ,
Advances in Intelligent Systems and Computing 898,
https://doi.org/10.1007/978-981-13-3393-4_28

1 Introduction

Software quality is much more important for the success of the software. The rate of software crisis is increasing day by day due to lack of testing process. The manual testers work hard on the software to maintain the quality [1]. But due to human error, software does not get that much efficiency which is required to make software error-free.

So the keyword-driven framework helps to maintain the test cases and test data individually. The keyword-driven framework is a functional testing framework also called table-driven framework or action word framework. This partitions the test case into four parts. The first part is tested setup part, second is test object, the third one is action, and the last one is test data [2]. This division can be done and maintained with the help of the excel sheet, and the brief description is given below:

Test step: This describes the step of the test which is going to perform on the test object. It can also say that it describes the flow of the test.

Test object: This section defines the Web elements or the objects present on the Web page such as user ID, password.

Action: This defines the name of the action that is going to perform on the objects like click, input, select.

Test data: This defines the data values passed on the textboxes or the data values which is used for the testing purpose.

The basics of the keyword-driven framework are to separate the coding part from the test case and the test step. By separating this, nontechnical person can also easily understand the automation process. With the help of this framework, a manual tester can also write their automation script and automate the process. The keyword can be split into base and user keyword in Fig. 1.

The lists of the components which we require to achieve the above tasks are:

Excel sheet: This contains the overall data required for the test cases, test step, test objects, test actions.

Keyword function library: In this, there are lots of files which are used for the automation. These files store the working of the actions and all actions callable from this file.

Object repository: These are the property files which are stored in the HTML elements present in the Web application. This file is linked with objects in the test.

Datasheet: This is an excel file used to keep data values required by the objects to perform actions on it.

Execution engine: All the processes are triggered from this path.

There are many advantages of this framework. In this, once the framework is set up by the hard code automation engineers, there is not any requirement of the technical persons. So when the test is set up, then any nontechnical person or manual testers can easily handle this easily. In this, the maintenance [4] of the test suit or the test setup is very easy to handle. Once you are clear about your project or the

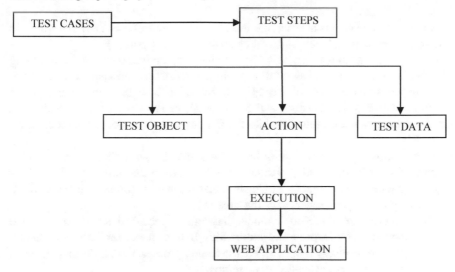

Fig. 1 Work flow of keyword-driven framework [3]

business requirement which is done on the client side, you can start your development of the framework. The reusability of the components is also increased because of making all the components file individually. The code is also reusable as it fetches all components from a different file.

Rest of the paper is organized as follows: Work related to this paper is discussed in Sect. 2. In Sect. 3, we present the proposed approach. Results and simulation are discussed in Sect. 4. Finally, paper is concluded in Sect. 5.

2 Related Work

In the existing system, we have a test script that belongs to a particular Web application and it has less flexibility of the code and nontechnical users or manual testers cannot understand the test script. They do not have technical knowledge of writing a test script and programming languages.

But this system overcomes this inability of manual testers and provides an interface which helps the testers to add their own test script in test suites [5].

Hui et al. predict a model "Linux keyword-driven distributed test automation framework (LKDT)" which is based on the examines of current automated testing frameworks; for software testing, automation it is combined with the actual project practice. This model estimates the framework feasibility and superiority [3].

There are many numbers of models exist for automated testing of the Web application to examine the use of keyword-driven testing. In the keyword-driven testing, the functionality of system under test is documented in a table and also in step-by-step instructions for each test. It requires the production of modular, reclaimable test element. These elements are met into test scripts. These test scripts can be split into many durable actions. The result will be analyzed by a report [6].

Takala et al. establish a generic open-source adapted framework for simplifying the combination of model-based test generators with test execution tools. This framework depends on the keyword concept. It is most associable in online testing, which requires that between generator and SUT, messages are passed continuously. The architecture is based on plug-ins, like the Eclipse-integrated development environment [7].

Automation of software testing techniques increase the effectiveness and efficiency of the software testing. Automation software testing requires a high data-driven amount of investment for buying the software and hardware. By using automation testing techniques, testers save the time [8].

Nguyen et al. explain the topic model that defines "technical aspects topics in the textual contents of bug reports and source files. In this, model correlates bug reports and corresponding buggy files" [9]. By the testing process, testers identify the bugs in the software which are needed to be removed.

3 Proposed Model

Our software is basically based on the keyword-driven framework. This framework helps nontechnical persons or manual testers to understand easily, and they can easily add the test cases. This also helps to use the code again and again. In this proposed methodology, we test the Web application and generate the bug report after testing all the test cases present in the test suites.

It will help in that Web application where the user interfaces change continuously or there is no stability of the GUI. In that case, the automation test users who use normal framework have to change test script accordingly. But with the help of the keyword-driven framework, we can easily tackle this type of changes, and with the small change, they can execute the test cases.

In the proposed approach, we are using keyword-driven framework for regression testing and generate the test result. The test result is having five fields. In the first field, we define the name of the test cases, in the second field, we define the keyword, the third field is about the execution, and the last one is for a result which defines the status of the test. This model has Java library file which defines all the Java files like action, objects, test cases. The Java library file works as a repository which is used to store all files required to set up for the framework.

It also has test data sheet which is an excel sheet that contains all the data required to test the Web application. This system fetches data from the test data sheet and used to test the Web application. The third most important thing required to test the Web application is selenium WebDriver. This is the most important part to automate the test because this trigger browser, and without this, we cannot hit the URL. This WebDriver supports multiple programming languages, but we are using Java for writing test script using selenium WebDriver.

The generated report is in HTML format, and we can easily see the report using this. This report is having multiple fields, and it will show the pass and fail test cases according to test their objects. We can generate a report using file mechanism of the Java.

Now the last one is the main driver script; these driver scripts trigger all the test cases. It will read the test step one at a time and execute step with their corresponding test cases and then log the status or the result of the test in the report or test report. There is a mailing API which is used to mail that generated report to the concerned team for the reference, and it will send to the developers so that if there is any bug found, then it will resolve by the developers. The mailing API has Java code which takes sender mail ID and password with the receiver mail IDs. And it will also take a port number which helps to make the connection. This saves the time of the bug locking system, and developers easily resolve that bug. This kind of feature is used by the company which requires high efficiency. The steps involved to perform keyword-driven testing are as follows:

Keyword-driven testing can be done in two ways, manually as well as automated. But generally, this form of testing is used in automation testing. It is a scripting technique which uses the data files to store the keywords used in the application which is being tested by the testers.

The aims of keyword-driven testing or keyword-based testing in automation are:

- Reduce the maintenance cost;
- It avoids the duplicated bugs;
- Increase the reuses of function scripting;
- It supports the testing in a better way and also increases portability;
- Accomplish the more testing result with less effort.

In keyword-driven testing, in the initial stage of the development process, we create simple test cases for checking the functionality of the software and we can perform testing activities on software application module-by-module. The easiest way to perform keyword-driven testing is to record them. After the recording activity, the test cases will be modified or enhanced according to the requirements of the customer.

Each keyword in keyword-driven testing will link with at least single command, function, or test scripts, which are used to perform the actions related to that particular keyword. When the different test cases are executed in automation testing, the test automation framework calls a test library from where the keywords will be interpreted.

The keyword-driven testing involves the different activities which are given as:

Step 1 Identifying the keywords, i.e., low-level keywords or high-level keywords;
Step 2 Implementing the keywords as executable;
Step 3 Create the test cases;
Step 4 Create the driver scripts;
Step 5 At last, execute the test scripts (automation).

Table 1 Test results

TC name	Keyword	Description	Execute	Result
TC_01	Login	Login to application	Yes	Pass
	createOrd	Create a new order	Yes	Pass
	srchOrd	search for an order	Yes	Fail
	modifyOrd	Modify the order	No	NA
	srchOrd	search for an order	No	NA
	Logout	Logout from application	Yes	Not run
TC_02	******	*****	**	**
	*****	*****	**	**

*Signifies that the similar reports are generated as given in TC_01

The different tools used for keyword-driven testing are given below:

- HP QTP
- Selenium.

4 Result and Simulation

Firstly it will test all the test cases, generate the bug report after testing, and then the mail is sent to the concern team.

This report also contains the time of starting the test and ending test, as well as, it will capture the screen in between of the execution of test so that debugging becomes easy. The format of the report is like excel sheet as shown above, but it is in HTML format and contains all test steps with the number of the test cases.

It will read test setup file line by line and execute accordingly. The report looks like as given in Table 1. There are many test cases executed automatically and report is generated. The result shows that a test case passed or failed in result column.

5 Conclusion

We proposed a keyword-driven framework approach with the help of selenium automation testing tool in which we automatically tested the whole Web application and generated the bug report automatically. In keyword-driven framework or testing, we used a particular keyword for specific functionality instead of writing the multiple steps. It is important to design the test suits in a proper and efficient way.

We used the data files to store the keywords used in the application which is being tested automatically. In this approach, a bug report and its graphical representation are generated automatically which reduce the testers or developers effort and time. Our results show that the given approach is better as compared to existing approaches.

References

1. Jeong, G., Kim, S., Zimmermann, T.: Improving bug triage with bug tossing graphs. In: Proceedings of Joint Meeting of the European Software Engineering Conference and the ACM SIGSOFT International Symposium on Foundations of Software Engineering, pp. 111–120 (2009)
2. Yang, G., Zang, T., Lee, B.: Towards semi-automatic bug triage and severity prediction based on topic model and multi-feature of bug reports. In: 2013 IEEE 38th Annual Conference on Computer Software and Applications (COMPSAC), pp. 97–106. IEEE (2014)
3. Hui, J., Yuqing, L., Pei, L., Jing, G., Shuhang, G.: LKDT: a keyword-driven based distributed test framework. In: International Conference on Computer Science and Software Engineering, pp. 719–722 (2008)
4. Sun, C., Lo, D., Khoo, S.-C., Jiang, J.: Towards more accurate retrieval of duplicate bug reports. In: Proceedings of the 2011 26th IEEE/ACM International Conference on Automated Software Engineering (2011)
5. Panichella, A., Dit, B., Oliveto, R., Di Penta, M., Poshyvanyk, D., De Lucia, A.: How to effectively use topic models for software engineering tasks? An approach based on genetic algorithms. In: Proceedings of the 2013 International Conference on Software Engineering, pp. 522–531. IEEE Press (2013)
6. Bhattacharya, P., Neamtiu, I.: Fine-grained incremental learning and multi-feature tossing graphs to improve bug triaging. In: 2010 IEEE International Conference on Software Maintenance (ICSM), pp. 1–10. IEEE (2010)
7. Takala, T., Maunumaa, M., Katara, M.: An adapter framework for keyword-driven testing. In: Ninth International Conference on Quality Software, Department of Software Systems, Tampere University of technology, Finland, pp. 201–210 (2009)
8. Cubranic, D.: Automatic bug triage using text categorization. In: Seke 2004: Proceedings of the Sixteenth International Conference on Software Engineering & Knowledge Engineering (2004)
9. Nguyen, A., Nguyen, T., Al-Kofahi, J., Nguyen, H., Nguyen, T.: A topic-based approach for narrowing the search space of buggy files from a bug report. In: 2011 26th IEEE/ACM International Conference on Automated Software Engineering (ASE), pp. 263–272 (2011)

Urban Bus Arrival Time Prediction Using Linear Regression and Kalman Filter—A Comparison

Neeraj Ramkumar and Archana Chaudhari

Abstract The study aims to use two statistical processes to predict the arrival time of a bus in the highly dynamic traffic conditions of Mumbai. The paper provides a comparison of regression and Kalman filter as an attempt to model the travel time for a bus. GPS data collected from the bus during field study was used for training and validation of both the models. The Kalman filter model is leveraged to provide real-time information and is used to exploit the correlation between the test bus and previous buses plying along the same route and is shown to perform better among the two for forecasting travel time.

Keywords Travel time prediction · Kalman filter · Regression

1 Introduction

Traffic congestion is a major problem in densely populated cities like Mumbai in developing countries. Majority of the city populace use a combination of different transport systems available like local trains, metro railways and buses to reach their destination. In such a scenario, an arrival time estimate would be desirable for trip planning and would be a time saver. Additionally, reliable estimates also have the potential of attracting new users for the transportation system.

The Brihanmumbai Electric Supply and Transport (BEST), the public bus service provider in Mumbai, has an estimated ridership of 26,00,000 people per day [1]. A bus arrival time prediction system would be extremely beneficial to all the consumers. This paper aims to provide a comparison of a multilinear regression model and a Kalman filtering model for predicting bus arrival times. The report outlines both the methods used and the considerations and input data used for the models.

N. Ramkumar (✉) · A. Chaudhari
Electronics and Telecommunication Department, Dwarkadas J. Sanghvi College of Engineering, Mumbai 400056, India
e-mail: neerajr44@live.com

A. Chaudhari
e-mail: archana.chaudhary@djsce.ac.in

© Springer Nature Singapore Pte Ltd. 2019
J. Wang et al. (eds.), *Soft Computing and Signal Processing* ,
Advances in Intelligent Systems and Computing 898,
https://doi.org/10.1007/978-981-13-3393-4_29

Due to increased advances in the telecommunication and easy availability of information technology, new ways of integrating real-time information in passenger information system (PIS) have come up [2]. This study also uses location data from a GPS module fitted inside the buses as input for the models used for the prediction. Several cities follow a similar methodology and make the live location data available to the passengers and provide ride time estimates using this data. However, this technology and service are not yet commonplace in India and this creates a need for providing trip time estimates for dynamic traffic conditions that are characteristic of densely populated cities in India [3].

2 Literature Review

The approaches used to solve the problem of travel time estimation usually are based on historical average models, regression models, exponential smoothing models, Kalman filtering models and ANNs.

Historical average models usually involve the usage of average time or average speed of the vehicles in a stretch of the route. Lin and Zeng [4] made certain assumptions about schedule adherence and dwell times and used historical data to provide real-time information about the time of arrival. Some studies also used speed data from GPS [5, 6]. Generally, historical data models require large amounts of data storage infrastructure and are computationally expensive and additionally are unsuited for large cities with dynamic traffic conditions [7].

Regression models have been extensively used for travel time prediction. A variety of predictor variables including the no. of bus stops, average speed, distance, dwell time, passenger count, departure and arrival times have been used in previous research [8, 9]. Generally, historical average and regression models were used in studies as a comparison for other models. Regression models have been outperformed by other models in both Jeong and Rilett [10] and Ramakrishna et al. [11].

Machine learning techniques like ANNs are extremely good at dealing with complex nonlinear relationships between predictor and response variables and have thus been used for travel time estimation by Jeong and Rilett [10], Ramakrishna et al. [11], Chien et al. [12] and Chen and Chien [13]. SVM also belongs to this category, and support vector regression has been used by Bin et al. mainly due to its generalization ability and the fact that unlike ANN and SVMs are not amenable to the overfitting problem [14]. The disadvantages of ML methods include the computational expense in training these models and the necessity to periodically retrain the models to accommodate the changing nature of road traffic due to external influences. Another disadvantage of these models is that the results are location-dependent [7].

Kalman filter has been used to predict travel times extensively [15–17] due to its ability to handle noise and using dynamic real-time information to continuously update the state variable. The results from Kalman filter are promising, and the constant updating feature makes it suitable for the dynamic traffic conditions of India.

3 Data Collection

The study route selected was from Mulund, Mumbai to Upvan, Thane (Route 110). This route was selected due to the variation of traffic densities along the route and since this is one of the busiest routes in Mumbai. The data collected included the latitude and longitude information along the route, dwell times, arrival and departure times and geo-coordinates of the bus stops. For model training purposes, data for 20 trips was collected. To validate the model, data from four trial runs was used.

Data processing was carried out by dividing the whole route into segments where each segment was defined as the part of the route between two stops. The geo-coordinate information was converted to distance information from the last bus stop. Given geo-coordinates of two points, $P_1(lat_1, lon_1)$ and $P_2(Lat_2, Lon_2)$, the distance between the two points D_{12} was calculated as:

$$D_{12} = \cos^{-1}(\sin(lat_1) * \sin(lat_2) + \cos(lat_1) * \cos(lon_1 - lon_2)). \tag{1}$$

Travel time was forecasted for each data point in the current segment from the current location to the end of that segment. This approach has multiple advantages because many buses in Mumbai have common segments along their routes and data for the Kalman Filtering model would be very recent because of this as opposed to running a Kalman filter model trip-wise for the whole route in Table 1.

4 Algorithms

4.1 Multilinear Regression

Linear regression is used to predict a continuous-valued output based on inputs about an explanatory variable. When multiple explanatory variables are used to predict a response variable, it is called multiple linear regression. As a continuous time value was to be predicted, regression can be used for arrival time prediction. It is a parametric approach which assumes a linear relation between the input and the output parameters and performs operation on a dataset where the target values are already defined [18]. In the case of travel time prediction, a linear regression model was fitted to one segment (stop-to-stop) of the route to predict the travel time to the

Table 1 Table format post the processing of raw geo-coordinate data

Distance (in m)	Time elapsed	Speed (in mph)
1.60934	0.6	2.76
3.21868	2.04	3.5
4.82802	2.8	4.18

end of the segment from any point in the segment. The explanatory variables used are the distance to the bus stop and the average speed of the bus.

$$T_b = \beta_0 + \beta_1 * \text{DistanceLeft} + \beta_2 * \text{AverageSpeed}. \tag{2}$$

T_b Time to next bus stop.

The categorical variables to denote weather, the day of the week and peak-hour slots were considered, but the regression models thus developed did not show sufficient difference to warrant their inclusion in the model. Dwell time was used as a sufficient proxy for passenger demand as high passenger demand will lead to longer dwell times as it will take longer to board the bus. Dwell time was separately calculated by taking an average of dwell times during peak-hour slots and non-peak-hour slots. Schedule adherence was not considered for the development of the model.

4.2 Kalman Filter

Kalman filter is a recursive algorithm that provides a current estimation of state variable in the presence of uncertain data or error-filled real-time measurements [19]. This property of the filter makes travel time estimation an ideal use case for it and has been used extensively in previous research in a variety of ways [15–17]. This paper proposes a method to estimate the travel time to the bus stop at any point in the route by using a Kalman filter. The Kalman filter is a linear optimal filter that assumes that the process and measurement noise are Gaussian. It relies on the Gaussian property of multiplication of two Gaussians results in another Gaussian. For a Kalman filter, a mathematical model of the system and equation for the transformation of measurements to state variable are necessary. These are given by:

$$X_k = F_k * X_{k-1} + B_k * U_k + w_k. \tag{3}$$
$$Z_k = H * X_k + v_k. \tag{4}$$

X_k is the current state of the system, and F_k is the prediction matrix used to obtain the current state estimate from the previous state of the system.
w_k, v_k are process noise and measurement noise, respectively, and are assumed to be zero mean white Gaussian noise.
H_k is the transformation matrix for modelling the sensor measurement Z_k.

For systems with external influence, B_k and U_k are the control matrix and vector. The Kalman filter works in two steps:

Prediction step: In this step, an estimate at the current time step is acquired from the best estimate at the previous time step. The priori estimate $\left(X_k^p\right)$ and covariance $\left(P_k^p\right)$ of state variable are:

$$X_k^p = F_k * X_{k-1} + B_k * U_k. \tag{5}$$

$$P_k^p = F_k * P_{k-1} * F_k^T + Q_k. \tag{6}$$

Innovation step: In this step, the Kalman gain is calculated by comparing the estimate calculated in the previous step and the measurement at current time step. Using the Kalman gain, the current best estimate is calculated.

Kalman gain is calculated as:

$$K = P_k * H_k^T \left(H_k * P_k * H_k^T + R_k\right)^{-1}. \tag{7}$$

The posteriori best estimate (X_k) and error covariance (P_k) are calculated as:

$$X_k = X_k^p + K(Z_k - H_k * X_k). \tag{8}$$

$$P_k = P_k^p - K * H_k * P_k^p. \tag{9}$$

To estimate the travel time using Kalman filter, the route is divided into segments where each segment is the part of the route in between bus stops. At each data point on the route, the travel time to the end of the segment is estimated by using the data from the previous n buses that passed through the same segment. The state variable at each data point is assumed to be the travel time to the end of the segment from that data point. The prediction matrix for the ith bus along the route (F_i) for segment s is calculated as:

$$F_i = \frac{TT_{i-1}^{s-1}}{TT_i^{s-1}}. \tag{10}$$

where TT_{i-1}^{s-1} denotes total time to cover previous segment $(s-1)$ by the previous bus $(i-1)$. For the ith bus going along segment s, at each data point, the priori time estimate $\left(X_i^p\right)$ and error covariance $\left(P_i^p\right)$ are calculated as:

$$X_i^p = F_i * X_{i-1}. \tag{11}$$

$$P_i^p = F_i * P_{i-1} * F_i^T + Q_i. \tag{12}$$

The Kalman gain (K) and measurement Z_i are calculated as:

$$K = P_i * H_i^T \left(H_i * P_i * H_i^T + R_i\right)^{-1}. \tag{13}$$

$$Z_i = F_i * TT_{i-1}^s - X_{elapsed}. \tag{14}$$

where $X_{elapsed}$ is the elapsed time since the start of the segment at that data point.

The posteriori best time estimate (X_i) and error covariance (P_i) are calculated as:

$$X_i = X_i^p + K(Z_i - H_i * X_i). \tag{15}$$

$$P_i = P_i^p - K * H_i * P_i^p. \tag{16}$$

The algorithm is repeated for all the buses that plied on the same segment before the current bus. Unlike previous studies [3, 17], this method does not rely on dividing the route into segments of constant length. Q and R are assumed to be constants throughout the running of the Kalman filter. However, they can be different for each iteration. R is calculated from the data collected by the GPS unit.

5   Results and Analysis

The performance was measured over four trial runs, and metrics used for comparing the models are MAPE and RMSE. MAPE is a useful measure of forecasting accuracy which measures the deviation from actual data in terms of percentage. It has been used as a performance metric in previous studies [17, 20]. The formula for MAPE is:

$$MAPE = \frac{100}{n} \sum \frac{X_t - F_t}{X_t}. \tag{17}$$

It has the advantage of easy interpretability, and it also is scale-independent. However, due to certain limitations of the metric regarding bias for forecasts that are too low, RMSE is also considered as a metric for comparison in this paper. It is better for showing bigger deviations since the errors are squared. RMSE is given by:

$$RMSE = \sqrt{\frac{\sum (X_t - F_t)^2}{n}}. \tag{18}$$

Here,

n   is the total no. of data points that were collected along the route.
X_t is the actual value at data point t.
F_t is the forecasted value at data point t.

For linear regression, average RMSE and MAPE values were 18.52 and 33.263. A MAPE of 40 is considered as reasonable performance for a model that predicts travel time [20]. However, the regression model predicted negative values for certain data points when actual values were close to 0. To correct this, log-log transformation was used to ensure that the dependent variable is always positive. This resulted in average RMSE and MAPE values of 15.27 and 27.35. A considerable increase in performance was thus achieved, and the log-log transform also denoted that the relation between the dependent variables was multiplicative rather than additive. The histogram plot of residuals in Figs. 1 and 2 is also a near-symmetric bell-shaped distribution indicating that the normality assumption in regression is likely to be true.

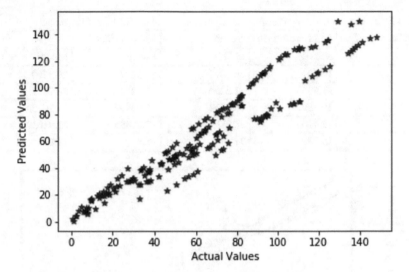

Fig. 1 Actual versus predicted time values for log-log transformed regression model

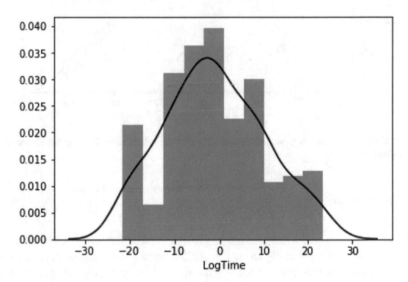

Fig. 2 Residuals graph for log-log transformed regression model

For the first trial run, data from the last five trips in the training set was used as inputs while running the Kalman filter model and for the second trial run data from the last four trips in the training set and the first trial run and so on for the remaining trial runs. As shown in Table 2, the Kalman filter model outperformed both the models mentioned above with an average RMSE of 8.43 and MAPE of 19.84. Figure 3 shows the difference between predicted time values and measured time values to the bus stop from each data point along the route for a trial run.

Table 2 Results from log-log transformed regression model and Kalman filter model

Trial run no.	Log-transformed regression		Kalman filter	
	RMSE	MAPE	RMSE	MAPE
1	15.6017	29.1892	7.52432	20.5760
2	21.5399	32.2254	13.1231	24.7871
3	10.5397	19.8734	6.03197	14.10253
4	13.4109	28.0893	7.06435	19.91039
Avg.	15.27	27.35	8.43	19.84

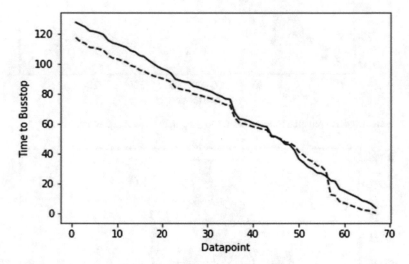

Fig. 3 Actual (*solid line*) and predicted (*dashed line*) time values using Kalman filter versus data points along route

6   Conclusion

In the present study, travel time has been estimated by linear regression algorithm and by Kalman filter algorithm. By using the performance parameters of mean absolute percentage error and root mean square error, the Kalman filter algorithm was shown to perform better. For future work, integrating the two algorithms into a hybrid model can be considered. The results of the linear regression model can act as an input to the Kalman filtering model and be compared to the base Kalman filtering model. Additionally, extensive data collection over a long period of time is required in a variety of traffic conditions and at different times of the year during different seasons so that more robust models can be built.

References

1. Hindustan Times Article, https://www.hindustantimes.com/mumbai-news/better-buses-best-solution-how-to-revive-mumbai-s-bus-network/story-cKHqtW71eV2tM5IFOAFMrO.html
2. Balasubramaniam, P., Rao, K.R.: An adaptive long-term bus arrival time prediction model with cyclic variations. J. Public Transp. **18**(1) (2015)
3. Vanajakshi, L., Subramanian, S.C., Sivanandan, R.: Travel time prediction under heterogeneous traffic conditions using global positioning system data from buses. IET Intell. Transp. Syst. **3**(1), 1–9 (2008)
4. Lin, W.H., Zeng, J.: Experimental study of real-time bus arrival time prediction with GPS data. Transp. Res. Rec. **1666**, 101–109 (1999)
5. Weigang, L., Koendjbiharie, W., Juca, R.C., Yamashita, Y., Maciver, A.: Algorithms for estimating bus arrival times using GPS data. In: The IEEE 5th International Conference on Intelligent Transportation Systems, pp. 868–873 (2002)
6. Sun, D., Luo, H., Fu, L., Liu, W., Liao, X., Zhao, M.: Predicting bus arrival time on the basis of global positioning system data. In: Transportation Research Record. National Academy, Washington, DC (2007)
7. Altinkaya, M., Zontul, M.: Urban bus arrival time prediction: a review of computational models. Int. J. Recent Tech. Eng. **2**(4) (2013)
8. Patnaik, J., Chein, S., Bladihas, A.: Estimation of bus arrival times using APC data. J. Public Transp. **7**(1), 1–20 (2004)
9. Abdelfattah, A.M., Khan, A.M.: Models for predicting bus delays. Transp. Res. Rec. **1623**, 8–15 (1999)
10. Jeong, R., Rilett, L.: The prediction of bus arrival time using AVL data. In: 83rd Annual General Meeting, Transportation Research Board, National Research Council, Washington D.C., USA (2004)
11. Ramakrishna, Y., Ramakrishna, P., Sivanandan, R.: Bus travel time prediction using GPS data. In: Proceedings Map India (2006)
12. Chien, S.I.J., Ding, W., Wei, C.: Dynamic bus arrival time prediction with artificial neural networks. J. Transp. Eng. **128**(5), 429–438 (2002)
13. Chen, M., Chien, S.: Dynamic freeway travel-time prediction with probe vehicle data, link based versus path based. Transp. Res. Rec. 157–161 (2001)
14. Bin, Y., Zhongzhen, Y., Baozhen, Y.: Bus arrival time prediction using support vector machines. J. Intell. Transp. Syst. **10**(4), 151–158 (2006)
15. Chien, S.I.J., Kuchipudi, C.M.: Dynamic travel time prediction with real-time and historic data. J. Transp. Eng. **129**(6), 608–616 (2003)
16. Yang, J.-S.: Travel time prediction using the GPS test vehicle and Kalman filtering techniques. In: American Control Conference, Portland, Oregon, USA (2005)
17. Vanajakshi, L.: Development of a real time bus arrival time prediction system under Indian traffic conditions. https://coeut.iitm.ac.in/APTS_Finalreport_2016.pdf
18. Draper, N.R., Smith, H., Pownell, E.: Applied Regression Analysis, vol. 3. Wiley, New York (1966)
19. Kalman, R.E.: A new approach to linear filtering and prediction problems. Trans. ASME-J. Basic Eng. 82(Series D), 35–45 (1960)
20. Brooks, P.: Metrics for IT Service Management. Van Haren Publishing, Netherlands (2006)

Spark Streaming for Predictive Business Intelligence

M. V. Kamal, P. Dileep and D. Vasumati

Abstract Apache spark can process the data in real time with the test mining and natural language processing. The business intelligence can be improved by collecting and processing the data from Web in real time. Process mining collects the data from event logs in process discovery and then diagnosis the difference between the observed and reality through event logs and extended the data of the event. Dealing with huge data process mining finds difficulty in processing. Spark handles the data processing speed and real time. It receives the input data and segregated into batches that put up in processing. The incoming data appended to the already existing data for processing. It identifies the problems and quickly reports generation of processing data.

Keywords Spark · Process mining · Natural language processing Business intelligence

1 Introduction

The business intelligence achieves when dealing with huge amount of data. Hadoop is an open-source software, and it collects the data and stores in HDFS and starts processing with the tools in Hadoop [1]. Hadoop takes more time for processing, and it is not able to process the data in real time. Spark processes the real-time

M. V. Kamal (✉)
Jawaharlal Nehru Technological University, Hyderabad, Hyderabad, India
e-mail: kamalmv@gmail.com

P. Dileep
Andhra University, Vishakapatanum, India
e-mail: dileep_p505@yahoo.co.in

D. Vasumati
Department of CSE, JNTUCEH, Jawaharlal Nehru Technological University, Hyderabad, Hyderabad, India
e-mail: roshan44@gmail.com

© Springer Nature Singapore Pte Ltd. 2019
J. Wang et al. (eds.), *Soft Computing and Signal Processing*,
Advances in Intelligent Systems and Computing 898,
https://doi.org/10.1007/978-981-13-3393-4_30

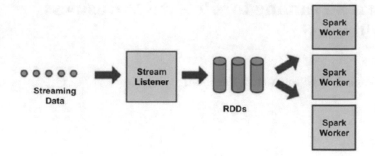

Fig. 1 Spark architecture

huge data. Spark receives the data and segregated into live streams. The data are put into the spark streaming. The input data are divided into batches and put that into the spark engine for processing. The batches of processed data are getting a result. The available data is processing and then the discovered data is appended into the processing. So the real-time processing result helps in achieving the business intelligence [2]. Spark has SQL structure and query language can able to analyze the live data.

Spark has resilient distributed dataset for processing, but it is not allowing to append the data the currently discovered data so the operations are made in streams. It is a fundamental concept of spark streaming. The streams are resilient distributed dataset with elements in Fig. 1.

2 Literature Survey

2.1 Keith Gutfreund [3]

This paper proposes the test mining and natural language processing of apache spark. It is capable of handling unstructured data in real time. The processed patterns achieve customer relationship management [3].

2.2 Wil van der Aalst [4]

Process mining concentrates on process-centric analysis, and the process discovers from the event log. The difference between the observed and reality process is checked by conformance checking, and process of the model is enhanced [4].

2.3   Ramkrushna et al. [5]

Spark is an open-source software for processing the huge data. The data are processed in real time so the patterns are generated out of clear glimpse and processes are performed better than MapReduce [5].

2.4   Olivier Caya and Adrien Bourdon [2]

The value of the conceptual framework is created from the business intelligence and analytics usage in competitive sports and identifies the value creation at each level using spark [2].

2.5   Mudasir Ahmad Wani and Suraiya Jabin [6]

In the last decade, business intelligence faces main problem with the big data [6]. When the data is huge for processing, the open-source software Hadoop is used to store and process. If the data is receiving in real time, spark can process the data.

2.6   Avid Machine Learning [7]

Concordance gives the perspective of given word. Before word removing matching of concordance is done; otherwise, the extracted word is not around the looking of the full sentence [8].

2.7   James Pustejovsky [9]

In n-gram model starting, the process in random state under probability distribution and the states are changed from one another depends upon probability distribution. Probability of next word is based on its previous words [10].

2.8   Kai-Wei Chang [11]

The word sequences are probability distribution in sentences which are used to generate strings. The n-gram model assumes that every word is based on its last n − 1 word [1].

3   Existing System

Process mining is a grouping of the event logs and process models. The data collected from the Web client and Web server. The client and server interactions take place for analyzing the meaning patterns. The process models are extracted from the data usage. It analyzes the outcomes of the process in different parts based on the Web metrics interest. The process model changes arise to the changes arises in the user. The detailed look made by the clickstream.

1.   The final state reaching on Web site is maximized.
2.   The sales expectations also increased in every visit.

   Process mining is divided into three types: process discovery, conformance checking, and model enhancement shown in Fig. 2.

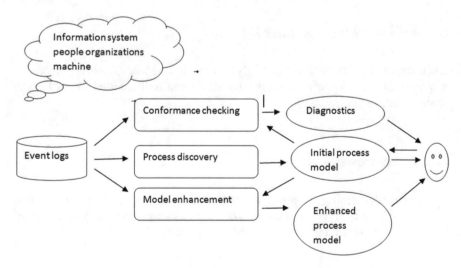

Fig. 2   Process mining

3.1 Process Discovery

It starts with an event log. It is automatic learning for models from event logs. Events which are correlated to the particular process instances are ordered. Process models which are discovered mainly used to make discussion of the problems arising among stakeholders. Viewing the actual process finds its problems, and generating improves idea for project. The process model which is discovered is acting as a template.

3.2 Conformance Checking

It also starts with the process model and event log. It discovers the difference between the process-modeled behavior and observed behavior in the event log. It is used for checking the documented quality process. Identifies the cases which are deviated and understands common things takes place in it. In the process, fragments will be identified with occurance of high side deviations. Further, the quality of process model will be judged and the same has been discovered. Genetic algorithm evaluates the models as if created new one by using conformance checking [12–15]. It is also a start of a model enhancement. It is used for many reasons for checking the various reasons from evaluating the process model discovery to its auditing and its compliant monitored.

3.3 Model Enhancement

It improves or extends the existing process model by using an event log. The log corrects the non-fitting process model. It also corrected by the diagnostics given by model alignment. Event log has information about case, resources, and timestamps. It analyzes the time waited in between the activities. The main bottlenecks are identified by calculating the time difference between the related events and also compute the variance between the events [16]. Event information discovers the certain groups executing the connected events. Clustering constructs social networks and analyzes the performance of the resource. Classification technique analyzes the deciding points in the process model. Process mining is not for an offline process, but it is also useful for making predictions at runtime.

4 Proposed System

Spark is open-source software for processing the real-time data. It distributes data collection; data are stored in RAM. It is evaluated by the most process engine [17].

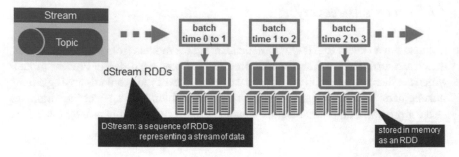

Fig. 3 Series of RDD

It distributes data as in-memory RDD's resilient distributed dataset. Hadoop also can able to store and access the big data, but it stores the data and then only starts processing. It has more time for processing the data. Spark will process the data in speed manner in real-time process. Spark has abstractions of built-in programs, and libraries are fit by machine learning, SQL—structured query language, and ETL—extract transform load algorithms.

Spark can directly access to the distributed data, operations performed on RDD and abstractions provided for viewing RDD's as a relational database. Operations performed by SQL.

Listing 1.1. Spark deploying the data and the multiple machines using resilient distributed dataset.

```
>>>rows = sqlContext.sql('select * from t_table').take(2)
>>> for x in rows:
print x
Row(Message=u'onsider the ethical plagiarism [Reference: 141031-003226] We are
escalating …
Row(Message=u'onsider the ethical plagiarism [Reference: 141031-003226] Incident
No: 141031-003226 Mail forwarded from …
    Command took 5.03s
```

Spark can operate the data which receives in real-time. The Spark resilient distributed data set is unalterable as shown in Fig. 3. This will not allow to add new RDD which was in existence. The freshly arriving data are as data streams. Spark has an abstraction and also programming engine both of it changes the streamed data into sequence of RDD is DStream—discretized stream [16, 18–21].

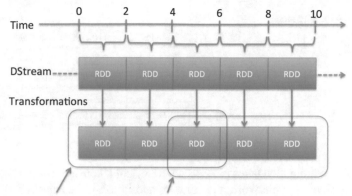

Batch Length: 2 seconds, Window Length: 6 seconds, Slide Interval: 4 seconds

Fig. 4 Transformations of DStream

The series of RDD are creating a DStream. The DStream are processed separately or combined with an RDD.

```
sc = StreamingContext(sc, 60)

rec= sc.socketTextStream(host,port)

fRDD = rec.map(lambda x: x.split('\t'))

textRDD = fRDD.map(lambda x: x[10])

wRDD=tRDD.flatMap(lambda x: x.split(' '))

pair = wRDD.map(lambda x: (x,1))

wordcount = pair.reduceByKey(lambda x,y: x+y)

wordcount.print()
```

Read the CRM data and process it for every minute. Separate the delimited records into fields. The operation starts with the data on text field and finds the frequency occurrence of words in the batches of RDD in Fig. 4. This is achieved in a speed manner and in real-time basis for every 60 s. Business intelligence can achieve in process the real-time data. If the data is created, it is put up in the process.

4.1 Natural Language Processing

The language is understandable and its operations takes place with translation program. The concordance is implemented for extracting the sentences which have the key phrases. The examination also happens with its preceding and following of key

Fig. 5 NL processing steps

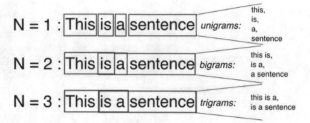

phrases. The examination takes place as unigram, and it considers only single word. Bigram considers double word and trigram, and it considers triple word [22–24] and then n-grams for multiword. For identifying the important trends, n-grams are examined in Fig. 5.

```
import ree
def makeNGrams(n,text):
splitwords = text.split(" ")
filteredwords=[]
for i in splitwords:
if ree.match("[a-zA-Z]+", i):
filteredwords.append(i)
splitwords = filteredwords
ngrams = []
for j in xrange(len(wrd) -(n-1)):
gram = "
for k in xrange(n):
gram += words[j+k] + ' '
ngrams.append(gram.strip())
return ngrams
```

4.1.1 Concordance

The most frequent words are automatically grabbed and applied with concordance for finding the text surrounded that words. W is the width of the words, and L is the number of results predicted to show. Recurrent occur bigram passes to the sentence as concordance shows the framework around the bigram.

ID	Output
141002-001480	that you are having trouble with your purchase. In order to
140929-007053	scussed I am having trouble with advanced author searches
140929-003433	hear you are having trouble with access to Science Direct.

Fig. 6 Concordance output

```
tword = orderList[len(orderList)-1][0]
print tword
print wordtext.concordance(tword, W=80, L=10)
tngram = returnedwords_collocations[1]
concordancesen = concordancesentencebysentence(tngram, unprocessedtext)
list = len(concordancesen )
for indx in range(0, listlen):
    print('-----')
    print(concordancesen[indx])
```

Concordance Output:
The text of the string is broken into n-grams. Discarding the non-letters is by splitting the word. The n-grams are again converted into space-separated words shown in Fig. 6. The list of n-gram strings is returned.

5 Conclusion

The huge data processed in real time by using spark ecosystem. The SQL queries are used to process the data. Frequently analyzed data are unstructured data. The text mining and natural language processing process the data. The data are processed as a batch if stream and currently discovered data are appended with it. Business intelligence is improved by collecting and processing huge data in real time automatically.

References

1. Grover, M., Malaska, T., Seidman, J., Shapira, G.: Hadoop Application Architectures, 1st edn. O'Reilly, Sebastopol, CA (2015)
2. Caya, O., Bourdon, A.: A framework of value creation from business intelligence and analytics in competitive sports. In: 2016 49th Hawaii International Conference on System Sciences (HICSS)

3. Gutfreund, K.: Big data techniques for predictive business intelligence. J. Adv. Manage. Sci. **5**(2) (2017)
4. Van der Aalst, M.: Using Process Mining to Bridge the Gap between BI and BPM. Eindhoven University of Technology, The Netherlands
5. Maheshwar, R.C., Haritha, D., Haritha, D.: Survey on high performance analytics of big-data with apache spark. In: Advanced Communication Control and Computing Technologies (ICACCCT) (2016)
6. Wani, M.A., Jabin, S.: Big Data: Issues, Challenges and Techniques in Business Intelligence
7. Avid Machine Learning Natural Language Processing—Concordance CG 5 August 2017
8. White, T.: Hadoop: The Definitive Guide, 3rd edn, pp. 11–12. O'Reilly, Sebastopol, CA (2012)
9. Pustejovsky, J.: Computational Linguistics, Brandeis University, 23 January 2015
10. Bhandarkar, M.: MapReduce programming with apache hadoop. In: Proceedings of 2010 IEEE International Symposium on Parallel & Distributed Processing, Atlanta, GA, April 2010
11. Chang, K.-W.: N-gram, CS @ University of Virginia (2016)
12. Karthika, I., Gokulraj, P., Saravanan, S.: Prediction of sales using big data analytics. J. Adv. Chem. **12**(20)
13. Karthika, I., Priyadharshini, S.: Survey on location based sentiment analysis of twitter data. IJEDR **5**(1) (2017). ISSN 2321-9939
14. Saravanan, S., Venkatachalam, V.: Advance map reduce task scheduling algorithm using mobile cloud multimedia services architecture. IEEE Dig. Explore pp. 21–25 (2014)
15. Saravanan, S., Venkatachalam, V.: Enhanced bosa for implementing map reduce task scheduling algorithm. Int. J. Appl. Eng. Res. **10**(85), 60–65 (2015)
16. Manning, C., Schütze, H.: Foundations of statistical natural language processing, 2nd edn. MIT Press, Cambridge, MA (2000)
17. Zaharia, M., Chowdhury, M., Franklin, M.J., Shenker, S., Stoica, I.: Spark: cluster computing with working sets. In: Proceedings of HotCloud 2010, 2nd USENIX Workshop on Hot Topics in Cloud Computing, Boston, MA, 2010
18. Karau, H., Konwinski, A., Wendell, P., Zaharia, M.: Learning Spark, Lightning-Fast Big Data Analysis, 1st edn. O'Reilly, Sebastopol, CA (2015)
19. Turing, A.M.: Computing machinery and intelligence. Mind **59**(236), 433–460 (1950)
20. IBM Press Release. 701 Translator (Online), 8 Jan 1954. http://www-03.ibm.com/ibm/history/exhibits/701/701_translator.html
21. Manning, C., Schütze, H.: Foundations of Statistical Natural Language Processing, 2nd ed. MIT Press, Cambridge, MA, chap. 1.4.5, pp. 31–34 (2000)
22. Jurafsky, D., Martin, J.H.: Speech and Language Processing: An Introduction to Natural Language Processing, Computational Linguistics and Speech Recognition, 2nd edn. Prentice Hall, Upper Saddle River, NJ (2008)
23. Karthika, I., Porkodi, K.P.: Fraud claim detection using spark. Int. J. Innov. Eng. Res. Technol. **4**(2) (2017). ISSN 2394-3696
24. Karthika, I., Porkodi, K.P.: Automatic monitoring and controlling of weather condition using big data analytics. Int. J. Adv. Res. Comput. Commun. Eng. **6**(1) (2017)

Memory Allocator for SMP and NUMA-Based Soft Real-Time Operating System Using MemSimRT

Vatsalkumar Shah and Apurva Shah

Abstract As we have entered into an era of high-performance computing, the demand for multi-core architecture has increased. NUMA architecture-based systems are the outcome of this tendency and offer an organized scalable design. However, existing dynamic memory allocators are not capable of performing on a multiprocessor architecture and do not comply with real-time system requirements as well. Researches have proved that the existing memory allocators for any operating systems which support NUMA architecture are not suitable for real-time applications. Hence, in this paper, a new dynamic memory allocator has been proposed which is having consistent execution time and the acceptable fragmentation not only for SMP architecture but also for NUMA-based soft real-time systems.

Keywords SMP · NUMA · RTOS

1 Introduction

This memory management allocator is especially designed and implemented RTOS which supports both SMP and NUMA architecture. The main criteria of the real-time operating system (RTOS) are to complete the task in given time limit which is also called the deadline. RTOS has various features [1–3], but a core part of each operating system is the memory management. Memory management means allocation or deallocation of memory blocks on demand.

Since last three to four decades, a huge amount of improvement has been done for the different embedded devices. Embedded devices maybe smartphones, may be

V. Shah (✉)
Birla Vishvakarma Mahavidyalaya Engineering College, Vallabh Vidyanagar, Gujarat, India
e-mail: vatsal.shah@bvmengineering.ac.in; vatsal.shah-cse@msubaroda.ac.in

V. Shah · A. Shah
CSE Department, M. S. University, Baroda 390001, Gujarat, India
e-mail: apurva.shah-cse@msubaroda.ac.in

© Springer Nature Singapore Pte Ltd. 2019
J. Wang et al. (eds.), *Soft Computing and Signal Processing*,
Advances in Intelligent Systems and Computing 898,
https://doi.org/10.1007/978-981-13-3393-4_31

tablets, etc. But for improving efficiency, uniprocessor is not enough to fulfill the demand [4, 5].

In the twenty-first century, as we have entered into an era of high-performance computing, the demand for multi-core architecture has gained momentum. NUMA architecture-based systems are the outcome of this tendency and offer an organized scalable design. However, existing dynamic memory allocators are not capable of performing on a multiprocessor architecture and do not comply with real-time system requirements as well [4, 5].

2 Related Works

For general purpose operating system as well as for real-time operating system, different memory allocators are available. These algorithms are classified as traditional and non-traditional algorithms [1–3]. Traditional algorithms are (1) sequential fit algorithm, (2) buddy allocator algorithm, (3) indexed fit, (4) bitmapped fit. And non-traditional algorithms are (1) Doug Lea (Dlmalloc), [5] (2) half-fit, [6] (3) TLSF, (4) tcmalloc [7].

These algorithms have already been explained in detail in my previous paper title as "Proposed Memory Allocation Algorithm for NUMA based Soft Real-Time Operating System," published in International Conference On Emerging Technologies In Data Mining And Information Security (IEMIS 2018).

3 Objectives

There are two main objectives [8–10] for this proposed allocator. These objectives are already explained in detail in my previous paper as mentioned above.

(1) Less memory fragmentation.
(2) Consistent execution time.

4 Proposed Algorithm

The general description of these proposed allocators is already discussed in the previous paper as mentioned above (Fig. 1).

Hence, this section will show the actual pseudocode for how blocks are arranged, how to allocate memory block as per request, and how to search remote node.

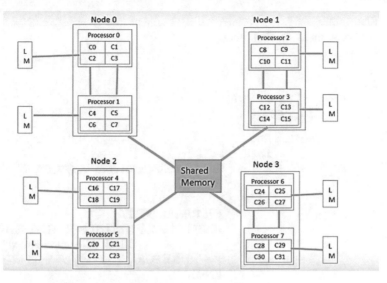

Fig. 1 Proposed algorithm structure [11]

4.1 Pseudocode for Arrangement of Blocks

```
BEGIN
IF  Block Size <= 512 bytes THEN
        Hashing data structure where each key is multiple of 4 up to 512 bytes
        At each key, link list of 64 nodes of same Size
ELSE IF Block Size > 512 bytes AND Block Size  <= 2 Mb THEN
        Create Two level list
        Primary Index which Stores range of 2^{PI} to 2^{PI+1} -1 where PI ∈ [9, 21]
        Each primary index is divided in ranges by 2^{range} , where range =6
ELSE IF Block Size > 2 Mb THEN
        Block will be arranged in descending order of Size
ENDIF
END
```

4.2 Pseudocode for Block Allocation

RB = Requested block size,
PI = Primary index,
SI = Secondary index,
range = divides the primary level ranges in a number of ranges linearly
ISL = Index of segregated list which holds the free block tree,
FR = Fragmentation
RS = Number of request satisfied (initialize with 0),
RN = Root node of AVL tree,

BS = Block size,
MAB = Maximum available blocks.


```
BEGIN
        IF RB <= 512 bytes THEN
                PI = ⌊(RB−1)/4⌋
                WHILE true
                        IF SI > -1 THEN
                                CALL smallBlockAllocation(PI, SI, RB)
                                BREAK
                        ELSE
                                INCREAMENT PI
                                IF PI EQUAL 9 AND SI EQUAL -1 THEN
                                        PRINT "Block Allocation Failed"
                                        RETURN
                                ENDIF
                        ENDIF
                ENDWHILE
        ELSE IF RB > 512 b AND RB <= 2 Mb THEN
```

$$ISL\,(PI, SI) = \begin{cases} PI = \lfloor \log_2 RB \rfloor & where\ PI \in [9, 21] \\ SI = \left\lfloor \dfrac{(RB - 2^{PI})}{2^{PI-range}} \right\rfloor & where\ SI \in [0, 63] \end{cases}$$

```
                CALL normalBlockAllocation(PI, SI, RB)
        ELSE
                Blocks are arranged in descending order of Size
                PI index starts with 0 up to MAB.
                CALL largeBlockAllocation(PI, RB)
        ENDIF
END

smallBlockAllocation (PI, SI, RB)
BEGIN
        PRINT "Small Block Allocated"
        Compute FR as (PI+1)*4 − RB
        DECREMENT SI
        INCREMENT RS
END
```

```
normalBlockAllocation (PI, SI, RB)
BEGIN
        WHILE true
                IF RN >= RB THEN
                        Allocate RN;
                        PRINT "Normal Block Allocated"
                        BREAK

                ELSE
                        SET RN as Right Child of RN
                        IF RN reach to Leaf node AND RN <= RB THEN

                                INCREMENT SI
                                IF SI EQUAL 2^range -1 THEN
                                        INCREMENT PI
                                        IF PI EQUAL 21 AND SI EQUAL 2^range
                        -1 THEN
                                        PRINT "Block Allocation Failed"
                                                RETURN
                                        ENDIF
                                ENDIF
                        ENDIF
                ENDIF
        ENDWHILE
        Balance AVL tree to maintain level -1, 0, +1
        Compute TR as RN –RB
        INCREMENT RS
END

largeBlockAllocation(PI, RB)
BEGIN
        IF RB <= BS at PI^th index THEN
                Divide block in to RB and (BS at PI^th index - RB)
                Compute BS at PI^th index as BS at PI^th index - RB
        ELSE
                INCEREMENT PI
                IF PI > MAB THEN
                        PRINT  "Block Allocation Failed"
                        RETURN
                ENDIF
        ENDIF
END
```

4.3   Pseudocode for Remote Node Search

Mem_{ui} = Memory utilization of ith node
Mem_{us} = Memory utilization of self node
T_U = upper limit of threshold,
T_L = lower limit of threshold,
U and L are constants. (U >1 and L < 1)
Here U = 1.3 and L = 0.7

Mem_{u_avg} = Average Memory utilization of all available node including shared memory

```
BEGIN
        Calculate Memory Utilization of each node
        Find Average Memory Utilization.
```

$$Mem_{u\_avg} = \frac{Mem_{u1} + Mem_{u2} + Mem_{u3} + \cdots + Mem_{un}}{n}$$

```
        Find Upper and Lower Threshold Values
```
$T_U = H \times Mem_{u\_avg}$
$T_L = L \times Mem_{u\_avg}$

```
        Sort node in ascending order of utilization
        Categorize each node Ideal, lightly loaded, Average Loaded and Heavily
Loaded
        IF Self node is Ideal Node THEN
                Use local memory of self node for utilization.
        ELSE IF Ideal Node is available THEN
                Use local memory of ideal node for utilization
        ELSE IF Lightly Loaded Node is available THEN
                IF Memory utilization of Lightly Loaded Node <= Mem_us THEN
                        Use local memory of Lightly Loaded node for utilization.
                ELSE
                        Use local memory of self node for utilization.
                ENDIF
        ELSE IF Average Loaded Node is available THEN
                IF Memory utilization of Average Loaded Node <= Mem_us THEN
                        Use local memory of Average Loaded node for utilization.
                ELSE
                        Use local memory of self node for utilization.
                ENDIF
        ELSE
                Use local memory of self node for utilization.
        ENDIF
END
```

5 Results

There are five different test cases shown in Tables 1 and 2.

1. **SMP**

Case 1: Existing and proposed allocators allocate from local memory.

2. **NUMA**

Case 2: Existing from local and proposed follow Local -> Shared -> Ideal.
Case 3: Existing From local and proposed From ideal.
Case 4: Existing and proposed both from ideal.
Case 5: Existing and proposed follow Local -> Shared -> Ideal.

Table 1 All algorithms execution time for all test cases

Parameter		Execution time (ms)			
Test no.	Cases	Dlmalloc	tcmalloc	TLSF	Proposed
Case 1	Best case	287.8581	330.3003	268.598	234.6128
	Average case	1904.826	2890.503	1461.272	1067.995
	Worst case	3204.577	4352.133	2153.912	1847.152
Case 2	Best case	326.2426	410.8068	290.2026	374.3901
	Average case	2013.324	2988.474	1522.335	2303.212
	Worst case	3202.561	4348.651	2049.025	3361.854
Case 3	Best case	338.0589	420.5202	296.4418	245.4583
	Average case	2054.716	2995.241	1539.964	1115.835
	Worst case	3283.047	4288.159	2064.704	1785.676
Case 4	Best case	374.8572	444.5905	319.6948	249.566
	Average case	2110.58	3240.527	1530.277	1149.484
	Worst case	3277.428	4467.102	2172.658	1860.321
Case 5	Best case	528.5204	636.5573	480.8449	385.8492
	Average case	2928.034	4166.439	2541.517	2252.019
	Worst case	4231.555	5107.635	3684.495	3367.702

Table 2 All algorithms fragmentation for all test cases

Parameter		Fragmentation (%)			
Test no.	Cases	Dlmalloc	tcmalloc	TLSF	Proposed
Case 1	Best case	43.6472	29.684	22.4791	17.5031
	Average case	52.3926	35.157	27.0205	22.0902
	Worst case	60.4389	43.6719	32.0433	26.9948

(continued)

Table 2 (continued)

Parameter		Fragmentation (%)			
Test no.	Cases	Dlmalloc	tcmalloc	TLSF	Proposed
Case 2	Best case	43.5037	36.6849	21.5874	10.5141
	Average case	52.5491	44.4944	29.5867	15.6241
	Worst case	61.4645	43.3702	36.5828	20.5572
Case 3	Best case	45.2763	33.6326	24.0237	15.4697
	Average case	53.8153	44.9067	30.2955	19.5226
	Worst case	60.4389	43.6719	32.0433	26.9948
Case 4	Best case	35.2178	26.3902	19.2872	14.5781
	Average case	43.0248	35.5497	26.3408	19.7854
	Worst case	52.6015	43.6602	35.9747	25.1212
Case 5	Best case	31.4948	23.8764	17.5356	10.4119
	Average case	38.895	31.6255	22.913	15.0111
	Worst case	47.3835	38.2598	30.8472	19.1314

6　Conclusion

As shown in this paper, an allocator which has been proposed for SMP, as well as NUMA-based soft RTOS, provides consistent execution time and acceptable fragmentation as compared to other existing algorithms.

References

1. Diwase, D., Shah, S., Diwase, T., Rathod, P.: Survey report on memory allocation strategies for real time operating system in context with embedded devices. Int. J. Eng. Res. Appl. 2(3), 1151–1156 (2012)
2. Shah, V., Shah, A.: An analysis and review on memory management algorithms for real time operating system. Int. J. Comput. Sci. Inf. Secur. (IJCSIS) 14(5) (2016)
3. Kim, S.: Node-Oriented Dynamic Memory Management for Real-Time Systems on ccNUMA Architecture Systems (2013)
4. Liu, J.W.S.: Real-Time System. Person Education, pp. 20–40
5. Lea, D.: A Memory Allocator. http://g.oswego.edu/dl/html/malloc.html. Unix/Mail December, 1996
6. Ogasawara, T.: An algorithm with constant execution time for dynamic storage allocation. In: RTCSA '95: Proceedings of the 2nd International Workshop on Real-Time Computing Systems and Applications, pp. 21–25, Washington, DC, USA. IEEE Computer Society (1995)
7. Sanjay Ghemawat, P.M.: Tcmalloc: Thread-Caching Malloc (2010)
8. Werstein P., Situ, H., Huang, Z.: Load balancing in a cluster computer. In: Seventh International Conference on Parallel and Distributed Computing, Applications and Technologies (PDCAT'06) (2006)
9. Shah, V., Patel, K.: Load balancing algorithm by process migration in distributed operating system. Int. J. Comput. Sci. Inf. Technol. Secur. (IJCSITS) 2(6) (2012). ISSN: 2249–9555

10. Shah, V., Shah, A.: Critical analysis for memory management algorithm for NUMA based real time operating system. In: IEEE Xplorer in December 2017
11. Shah, V., Shah, A.: Proposed memory allocation algorithm for NUMA based soft real time operating system. In: International Conference on Emerging Technologies in Data Mining and Information Security (IEMIS 2018)

Comparative Study of AODV And OLSR Routing Protocols

Meenal Jain and Manoj Kumar Pal

Abstract In this paper, we have implemented two routing algorithms over Mobile Ad-hoc Network (MANET) for wireless network. First is Ad-hoc On-Demand Distance Vector (AODV) routing protocol and second is Optimized Link State Routing Algorithm (OLSR) routing protocol. These algorithms enable the way for optimization of the routing parameters. These algorithms take into account the merits and demerits of both the protocols in the mobile network. Herein, we used Network Simulator- ns-2 and Xilinx (version 10.1) for Very High-Speed Integrated Circuit (VHSIC) hardware description language implementation of AODV and OLSR routing algorithms in wireless ad-hoc network. Experimental analysis has been done to compare the performance of both algorithms. In ns-2 software, the algorithm aims at improving the performance parameters of routing such as delay, throughput, jitter, and also added OLSR patch to run OLSR routing algorithm. In Xilinx software, algorithm aims at improving the performance parameter of routing such as macrocells, product term, register used, pin, functional block and memory used.

Keywords Ad-hoc · AODV · ns-2 · OLSR · VHDL

1 Introduction

Wireless network involving everyday due to its compatibility in both ad-hoc as well as self-emerging infrastructure along with fixed wireless router access point and base station. Wireless network is very affordable operational efficient which provides

M. Jain (✉) · M. K. Pal
Department of Electronics and Communication Engineering, M.I.T.R.C Alwar,
Alwar 301001, India
e-mail: meenaljainece@gmail.com

M. K. Pal
e-mail: manoj309.pal@gmail.com

© Springer Nature Singapore Pte Ltd. 2019
J. Wang et al. (eds.), *Soft Computing and Signal Processing*,
Advances in Intelligent Systems and Computing 898,
https://doi.org/10.1007/978-981-13-3393-4_32

mobility and convenience to the user. It is used in Bluetooth, Wi-Fi, mobile phones, etc. Everyday research is going on in ad-hoc network and this attract researchers to do research in this field. It is an independent, self-configuring and without pre-define infrastructure network containing mobile routers and wireless channel forming random graph. The property of having infrastructure less makes it different from fixed network and useful in nodes mobility and designing new path between nodes. This also limits it from using in fixed network. Fundamentally, routing includes two activities—one is deciding optimal routing paths and other is transporting data. In wireless network, there are many categories of routing algorithms such as reactive, proactive, geographical, geocasting, hierarchical, multipath, power-aware and hybrid routing protocols. This paper focuses on reactive AODV routing protocol and proactive OLSR routing protocol [1–3].

First of all, undergone the wireless networks and types of these networks. In this paper, a survey of routing algorithm for wireless networks is described [1]. It represents working, types, uses of routing algorithm and their applications [2]. It describes the importance of routing protocols and network performances such as throughput, delivery ratio and end-to-end delay have analysed [3]. Comparison between AODV and DSDV routing algorithms in terms of their working principle [4]. This paper describes different types of internal and external attacks which affect the AODV protocol and preventing method to get rid off [5]. It describes working process of OLSR protocol and message transmitted from source to destination in a secure manner, calculation of delay, throughput in ns-2 and how to reduced overhead flooding [6–8]. AODV and OLSR are been compared in terms of their working principle, packet delivery ratio, end-to-end packet delay and normalized routing load [9–11]. Performance analysis of DSDV Dynamic Source Routing, (DSR), AODV, OLSR and Temporally Ordered Routing Algorithm (TORA) on their basis of packet delivery ratio, control overhead, end-to-end delay, network throughput, packet loss in MANET Optimized Network Engineering Tools, (OPNET), Wireless Personal Area Network (WPAN), highway and city scenarios [12–15].

In this paper, software and hardware implementation will be performed by ns-2 and VHSIC hardware description language. ns-2 is a real-time platform to implement routing algorithm. ns-2 works on Tool Command Language (TCL) script and worked at packet level. This software gives knowledge about how the nodes work in real time. Hardware Description Language (HDL) is a hardware description language that can be utilized to show a digital system at numerous level of abstraction running from the algorithmic level to gate level.

2   Working of Routing Algorithms

In this paper, basically two routing algorithms are implemented—reactive AODV [1–5, 9–15] routing algorithm and proactive OLSR [6–8].

2.1 Ad-hoc On-Demand Distance Vector (AODV)

It has some different types of control packets named as route request, route reply, HELLO message and route reply where route reply is unicasted and route request and HELLO message are broadcasted. AODV sends message periodically to get the newly arrived neighbours. Whenever the source wants to transmit the data to destination, AODV routing protocol always check the table to find whether there is any active route is present towards destination or not. If it finds any active route, then data is transmitted to the next node according to route else route request packet is broadcasted and time is set. This procedure is repeatedly performed until the route reply is received at destination. When route request is obtained at destination, AODV routing protocol forwards route reply to source after checking the reverse path. If any path is detected it send the information otherwise route is disconnected and send RERR message to whole path [4, 15].

2.2 Optimized Link State Routing (OLSR) Protocols

OLSR protocol is a proactive routing protocol. In this protocol, whenever the routes are needed they are always available immediately. Multipoint relay is used to lower the number of broadcasts by having reduced broadcast in some areas of network likewise also lessening the possible overhead in the network. OLSR protocol utilizes two sorts of control messages—one is HELLO and another is TC. HELLO messages are utilized for searching the data about the hosts, neighbours as well as the connection status. Using HELLO message, the MPR selector set is developed which depicts neighbours has picked this host to behave as MPR and with this data, the host can compute its own set of the MPRs. The HELLO messages transmit just a single hop away, however, the TC messages are broadcasted all over the network periodically and TC messages can only be forward by the MPR host [6, 7].

2.3 Comparative Study Between AODV and OLSR Routing Protocol

Basically, AODV and OLSR both are used to avoid the time consumption but they are different in certain properties and these differences have shown in Table 1. Table 1 shows the comparison between AODV and OLSR routing protocols based on some basic properties.

Table 1 Difference between AODV and OLSR routing algorithm

Performance constraints	OLSR	AODV
Classification	Table-driven or proactive	On-demand or reactive
Protocol type	Link state routing	Distance vector
Route maintained	Route table	Route table
Requires sequence	No	Yes
Bandwidth usage	More bandwidth	Less bandwidth
Resource usage	More resource uses	Less resource uses
Security considerations	Less secure	More secure

Table 2 ns-2 simulation parameters

Simulation parameter	Value
Simulator	ns-2.34
Simulation area	for AODV-800*800 for OLSR-1000 * 1000
Number of nodes	3, 4, 5, 7
Routing protocol	AODV, OLSR
Mobility	Random
Channel type	Wireless
Antenna model	Antenna/omni antenna
MAC type	MAC/802_11

3 Experimental Results

There are two types of methodologies to implement AODV and OLSR routing protocols. (1) Network simulator based on real time system. (2) Hardware implementation performed on Xilinx software in VHDL.

Real-Time System (Network Simulator): It works on the network simulator 2.34 in Ubuntu operating system for implementation of AODV and OLSR routing protocols. Table 2 shows the simulation parameters.

Hardware Implementation: It works on the Xilinx 10.1 software for VHDL implementation of AODV and OLSR routing protocols. There are certain steps to implement both routing protocols. These steps are defined below.

Algorithm 1: AODV routing algorithm

Step 1: In this algorithm, four messages are RREP, RRER, RREQ and HELLO to define input and clk & reset to initiates network.

Step 2: Define the nodes and how the packet traverse in the network.

Step 3: With the help of state machine diagram, VHDL code is done for AODV routing algorithm.

Algorithm 2: OLSR routing algorithm

Step 1: In this algorithm, three messages are HELLO, TC and MID to define input condition and clk & reset to initiate the topology of OLSR routing algorithm.
Step 2: Take the advantage of MID message to avoid the flooding in whole network.
Step 3: With the help of state machine diagram, VHDL code is done for OLSR routing protocol.

4 Results and Discussion

4.1 ns-2 Simulation Results

Delay: It is defined as the time taken to transfer data from source to destination.
Throughput: It is the rate of data packets successfully transmitted in a unit of time in the network during the simulation.
Jitter: Jitter (noise) is the difference between subsequent periods of time for a given task in a real-time operating system. So, the comparison between ns-2 results concluded that OLSR routing protocol is better than AODV routing protocol.

4.2 VHDL Simulation Results for AODV Routing Protocol

Experimental results of AODV routing protocols reveal that number of macrocells of 47% of total cells (15 out of 32), number of product term available is 30% of total pterm (33 out of 112), number of register utilized is 29% of total register (9 out of 32), number of pins used is 28% of total pins (9 out of 33), number of functional block used is 15% of total block (12 out of 80) and memory deploy is 70912 kilobytes as shown in Figs. 1 and 2, respectively. In this waveform RREQ, clk and reset act as an input. RREP, Destination, out_ack and RRER act as output. This shows when the RREQ message is transferred, clk and reset become high. But if RREP message could not transfer from destination to source due to path breakage, then both Dest

Fig. 1 Input-output view for AODV routing algorithm

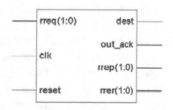

Fig. 2 Register transfer level schematic of AODV routing algorithm

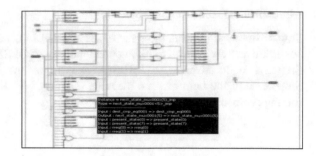

Fig. 3 Technology schematic of AODV routing algorithm

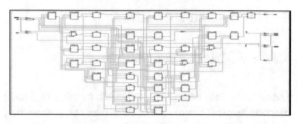

Fig. 4 Simulation waveform of AODV routing algorithm

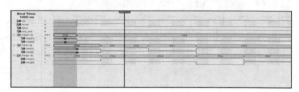

and out_ack become zero and RRER message becomes high. Figures 3 and 4 show the simulation waveform and technology schematic of AODV routing algorithm.

4.3 VHDL Simulation Results for OLSR Routing Algorithm

Experimental results of OLSR routing algorithm reveal that number of macrocells of 44% of total cells (32 out of 154), number of pterm avail is 19% of total pterm (21 out of 112), number of register utilized is 41% of total register (13 out of 32), number of pins used is 31% of total pins (10 out of 33), number of functional block used is 13% of total block (10 out of 80) and memory deploy is 129756 kilobytes as shown in Figs. 5 and 6, respectively. In this waveform HELLO, clk and reset behave as an input. Dest, OUT_ACK, TC and MID behave as an output. This shows that reset and clk become high. TC message forwarded to the network with MID message, if path is not found then OUT_ACK and Dest become low. Comparing both the routing algorithms concludes that OLSR algorithm is better than AODV routing algorithm in both environment real-time and hardware implementation. Figures 7 and 8 show the simulation waveform and technology schematic of OLSR routing algorithm.

Fig. 5 Input-output view for
OLSR routing algorithm

Fig. 6 Register transfer
level schematic of OLSR
routing algorithm

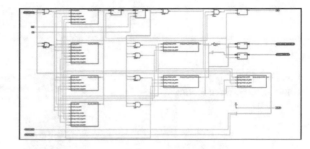

Fig. 7 Technology
schematic of OLSR routing
algorithm

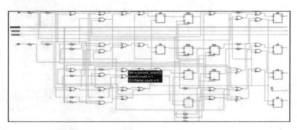

Fig. 8 Simulation waveform
of OLSR routing algorithm

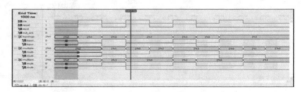

4.4 *Compararison Between AODV and OLSR Routing Algorithms*

Figures 9, 10 and 11 show the comparison among routing algorithms in terms of delay, jitter and throughput, respectively.

5 Future Scope

This paper presents two different kinds of ad-hoc routing algorithms those are compared on the basis of throughput, jitter and delay. A simulation-based result can be enhanced by different strategies such as fuzzy algorithm. OLSR routing protocol

Fig. 9 Delay comparison between routing algorithms

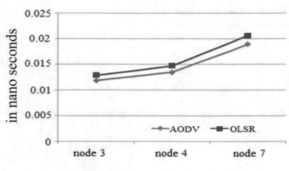

Fig. 10 Throughput comparison between routing algorithms

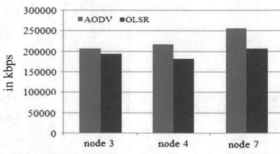

Fig. 11 Jitter comparison between routing algorithms

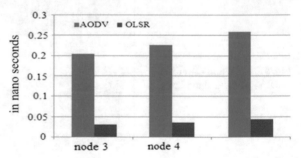

uses more registers and pins as compared to AODV routing protocol in hardware implementation, which can be improved further.

References

1. Alotaibi, E., Mukherjee, B.: A survey on routing algorithms for wireless ad-hoc and mesh networks. Comput. Netw. **56**(2), 940–965 (2012)
2. Ade, S.A., Tijare, P.A.: Performance comparison of AODV, DSDV, OLSR and DSR routing protocols in mobile ad hoc networks. Int. J. Inf. Technol. Knowl. Manag. **2**(2), 545–548 (2010)
3. Gowrishankar, S., Basavaraju, T.G., Singh, M., Sarkar, S.K.: Scenario based performance analysis of AODV and OLSR in mobile ad hoc networks. In: Proceedings of the 24th South East Asia Regional Computer Conference, vol. 15, no. SP4 (2007)

4. Hajare, P.A., Tijare, P.A.: Secure optimized link state routing protocol for ad-hoc networks. Int. J. Comput. Sci. Inf. Technol. **3**(1), 3053–3058 (2012)
5. Huhtonen, A.: Comparing AODV and OLSR routing protocols. In: Telecommunications Software and Multimedia, pp. 1–9 (2004)
6. Jagdale, B.N., Patil, P., Lahane, P., Javale, D.: Analysis and comparison of distance vector, DSDV and AODV protocol of MANET. Int. J. Distrib. Parallel Syst. **3**(2), 121 (2012)
7. Lokanath, S., Thayur, A.: Implementation of AODV protocol and detection of malicious nodes in MANETs. Int. J. Sci. Res. (IJSR)
8. Gupta, M., Kaushik, S.: Performance comparison study of aodv, olsr and tora routing protocols for manets. IJCER **2**(3), 704–711 (2012)
9. Mohan, B.A., Sarojadevi, H.: Study of scalability issue in optimized link state routing (OLSR) protocol. Int. J. Adv. Technol. Eng. Res. **2**(4), 40–44 (2012)
10. Perkins, C., Belding-Royer, E., Das, S.: Ad Hoc On-demand Distance Vector (AODV) Routing, p. 3561. No, RFC (2003)
11. Srivastava, P., Verma, S., Singh, P., Grag, B., Rastogi, A.: A concept paper on comparative study for fpga implemented adhoc routing algorithm. ICM **I**, 2014 (2014)
12. ShubhrantJibhkate, S.K., Kamble, A., Jeyakumar, A.: AODV and OLSR based routing algorithm for highway and city scenarios. Int. J. Adv. Res. Comput. Commun. Eng. **4**(6), 275–280 (2015)
13. Vijaya, I., Mishra, P.B., Rath, A.K., Dash, A.R.: Influence of routing protocols in performance of wireless mobile adhoc network. In: 2011 Second International Conference on Emerging Applications of Information Technology (EAIT), pp. 340–344 (2011, February)
14. Dabreand, U.A., Khan, J.A.: Analysis of delay and throughput using olsr for security of manet. Int. J. Res. Emerg. Sci. Technol. **2**(1) (2015)
15. Upadhyay, A., Phatak, R.: Performance evaluation of AODV DSDV and OLSR routing protocols with varying FTP connections in MANET. IJRCCT **2**(8), 531–535 (2013)

Analysis of MRI-Based Brain Tumor Detection Using RFCM Clustering and SVM Classifier

Venkateswara Reddy Eluri, Ch. Ramesh, Siva Naga Dhipti and D. Sujatha

Abstract The infected tumor area from a magnetic resonance image can be segmented, detected, and extracted accurately by the radiologist experts only through the experience. The complexities and limitations involved in this process are investigated/overcome through distributed rough fuzzy C-means (DRFCM). The support vector machine (SVM)-based classifier improves the accuracy and quality of the segmented tissue. Typically, the best clustering process makes the index values of XB, DB, and RAND as minimum as possible. The performance and quality analysis of the proposed method have been evaluated based on the accuracy, specificity, sensitivity, and also the similarity index of dice coefficient.

Keywords Clustering · Classification · Distributed fuzzy C-means · SVM
Cluster index · Segmentation · Brain tumor

1 Introduction

This examination keeps an eye on the issues of division of strange personality tissues and custom tissues, for instance, dim issue, white issue, and cerebrospinal fluid from alluring resonation pictures using feature extraction framework and support vector machine classifier.

The tumor is, in a general sense, an uncontrolled improvement of threatening cells in any bit of the body, while a cerebrum tumor is an uncontrolled advancement of damaging cells in the mind. A cerebrum tumor can be merciful or risky. The kind

V. R. Eluri (✉) · D. Sujatha
Department of CSE, Malla Reddy College of Engineering and Technology, Hyderabad, Telangana, India
e-mail: evr.eluri@gmail.com

Ch. Ramesh
Department of CSE, Rayalaseema University, Kurnool, Andhra Pradesh, India

S. N. Dhipti
Nalla Malla Reddy Engineering College, Divya Nagar, Hyderabad, India

© Springer Nature Singapore Pte Ltd. 2019　　　　　　　　　　　　　　　319
J. Wang et al. (eds.), *Soft Computing and Signal Processing* ,
Advances in Intelligent Systems and Computing 898,
https://doi.org/10.1007/978-981-13-3393-4_33

cerebrum tumor has consistency in structure and does not contain dynamic (growth) cells, though harmful mind tumors have non-consistency (heterogeneous) in structure and incorporate dynamic cells. The gliomas and meningioma are the cases of poor-quality tumors, delegated kind tumors and glioblastoma and astrocytoma which are class of high-review tumors, named harmful tumors.

2 Related Work

Amid the most recent decade, the measure of datasets has expanded, and with it, additionally the dimensional of each question put away bunch investigation.

Endeavors to discover specific groups of similar articles and the multifaceted nature of the errand rely upon numerous components that can be the quantities of items to the group and their portrayal in the dataset just to refer to a few [1].

Clustering such massive set of images makes running even the most ordinary clustering algorithm not trivial [2]. The time complexity explodes over a certain size requiring more powerful machines to complete the task. The standard approach for huge information handling is spoken to by the appropriated calculation [3].

Subsequently, it is required an approach to seriously part of the dataset, a conveyed handling structure to process them independently and efficiently, and different methods to refine the bunching [4].

The location of a cerebrum tumor at a beginning time is a fundamental issue for giving enhanced treatment [5]. Once a cerebrum tumor is clinically suspected, radiological assessment is required to decide its area, its size, and effect on the encompassing territories. By this data, the best treatment, surgery, radiation, or chemotherapy are chosen. It is apparent that the odds of survival of a tumor-contaminated patient can be expanded substantially if the tumor is recognized, precisely in its beginning time. Accordingly, the investigation of cerebrum tumors utilizing imaging modalities has picked up significantly in the radiology center [6, 7].

3 Methodology

This area displays the materials, the wellspring of cerebrum MR picture dataset, and the calculation used to perform mind MR tissue division. Figure 1 gives the streaming chart of the count. As test pictures, distinctive MR pictures of the mind were utilized, including T1-weighted MR pictures with repetition time (TR) of 1740 and echo time (TE) of 20, T2-weighted MR pictures with repetition time (TR) of 5850 and echo time (TE) of 130, and FLAIR-weighted MR pictures with repetition time (TR) of 8500 and echo time (TE) of 130. These test pictures were gained utilizing a 3 T Siemens Magnetom Spectra MR machine. The aggregate quantities of cuts for all channels were 15, which prompt a sum of 135 pictures at nine cuts or pictures for

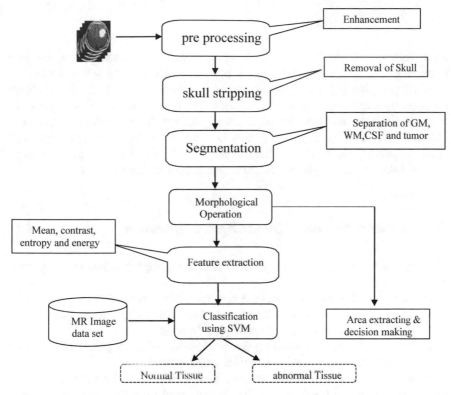

Fig. 1 Steps used in the proposed algorithm

each patient with a field of the perspective of 200 mm, an interslice hole of 1 mm, and voxel of size 0.78 mm × 0.78 mm × 0.5 mm.

The proposed strategy is connected to cerebrum MRI of 512 × 512 pixel measure informational index and later changed over to grayscale for additionally handling. The accompanying area demonstrates the preparing of the calculation.

3.1 Preprocessing

The basic errand of preprocessing is to improve the idea of the MR pictures and make it in a casing suited for moreover getting ready by human or machine vision structure. Also, preprocessing modifies certain parameters of MR pictures, for example, upgrading the flag to-clamor proportion, and enhances the visual appearance of MR picture, emptying the immaterial disturbance and undesired parts beyond anyone's ability to see, smoothing the inner bit of the locale, and securing its edges.

3.2 Skull Stripping

Skull stripping is a vital procedure in biomedical picture investigation, and it is required for the compelling examination of mind tumor from the MR pictures.

Skull stripping is the way of taking out all non-brain tissues in the cerebrum pictures. By skull stripping, it is conceivable to evacuate extracerebral tissues, for example, fat, skin, and skull in the cerebrum pictures. There are a few procedures accessible for skull stripping; a portion of the common methods is programmed skull stripping utilizing picture shape, skull stripping in light of division and morphological task, and skull stripping given histogram examination or limit esteem.

3.3 Segmentation and Morphological Operation

The division of the tainted brain regions in MRI is accomplished through the accompanying strides:

In the initial step, the preprocessed cerebrum MR picture changed over into a binary image with a threshold for the cut-off of 128 being selected. The pixel values greater than the selected threshold are mapped to white.

In the second step, remembering the ultimate objective to get rid of the white pixel, a deterioration undertaking of morphology is used.

At long last, the dissolved regions and the first picture are both partitioned into two similar areas and the dark pixel locale extricated from the disintegrate task is considered a cerebrum MR picture veil. In this study, hybrid rough fuzzy C-means algorithm is used for effective segmentation of brain MR image.

3.4 Feature Extraction

It is the way of gathering more elevated amount data of a picture, for example, shape, surface, shading, and complexity. The surface examination is an essential parameter of human visual observation and machine learning framework. It is utilized viably to enhance the exactness of determination framework by choosing notable highlights. Haralick et al. presented a standout among the most broadly utilized picture investigation use of gray-level co-event matrix (GLCM) and surface component. This method takes two stages for highlight withdrawal from the medicinal pictures. In initial step, the GLCM is processed, and in the other, the texture features based on the GLCM are calculated. Because of the multifaceted structure of enhanced tissues. Textural discoveries and examination could enhance the determination, different phases of the (tumor arranging), and treatment reaction evaluation. The insights include an equation for a portion of the essential highlights which is recorded beneath.

(1) Mean (M): The mean of an image represented by 'M' is used to describe the intensity of individual pixels in the image. Equation (1) is used to determine the mean of the image [8–10].

$$M = \left(\frac{1}{mxn}\right) \sum_{x=0}^{m-1} \sum_{y=0}^{n-1} f(x, y) \tag{1}$$

(2) Standard Deviation (SD): The standard deviation is also referred to the second central moment that describes the probability distribution of the population. The higher the value, the better the clarity in the image. Equation 2 is used to determine the SD [10].

$$SD(\sigma) = \sqrt{\left(\frac{1}{mxn}\right) \sum_{x=0}^{m-1} \sum_{y=0}^{n-1} (f(x, y) - M)^2} \tag{2}$$

(3) Entropy (E): Entropy is figured to describe the irregularity of the textural picture. Equation 3 is used to determine the entropy of an image [11].

$$E = - \sum_{x=0}^{m-1} \sum_{y=0}^{n-1} f(x, y) \log_2 f(x, y) \tag{3}$$

(4) Skewness (Sk(X)): Skewness is defined as the gauge of symmetry or the short of symmetry. Equation 4 is used to determine the skewness denoted by S(X) [12].

$$\underset{k}{S}(X) = \frac{1}{(mxn)} \frac{\sum |(f(x, y) - M^3|}{SD^3} \tag{4}$$

(5) Contrast (Con): Contrast is defined as the measure of the strength of a pixel in the image with its neighborhood pixels in the image. Equation 5 is used to calculate the contrast of the image [11, 12].

$$C_{on} = \sum_{x=0}^{m-1} \sum_{y=0}^{n-1} (x - y)^2 f(x, y) \tag{5}$$

(6) Inverse Difference Moment (IDM): Inverse difference moment is also referred as homogeneity and defined as a measure of local homogeneity of an image. The IDM value ranges from one to many based on the textures in the image(s) in Figs. 1, 2, and 3. IDM is computed using Eq. 6 and shown in Table 1.

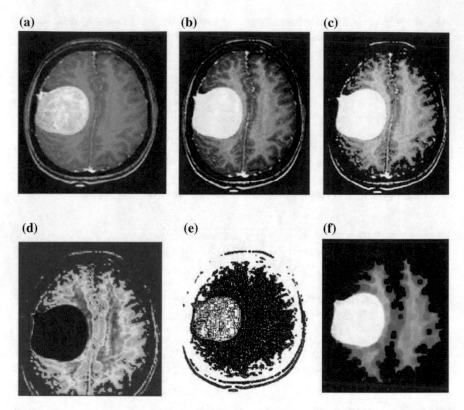

Fig. 2 Experimental results **a** original image, **b** enhanced image, **c** skull-stripped image, **d** wavelet decompose image, **e** intense segmented image, **f** tumor region

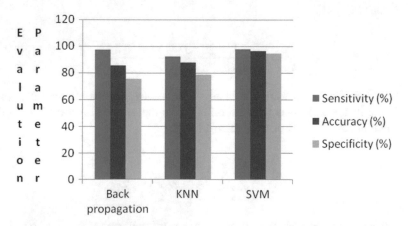

Fig. 3 Comparative analysis of classifier

Table 1 Comparison of accuracies in different classifiers

Number of test images (normal = 67, abnormal = 137)

Evaluation parameter	Back propagation	KNN	Proposed classifier (SVM)
Sensitivity (%)	97.4	92.53	97.74
Accuracy (%)	85.57	88.07	96.66
Specificity (%)	75.54	78.79	94.54

$$IDM = \sum_{x=0}^{m-1} \sum_{y=0}^{n-1} \frac{1}{1 + (x+y)^2} f(x, y) \tag{6}$$

4 Experimental Results

To validate the performance of our algorithm, we used benchmark dataset which is collected from brain Web dataset (http://brainweb.bic.mni.mcgill.ca/cgi/brainweb1). The images used for our analysis mostly included T2-weighted modality with 3 mm slice thickness, 1% noise, and 20% intensity non uniformity.

5 Conclusions

In this examination, utilizing MR pictures of the mind, we portioned cerebrum tissues into normal tissues, for example, white issue, dim issue, cerebrospinal liquid (foundation), and tumor-tainted tissues. Fifteen patients tainted with a glial tumor, inconsiderate and harmful stages, aided this examination. We utilized preprocessing to enhance the flag to-clamor proportion and to wipe out the impact of undesirable commotion. We used a skull stripping calculation given limit procedure to improve the skull stripping execution.

Besides, we utilized rough fuzzy C-means to section the pictures and bolster vector machine to arrange the tumor organize by breaking down element vectors and territory of the tumor. In this examination, we researched surface-based and histogram-based highlights with an ordinarily perceived classifier for the characterization of mind tumor from MR cerebrum pictures. From the test comes about performed on the distinctive images, obviously the investigation for the cerebrum tumor recognition is quick and exact when contrasted and the manual location performed by radiologists or clinical specialists.

The different execution factors likewise demonstrate that the proposed calculation gives better come about by enhancing specific parameters, for example, mean,

MSE, PSNR, exactness, affectability, specificity, and dice coefficient. Our experimental results demonstrate that the proposed work can help in the precise and timely detection of cerebrum tumor along with the identification of its exact location. In this manner, the proposed method is huge for mind tumor location from MR pictures. The results accomplished 96.51% exactness showing the effectiveness of the proposed method for distinguishing normal and abnormal tissues from MR pictures. Our outcomes prompt the conclusion that the proposed technique is appropriate for coordinating clinical choice emotionally supportive networks for essential screening and finding by the radiologists or clinical specialists.

References

1. Smith, T.F., Waterman, M.S.: Identification of common molecular subsequences. J. Mol. Biol. **147**, 195–197 (1981)
2. May, P., Ehrlich, H.C., Steinke, T.: ZIB structure prediction pipeline: composing a complex biological workflow through web services. In: Nagel, W.E., Walter, W.V., Lehner, W. (eds.) Euro-Par 2006.LNCS, vol. 4128, pp. 1148–1158. Springer, Heidelberg (2006)
3. Fahad, A., Al Shatri, N., Tar, Z., Al Mari, A.: A survey of clustering algorithms for bigdata: taxonomy and empirical analysis. IEEE Trans. Emerg. Top. Comput. **2**(3), 267–279 (2014)
4. Guha, S., Rastogi, R., Shim, K.: Cure: an efficient clustering algorithm for large databases. In: Proceedings of ACMSIGMOD Record, vol. 27, no. 2, pp. 73–84 (2008)
5. Chen, M.S., Han, J., Yu, P.S.: Data mining: an overview from a database perspective. IEEE Trans. Knowl. Data Eng. **8**(6) (2007)
6. Zait, M., Messatfa, H.: A comparative study of clustering methods. Futur. Gener. Comput. Syst. **13**, 149–159 (2005)
7. Chen, M.S., Han, J., Yu, P.S.: Data mining: an overview from a database perspective. IEEE Trans. Knowl. Data Eng. **8**(6) (2005)
8. Venkateswara Reddy, E., Reddy, E.S.: Image segmentation using rough set based fuzzy KMeans clustering algorithm. Global Journals Inc, USA, GJCST, Vol. 13, Issue 6, Version 1.0, pp. 23–28 (2013)
9. Venkateswara Reddy, E., Reddy, E.S.: Image segmentation using rough set based fuzzy c means clustering algorithm. In: International Journal of Computer Applications, USA, (IJCA), Vol. 74. No. 14, pp. 23–28 (2014)
10. Venkateswara Reddy, E., Reddy, E.S.: A comparative study of color image segmentation using hard, fuzzy, rough set based clustering techniques. Counc. Innov. Res. IJCT, **11**(8), 2873–2878 (2013)
11. Emam, W.M., Saad, A.E.H.A., Riad, R., et.al.: Morphometric study and length- weight relationship on the squid Loligo forbesi (Cephalopoda: Loliginidae) from the Egyptian Mediterranean waters. Int. J. Env. Sci. Eng. (ijese) **5**, 1–13 (2014)
12. Li, B., Lai, Y.-K., Rosin, P.L.: Example-based image colorization via automatic feature selection and fusion, Neuro Comput. **266**, 687–698 (2017)

Image Retargeting Using Dynamic Load Balancing-Based Parallel Architecture

Ganesh V. Patil and Santosh L. Deshpande

Abstract Nowadays, enormously expanding use of mobile gadgets for capturing images is getting overwhelming response. This fact results in a tremendously increasing usage of digital images. To maintain the quality of vastly pervading digital images on variable sized display contraptions becomes a pensive task for a Web administrator. We are providing a three-leveled image retargeting approach on a parallel architecture with ranking-based dynamic load balancing (RBDLB). Image retargeting is both computational and memory intensive task. Static load balancing cannot offer equity to image retargeting errand as incoming image jobs are required to be processed at dynamic time. The motive of thought process of this undertaking is to provide a good response time and efficient resource utilization in a task of image retargeting without compromising quality of image.

Keywords Image retargeting · Image resizing · Quantization · Compression
Dynamic load balancing

1 Introduction

The evolution of smart handheld devices with Internet facility, good quality graphics, good processing capabilities results in high-rate mobile data consumption. The rate of capturing images on mobile devices is increasing at a high rate. According to worldwide survey in 2017, more than 1.2 trillion images are captured using the smart mobile devices [1]. As per growing user demands, display sizes of these handheld gadgets are rapidly changing. The resultant scenario produced different handheld gadgets with variable display sizes. Different sized communication gadgets demand images with multiple resolutions which should be appropriate for respective display

G. V. Patil (✉) · S. L. Deshpande
Vishveshwaraya Technical University Belgaum, Belgaum, Karnataka, India
e-mail: pganeshv@gmail.com

S. L. Deshpande
e-mail: sld@vtu.ac.in

© Springer Nature Singapore Pte Ltd. 2019
J. Wang et al. (eds.), *Soft Computing and Signal Processing* ,
Advances in Intelligent Systems and Computing 898,
https://doi.org/10.1007/978-981-13-3393-4_34

327

sized device. This dynamically changing demand overburdens the efforts of Web administrator. It makes a pensive task for a Web administrator to provide appropriate sized image for a respective gadget with good image resolution. When a small image is to be displayed on a gadget with a large-sized display unnecessarily computational power is utilized for image scaling. This overall process of image scaling is called image retargeting. Image retargeting is used to convert an image into different sizes to display it on screen of variable sizes. In proposed work, we are providing three-leveled image retargeting technique based on image resizing, quantization, and compression.

Existing image retargeting techniques are classified into two types [2], i.e., subjective and objective. Subjective methods of image retargeting are mostly time-consuming and require maximum user participation, and its large combination of result increases complexity. In contrast to subjective methods, objective methods are fast and automatic.

If we consider pixel-wise mapping, then image retargeting algorithms are basically classified into two classes—i. discrete and ii. continuous. In discrete method of image retargeting, less important pixels are removed to get final retargeted image. Continuous retargeting obtains optimal pixel size to get a targeted sized image [3].

Image retargeting can also be achieved by image resizing. The most commonly used image resizing techniques are homogeneous resizing, cropping, letterboxing/pillarboxing. Common limitations of these methods are stretching or squishing, loss of image contents, and waste of limited screen space [4].

Liang et al. [2] have given an objective metric for accessing the quality of retargeted images. Objective metric is based on the preservation of salient region, global structure, aesthetics, and symmetry.

Resizing technique proposed by Lin et al. [4] has given accumulative energy-based seam carving for image resizing. This experimental work gives a parallel computing environment to resize the image. This technique is based on seam carving [5] method of image resizing. Seam carving is image resizing technique based on internal details of image. A seam consists of the adjacent pixels of the image. The process of seam carving is to determine which pixels should be removed or inserted. The seam deletion or insertion is based on the energy level of seam.

Hua et al. [6] present image strip scaling method of image retargeting. In this experimental work, image energy map is constructed using the gradient information of image and then image is divided into several energy strips.

Zhu et al. [7] have put light on key problems in content-aware image retargeting experimental work; they are:

i.   Human attention-based correlation of image areas using importance map.
ii. Generation of global optimal operator sequence based on image content property.

Zhang et al. [3] proposed the alignment algorithm with three-level retargeting process. Quality of image retargeting is measured by using a combination of fidelity measures and inconsistency detection metrics.

Zhang et al. [8] proposed a new image retargeting method which is a combination of both seam caving and warping. This technique utilizes the advantages of both seam carving and warping and tries to avoid the limitation of both.

Today, rate of handheld devices for image capturing is leaning toward the extreme end. Variability in display sizes with different processing capabilities overburdens the Web administrator while maintaining the mapping of display size with the image size. This problem is solved by image retargeting algorithms. But rate of incoming images toward Web administrator for processing and rate of image retargeting required is so large that the task of image retargeting becomes highly computation and memory intensive. To provide an efficient image retargeting with available set of resources is becoming a challenging task for a Web administrator. Web administrator is always persisting to achieve efficient image processing with good response time and efficient resource utilization. Today, pervading use of digital images in social community sites, blogs makes image retargeting as an unavoidable task. The rate of image capturing and processing is increasing at dynamic speed. In this case, static load balancing cannot provide justice to image retargeting. As the computational load is changing at dynamic time, we have proposed a dynamic load balancing-based parallel image retargeting technique. The main objective of this work is to provide a well-balanced and efficient image retargeting with maximum resource utilization.

2 Methodology

We have proposed a three-level image retargeting approach as shown in Fig. 1. They are: 1. resizing, 2. quantization, and 3. compression.

2.1 Resizing

To match with the display size of targeted device, image needs to be resized. To accomplish this task of image resizing, we are using ImageMagick library's resize operation to scale the image into different dimensions. Internally, resize operation is done using Lanczos [9] filter. Resize operation does the task of image resizing to have images with multiple dimensions. These dimensions are later used to serve the client requesting the image.

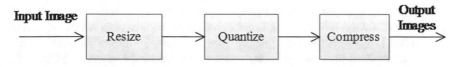

Fig. 1 Technique of image retargeting

2.2 Quantization

Capacity of human eye to distinguish the different combinations of colors is very small. Taking an advantage of this fact quantization [10–12] process does the palletization of similar colors in a color space of image. Storage efficiency is achieved by color grouping. Storage required for an image is directly proportional to number of colors in an image. If the palette size is small, then the resulting image is also small. We have used pngquant [13] tool for quantization of image. This tool is used for the palletization of image using homogeneous color grouping which results in storage optimization. This process minimizes the storage by considering the limitation of visual differentiation made by the human eye. So it does not compromise the image quality.

2.3 Compression

This step involves advpng [14] which does the image compression without loss of visual quality of image. This tool compresses the .png file to reduce its size without remarkable change in image contents.

In this way, we are achieving the image retargeting using three-step image processing. Image resizing reduces the size of image, quantization reduces the storage requirement, and compression achieves lossless compression of image.

3 Architecture of Proposed Work

The proposed system uses master–slave architecture to process image through three different levels, i.e., resize, quantization, and compression. Golang library is used for the implementation of proposed work in Fig. 2.

HTTP protocol is used for internode communication. We use ImageMagick's [15] tools for image resizing, pngquant [13] for quantization or palletization of PNG images, and advpng [14] for lossless compression.

1. Dynamic state information is collected by master node from all slave nodes in a network. Master collects system states from the workers and serves it to the Prometheus stats collector. Load information visualization is done by using Grafana [16]. Prometheus module provides required state information to Grafana. Master collects the basic attributes required for load calculation, i.e., CPU idle percentage, free memory percentage, and communication latency.
2. Master automatically starts a HTTP server. Masters monitor maintains list of nodes connected, their address, and status information.
3. The master fetches the images stored files at the central location and creates job object of each image.

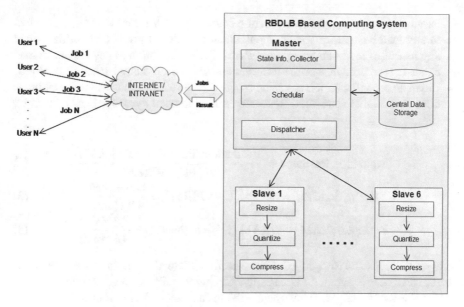

Fig. 2 Architecture of proposed work

4. In parallel image processing, information of all available nodes with their status information is given to a scheduler. Based on dynamic status information, job scheduling is done by a scheduler. The master node then assigns job to the node selected by the scheduler.
5. The results can be viewed in real time in the second tab of the Web dashboard. A CSV of the same data is also generated.
6. Master also monitors the central storage for new incoming input images to a system.

4 Scheduling

To achieve better efficiency and maximum resource utilization, we are using ranking-based dynamic load balancing (RBDLB)-based master–slave architecture. RBDLB computing system consists of set of computing nodes (i.e., from CN1 to CN N). For our experimental work of image retargeting, $CN = 6$. As we are using master–slave architecture, master node acts as a responsible coordinator of the system.

Master node contains state info collector, RBDLB scheduler, and dispatcher as main modules. Every slave node contains resize, quantization, and compression module.

- State info collector collects the dynamic state information from all nodes periodically. The research work done by Patil and Deshpande [17] gives a load balancing based on the dynamic information of resources, i.e., CPU and RAM as prime factors of load balancing. Total computational capacity of node, CPU utilization at any instant of load balancing, total RAM, RAM utilization at any instant of load balancing are used as a prime information blocks for dynamic scheduling of job.

- Based on the information collected by the state info collector, resource ranking is done by RBDLB scheduler. Resource ranking is done by using Eqs. 1, 2, and 3 [18].

$$\text{Evaluation Resource} = \frac{\text{EvaluationCPU} + \text{EvaluationRAM}}{\text{WCPU} + \text{WRAM}} \tag{1}$$

$$\text{EvaluationCPU} = \text{WCPU}(1 - \text{CPUload}) * \frac{\text{CPUspeed}}{\text{CPUmin}} \tag{2}$$

$$\text{EvaluationRAM} = \text{WRAM} \, (1 - \text{RAMusage}) * \frac{\text{RAMsize}}{\text{RAMmin}} \tag{3}$$

where WCPU = the weight of CPU speed; CPUload = the instantaneous CPU load; CPUspeed = actual CPU capacity; CPUmin = CPU speed at minimum level; WRAM = RAM weight; RAMusage = the instantaneous RAM usage; RAMsize = total capacity of RAM; and RAMmin = RAM size at minimum level.

Image processing involves both CPU and RAM usage at maximum extent. So we have given same weight to CPU and RAM, i.e., WCPU = WRAM.

- After calculating ranks by RBDLB scheduler, dispatcher node dispatches the tasks to a computing node with higher rank.

5 Experimental Setup

The proposed architecture is based on single master coordinated system as shown in Fig. 2. The specification of nodes used for this experimentation is given in Table 1.

Table 1 Node specification details

Sr. no.	Name of node	Specification details
1	Master	Dell Vostro 1014, CPU Intel Core 2 Duo, 2.10 GHz, RAM 2 GB, OS: Windows 7(64 bit)
2	Slave	Dell Optiplex 380, CPU Intel Core 2 Duo, 2.93 GHz, RAM 2 GB, OS: Windows 7(32 bit)
3	Network	Gigabit network

Table 2 Software details

Sr. no.	Software name	Purpose
1	Prometheus	Dynamic state information collection
2	Grafana	Load visualization
3	ImageMagick	Image resization
4	Pngquant	Image quantization
5	Advpng	Image compression

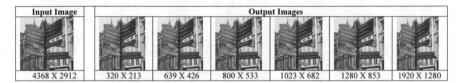

Input Image	Output Images					
4368 X 2912	320 X 213	639 X 426	800 X 533	1023 X 682	1280 X 853	1920 X 1280

Fig. 3 Sample output images with different sizes

6 Results and Discussions

Table 1 shows the results obtained by using the parallel master–slave architecture of image retargeting. Table 2 gives the time required for image targeting level-wise. The size of the original image and targeted image is compared with required computational and communication time in seconds. The following observations are drawn from results given in Table 3:

1. Computational time depends not only on size of image but also on the image contents.
2. Parallel processing also requires considerable amount of time for intercommunication between nodes.
3. Resize, quantize, and compress time are not uniform for images. In some cases, quantize time is greater than other two, in some cases, resize time is greater, and in remaining cases, compression time is greater.
4. In eighth result of Table 3, the communication time is equal to computational time. It depends on the content complexity of image. In process of image retargeting image size along with image contents are key factores of computation.

Table 4 shows the results obtained by Lin et al. in [4] for the seam carving-based image resizing using parallel architecture on Octa Core 3.40 Ghz CPU and 32 GB RAM.

By comparing Tables 3 and 4, it is proven that our results are quite effective than the same shown in Table 4. The experimental setup used for current experimentation i.e shown in Table 1 is also of lower configuration than the same required by Lin et al. in [6]. Figure 3 shows the sample output images with different sizes after image retargeting of sample input image.

Table 3 Time required for image retargeting

Sr. no.	Original size (pixels)		Targeted size (pixels)		Resize time (s)	Quantize time (s)	Compress time (s)	Total computational time (s)	Total time required (s)	Communication time (s)
	Width	Height	Width	Height						
1	3072	2048	320	213	7.05	9.18	5.48	21.73	27.00	5.26
2	3072	2048	639	426	2.09	2.27	1.89	6.27	7.00	0.72
3	3072	2048	1023	682	14.91	6.11	10.64	31.67	38.00	6.32
4	2912	4368	320	213	13.70	8.76	11.95	34.42	42.00	7.57
5	2912	4368	639	426	9.98	10.30	7.68	27.97	36.00	8.02
6	2912	4368	1023	682	18.61	10.17	21.48	50.27	59.00	8.72
7	3840	5760	320	213	22.11	8.82	11.91	42.85	48.00	5.14
8	3840	5760	639	426	0.13	0.17	0.10	0.42	14.00	13.57
9	3840	5760	1023	682	19.05	11.60	12.34	43.01	61.00	17.98

Table 4 Results obtained by Lin et al. in [5]

Sr. no.	Image size	Target size	Time required (s)
1	400 × 400	300 × 300	0.684
2	800 × 600	600 × 600	5.373
3	1024 × 768	768 × 768	11.476

References

1. World Wide Survey. https://mylio.com/true-stories/tech-today/how-many-digital-photos-will-be-taken-2017-repost. Accessed 12 March 2018
2. Liang, Y., Liu, Y.-J., Gutierrez, D.: Objective quality prediction of image retargeting algorithms. IEEE Trans. Visual Comput. Graph. **23**(2), 1–13 (2017)
3. Zhang, Y., Ngan, K.N., Ma, L., Li, H.: Objective quality assessment of image retargeting by incorporating fidelity measures and inconsistency detection. IEEE Trans. Image Process., 1–14. (This article has been accepted for publication but not yet published)
4. Lin, Y., Niu, Y., Lin, J., Zhang, H.: Accumulative energy based seam carving for image resizing. In: Presented in 17th International Conference on Parallel and Distributed Computing, Applications and Technologies (PDCAT), pp. 366–371 (2016)
5. Avidan, S., Shamir, A.: Seam carving for content aware image resizing. ACM Trans. Graph. **27**(3), 1–9 (2008)
6. Hua, S., Wei, H., Su, T.: Fast image retargeting based on strip dividing and resizing. J. Syst. Eng. Electron. **25**(6), 1072–1081 (2014)
7. Zhu, L., Chen, Z.: Fast genetic multi-operator image retargeting. In: Presented in IEEE Conference on Visual Communications and Image Processing (VCIP), November 2016
8. Zhang, L., Li, K., Qu, Z., Wang, F.: Seam warping: a new approach for image retargeting for small displays (2015)
9. Lancoz. http://www.imagemagick.org/Usage/filter/#lanczos. Accessed 6 Nov 2017
10. Image Quantization. https://en.wikipedia.org/wiki/Quantization_(image_processing). Accessed 6 Nov 2017
11. Image Compression (2017). https://en.wikipedia.org/wiki/Image_compression. Accessed 6 Nov 2017
12. Image Resizing. http://www.imagemagick.org/Usage/resize/. Accessed 6 Nov 2017
13. Pngquant. https://pngquant.org. Accessed 6 Nov 2017
14. Advpng. http://www.advancemame.it/doc-advpng.html. Accessed 6 Nov 2017
15. Image Magicks Resize. http://www.imagemagick.org/Usage/resize/. Accessed 6 Nov 2017
16. Grafana. https://grafana.com. Accessed 6 Nov 2017
17. Patil, G., Deshpande, S.L.: Distributed rendering system for 3D animation with blender. In: IEEE International Conference on Advances in Electronics, Communication and Computer Technology, pp. 92–98, December 2016
18. Li, M., Baker, M.: The Grid Core Technologies, Chapter 6, p. 252. Wiley

Exploring the Nexus Between Mobile Phone Penetration and Economic Growth in 13 Asian Countries: Evidence from Panel Cointegration Analysis

Ujjal Protim Dutta, Abhishek Gupta and Partha Pratim Sengupta

Abstract This study endeavours to examine the nexus between mobile phone penetration and economic growth for 13 Asian nations over the period of 1990–2016. To attain this objective, we have initially employed LLC and IPS panel unit root tests to investigate whether the variables are stationary or not. Having found a same order of integration for each of the variables, we applied Pedroni's panel cointegration method to investigate the cointegrating relationship between the variables. The outcome of the study reveals the presence of a cointegrating relationship between the variables. Accordingly, we applied FMOLS and DOLS estimators to estimate the mobile penetration retention coefficient. The results of both methods confirmed the significant effect of mobile phone penetration on economic growth. Thus, besides investment in human and physical capital, governments of these countries can enhance their economic growth by harnessing the potential of mobile phone penetration.

Keywords Mobile phone penetration · Economic growth · Panel data
Cointegration

1 Introduction

Mobile and Internet users all around the world have proliferated since the 1990s. World's Internet users and mobile phone subscribers soared from three million and 11 million to more than three billion and six billion, respectively, in 2014 [1]. Due

U. P. Dutta (✉) · A. Gupta · P. P. Sengupta
Department of Humanities and Social Sciences, National Institute of Technology Durgapur,
Durgapur 741302, West Bengal, India
e-mail: ujjaldtt.06@gmail.com

A. Gupta
e-mail: aguptaindia2012@gmail.com

P. P. Sengupta
e-mail: pps42003@yahoo.com

© Springer Nature Singapore Pte Ltd. 2019   337
J. Wang et al. (eds.), *Soft Computing and Signal Processing* ,
Advances in Intelligent Systems and Computing 898,
https://doi.org/10.1007/978-981-13-3393-4_35

to such a deep penetration of telecommunication services in the last few decades, we observe the arrival of companies attracting extensive investment from within India and also abroad [2, 3]. Mobile technology and Internet have penetrated deep into the developed and developing nations, into the lives of rural and urban society. The developing and poor nations which saw the diffusion of mobile technology and Internet services into their country towards the end of the twentieth century also observed an impressive rate for adoption of such technologies [1]. For instance, one may note that in sub-Saharan Africa mobile connections have increased 10 times compared to landlines which dramatically rose by 49% annually between 2002 and 2007.

Many countries around the world use these services for the welfare of the citizens. These countries have developed systems such that governments and citizens can interact with each other on major issues. For instance, in Japan, we see government notifies the public through mobile phones seconds before a major earthquake is about to hit the island nation. In countries of Africa, citizens are able to view real-time reports of violent confrontations that are shared by their fellow citizens themselves [4]. The agriculture economy-based country such as India which is mainly dependent on monsoon for the cultivation has developed a system where the farmers get a notification about monsoon period and related details about cropping period and harvesting.

The governments of almost all countries are striving hard to evolve this mere tool of communication into a major platform for service delivery. Their endeavours have the potential to bring revolution in this sector as has been observed in the present scenario where mobile phones are no more considered as a mere mode of reducing communication and coordination cost but also seen as a medium that can uplift the standard of life through advanced applications and services [4]. These services give them access to better and innovative methods to mitigate their losses. The citizens of a nation can now collect information about government policy, banking services availability, credit availability through Internet connection available on their phones. Some researchers suggest that the reduction in cost of communication through mobile phones can have tangible economic benefits. This may lay direct benefit to agriculture, producer welfare, labour market efficiency and consumer welfare based on the circumstances within an economy and requirements of the country [5–8]. Another research by Lam and Shiu, 2010 [9], shows a direct relationship between real GDP and development of telecommunication in the developed countries.

The aim of our research is to study the effects of mobile phone penetration on economic growth of 13 selected Asian countries over the period of 1990–2016. The study is important for two reasons: firstly, as per the new economic order [1], by 2020, among the largest seven countries, in terms of purchasing power parity, four will be from Asia (China, India, Japan and Russia). In addition to this, another research suggests that mobile communication technology plays a significant role in global economic growth [9].

2 Literature Review

2.1 Diffusion of Mobile Telephony

As stated by Gupta and Jain, 2012 [10], two major factors responsible for the diffusion of mobile subscription in any economy are good competitive practices among the firms and effective government intervention. The Indian mobile telephony market got a boost from both the efficient service providers and radical government that led to higher investment inflows towards the sector and sustained high diffusion rate. As discussed by Andranaivo and Kpodar, 2012 [11], for African countries, mobile phone diffusion can be an important tool to bridge the financial infrastructure gap. Africa appears to be lagging in number of ATMs, bank branches and bank's outreach to poor, financial inclusion and even in the per cent of people who owns a mobile. There are a number of government initiatives that are based on mobile connectivity and delivery of service in the mode of Direct Benefit Transfer.

In China, Internet was accessible to people by 1994, and since then, with a series of reforms in the industry, the sector grew from state-controlled to market-controlled. Since then, we have observed a significant surge in the numbers of companies providing telecom services, which rose to 4400 till 2008 (the major industry consolidation and government reform were done in 2008) [12–15].

2.2 Nexus Between Mobile Phone and Growth

The role of mobile telecommunication services on total factor productivity (TFP) is well recognized in the existing literature. During the last two decades, due to the privatization, advancements in telecommunications technology and market liberalization, the world is experiencing a faster growth rate. Especially, the developing and underdeveloped countries have experienced a rapid growth in the same period [9]. Information and communication technology (ICT) was a theme of earlier interests to study its wide impact on economic growth where the focus of researchers laid on gadgets such as television, radio and other communicative mediums. But since the rise of mobile telephony and its widespread diffusion in societies all around the world, the focus has shifted towards mobile telephony's impact on the economic growth [11]. As said by Tcheng and others, 2007 [16], ICT has become a factor for the economic growth. They further stated that ICT, as the engine of economic growth, is considerably related to any business activity, reduces cost for the users and helps in the development of new and innovative products.

It was earlier assumed that only the European and high-income countries would be able to mobilize resources for the telecommunication sector, but later it was found that the development of network, telephone services and the Internet is provided with externalities that led to higher economic growth path even for middle-income and developing countries as well [11]. Similarly, Madden and Savage [17], discussed the relationship of real GDP and telecommunication investment in Central and Eastern Europe (CEE) and stated a bidirectional trend. Meanwhile, a further research by Dutta [18] suggested that the effect of telecommunication infrastructure on economic growth was higher and unidirectional. Another research by Chakraborty and Nandi [19], stated bidirectional relationship between telephony and GDP for both short term and long term. He further stated that when countries under study are divided based on the degree of privatization, it was found that bidirectional relationship was valid for nations with higher extent of privatization whereas for the nations with lower extent of privatization the trend was found to be unidirectional with causality running from teledensity to GDP.

3 Materials and Methods

3.1 Framework

Based on the existing literature, along with labour and capital stock, we have also considered mobile phone penetration as a source of growth. To understand the influence of mobile phone penetration on economic growth, we develop the following aggregate production function model:

$$y = f(L, K, M) \tag{1}$$

where y is the real GDP per capita, L is the labour input, K is the capital stock, and M is the mobile phone penetration. We have converted Eq. (1) into a log-linear form which can be specified as follows

$$lny = \beta_0 + \beta_1 lnL + \beta_2 lnK + \beta_3 lnM \tag{2}$$

where the coefficients β_i, $(i = 1 \ldots n)$ relate to GDP per capita labour input, capital stock and mobile phone penetration.

3.2 Data

The study used annual macroeconomic data of 13 Asian countries over the period of 1990–2016. The 13 Asian countries consist of India, China, Pakistan, Sri Lanka, Bangladesh, Indonesia, Vietnam, Japan, Nepal, Philippines, Thailand, Bhutan and Tajikistan. We have used annual data on GDP per capita (constant 2010 US$) for economic growth. For labour input and capital stock, the study used ILO estimated "labour force participation rate, (% of total population ages 15+)" and "gross fixed capital formation (constant 2010 US$)", respectively. We have used "mobile cellular subscriptions" for mobile phone penetration. We have obtained the data for all the selected variables "from World Development Bank Indicators of World Bank".

3.3 Unit Root Test

To confirm the stochastic properties of the variables along with their orders of integration, the study employs ""Levin–Lin–Chu (LLC)" [20] and "Im–Pesaran–Shin (IPS)"" [21] unit root tests. These panel unit root tests have more advantage than the traditional time series approach as it overcomes the problems of size and power of the individual time series techniques. In addition to this, according to Baltagi [22], the statistics of the panel unit root test conveniently lead to a standard normal distribution. The LLC and IPS unit root tests are nothing but an extension of unit root test as specified in augmented Dickey–Fuller, for time series approach. LLC test performs on the basis of null hypothesis of the variables suffering from unit root against the alternative of no unit root in the series. Moreover, it assumes the same autoregressive process across the cross sections. On the other hand, IPS test permits heterogeneity autoregressive process across the cross section.

Structure of both panel unit root tests can be expressed with the help of the following augmented Dickey–Fuller regression equation.

$$\Delta y_{i,t} = \alpha_i + \rho y_{i,t-1} + \sum_{j=1}^{p_i} \alpha_j \Delta y_{i,t-j} + \varepsilon_{i,t} \tag{3}$$

3.4 Panel Cointegration Test

Like IPS panel unit root test, Pedroni cointegration technique also allowed heterogeneity across individual panel member. By allowing heterogeneity across the panel members, this cointegration technique has a great advantage as it is unrealistic to assume common cointegrating vectors across the panel members. Pedroni's [23] panel cointegration technique can be started with the help of the following regression specification:

$$y_{i,t} = \alpha_i + \delta_i t + \beta_1 X_{1,i,t} + \beta_2 X_{2,i,t} + \cdots + \beta_n X_{n,i,t} + \varepsilon_{i,t} \tag{4}$$

where α_i is country-specific effects, δ_i is deterministic time trend, and $\varepsilon_{i,t}$ denotes estimated residuals.

The cointegration test is performed on the basis of the regression of the residuals obtained from Eq. (2). Accordingly, the following regression equation is performed on the estimated residuals ($\varepsilon_{i,t}$) to check the null hypothesis that variables are not cointegrated.

$$\varepsilon_{i,t} = \rho_i \varepsilon_{i,t-1} + \mu_{i,t} \tag{5}$$

Pedroni's [23] suggested two types of cointegration approaches: "within dimension approach" and "between dimension approach". The first one includes four test statistics, namely "panel v", "panel rho", "panel (PP) statistic" (nonparametric) and "panel ADF statistic (parametric)". On the other hand, second one "includes three statistics," namely "group rho", "group PP statistics" (nonparametric) and "group ADF statistics (parametric)". In case of "within dimension approach", Pedroni's [23] panel cointegration technique assumed null hypothesis of $H_0 : \rho_i = 1$ *for all i* against the alternative $H_1 : \rho_i < 1$ *for all i*. On the other hand, in case of "between dimension approach", it assumed $H_0 : \rho_i = 1$ *for all i* against the alternative of $H_1 : \rho_i < 1$ *for atleat one i*.

3.5 The FMOLS and DOLS Tests

To estimate fully modified OLS and dynamic OLS, we have considered the following regression equation:

$$GDP_{i,t} = \alpha_i + x_{i,t}\beta + u_{i,t} \tag{6}$$

where α_i is country-specific effects, $GDP_{i,t}$ is the gross domestic product (an I (1) process), β is the vector of slopes $(k, 1)$ dimension, and $u_{i,t}$ are stationary disturbance terms. In the above specification, $x_{i,t}$ are assumed to be $(k, 1)$ vector of explanatory variables. It is assumed that $x_{i,t}$ $(k, 1)$ vector of independent variables are integrated processes of order one for all cross-sectional items, where

$$x_{i,t} = x_{i,t-1} + \varepsilon_{i,t} \tag{7}$$

The fully modified OLS estimator is obtained by making serial correlation and endogeneity correction to the ordinary least square estimator. Accordingly, the resulting FMOLS estimator can be specified in the following manner:

$$\hat{\beta}_{FM} = \left(\sum_{i=1}^{N} \sum_{t=1}^{T} \left(X_{i,t} - \bar{X}_i \right)^2 \right)^{-1} \sum_{i=1}^{N} \left(\sum_{t=1}^{T} \left(X_{i,t} - \bar{X}_i \right) \widehat{GDP}_{i,t}^* - T\hat{\delta}_{\varepsilon u} \right) \tag{8}$$

where

$\widehat{GDP}_{i,t}^*$ is the transformed variable of $GDP_{i,t}$ for making endogeneity correction and $\hat{\delta}_{\varepsilon u}$ denotes serial correlation correction term.

Kao and Chiang [24] have extended the DOLS estimator to panel analysis. This estimator is used to achieve serial correlation and endogeneity correction. According to Lean and Smyth [25], this method has a possibility of controlling endogeneity problem in the model which augments the cointegration regression with lead and lagged differences of mobile phone penetration and other independent variables. Thus, DOLS estimator offers a robust correction of endogeneity in the regressors [26].

We can obtain the DOLS estimator from the following regression equation:

$$GDP_{i,t} = \alpha_i + x_{i,t}\beta + \sum_{k=-p_1}^{p_2} \delta_k \Delta GDP_{i,t-k} + \sum_{k=-q_1}^{q_2} \lambda_{ik} \Delta x_{i,t-k} + u_{i,t} \tag{9}$$

where

α_i is country-specific effects, λ_{ik} is the coefficient of a lead or lag of first differenced explanatory variables, and $u_{i,t}$ is the error terms following $I(0)$ process.

4 Empirical Findings

4.1 Results of Unit Root Test

We employ LLC and IPS tests to all the selected variables. Results of the tests are presented in Table 1. It is observed from the results that for all the variables null of unit root cannot be rejected at the levels. However, they become stationary at first differences. Thus, we can conclude that all the variables are non-stationary at the levels and are stationary at the first difference.

Table 1 LLC and IPS test results

Variables	Test	Level	First difference
GDP	LLC	−0.494	−4.748*
	IPS	3.224	−5.784*
Capital	LLC	−1.302	−7.676*
	IPS	1.663	−7.814*
Labour	LLC	−0.785	−2.737*
	IPS	0.482	−6.538*
Mobile penetration	LLC	−1.511***	−4.080*
	IPS	0.791	−3.282*

Note *, ** and *** denote rejection of null hypothesis at one, five and ten per cent levels

Table 2 Pedroni's cointegration test results

Dimension	"Test statistics"	Statistics	Probability
Within dimension	"Panel v statistic"	0.328	0.371
	"Panel ρ statistic"	−0.343	0.365
	"Panel t statistic: (nonparametric)"	−2.040**	0.020
	"Panel t statistic (ADF): (parametric)"	−3.073*	0.001
Between dimension	"Group ρ statistic"	1.051	0.853
	"Group t statistic: (nonparametric)"	−1.836**	0.033
	"Group t statistic (ADF): (parametric)"	−4.048*	0.000

Note *, ** and *** denote rejection of no cointegration at 1, 5 and 10% levels of significance

4.2 Results of Pedroni's Cointegration Technique

Use of this test allowed us to test the cointegration among the selected variables. In case of "within dimension approach", two of four statistics reject the null hypothesis of no cointegration at 5 and 1% levels (see Table 2). Similarly, in case of "between dimension approach", two of three statistics reject the null hypothesis of no cointegration at 5 and 1% levels. Four of seven test statistics proposed by Pedroni reject the null hypothesis that variables are not cointegrated (see Table 2). Thus, we can summarize that there is a significant cointegrating relationship between mobile phone penetration and economic growth.

4.3 Results of FMOLS and DOLS Estimators

Having found cointegration among the selected variables, it is useful to estimate the explanatory variables' retention coefficients with the help of panel cointegrating estimator. "To attain this objective, the study employed FMOLS and DOLS panel estimators". The study used both the versions (pooled and group) of the FMOLS and DOLS. Table 3 reports the results of the FMOLS and DOLS tests. It is clearly observed from the results that the coefficients of mobile penetration (0.144 and 0.050) are positive. Moreover, the coefficients are statistically significant at one per cent level for both pooled and grouped versions of FMOLS. Similarly, in case of DOLS, the coefficients of mobile penetration (0.127 and 0.072) are statistically significant at 1 and 5% levels of significance (see Table 3). Thus, the outcomes of the study indicate the significance of mobile penetration on economic growth of the selected Asian countries. Our results are consistent with the results of Qiang et al. [27].

Table 3 FMOLS and DOLS tests

Variable	FMOLS		DOLS	
	Pooled	Grouped	Pooled	Grouped
Capital	2.498*	2.713*	3.466*	3.409*
	(0.227)	(0.242)	(0.242)	(0.288)
Labour	−0.598	1.463***	−1.839*	−6.372*
	(0.545)	(0.791)	(0.648)	(0.614)
Mobile penetration	0.144*	0.050*	0.127*	0.072**
	(0.018)	(0.016)	(0.019)	(0.028)
R-squared adj.	0.963		0.990	

Note *, ** and *** indicate the significance of the variables at one, five and ten per cent, respectively. Numbers in the parenthesis indicate standard error

5 Conclusion

The study is an endeavour to explore the nexus between "mobile phone penetration and GDP per capita for 13 Asian nations." To attain this objective, the study used advanced panel data framework to take into account of the possible endogeneity problems. We have initially employed "IPS and LLC unit root tests to investigate whether the variables (labour, mobile penetration and capital) are stationary or not. Having found the same order of integration, i.e. I (1), we applied Pedroni's panel cointegration technique to investigate the cointegrating relationship between the variables". Out of seven test statistics, four rejected the null hypothesis that variables are not cointegrated. Accordingly, we applied FMOLS and DOLS estimators to measure the mobile penetration retention coefficient. The results of both the methods confirmed the effect of mobile penetration on economic growth of 13 selected Asian nations over the period of 1990–2016.

The results of the study have important policy implication for the Asian nations. Governments of these countries should encourage domestic as well as foreign investment in this sector. In addition to this, lowering the cost of communication is another critical point for the expansion of mobile phone and growth.

References

1. Jorgenson, D.W., Vu, K.M.: The ICT revolution, world economic growth, and policy issue. Telecommun. Policy **40**, 383–397 (2016)
2. Gruber, H.: Competition and Innovation: the diffusion of mobile telecommunication in Central and Eastern Europe. Inf. Econ. Policy **13**, 19–34 (2001)
3. Gruber, H., Verboven, F.: The diffusion of mobile telecommunications services in the European Union. Eur. Econ. Rev. **45**, 577–588 (2001)

4. Jenny, C.A., Isaac, M.M.: Mobile phones and economic development in Africa. J. Econ. Perspect. **24**, 207–232 (2010)
5. Jensen, R.T.: The digital provide: information (technology), market performance and welfare in the South Indian fisheries sector. Quart. J. Econ. **122**, 879–924 (2007)
6. Jenny, C.A.: Information from markets near and far: mobile phones and agricultural markets in Niger. Am. Econ. J.: Appl. Econ. **2**, 46–59 (2010)
7. Jenny, C.A.: Does digital divide or provide? The impact of mobile phones on grain markets in Niger. BREAD Working Paper 177 (2008)
8. Klonner, S., Patrick, N.: Does ICT Benefit the Poor? Evidence from South Africa (2008). http://privatewww.essex.ac.uk/~pjnolen/KlonnerNolenCellPhonesSouthAfrica.pdf
9. Lam, P.L., Shiu, A.: Economic growth, telecommunications development and productivity growth of the telecommunications sector: evidence around the world. Telecommun. Policy **34**, 185–199 (2010)
10. Gupta, R., Jain, K.: Diffusion of mobile telephony in India: an empirical study. Technol. Forecast. Soc. Chang. **79**, 709–715 (2012)
11. Andrianaivo, M., Kpodar, K.: Mobile phones, financial inclusion, and growth. Rev. Econ. Inst. **3**, 30 (2012)
12. Loo, B.P.Y.: Telecommunication reforms in China: towards an analytical framework. Telecommun. Policy **28**, 697–714 (2004)
13. Xia, J.: Linking ICTs to rural development: China's rural information policy. Gov. Inf. Q. **27**, 187–195 (2010)
14. Chen, X., Gao, J., Tan, W.: ICT in China: a strong force to boost economic and social development. In: Berleur, J., Avegrou, C. (eds.) Perspectives and Policies on ICT in Society, IFIP International Federation for Information Processing, vol. 179, pp. 27–36. Springer, Boston (2005)
15. Xia, J.: The third-generation-mobile (3G) policy and deployment in China: current status, challenges, and prospects. Telecommun. Policy **35**, 51–63 (2011)
16. Tcheng, H., Huet, J.M., Romdhane, M.: Les enjeux financiers des telecoms en Afrique subsaharienne. Technical Report, Paris, InstitutFrancais des relations Internationales (2010)
17. Madden, G., Savage, S.J.: CEE telecommunications investment and economic growth. Inf. Econ. Policy **10**, 173–195 (1998)
18. Dutta, A.: Telecommunications and economic activity: an analysis of Granger causality. J. Manag. Inf. Syst. **17**, 71–95 (2001)
19. Chakraborty, C., Nandi, B.: Privatization, telecommunications and growth in selected Asian countries: an economic analysis. Commun. Strat. **58**, 277–297 (2003)
20. Levin, A., Lin, C.F., Chu, C.S.J.: Unit root tests in panel data: asymptotic and finite-sample properties. J. Econom. **108**, 1–24 (2002)
21. Im, K.S., Pesaran, H.M., Shin, Y.: Testing for Unit roots in heterogeneous panels. J. Econom. **115**, 53–74 (2003)
22. Baltagi, B.H.: Economic Analysis of Panel Data, 2nd edn. Wiley, New York (2001)
23. Pedroni, P.: Critical values for cointegration tests in heterogeneous panels with multiple regressors. Oxford Bull. Econ. Stat. **61**, 653–670 (1999)
24. Kao, C., Chiang, M.: On the estimation and inference of a cointegrated regression in panel data. In: Baltagi, B.H., Fomby, T.B., Hill, R.C. (eds.) Nonstationary Panels, Panel Cointegration, and Dynamic Panels. Advances in Econometrics, vol. 15, pp. 179–222. Emerald Group Publishing Limited (2001)
25. Lean, H.H., Smyth, R.: CO2 emissions, electricity consumption and output in ASEAN. Appl. Energy **87**, 1858–1864 (2010)
26. Afonso, A., Jalles, J.T.: Revisiting Fiscal Sustainability: Panel Cointegration and Structural Breaks in OECD Countries (2012). https://papers.ssrn.com/sol3/papers.cfm?abstract_id=2128484
27. Qiang, C., Rossotto, C., Kimura, K.: Economic impact of broadband. In: Information and Communications for Development 2009: Extending Reach and Increasing Impact, vol. 3, pp. 35–50 (2009)

Subjective Answer Grader System Based on Machine Learning

Avani Sakhapara, Dipti Pawade, Bhakti Chaudhari, Rishabh Gada, Aakash Mishra and Shweta Bhanushali

Abstract According to experts, a good test paper should have a combination of objective and subjective questions. But the current online examinations mainly consist of only objective questions. This is because an accurate computerized grading is possible for such questions. But achieving an accurate computerized grading for the subjective questions is still a matter of concern. To address this problem, in this paper, we have designed and implemented a machine learning-based subjective answer grader system (SAGS) using two algorithms, namely latent semantic analysis (LSA) and information gain (IG) for the generation of grades. We have proposed the enhancement of these algorithms through synonym replacement using WordNet, and the accuracy of these algorithms is measured by comparing the generated scores with the scores given by human evaluators.

Keywords Subjective answer evaluation · Machine learning
Latent semantic analysis (LSA) · Information gain (IG) · WordNet

A. Sakhapara (✉) · D. Pawade · B. Chaudhari · R. Gada · A. Mishra · S. Bhanushali
Department of IT, K.J. Somaiya College of Engineering, Mumbai, India
e-mail: avanisakhapara@somaiya.edu

D. Pawade
e-mail: diptipawade@somaiya.edu

B. Chaudhari
e-mail: bhakti.chaudhari@somaiya.edu

R. Gada
e-mail: rishabh.bg@somaiya.edu

A. Mishra
e-mail: aakash.mishra@somaiya.edu

S. Bhanushali
e-mail: shweta29@somaiya.edu

© Springer Nature Singapore Pte Ltd. 2019   347
J. Wang et al. (eds.), *Soft Computing and Signal Processing* ,
Advances in Intelligent Systems and Computing 898,
https://doi.org/10.1007/978-981-13-3393-4_36

1 Introduction and Related Work

In the current online examination system, only objective-type questions are assessed using computer-assisted assessment (CAA). However, CAA lacks the capability of evaluating descriptive answers. Most of the examinations have a combination of objective and subjective questions for the evaluation of the students. Therefore, the CAA system must be capable of evaluating the descriptive answers as well [1]. CAA system helps to avoid human errors and enables instant generation of result. Most of the earlier attempts at CAA were either only applicable for objective answer evaluation (MCQ-based marking) or used strict keyword matching (keyword analysis) which provided poor results. Our objective is to design a system which will evaluate short subjective answers using a combination of keyword matching, semantic matching, machine learning and statistical techniques.

Many researchers have worked in this area. Li et al. [2] have presented the KNN algorithm for automatic essay scoring. In this, each essay is represented by the vector space model (VSM) by calculating the term frequency and inverse document frequency (TF-IDF) weights. The similarity of essays is calculated using cosine similarity. Himani et al. [3] have implemented a prototype for subjective answer evaluation based on latent semantic analysis (LSA) algorithm. The LSA algorithm uses a training dataset consisting of pre-graded answers to calculate the score of the student's answer. Arun et al. [4] have implemented an enhanced NLP [5] -based answer evaluation system. This method assesses text by computing a score based on an explicit concept match between the student's answer and teacher's answer (i.e. reference). Syamala Devi and Himani [6] presented the paper in which several machine learning techniques like latent semantic analysis (LSA) [7, 8], bilingual evaluation understudy [9] and maximum entropy [10] are explored for evaluating the subjective answer. Md. Monjurul Islam et al. [11] have developed a system for essay grading using generalized latent semantic analysis (GLSA). In GLSA, n-gram by document matrix is created instead of a word by document matrix of LSA. GLSA preserves the word proximity, but it takes more processing time than LSA. There is also not much difference in the accuracy between the two algorithms which makes LSA superior to GLSA. The summary of different techniques for the evaluation of subjective answers is given in Table 1, and the comparison of the algorithms is given in Table 2.

From the literature survey, it is found that LSA is the most widely used algorithm. Thus in this paper, LSA algorithm is used as the basis of comparison with the other algorithms.

Table 1 Survey of subjective evaluation techniques

Technique	Parameters used	Dataset description	Accuracy %
K-nearest neighbour (KNN) [2]	TF-IDF term—weighting, document frequency, number of nearest neighbour, cosine similarity	CET4 essays in the Chinese learner English corpus (CLEC)	76
Enhanced NLP [4]	Keyword analysis, information extraction, semantic matching	Dataset released by Rada Mihalcea and Michael Mohler	80
Latent semantic analysis (LSA) [7]	TDF matrix, singular value decomposition (SVD), cosine similarity	Database management systems	59–88
Generalized latent semantic analysis (GLSA) [11]	N-gram by document matrix, singular value decomposition (SVD), cosine similarity	Database engineering course for class of 61 students	34–91

Table 2 Comparison of algorithms for subjective answer evaluation

Algorithm	Semantic study	Word order consideration	Syntactic importance	Processing time
GLSA	Yes	Yes	Yes	Maximum
LSA	Yes	No	No	Intermediate
KNN	Yes	No	No	Intermediate

2 Proposed Subjective Answer Grader System (SAGS)

SAGS uses machine learning [12] to evaluate the subjective answers consisting of around 60–100 words. It generates a score between 0 and 3 marks for the input answer which is to be graded. The working of the SAGS is shown in Fig. 1.

The working of the SAGS can be mainly divided into two phases: training phase and grading phase.

Training phase consists of the following steps:

1. **Store training dataset**: Kaggle dataset [13] consists of some biology questions and their pre-graded answers along with actual grader scores within the range of 0–3 marks. The training dataset for SAGS consists of two biology questions from Kaggle dataset, where one biology question has around 1200 pre-graded answers and another question has around 900 pre-graded answers. These pre-graded answers and its actual grader scores form the training dataset of SAGS.

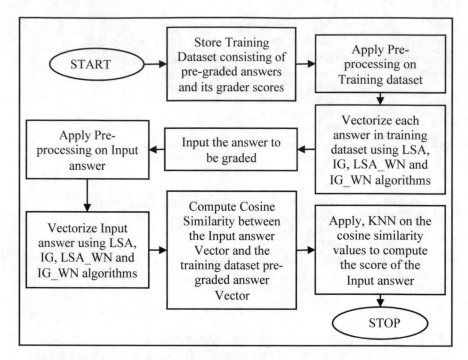

Fig. 1 Working of subjective answer grader system (SAGS)

2. **Apply pre-processing to training dataset**: Each pre-graded answer in the training dataset is pre-processed. The text is tokenized, stop words are removed, and words are auto-corrected and are reduced to their root form through stemming [14].
3. **Vectorize training dataset**: Each pre-graded answer in the training dataset is vectorized using the four algorithms, namely latent semantic analysis (LSA), information gain (IG), latent semantic analysis with WordNet enhancement (LSA_WN) and information gain with WordNet enhancement (IG_WN).

Grading phase consists of the following steps:

1. **Input the answer to be evaluated**: A subjective answer consisting of around 60–100 words is inputted to the SAGS which is to be graded.
2. **Apply pre-processing to input answer**: The same pre-processing steps as described in Step 2 of the training phase are applied to the input answer. The input answer is tokenized, stop words are removed, and words are auto-corrected and stemmed.
3. **Vectorize input answer**: Similar to Step 3 of training phase, the input answer is vectorized using the four algorithms, viz., LSA, IG, LSA_WN and IG_WN.
4. **Compute cosine similarity**: Cosine similarity is calculated between the input answer vector and the training dataset vectors.

5. **Compute the score for the input answer**: Using K-nearest Neighbours (KNN) algorithm with the value of k = 2, the top two nearest pre-graded answers based on cosine similarity values are selected. The average of the grader score of these top two selected pre-graded answers is the score computed for the input answer.

3 Basic and Enhanced LSA and IG Algorithms

After applying pre-processing to the pre-graded and input answers, we have applied latent semantic analysis (LSA) and information gain (IG) algorithms. We have tried to enhance these algorithms by combining the power of WordNet with it.

3.1 Latent Semantic Analysis (LSA)

Latent semantic analysis (LSA) is used to identify relationships between the pre-graded answers in the training dataset and the terms present in it by inferring meaning related to the pre-graded answers and the terms. It is a technique used for clustering similar pre-graded answers together. The steps of the algorithm are:

1. **Construct a term frequency matrix**: The pre-graded answers in the training dataset are represented in the form of a matrix ($TF_{training\_dataset}$) where each row represents one unique term in the training dataset and each column represents one pre-graded answer from the training dataset. Each cell represents the frequency of the term referred by the row, present in the pre-graded answer which is indicated by the column. Similarly, a separate one-column matrix (TF_{input_answer}) is constructed for the input answer which is to be graded.

2. **Apply SVD to the term frequency matrix**: The ($TF_{training\_dataset}$) term frequency matrix for the training dataset, generated in Step 1, is decomposed into the product of three matrices (X, Y and Z) such that the product of these three matrices is same as the term frequency matrix. This technique is called as singular value decomposition (SVD) and is represented by Eq. (1).

$$TF_{training\_dataset} = [X][Y][Z]^{T} \qquad (1)$$

where X represents the term eigenvector values, Y represents the diagonal matrix having scaling values, and Z represents the pre-graded answer eigenvector values.

3. **Apply k-dimension reduction to the X, Y and Z matrices**: The X, Y and Z matrices are reduced to k-dimension by retaining first k columns of X and Z matrices (X_k and Z_k) and first k rows and columns of matrix Y (Y_k). The remaining values in all the three matrices are truncated. After truncation, each row of matrix Z_k corresponds to the vector of each pre-graded answer in the training dataset. It is represented by Eq. (2).

$$Vector_{pregraded\_answer} = Each\ row\ in\ Z_k matrix \tag{2}$$

4. **Compute k-dimensional vector for the input answer**: The k-dimensional vector for the input answer is calculated using the term frequency matrix (TF_{input_answer}) of the input answer as shown in Eq. (3).

$$Vector_{input\_answer} = \left[TF_{input\_answer}\right]^T [X_k][Y_k]^{-1} \tag{3}$$

3.2 Information Gain (IG)

Information gain (IG) [15] is mostly used for feature selection in text categorization. In SAGS, features refer to the terms present in the training dataset and class labels correspond to the grader scores ranging from 0 to 3 marks; i.e. there are four classes in SAGS. IG(T, S) gives the degree to which term T affects the probability of the corresponding pre-graded answer being assigned score S. This algorithm works on eliminating features with low information gain with respect to the output score. The steps to be carried out are:

1. **Construct a term frequency matrix**: A ($FC_{training\_dataset}$) term frequency matrix is constructed for the pre-graded answers of the training dataset. In this matrix, rows represent terms and columns represent pre-graded answers. Similarly, a separate matrix (FC_{input_answer}) is constructed for the input answer to be graded.
2. **Calculate the information gain**: Information gain between each term (T) and each grader score class (S) is calculated using Eq. (4).

$$IG(T, S) = \sum_{s \in S} \sum_{t \in T} p(t, s) log\left(\frac{p(t, s)}{p(t)p(s)}\right) \tag{4}$$

where p(t, s) is the joint probability function of T and S, and p(t) and p(s) are the marginal probability distribution functions of T and S, respectively. The same procedure is repeated for the input answer as well.
3. **Select k-best terms**: The terms with maximum information gain are selected, and other terms are eliminated from the term frequency matrix ($FC_{training\_dataset}$). This matrix with k-best terms is represented as $FC_{training\_dataset}^k$. The vector of each pre-graded answer in the training dataset is given by Eq. (5).

$$Vector_{pregraded\_answer} = Each\ column\ in\ FC_{training\_dataset}^k \tag{5}$$

Similarly, k-best terms are selected from FC_{input_answer} to form the vector of input answer. It is represented by Eq. (6).

$$Vector_{input\_answer} = FC_{input\_answer}^k \tag{6}$$

3.3 Enhancement of LSA (LSA_WN) and IG (IG_WN) Using WordNet

WordNet [16] is a lexical database of English language. Nouns, verbs, adjectives and adverbs are grouped into sets of cognitive synonyms (synsets), each expressing a distinct concept. Existing algorithms like LSA and IG can be enhanced by using WordNet. Terms in the training dataset are replaced by their contextual synonyms to reduce the number of overall terms and aggregate terms having the same meanings. The WordNet enhancement procedure to be integrated with LSA and IG is as follows:

1. **Read the terms**: After applying the pre-processing steps to the terms of the pre-graded answers of the training dataset, these terms are read by SAGS.
2. **Construct an internal dictionary and add terms to it**: In this step, for each term of the pre-processed training dataset, the term is looked up in an internal dictionary. This dictionary consists of unique words from the entire training dataset; i.e. no two words will mean the same thing in it. Initially, this dictionary is empty. If the term under consideration is found in the internal dictionary, then the term is kept as it is and the next term is read. But if the term is not found in the internal dictionary, then

   - Synset for the term is generated and
   - It is checked whether any synonym of the term is present in the internal dictionary.
   - If present, the term is replaced with the synonym word of the synset, which is found in the internal dictionary. Otherwise, if no synonym word from the synset is found in the internal dictionary, then the term is added to the internal dictionary.

   After performing synonym replacement, the terms of the pre-graded answers of the training dataset are compressed to unique terms which in turn provides more accurate frequency counts of terms.
3. **Apply LSA or IG algorithm**: LSA or IG algorithm is applied to the set of the compressed unique terms generated in Step 2.
   Steps 1–3 are repeated for the input answer as well.

4 Results and Discussions

For experimentation purpose, we have considered total of 2100 pre-graded answers along with the actual grader scores for two biology questions, from Kaggle dataset. Eighty percentage of these pre-graded answers is used for training purpose, and 20% is used for testing purpose. The grading score computed by the four algorithms for five sample pre-graded answers from Kaggle dataset for Question 1 (Q1) is given in Table 3.

Table 3 Grading scores computed by different algorithms for sample answers of Q1

Answers	Human grade	LSA	LSA_WN	IG	IG_WN
A1	1	0	1	0	1
A2	3	2	1	1	3
A3	3	1	2	1	3
A4	2	1	1	0	2
A5	0	1	0	0	0

Table 4 Accuracy and correlation coefficient of LSA, IG, LSA_WN and IG_WN algorithms

	Accuracy%		Correlation coefficient	
Algorithm	Question 1 (Q1) (%)	Question 2 (Q2) (%)	Question 1 (Q1)	Question 2 (Q2)
LSA	44	37.7	0.31	0.17
IG	16	13.11	0.23	0.11
LSA_WN	81.5	78.7	0.56	0.62
IG_WN	83	85.24	0.78	0.84

On analysing the scores computed by the four algorithms, it was observed that integration of WordNet, in which the terms are replaced by its synonyms, increased the accuracy of 10% of the answers of the dataset. LSA_WN and IG_WN helped to determine more accurate scores for the answers which were under-marked by the basic LSA and IG algorithms.

The performance of all the four algorithms is measured in terms of accuracy and correlation coefficient. The accuracy measurement and the correlation coefficient values of the four algorithms are presented in Table 4. It is observed that the accuracy of basic LSA and IG algorithms is very low. But by incorporating WordNet with LSA and IG algorithms (LSA_WN and IG_WN), the accuracy of both the algorithms is increased to a huge extent. The highest accuracy is provided by the information gain with WordNet (IG_WN) algorithm. The correlation coefficient is used to measure the relationship between the scores computed by each algorithm with the actual human evaluation score. It is observed that, out of all the four algorithms, IG_WN depicts a high correlation value with the human evaluation score. Thus, it is inferred that the scores computed by the IG_WN algorithm are much more closer to the human evaluation score, as compared to the other three algorithms.

5 Conclusion and Future Work

This paper describes a system for the evaluation of subjective answers using machine learning techniques with LSA and IG algorithms. Further, LSA and IG algorithms are enhanced, by applying term conversion using lexicon like WordNet to deal with the problems of polysemy and synonymy. The accuracy of these algorithms is measured by comparing the computed scores of the answers with the human evaluation scores. After the enhancement of LSA and IG with WordNet, the accuracy of LSA and IG is increased from less than 40% to more than 75%. Information gain with WordNet (IG_WN) gives the best accuracy of around 83%. The system can be evolved further to deal with long answers, formulas, diagrams and equations. Also, a feedback system can be implemented using which students can get a detailed evaluation report which will highlight the answer text where the students lost marks. It will also be interesting to observe the behaviour of other algorithms with the inclusion of WordNet.

References

1. Pooja, K., Amitkumar, M., Kavita, D., Tejaswini, D.: Online examination with short text matching. In: IEEE Global Conference on Wireless Computing and Networking, pp. 56–60, Lonavala, India (2014)
2. Li, B., Lu, J., Yao, J., Zhu, Q.: Automated essay scoring using the KNN algorithm. In: IEEE Proceedings of the International Conference on Computer Science and Software Engineering, pp. 735–738, Hubei, China (2008)
3. Himani, M., Syamala Devi, M.: Subjective evaluation using LSA technique. Int. J. Comput. Distrib. Syst. (2013)
4. Arun, P., Parshu, D., Karma, W., Kesang, W., Uttar, R., Yeshi, J.: Automatic answer evaluation: NLP approach. In: ResearchGate, pp. 1–5 (2016). https://doi.org/10.13140/RG.2.1.2563.6726
5. James, A.: Natural language processing. In: Encyclopedia of Computer Science, 4th edn., ACM Digital Library, pp. 1218–1222. Wiley, Chichester (2003)
6. Syamala Devi, M., Himani, M.: Machine learning techniques with ontology for subjective answer evaluation. Int. J. Nat. Lang. Comput. 5(2), 1–11 (2016)
7. Thomas, L.: Latent semantic analysis. In: Major Reference Works, Wiley Online Library (2006)
8. Nicholas, E., Xiaoni, Z., Victor, P.: Latent semantic analysis: five methodological recommendations. Eur. J. Inf. Syst. 21(1), 70–86 (2010) (Special Issue on Quantitative Methodology)
9. Kishore, P., Salim, R., Todd, W., Wei-Jing, Z.: BLEU: a method for automatic evaluation of machine translation. In: Proceedings of the 40th Annual Meeting of the Association for Computational Linguistics, pp. 311–318, Philadelphia (2002)
10. Jana S.: Using a MaxEnt classifier for the automatic content scoring of free-text responses. In: Proceedings of the 30th International Workshop on Bayesian Inference and Maximum Entropy Methods in Science and Engineering, pp. 41–48. AIP Press (2011)
11. Md. Monjurul, I., Latiful Hoque, A.S.M.: Automated essay scoring using generalized latent semantic analysis. J. Comput. 7(3), 616–626 (2012)
12. Fabian, P., Gael, V., Alexandre, G., Vincent, M., Bertrand, T., Olivier, G., Mathieu, B., Peter, P., Ron, W., Vincent, D., Jake, V., Alexandre, P., David, C., Matthieu, B., Matthieu, P., Edouard, D.: Scikit-learn: machine learning in python. J. Mach. Learn. Res. (JMLR) 12, 2825–2830 (2011)
13. Kaggle Dataset, The Hewlett Foundation: Short Answer Scoring. https://www.kaggle.com/c/asap-sas. Accessed 11 April 2018
14. Martin, P.: An algorithm for suffix stripping. Program 14(3), 130–137 (1980). https://doi.org/10.1108/eb046814
15. Harun, U.: A two-stage feature selection method for text categorization by using information gain, principal component analysis and genetic algorithm. Knowl. Based Syst. 24(7), 1024–1032 (2011) (Elsevier)
16. George, M.: WordNet: a lexical database for English. Commun. ACM 38(11), 39–41 (1995)

Lesion Classification Using Convolutional Neural Network

Mayank Sharma and Aishwarya Bhave

Abstract Malignant melanoma is uncommon in India as compared to the Western nations. However, its growth in recent years has been significant. Early detection of malignant skin lesions can help in proper cure. All recent works on automated classification of skin lesions generate a set of features based on the lesion segment such as lesion diameter and texture. The lesions are then classified into malignant and benign classes based on these features. In our work, we use convolutional neural networks (CNNs) with LeNet architecture in order to automate the feature extraction and selection process. We classify skin lesions in binary class of malignant and benign using ISBI 2016 and PH2 data set with an accuracy of 75% and 97.91%, respectively.

Keywords Skin lesion · Skin lesion classification · Convolutional neural network · Deep learning · Malignant melanoma

1 Introduction to Skin Lesion Classification

Skin cancer consists of aberrant growth of skin cells along with rapid dispersion of such cells throughout the human body. There are various types of skin cancers amongst which malignant melanoma is the most prominent and dangerous one. Malignant melanoma is significantly observed in countries like Australia and New Zealand [1, 2], America [3] and European countries [4]. In America alone, over 9000 deaths from melanoma are anticipated in 2018 [3]. The task of skin lesion classification [5] deals with the following objectives: given an image of a skin lesion [6], classify it as either malignant melanoma or benign melanoma. Malignant melanoma varies in various characteristics such as asymmetricity, border, colour, and texture

M. Sharma (✉) · A. Bhave
National Institute of Technology Raipur, Raipur, India
e-mail: mayanksharma.nitrr@gmail.com

A. Bhave
e-mail: aishwaryabhave54@gmail.com

© Springer Nature Singapore Pte Ltd. 2019
J. Wang et al. (eds.), *Soft Computing and Signal Processing*,
Advances in Intelligent Systems and Computing 898,
https://doi.org/10.1007/978-981-13-3393-4_37

from benign melanoma [7]. In this paper, we propose a method for skin lesion classification using LeNet [8], a basic but powerful CNN architecture, as an alternative to other computationally extensive deep learning architectures.

2 Literature Review

Recent works in skin lesion classification include resource-exhaustive and time-taking manual extraction of features by various methodologies such as "ABCD rule of dermatology" [9], "7-point checklist" [10] and "The CASH algorithm" [11]. Machine learning algorithms such as K-nearest neighbours [7, 12], support vector machine [12–16] and fuzzy classification techniques [17] have been applied over extracted features of image. Recently, FCN [18] has been applied over ISBI 2016 [1] skin lesion image data set. In this work, we try to automate this feature extraction [19] and selection process in neural networks by using a basic architecture of convolutional neural network, LeNet architecture [8] which is resource efficient and at the same time produces accurate results.

3 Data Gathering and Preprocessing

We have used two data sets, PH2 dermoscopy image database from Pedro Hispano Hospital [19] and The ISBI 2016 challenge data set for skin lesion analysis towards melanoma detection from International Skin Imaging Collaboration (ISIC) [20].

PH2 data set consisted of 200 lesion images out of which 160 where benign and 40 were malignant. The ISBI data set had 900 training and 379 testing images. Out of 900 training images, 719 were benign, 176 were malignant and 5 were unlabelled. Out of 379 testing images, 303 images were benign, 73 were malignant and 3 were unlabelled. The preprocessing was done in three stages: (1) data set balancing and image augmentation and (2) downsizing image (Tables 1, 2).

Table 1 Description of binary classification using PH2 data set

Data label	Training	Testing	Total
Benign	2304	576	2880
Malignant	2304	576	2880
Total	4608	1152	5706

Table 2 Description of binary classification using ISBI 2016 data set

Data label	Training	Testing	Total
Benign	1974	765	2739
Malignant	1974	765	2739
Total	3948	1530	5478

3.1 Data set Balancing and Image Augmentation

Since the original data sets were not balanced and the images in each class were not sufficient for training, additional images were derived from the original images using operations such as rotation (The images are rotated by 90°, 180° and 270°), flip (the images are flipped along the horizontal axis and the vertical axis using OpenCV) and shear (the images are sheared at 30° both vertically as well as horizontally). The PH2 data set size increased from 200 images to 5706 images after the application of these operations. Each class had 2880 images approximately thus balancing the data set. The ISBI 2016 data set size increased from 1279 images to 5478 images after the application of these operations. Each class had 2739 images approximately thus balancing the data set.

3.2 Downsizing Image

The images in PH2 data set were originally of the size 767 × 576, and those in ISBI 2016 data set were originally of the size 1022 × 767. These were downsized to a size of 120 × 160 in order to reduce the computational time in training as well as testing.

4 Methodology Used

Convolutional neural network [21] or CNN is a special type of deep neural networks that deals with the phenomena such as localization of the receptive field in high volume data, copying of weights forward as well as image subsampling using various kernels in each convolution layer. Convolution is a mathematical operation that employs feature extraction over the image.

Pseudocode for 7-layer Model

```
num_classes = 4
epochs = 20
dropout =0.7
```

```
learning_rate = 0.001
batch_size = 32
model = Sequential()
model.add(Lambda(lambda x: x/127.5 - 1., input_shape=(120, 160, 3),
        output_shape=(120, 160, 3)))
model.add(Conv2D(32, (3, 3), input_shape=(120, 160, 3)))
model.add(Activation('relu'))
model.add(MaxPooling2D(pool_size=(2, 2)))
model.add(Conv2D(32, (3, 3)))
model.add(Activation('relu'))
model.add(MaxPooling2D(pool_size=(2, 2)))
model.add(Conv2D(64, (3, 3)))
model.add(Activation('relu'))
model.add(MaxPooling2D(pool_size=(2, 2)))
model.add(Flatten())
model.add(Dense(64))
model.add(Activation('relu'))
model.add(Dropout(dropout))
model.add(Dense(num_classes))
model.add(Activation('softmax'))
rms = RMSprop(lr = learning_rate,)
model.compile(loss='binary_crossentropy',
            optimizer= rms,
            metrics=['accuracy'])
```

4.1 Model Architecture

LeNet [8] architecture was used primarily. This section describes in more detail the architecture of LeNet-5 and the CNN used in the experiments. LeNet-5 comprises seven layers, not counting the input, all of which contain trainable parameters (weights) [22]. The input is a 120×160 pixel image. The pixel values are normalized with respect to 255, and hence, black is associated with a pixel value of 0 and white is associated with a pixel value of 1.

5 Experimental Implementation

As part of data set collection, 200 lesion images from PH2 data set and 1279 images from ISBI 2016 data set were collected. These files were then rotated, flipped and sheared along with image augmentation to increase the size of data set to 5706 images with 2880 images of each class in PH2 data set and to 5478 images with

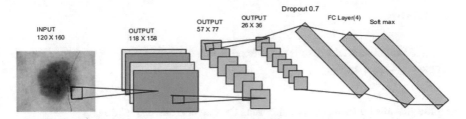

Fig. 1 LeNet-5 CNN structure for a 120×160 input image

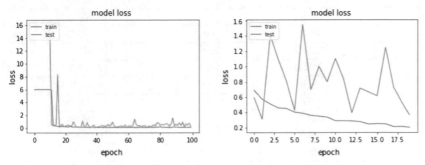

(a) Binary classification on PH2 Dataset
epoch = 100 Lr = 0.001 Dropout = 0.5

(b) Binary classification on ISBl Dataset
epoch = 20 Lr = 0.001 Dropout = 0.7

Fig. 2 Training and testing loss graphs for the best results so obtained on both the data sets

2739 images of each class in ISBI 2016 data set. The data sets consisted of images of size 767×576 and 1022×767, respectively, in PH2 and ISBI 2016 which were then resized to 120×160 size images for faster computation. A total of 4608 (2304 for each class) images were used for training, and 1152 (576 for each class) images were used for testing in PH2 data set, while 3948 (1974 for each class) images were used for training and 1530 (765 for each class) images for testing in ISBI 2016 data set. String labels of each class type are encoded using one hot encoding. Binary cross-entropy is used for binary class classification of benign and malignant lesion classes. A batch size of 32 images was used. The models were implemented and executed using the Keras wrappers with TensorFlow framework as the backend to run the models on the Nvidia GTX 960M GPU. Each of the models was trained using the RMSprop optimizer, and a suitable dropout was used to maintain a strong resistance to overfitting. The mean square error function was used for designing the loss function for all the models (Figs. 1 and 2).

6   Results

This section presents the results of the experiment using a LeNet-inspired architecture (Tables 3, 4, 5, 6 and 7).

Table 3 Results for binary classification using CNN architecture on PH2 data set

Epochs	Learning rate	Dropout	Accuracy
20	0.001	0.5	0.94792
20	0.001	0.7	0.94965
20	0.001	0.8	0.88976
20	0.0001	0.5	0.94357
20	0.0001	0.7	0.95573
50	0.001	0.5	0.966146
50	0.001	0.7	0.97656
50	0.0001	0.7	0.96094
100	0.001	0.5	**0.97917**
100	0.01	0.7	0.96354
100	0.0001	0.7	0.93056

Table 4 Confusion matrix for binary classification using CNN architecture on PH2 data set

Actual\Predicted	Benign	Malignant
Benign	TP = 572	FP = 4
Malignant	FN = 20	TN = 556

Table 5 Confusion matrix for binary classification using CNN architecture on ISBI 2016 data set

Actual\Predicted	Benign	Malignant
Benign	TP = 634	FP = 176
Malignant	FN = 228	TN = 582

Table 6 Results for binary classification using CNN architecture on ISBI 2016 data set

Epochs	Learning rate	Dropout	Accuracy
20	0.001	0.5	0.73642
20	0.001	0.7	**0.75062**
20	0.001	0.8	0.72716
20	0.0001	0.7	0.74815
50	0.001	0.7	0.74506
100	0.001	0.7	0.71975
200	0.001	0.7	0.69506

Table 7 Comparison with other works

Work	Accuracy
Lopez et al. [30]	0.81
Amelard et al. [16]	0.86
Abedini et al. [15]	0.91
Celebi et al. [13]	0.93
Ballerini et al. [7]	0.93
Cavalcanti and Scharcanski [12]	0.94
Bi et al. [18]	0.95
Abuzaghleh et al. [14]	0.97
Our work	0.979

7 Conclusion and Future Work

A comparison of LeNet-5 architecture model with previous works has been made. Experimental results show that the LeNet-5 (plain CNN architecture with seven layers) shows better results than FCN as well as other previous semi automated works. Future work can be done in accommodating medical images in transfer learning architecture to improve results over a large database of several disease images and in a short time. An unbalanced data set [31] can also be accommodated in future using Breiman's Random Forest [32].

References

1. Sneyd, M.J., Cox, B.: A comparison of trends in melanoma mortality in New Zealand and Australia: the two countries with the highest melanoma incidence and mortality in the world. BMC Cancer **13**(1), 372 (2013)
2. Stewart, B.W.K.P., Wild, C.P.: World Cancer Report 2014. Health (2017)

3. Siegel, R.L., Miller, K.D., Jemal, A.: Cancer statistics. CA Cancer J. Clin. **68**, 7–30 (2018). https://doi.org/10.3322/caac.21442
4. Balzi, D., Carli, P., Geddes, M.: Malignant melanoma in Europe: changes in mortality rates (1970–90) in European Community countries. Cancer Causes Control **8**(1), 85–92 (1997)
5. Celebi, M.E., et al.: A methodological approach to the classification of dermoscopy images. Comput. Med. Imaging Graph. **31**(6), 362–373 (2007)
6. Laskaris, N., et al.: Fuzzy description of skin lesions. In: Proceedings of SPIE, vol. 7627 (2010)
7. Ballerini, L., et al.: A color and texture based hierarchical K-NN approach to the classification of non-melanoma skin lesions. Color Medical Image Analysis, pp. 63–86. Springer, Dordrecht (2013)
8. LeCun, Y., et al.: Gradient-based learning applied to document recognition. In: Proceedings of the IEEE, vol. 86, no. 11, pp. 2278–2324 (1998)
9. Stolz, W., Riemann, A., Cognetta, A.B., Pillet, L., Abmayr, W., Holzel, D., Bilek, P., Nachbar, F., Landthaler, M., Braun-Falco, O.: ABCD rule of dermatoscopy: a new practical method for early recognition of malignant melanoma. Eur. J. Dermatol. **4**, 521–527 (1994)
10. Argenziano, G., Fabbrocini, G., Carli, P., Giorgi, V.D., Sammarco, E., Delfino, M.: Epiluminescence microscopy for the diagnosis of doubtful melanocytic skin lesions: comparison of the ABCD rule of dermatoscopy and a new 7-point checklist based on pattern analysis. Arch. Dermatol. **134**, 1563–1570 (1998)
11. Henning, J., Dusza, S., Wang, S., Marghoob, A., Rabinovitz, H., Polsky, D., Kopf, A.: The CASH (colour, architecture, symmetry, and homogeneity) algorithm for dermoscopy. J. Am. Acad. Dermatol. **56**(1), 45–52 (2007)
12. Cavalcanti, P.G., Scharcanski, J.: Automated prescreening of pigmented skin lesions using standard cameras. Comput. Med. Imaging Graph. **35**(6), 481–491 (2011)
13. Celebi, M.E., et al.: A methodological approach to the classification of dermoscopy images. Comput. Med. Imaging Graph. **31**(6), 362–373 (2007)
14. Abuzaghleh, O., Barkana, B.D., Faezipour, M.: Noninvasive real-time automated skin lesion analysis system for melanoma early detection and prevention. IEEE J. Trans. Eng. Health Med. **3**, 1–12 (2015)
15. Abedini, M.A.N.I., et al.: Accurate and scalable system for automatic detection of malignant melanoma. Dermoscopy Image Analysis, pp. 293–343 (2015)
16. Amelard, R., et al.: High-level intuitive features (HLIFs) for intuitive skin lesion description. IEEE Trans. Biomed. Eng. **62**(3), 820–831 (2015)
17. Stanley, R.J., et al.: A fuzzy-based histogram analysis technique for skin lesion discrimination in dermatology clinical images. Comput. Med. Imaging Graph. **27**(5), 387–396 (2003)
18. Bi, L., et al.: Dermoscopic image segmentation via multistage fully convolutional networks. IEEE Trans. Biomed. Eng. **64**(9), 2065–2074 (2017)
19. Mendona, T., et al.: PH 2-A dermoscopic image database for research and benchmarking. In: 2013 35th Annual International Conference of the IEEE Engineering in Medicine and Biology Society (EMBC). IEEE (2013)
20. International Skin Imaging Collaboration: Melanoma Project Website. https://isic-archive.com
21. Krizhevsky, A., Sutskever, I., Hinton, G.E.: Imagenet classification with deep convolutional neural networks. In: Advances in Neural Information Processing Systems (2012)
22. Agostinelli, F., et al.: Learning activation functions to improve deep neural networks (2014). arXiv:1412.6830
23. Bi, L., et al.: Automated skin lesion segmentation via image-wise supervised learning and multi-scale superpixel based cellular automata. In: 2016 IEEE 13th International Symposium on Biomedical Imaging (ISBI). IEEE (2016)
24. Goodfellow, I., et al.: Deep Learning, vol. 1. MIT press, Cambridge (2016)
25. Hartmann, W.M.: Dimension reduction vs. variable selection. In: International Workshop on Applied Parallel Computing. Springer, Berlin, Heidelberg (2004)
26. Simonyan, K., Zisserman, A.: Very deep convolutional networks for large-scale image recognition (2014). arXiv:1409.1556

27. Guyon, I., Elisseeff, A.: An Introduction to Feature Extraction, pp. 1–25. Berlin, Heidelberg, Springer (2006)
28. Peruch, F., et al.: Simpler, faster, more accurate melanocytic lesion segmentation through meds. IEEE Trans. Biomed. Eng. **61**(2), 557–565 (2014)
29. LaFleur-Brooks, M.: Exploring Medical Language: A Student-Directed Approach, 7th edn, p. 398. Mosby Elsevier, St. Louis, Missouri, US. ISBN 978-0-323-04950-4 (2008)
30. Lopez, A.R., et al.: Skin lesion classification from dermoscopic images using deep learning techniques. In: 2017 13th IASTED International Conference on Biomedical Engineering (BioMed). IEEE (2017)
31. Chawla, N.V.: Data mining for imbalanced datasets: an overview. Data Mining and Knowledge Discovery Handbook, pp. 875–886. Springer, Boston, MA (2009)
32. Breiman, L.: Random forests. Mach. Learn. **45**(1), 5–32 (2001)

'Learning to Rank' Text Search Engine Platform for Internal Wikis

Nilesh Sah and Harika Raju

Abstract A large number of companies maintain an internal workplace wiki to document specific goals or processes pertaining to different projects. These wikis can grow exponentially in terms of content size and hence must be supported with an efficient searching platform to facilitate fast lookup of the desired content. However, since these wikis contain highly sensitive matter, relying on external proprietary search engines such as Google, Bing is not possible. Companies, thus, rely heavily on existing open-sourced search engine platforms such as Lucene, Sphinx. Since the nature of the internal wikis can vary greatly, current user experience shows that the result produced by such search engine platform is often inaccurate. In this paper, we aim to present a search engine powered by 'Learning to Rank' system, having the capability to model its ranking algorithm according to the needs of the company.

Keywords Search engine · Learning to Rank · Hadoop · Machine learning

1 Introduction

This paper presents the design of an efficient text search engine, having the capability to mold its ranking algorithm depending upon the model of the internal wiki presented to it, resulting in accurate search results. Such a search engine platform will enable the companies to influence the relevance of the results produced according to their own needs. This customization is currently nonexistent or not supported properly in the majority of the open-sourced search engine platforms, e.g., Lucene, Sphinx.

In order for the search engine platform to determine a feasible configuration which best fits the wiki model, it is supported by a 'Learning to Rank' (L2R) machine

N. Sah · H. Raju (✉)
School of Computer Science & Engineering, Vellore Institute
of Technology, Vellore 632014, India
e-mail: harikaraju.g@gmail.com

N. Sah
e-mail: nilesh.sah13@gmail.com

© Springer Nature Singapore Pte Ltd. 2019
J. Wang et al. (eds.), *Soft Computing and Signal Processing* ,
Advances in Intelligent Systems and Computing 898,
https://doi.org/10.1007/978-981-13-3393-4_38

learning system [1]. L2R systems are based on relevance engineering in which we aim to identify certain features most important to a document and use these features to fine-tune the results and thus return an optimal ranking set.

The feasibility and the efficiency of the idea proposed in this paper regarding building a moldable searching platform are studied by building a search engine wrapper over the largest publicly accessible wiki, that is, the Wikipedia English corpus. Owing to the large nature of the Wikipedia corpus, which spans over 56 GB when uncompressed, most of the algorithms dealt in this paper will be discussed at a distributed computing level for faster data processing. We adopt Apache Hadoop as our standard implementation platform for the development of our in-house search engine with MapReduce being the de facto programming model.

2 Description

2.1 System Architecture

The architecture proposed for the project resembles the model for any other search engine. Figure 1 shows an overview of the different components that exist in our search engine design. As a one-time offline process, the entire wiki corpus is fed into the indexer for the generation of the index table. A trained L2R system serves as our ranking engine. The queries submitted by the users are sent to the search module for processing which communicates with the index database to pick the best possible matching results. These results are then sent to the ranking system to obtain an optimal order of display for them.

2.2 Learning to Rank (L2R)

'Learning to Rank' is an upcoming machine learning field based on relevance engineering which performs the task of ranking query-document pairs [2]. It is used intensely in the practice of information retrieval and data mining [3].

The input to an L2R system comprises of a set of queries and documents, where each query is associated with a number of documents. A label is assigned to each of the documents which denotes the relevance of the document with respect to the given query. This data is generally represented as vector of feature scores for every document. A mathematical model denoting the nature of the problem it addresses is as follows:

Consider $X = \{x_1, x_2, \ldots, x_m\}$ to be a set of query-document pairs which are associated with the label given by $Y = \{y_1, y_2, \ldots, y_m\}$ for $y_i \in \{1, \ldots, s\}$, where s denotes the maximum number of classification labels. Each x_i represents a query-

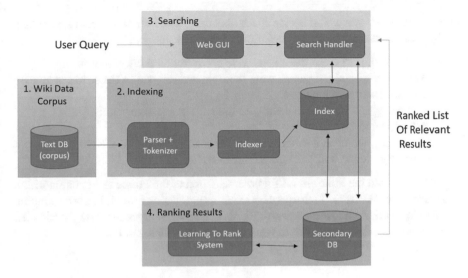

Fig. 1 System architecture for the proposed platform

document pair $\langle q, u \rangle$ with y_i denoting the relevancy score of the document u with respect to the query q.

Now, let $F = \{f_1, \ldots, f_n\}$ serve as the feature set such that $f_i : X \rightarrow R$. Thus, x_i can now be denoted as an n-dimensional vector $x_i = (f_1(x_i), \ldots, f_n(x_i))$. This X, Y pair serves as the training set for the model, and the aim is to predict the correct relevance score for an unseen query-document pair set X^i. The output of such an algorithm is a hypothesis function $H : X \rightarrow R$ which can be used for re-ranking documents.

Various libraries have been developed such as RankNet [4], RankBoost, RankSVM which address this problem based on different approaches: (1) point-wise approach, which treats the problem as a classification; (2) pairwise approach, which aims to identify the correct ordering set between any two documents for the same query; (3) listwise approach, which aims to optimize on the relative ordering of all the items in a list.

3 Methodology Adopted

3.1 Data Preprocessing

Most of the internal wikis maintained in corporations are written in wiki markup language and hence need to be parsed before extracting the required information from them. We use WikiClean library to effectively process the wiki markup language into a content stream which is then space tokenized. The individual tokens were then subjected to the following morphological operations:

(1) Case folding, wherein the tokens were converted into lowercase letters.
(2) Trimming special characters from the right and left end of the string to remove any unwanted punctuations.
(3) Replacing accented characters with their Latin character equivalents; e.g., 'é' gets replaced by 'e'.
(4) Stemming, which is a process of reducing an inflection of the word to its root form; e.g., 'cars' and 'car' get stemmed to 'car'.

We leveraged the Stanford NLP implementation of the Porter stemmer algorithm for carrying out task (4). Such a process helps in aggregating the results corresponding to different variants of the same word. Words which were not properly ASCII encoded containing non-decoded Unicode were ignored from the token list.

3.2 Indexing

Search engine indexing refers to the process of collecting data pertaining to different documents for use by search engines. Such an index aids in fast lookup for search queries. Without a search index, a considerable amount of effort would be required to process every search query.

We use the standard inverted index data structure for generating our search engine index for the wiki corpus. An inverted index works by essentially storing a mapping from a processed token to a list of documents containing the given token.

Our implementation of the indexing algorithm on Hadoop using MapReduce works as follows:

(1) Raw Wikipedia data is fed to the mappers, which is split into multiple parts internally by Hadoop, with every mapper receiving a part of the complete data to process.
(2) The data received by the mappers is first preprocessed to retrieve the actual content stream.
(3) For every processed token in the stream, the mapper emits a ⟨token, title⟩ pair to the reducer, where token corresponds to the processed term in the document and title corresponds to the title of the document.
(4) The reducer receives the token as a key and a list of all the document titles where the token occurred as the value. This list essentially serves as the inverted index for the given token. The output from the reducer thus serves as our index table entries.

```
Class Mapper
    method Map(docId key, doc value)
        title ← fetchTitle(value)
        tokenList ← parseContent(value)
        for all token ∈ tokenList do
            increment tokenCount[token]
        for all token ∈ tokenCount.keys do
            EMIT(token, <title, size(tokenList), tokenCount[token]>)

Class Reducer
    method Reduce(Token token, List values)
        titleList ← [ ]
        IDF ← log_e( size(values) / num of documents )
        for all value ∈ List values do
            TF ← value.tokenCount / value.tokenListSize
            Append(titleList, <TF, value.tokenCount, value.title>)
        EMIT(token, <IDF, titleList>)
```

Fig. 2 Pseudocode for indexing and TF–IDF value calculation

3.3 Feature Extraction

In order to come up with an efficient ordering for a set of documents, there must be a set of parameters over which we can differentiate between the different documents in order to determine which one is to be placed above the other. The major factors that we have taken into consideration in this paper include (1) term frequency–inverse document frequency and (2) PageRank.

We calculate TF and IDF values for every processed token during the indexing phase itself. Figure 2 depicts the combined pseudocode for indexing and TF–IDF value calculation when implemented on Hadoop using MapReduce.

$$IDF(t) = \log_e \left(\frac{Total\ number\ of\ documents}{Number\ of\ documents\ with\ the\ term\ t\ in\ it} \right)$$

$$TF(t) = \left(\frac{Number\ of\ occurence\ of\ the\ term\ t\ in\ the\ docuemnt}{Total\ number\ of\ terms\ in\ the\ document} \right)$$

PageRank (PR)

The state-of-the-art PageRank algorithm was developed by Google and is one of the first ranking algorithms to be used by the company. The algorithm works by analyzing the number and quality of links which are inbound or outbound to a given

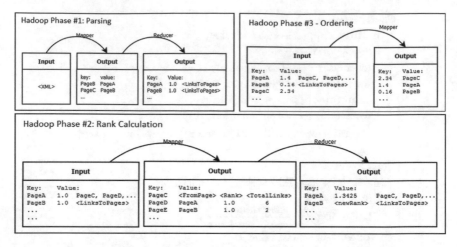

Fig. 3 Phases of PageRank calculation in Hadoop

page. It serves as a rough estimate regarding the importance of the Web site on the World Wide Web.

The mathematical equation expressing the PageRank is given as:

$$PR(u) = \frac{1-d}{N} + d * \sum_{v \in B_u} \frac{PR(v)}{L(v)}$$

$PR(u)$ denotes the PageRank for the page u,
$L(u)$ denotes the number of outbound links from the page u,
B_u denotes the set of all pages containing a link to the page u,
d denotes the damping factor, and
N denotes the total number of pages.

The implementation of the PageRank algorithm on Hadoop is as follows in Fig. 3:

(1) Raw Wikipedia data is first parsed and fed to the mappers. The mapper extracts the name of the article and identifies all the outbound links for the page. It emits data comprising of a pair of titles identified as the 'from page' and 'to page' for every link in the document.

(2) The reducer, for each page, accumulates the outbound link for that page and associates an initial PageRank of 1.0 to the page.

(3) The resulting data serves as the input for the next Hadoop job meant for calculating the PageRank by iterating multiple times over the generated data.

(4) The mapper of this job maps every outbound link with the key page along with rank and a total number of outbound links associated with the key page, and the reducer recalculates the rank for each of the pages along with restoring the original list of outbound links from the page. This job is repeated multiple times in order to arrive at a stable PageRank score.

(5) This generated output from the second job is then fed into a third Hadoop job which does a simple task of mapping the pages to their respective ranks and producing a sorted result.

3.4 Ranking Engine

The ranking engine is responsible for projecting an optimal ordering of the documents. For an enhanced search result, we use the RankLib implementation of the coordinate ascent algorithm for ranking our results. As opposed to RankSVM, Rank-Boost which operates pairwise, the runtime complexity of coordinate ascent algorithm is considerably better.

For our use case, the training set for the L2R model was prepared by querying the Wikipedia search API with a list of randomly sampled 7,110 frequently used unigrams and bigrams obtained from Norvig's n-gram dataset. Every query returned a list of 30 top matching documents according to the Wikipedia ranking algorithm. We labeled the first top five [1–5] documents with a score of 5, the next five [6–10] with a label of 4, and so on, resulting in a dataset of 21,330 labeled query-document pairs.

Every document was then replaced by its corresponding feature vector. The features selected for training the model were (1) query id, (2) covered query terms in the title, (3) title length, (4) covered query term ratio for the title, (5) length of the document, (6) PageRank score, (7) occurrence count of the query terms in the document, (8) sum of term frequency for the query term in the document, (8) minimum of the term frequency for the query term in the document, similarly (9) maximum of term frequency, (10) mean of term frequency, (11) sum of tf * idf, (12) min of tf * idf, (13) max of tf * idf, (14) mean of tf * idf, (15) IDF.

The model was trained with a configuration of (1) number of iterations: 25; (2) number of random restart: 5; (3) tolerance factor: 0.001.

3.5 Search Model

The search functionality of the proposed platform operates in the following manner,

(1) The user queries the search model with the required terms he wants to look up.
(2) The query terms undergo the same token preprocessing as described in Sect. 3.1, after which the search handler looks up the index table to fetch the documents associated with these query terms along with their feature metrics.
(3) Since the number of returned results can be huge, we filter out the documents on the basis of a simple heuristic given by *normalized term frequency + occurence count*. The first 1000 documents having greater heuristic value are chosen to be ranked by our ranking engine.

Table 1 Metrics calculated over the testing data

Type	Training data specification	Testing data specification	Metrics
Unigrams	6893 ranked list	278 ranked lists	NDCG@10: 0.8344
	165897 entries	5762 entries	NDCG@30: 0.9207
Bigrams	1425 ranked lists	6251 ranked lists	NDCG@10: 0.7701
	25100 entries	151014 entries	NDCG@30: 0.8875

(4)   The ranking engine receives the feature vector associated with each document in the list to be processed and ranks them depending upon the input model.

4   Evaluation

4.1   Experiment Setup

Amazon S3 service was utilized for storage of the Wikipedia data dump and other supporting artifacts [5]. The Hadoop jobs were executed on the distributed Amazon Elastic MapReduce (EMR) platform, using 4 c3.2xlarge instances. The overall runtime for the indexing and ranking jobs combined was approximately 32 h.

4.2   Efficiency Evaluation

In order to evaluate the efficiency of our proposed solution, we adopted metrics such as Discounted Cumulative Gain (DCG) and Normalized Discounted Cumulative Gain (NDCG) [6] to be calculated over our testing data, the result of which is tabulated in Table 1.

An ideal search engine produces a NDCG score of 1.0, which in our case evaluates to 0.8344 for the top 10 results and 0.9207 for the top 30 results in case of unigrams; for bigrams, the NDCG evaluation was 0.7701 for top 10 results and 0.8875 for the top 30 results. Table 2 lays down a one-to-one comparison of the top four results as produced by the search engine platform implemented by us over the one implemented by Wikipedia for certain search queries.

The manual testing of our search engine reflected promising performance as the required page would always show up in the top ten results. This enforces the fact that L2R models can capture the essence of relevant results, when trained properly.

Table 2 Comparison of search results between the proposed search engine system and the Wikipedia search engine

Search term	In-house search results	Wikipedia search results
Android	Android	Android
	Android (operating system)	Android Version History
	Android 18	Android (operating system)
	Android Nougat	Android Software Development
Sega	Sega	Sega
	Sega Genesis	Sega Genesis
	Sega Saturn	Sega CD
	Sega CD	Sega Hitmaker
Windows Vista	Windows Vista	Windows Vista
	Features new to Windows Vista	Windows Vista editions
	Comparison of Windows Vista and Windows XP	Features new to Windows Vista
	Windows Vista editions	Criticism of Windows Vista

5 Conclusion

We have presented a novel approach to formulate a search engine from scratch, which has the ability to morph its ranking algorithm by using 'Learning to Rank' systems. We make use of the coordinate ascent algorithm for training over the Wikipedia generated data, optimizing on the Expected Reciprocal Rank (ERR) metric, using the RankLib library. The entire platform was developed on a distributed system architecture. The performance of the system was satisfactory in terms of the results produced which supports our idea of building a moldable search engine.

In the future, we aim to improve the overall accuracy of the system by enhancing our feature set with more number of features such as BM25 score [5], click count, and content quality score. We also aim to integrate our proposed system with the existing open-sourced search engines such as Apache Lucene and Sphinx.

References

1. He, C., et al.: A survey on learning to rank. In: International Conference on Machine Learning and Cybernetics, vol. 3. IEEE (2008)
2. Li, H.: A Short Introduction to Learning to Rank (2011)

3. Liu, T.Y.: Learning to rank for information retrieval. Found. Trends Inf. Retr. **3**(3), 225–331 (2009)
4. Burges, C., et al.: Learning to rank using gradient descent. In: Proceedings of the 22nd International Conference on Machine learning. ACM (2005)
5. Boytsov, L., Belova, A.: Evaluating learning-to-rank methods in the web track adhoc task. In: TREC (2011)
6. Burges, C.J.C., Ragno, R., Le, Q.V.: Learning to rank with nonsmooth cost functions. In: NIPS, pp. 193–200 (2006)

Energy-Efficient Routing Protocols for Wireless Sensor Networks

Ch. Usha Kumari and Tatiparti Padma

Abstract The energy depletion is the major drawback in wireless sensor networks (WSNs). Sensor nodes use utilize tiny sized batteries, which can neither be replaced nor recharged. Hence, the energy must be optimally maintained in such battery-operated networks. One of the popular method of optimizing the energy efficiency is through clustering. In clustering methodology, all the sensor hubs are assembled into number of groups and each group is allocated with a group head called cluster head (CH). The CH consumes high energy than remaining sensor nodes (SNs) as they are responsible for data collection and data transmission. The cluster heads are usually overloaded with large number of sensing nodes and thereby leads to faster death. Thereby degrades the network execution. In this paper, we propose three different protocols—distributed energy-efficient clustering (DEEC), developed DEEC (DDEEC) and enhanced distributed energy-efficient clustering (E-DEEC). These protocols are compared with energy depletion of cluster head and lifetime. Simulations are carried by considering 500 sensor nodes and 30 cluster heads within a communication range of 200×200 m for 10,000 iterations. It is observed that number of nodes alive are 70% more in E-DEEC than DEEC and DDEEC. The packets transmitted to the BS are 80% more in E-DEEC than DEEC and DDEEC. The E-DEEC protocol enhances the network lifetime than DECC and DDEEC.

Keywords WSN · Cluster head · Distributed energy-efficient clustering DDEEC E-DEEC

Ch. Usha Kumari (✉) · T. Padma
GRIET, Hyderabad, Telangana, India
e-mail: ushakumari.c@gmail.com

T. Padma
e-mail: tatipartipadma@gmail.com

© Springer Nature Singapore Pte Ltd. 2019
J. Wang et al. (eds.), *Soft Computing and Signal Processing*,
Advances in Intelligent Systems and Computing 898,
https://doi.org/10.1007/978-981-13-3393-4_39

1 Introduction

WSNs are used in health monitoring, Internet of Things, robotics, medical applications and so on. Such system comprises lots of sensors, and all these nodes transmit information towards base station (BS). Each node is supplied with sensing unit, processing unit, communication unit and power unit. Sensor nodes have minimum amount of energy; therefore, the protocols should be designed carefully for efficient utilization of energy and for achieving longer lifetime of the network [1–3]. Parameters like initial energy, transmitted packets to the sink and CH selection and total nodes dead greatly influence the network lifetime [4–6]. Often SNs have a drawback of depleting battery life, so replacing of SNs is very complicated as they will spread randomly in the communication area [7]. Since nodes are placed distant from sink, direct communication is not possible because of less energy as direct communication requires more energy [8, 9]. Formation of clusters and selecting a cluster head selection greatly reduce battery consumption. This is one of the important technique for reduction in battery consumption.

2 Related Work

Sensor nodes in wireless networks are very small in size and are capable of gathering data and transmit the information to the CHs. The CHs transmit this information to the BS as shown in Fig. 1. Cluster formation can be implemented in two ways—homogenous networks and heterogeneous networks. Homogeneous networks are considered to be as the networks where the nodes have identical amount

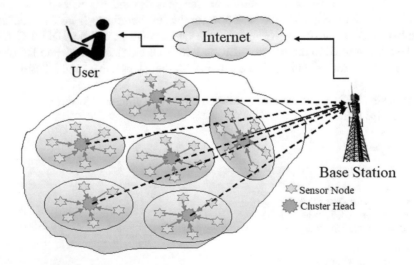

Fig. 1 Wireless sensor network architecture

of energy. There are various algorithms for homogeneous networks like low-energy adaptive clustering hierarchy (LEACH) [1, 10], hybrid-energy-efficient distributed clustering (HEED) [11] and power-efficient gathering in sensor information system (PEGASIS). Heterogeneous networks are the networks which have distinct energy levels for all the sensor nodes [12]. The various algorithms for heterogeneous networks are stable election protocol (SEP) [12], distributed energy-efficient clustering (DEEC), developed DEEC (DDEEC), enhanced DEEC (EDEEC) and threshold DEEC (TDEEC) [13].

HEED protocol described four essential goals in [14] for improving lifetime of network. In [15] DEEC, CHs are selected considering average energies and residual energies of nodes. Saini and Sharma [5, 10, 16] have offered E-DEEC which adds heterogeneity in network by presenting the super nodes having energy higher than the normal and the advanced nodes and respective probabilities. The proposed EDEEC enhances the lifetime and network stability.

3 System Model

The hardware energy dissipation shown in Fig. 2 which consists of a transmitter and receiver electronics. The energy dissipation model [1] uses free-space path loss (d^2), multipath channel models (d^4) and distance bewteen TX and RX. If the threshold is less than distance d_0, the free space (f_s) model is considered, else multipath (mp) model is considered. Let E_{elec} energy required by the electronics circuit ϵ_{fs} is the free-space energy loss, ϵ_{mp} multipath energy loss, d is the distance between source node and destination node and d_0 is crossover distance. The energy required for the sensor nodes for transmission of information to the base station over a distance d is given in Eq. (1) as:

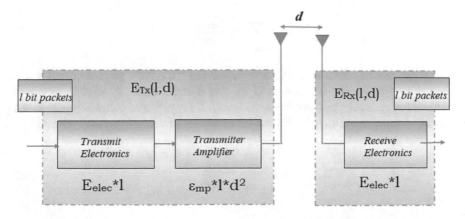

Fig. 2 Energy dissipation model

$$E_T(l, d) = \begin{cases} lE_{elec} + l\epsilon_{fs}d^2 & for \ d < d_0 \\ lE_{elec} + l\epsilon_{mp}d^4 & for \ d \geq d_0 \end{cases} \quad (1)$$

where, $d_0 = \sqrt{\dfrac{\epsilon_{fs}}{\epsilon_{mp}}}$. The energy consumed by the cluster head to receive message is given in Eq. (2)

$$E_{RX}(l) = lE_{elec} \quad (2)$$

4 Proposed Routing Protocols—DEEC, DDEEC and EDEEC

The DEEC is applied for heterogeneous WSNs. It uses two energy levels: residual and initial energy levels for CH selection. Higher energy SNs are selected as CHs.

Let n_i is number of iterations to CH for node S_i, the optimum number of cluster heads is denoted as $P_{op}N$, the probability of a SN S_i to become CH is P_i. The average energy for each round r is given in Eq. (3) as,

$$\bar{E_{av}}(r) = \frac{1}{N} \sum_{i=1}^{N} E_{avi}(r) \quad (3)$$

Cluster head selection probability is P_{chi} given in Eq. (4) as,

$$P_{chi} = P_{op} \left[\frac{E_{avi}(r)}{\bar{E}(r)} \right] \quad (4)$$

The cluster heads' average number is calculated for each round with probability of cluster head selection from [12] as in Eq. (5)

$$\sum_{i=1}^{N} P_{chi} = \sum_{i=1}^{N} P_{op} \frac{E_{avi}(r)}{E_r} = NP_{op} \quad (5)$$

Few nodes are eligible to become cluster heads which are having higher energy levels than normal nodes and denoted as G for that round r. Each SN selects a random number between zero and one for each round. If the obtained value is less than the threshold value as in Eq. (6), then it has probability of becoming a cluster head.

$$T_{ch}(S_i) = \begin{bmatrix} \dfrac{P_{chi}}{1 - P_{chi}(r \bmod \frac{1}{P_{ch}})} & S_i \in G \\ 0 & otherwise \end{bmatrix} \quad (6)$$

For two-level heterogeneous networks, the average probability for normal and advanced node is given as:

$$P_i = \begin{cases} \dfrac{P_{op}E_{avi}(r)}{(1+aq)\bar{E}(r)} & \text{if } S_i \text{ is the normal node} \\[4mm] \dfrac{P_{op}(1+a)E_{avi}(r)}{(1+aq)\bar{E}(r)} & \text{if } S_i \text{ is the advanced node} \end{cases} \tag{7}$$

The above Eq. (7) when extend to multilevel heterogeneous networks is given as in Eq. (8) [12]

$$P_i = \frac{P_{op}N(1+a)E_i(r)}{(N + \sum_{i=1}^{N} a_i)\bar{E}(r)} \tag{8}$$

DDEEC uses the residual energy for selection of cluster head. The difference between DEEC and DDEEC is in the probability equation defined in Eq. (9) which uses normal and advanced nodes [12].

$$P_i = \begin{cases} \dfrac{P_{op}E_{avi}(r)}{(1+aq)\bar{E}(r)} & \text{normal node} \\[4mm] \dfrac{P_{op}(1+a)E_{avi}(r)}{(1+aq)\bar{E}(r)} & \text{advanced node} \\[4mm] \dfrac{C \times P_{op}(1+a)E_{avi}(r)}{(1+aq)\bar{E}(r)} & \text{normal and advanced node} \end{cases} \tag{9}$$

EDEEC uses normal, adavanced and super nodes. Advanced are nodes having $'a'$ amount of high energy levels than normal nodes. Super nodes are having $'b'$ amount of higher energy levels. So the equation for P_i in EDEEC is given in Eq. (10) as:

$$P_i = \begin{cases} \dfrac{P_{op}E_{avi}(r)}{(1+q(a+rb)\bar{E}(r)} & \text{normal node} \\[4mm] \dfrac{P_{op}(1+a)E_{avi}(r)}{(1+q(a+rb)\bar{E}(r)} & \text{advanced node} \\[4mm] \dfrac{P_{op}(1+b)E_{avi}(r)}{(1+q(a+rb)\bar{E}(r)} & \text{Super node} \end{cases} \tag{10}$$

5 Simulations and Discussion

The stability period and lifetime of a network are the most important parameters for every sensor network. These parameters are investigated by considering number of dead and alive nodes as well as packets transmitted to the base station. Simulation parameters are given in Table 1. Figure 3 shows that all nodes are vanished at 4393 iterations in DEEC and at 3999 iterations in DDEEC, whereas in EDEEC the nodes

Table 1 Parameters for simulation

Parameters for simulation	Value
Operating range (N)	200×200
Sensor nodes (S_{ni})	200–500
Cluster heads (CH)	5–50
Initial energy of SNs	0.5 J
Total rounds	10,000
Range of communication	250 m
E_{elec}	40 nJ/bit
Crossover distance	90 m

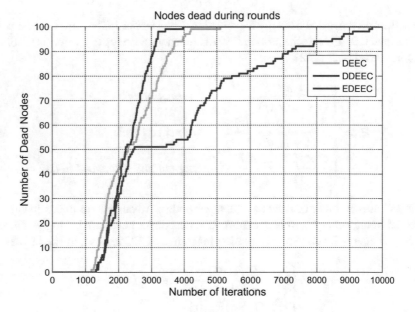

Fig. 3 Total number of dead nodes for 10,000 rounds

are alive even after more than 10,000 iterations. This shows packets transmitted to the BS are less in DEEC than EDEEC. Packets delivered to the BS are 32,00,000 for EDEEC, 11,00,000 for DDEEC and 84,275 for DEEC, respectively. This shows additional number of packets are transmitted to base station in EDEEC. The number of packets transmitted to BS is shown in Fig. 4.

For better performance of WSNs, the dead nodes should be as low as possible and this is attained by EDEEC. From Figs. 4 and 3 first nodes is dead at 1878, 1659, 1472, 1196 rounds for EDEEC, DDEEC and DEEC respectively. The thirteenth node is dead at 2555, 1799, 1661, 1289 for EDEEC, DDEEC and DEEC, respectively. All

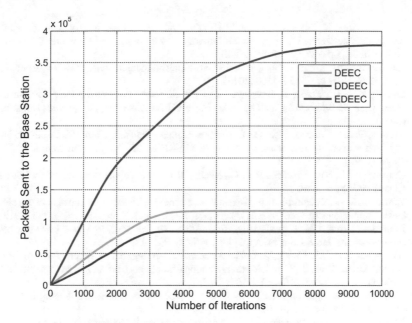

Fig. 4 Packets sent to the base station

nodes died at 5647, 3890 rounds. Packets transmitted to the base station in EDEEC, DDEEC, DEEC 236380, 79368 and 106514, respectively.

6 Conclusions

The stability period and the lifetime of the network are investigated by considering number of dead nodes, alive nodes and packets transmitted to the BS through MATLAB. Simulation is carried out for 200 nodes with 10,000 iterations. The performance is based on the energy depletion of SNs and death of nodes. Interference and node collisions are neglected in the simulation. The results show that nodes are dead faster in DEEC than in EDEEC by 53.8% for 10,000 iterations. Thus the stability period and the lifetime of the network increased notably.

References

1. Heinzelman, W.B., Chandrakasan, A.P., Balakrishnan, H.: An application-specific protocol architecture for wireless microsensor networks. IEEE Trans. Wirel. Commun. **1**(4), 660–670 (2002)
2. Xiangning, F., Yulin, S.: Improvement on LEACH protocol of wireless sensor network. In: International Conference on Sensor Technologies and Applications, 2007. SensorComm, pp. 260–264. IEEE (2007)

3. Ali, M.S., Dey, T., Biswas, R.: ALEACH: advanced LEACH routing protocol for wireless microsensor networks. In: International Conference on Electrical and Computer Engineering, 2008. ICECE, pp. 909–914. IEEE (2008)

4. Lin, H., Wang, L., Kong, R.: Energy efficient clustering protocol for large-scale sensor networks. IEEE Sens. J. **15**(12), 7150–7160 (2015)

5. Smaragdakis, G., Matta, I., Bestavros, A.: SEP: A Stable Election Protocol for Clustered Heterogeneous Wireless Sensor Networks. Boston University Computer Science Department (2004)

6. Krishna, Prasad, Vaidya, Nitin H., Chatterjee, Mainak, Pradhan, Dhiraj K.: A cluster-based approach for routing in dynamic networks. ACM SIGCOMM Comput. Commun. Rev. **27**(2), 49–64 (1997)

7. Kumari, Ch.U., Sasi Bhushana Rao, G., Madhu, R.: Erlang capacity evaluation in GSM and CDMA cellular systems

8. Taruna, S., Kohli, S., Purohit, G.N.: Distance based energy efficient selection of nodes to cluster head in homogeneous wireless sensor networks. Int. J. Wirel. Mob. Netw. **4**(4), 243 (2012)

9. Kour, H., Sharma, A.K.: Hybrid energy efficient distributed protocol for heterogeneous wireless sensor network. Int. J. Comput. Appl. **4**(6), 1–5 (2010)

10. Jung, S.M., Han, Y.J., Chung, T.M.: The concentric clustering scheme for efficient energy consumption in the PEGASIS. In: The 9th International Conference on Advanced Communication Technology, vol. 1, pp. 260–265. IEEE (2007)

11. Manjeshwar, A., Agrawal D.P.: TEEN: a routing protocol for enhanced efficiency in wireless sensor networks. In: null, p. 30189a. IEEE (2001)

12. Kumari, Ch.U., Ramya Krishna, M.: High performance wireless communication channel using LEACH protocols. Pak. J. Biotechnol. **13**, 52–56

13. Nguyen, L.T., Defago, X., Beuran, R., Shinoda, Y.: An energy efficient routing scheme for mobile wireless sensor networks. In: IEEE International Symposium on Wireless Communication Systems 2008 (ISWCS'08), pp. 568–572. IEEE (2008)

14. Younis, O., Fahmy, S.: HEED: a hybrid, energy-efficient, distributed clustering approach for ad hoc sensor networks. IEEE Trans. Mob. Comput. **3**(4), 366–379 (2004)

15. Heinzelman, W.R., Chandrakasan, A., Balakrishnan, H.: Energy-efficient communication protocol for wireless microsensor networks. In: Proceedings of the 33rd Annual Hawaii International Conference on System Sciences, p. 10. IEEE (2000)

16. Saini, P., Sharma, A.K.: E-DEEC-enhanced distributed energy efficient clustering scheme for heterogeneous WSN. In: 2010 1st International Conference on Parallel Distributed and Grid Computing (PDGC), pp. 205–210. IEEE (2010)

LSTM Based Paraphrase Identification Using Combined Word Embedding Features

D. Aravinda Reddy, M. Anand Kumar and K. P. Soman

Abstract Paraphrase identification is the process of analyzing two text entities (sentences) and determining whether the two entities represent the similar sense or not. This is a task of Natural Language Processing (NLP) in which we need to identify the sentences whether it is a paraphrase or not. Here, the chosen approach for this task is a deep Learning model that is Recurrent Neural Network-LSTM with word embedding features. Word embedding is an approach, from where we can extract the semantics of the word in dense vector representation. The word embedding models that are used for the feature extraction in Telugu are Word2Vec, Glove and Fasttext. These extracted feature models are added in the embedding layer of Long Short-Term Memory algorithm in order to classify the Telugu sentence pairs whether they are Paraphrase or not. The corpus for Telugu is generated manually from various Telugu newspapers. The sentences for word embedding model is also gathered from Telugu newspapers. This is the first attempt for paraphrase identification in Telugu using deep learning approach.

Keywords Paraphrase identification · Deep learning · RNN-LSTM
Word embedding model—Word2Vec · Glove · Fast-text · Corpus

1 Introduction

Paraphrase are the two text entities that are build with different syntactical words but convey the similar meaning or sense. Semantic uniformity is the main focused part in paraphrases rather than its syntax. The process of rewriting a text entity by preserving

D. Aravinda Reddy (✉) · M. Anand Kumar · K. P. Soman
Amrita School of Engineering, Center for Computational Engineering and Networking (CEN),
Coimbatore Amrita Vishwa Vidyapeetham, Coimbatore, TN, India
e-mail: cb.en.p2cen16004@cb.students.amrita.edu; d.aravind367@gmail.com

M. Anand Kumar
e-mail: m_anandkumar@cb.amrita.edu

K. P. Soman
e-mail: kp_soman@amrita.edu

© Springer Nature Singapore Pte Ltd. 2019   385
J. Wang et al. (eds.), *Soft Computing and Signal Processing*,
Advances in Intelligent Systems and Computing 898,
https://doi.org/10.1007/978-981-13-3393-4_40

the information of semantics in the different structure is known as paraphrasing. In the field of natural language processing, there exists many Paraphrasing application.

Non-paraphrases are the text entities which are not carrying semantic uniformity or equivalence between the sentences. A small amount of similarity exists in terms of syntax in the sentences, but they do not convey similar context. The structure of the sentences contains a mere portion of equivalence but the context differs.

Paraphrase identification is used in many fields. One of the main areas in which wide usage of this is plagiarism detection. Plagiarism, by the definition of Wikipedia, is the illegal appropriation and pirating the language, concept, ideology of another author and presenting them as one's own work. In the research community, the automatic detection of text re-usage plays the prominent role. But a major part of the work is focused only on the monolingual comparison (mostly English to English) whereas the largely unexplored part is a multilingual domain.

In a similar way, paraphrase identification is also used for extraction and retrieval of information, natural language generation, summarization, question answering and machine translation (MT). The information retrieving is major scope for data security and bio-medical field. Paraphrase identification is a crucial aspect which is almost emerging in such different fields.

The remaining section of the paper is systematically arranged as follows. Section 2 mentions some of the related works followed by a brief description of the paraphrase corpora for Telugu in Sect. 3. Section 4 explains about the methodology for paraphrase identification. The results of the experiment are given in Sect. 5. Section 6 has drawn the conclusion.

2 Related Works

This section represents the works in paraphrase identification so far. Natural Language Processing (NLP) is a field of artificial intelligence which deals with the human language. One of the complex task in NLP is Paraphrase identification. Several kinds of research have been under-went in this area and still many works are going on. In this area of research, the main focus is on the enhancing the existing problem. Some of the works which are related to paraphrase identification presented on different paraphrase data-set with different techniques in this section.

In this paper, [1] Bill Dolan et al. proposed "Support vector machine for paraphrase identification and corpus construction". In this proposed approach, the extraction of the clustered news article with annotated seed corpora to build the parallel corpus. This Large parallel corpus is induced into the support vector machine classifier for classification of data with the help of morphological and synonymy features. In [2] finch et al. proposes the semantic equivalence at the sentence level. In [3] Cordeiro et al. proposes the new metric of detecting paraphrases in unsupervised manner. In [4] Fernando et al. proposes the similar sense in between word meanings is also very helpful for the paraphrase identification. In [5] Socher et al. proposed Dynamic pooling and unfolding recursive auto-encoders for paraphrase detection, this work as an extension of recursive autoencoder (RAE). Recursive autoencoder is the recursive neural network, which learns the representation of words along with non-terminals in the parse tree recursively. In [6] Abraham et al. took the comparison study of statistical based and similarity based identification of paraphrases.

In [7] He et al. proposed "Multi-perspective sentence similarity modelling with convolutional neural network". They are mainly focused on an intrinsic feature. Sentence embedding is obtained by the convolutional neural network (CNN). In this, convolution is used for the whole version of embedding and every direction of embedding. In [8] Chitra et al. proposes fuzzy hierarchical clusters are processed with svm clusters to detect paraphrases using paraphrase recognizer.

In [9] Praveena et al. proposed an approach of chunking based Malayalam paraphrase identification using unfolding recursive auto-encoders. This method deals with unsupervised feature learning using recursive auto-encoders. In this approach, the extra features used are Glove, Word2vec and FastText word embedding with different dimensional vectors.

In [10] Mahalakshmi et al. proposed an approach for Paraphrase Detection for Tamil language using Deep learning algorithms. The real problem in paraphrase detection is that the semantics of the sentence has to be preserved while restating the sentence. Here, they have used unfolding recursive auto-encoders (RAEs) for leaning feature vectors for phrases in syntactic trees in an unsupervised way. In [11] Anand kumar et al. explains about the shared task on detecting paraphrase in Indian languages.

In [12] Aravinda Reddy et al. proposed an approach of "Paraphrase Identification in Telugu using Machine learning". In this method the chosen approach is machine learning with word share and tf-idf share features from the train data and predicted the test data with help of these features. For the classification in this model, various classifiers such as random forest classifier, support vector machines are used.

3 Telugu Paraphrase Corpora

3.1 Paraphrase Corpora for Telugu

Most of the Telugu data which is taken for the corpus is collected from the various newspaper manually. The entire corpus comprises of 4100 pairs of sentences in the combination of paraphrase and non-paraphrases as the same in [12]. In the phase of training we need good amount of data hence, the data is splitted into 3500 pairs of

Table 1 Paraphrase corpora for Telugu

	Paraphrase	Non-paraphrase	Total
Training dataset (Pairs)	1750	1750	3500
Test dataset (Pairs)	300	300	600

training sentences and 600 pairs of test sentences. In either of the training/testing data, the combination of both paraphrase and non-paraphrase exists. The Table 1 states about the paraphrases corpora for Telugu sentence pairs in the corpora.

4 Methodology

In this section, we discuss about the methodology that has been used to identify paraphrases in this task. The methodology which we followed is as given in Fig. 1.

4.1 Feature Extraction

4.1.1 Word Embedding for Telugu

Word embedding model is a modern text representation approach which is used in natural language processing. Word embeddings are the feature vectors for the words used in the neural network models in order to achieve state-of-art system in natural language processing. The word embedding model is a method in which the words in the corpus is mapped to low dimensional dense vector representation. The model will

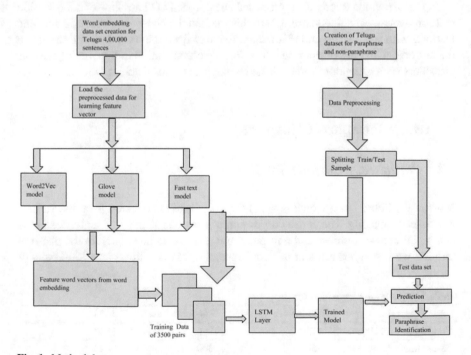

Fig. 1 Methodology

able to capture the semantic information of words. It is a major improvement from the bag of words model, where the model only represents word count and frequencies in a large sparse vectors distribution. In word embedding model, we will train the words from the large corpus and each word is a dense vector and a single point in word embedding space and the nearby words represent the same context. Mapping of the word to lower dimensional space.

$$\textbf{Word Embedding} = Word \longmapsto R^n \tag{1}$$

where

R^n N-dimensional space.

In word embedding model the main algorithm that are considered for this model are.

4.1.2 Word2Vec Model

In word embedding, this is one of the well known algorithm for learning words from the corpus. Word2vec model is a shallow neural network model in which words from the corpus are inputs to the model and get the output as the feature vector of the words in low dimensional space. Word2vec is a shallow learning model, but it converts text into a numerical format i.e., vectors which can understand by deep learning models. These feature vectors of each word represents a single point in embedding dimensional space. Grouping the similar word vectors together in vector space is the usefulness of word2vec. Word2vec model trains the words based on the other context words in the corpus. The model mainly considers any one of these two ways, either using context to predict a word (C-BOW model) or a word to predict the context (Skip-Gram Model). We have chosen a word to predict the context which is a skip-gram model to create dense vectors in our model.

$$Word2Vec\ model = \frac{1}{N} \sum_{n=1}^{N} \sum_{-i \le j \le i} \log P(W_{i+j}|W_n) \tag{2}$$

where

$P(W_{i+j}|W_n)$ Probability of finding the context words between the window $i + j$ when the word n is given in the skip gram model.

4.1.3 Glove Model

The glove is also known as Global vector word representation. In the word2vec model, the information about word co-occurrences isn't exploited fully. Whereas global vector model which is having benefits of both word2vec skip gram model for dense vector representation and matrix factorization for word co-occurrences. One of the matrix factorization method which is latent semantic analysis is used in order to capture statistical information of text corpus. This is achieved by decomposing large metrics to low-rank approximation. It is unsupervised way of learning word representation. The ratio of co-occurrence probabilities for various words can be examined with the word relationships.

$$P_{KW} = P(K|W) = \frac{X_{KW}}{X_K} \tag{3}$$

where

$P(K|W)$ Probability of Kth word appearing in the context of word (W).
$X_{KW} = X$ is the co-occurrence matrix in which number of times word k occurring in the word w context.

4.1.4 Fast-Text

The Facebook research team is developed word embedding model for efficient learning of word representations which is known as FastText in reference to jason brownlee blog. In the recent natural language processing tasks, this library has gained a lot of traction. FastText assumes that a word is created by n-gram characters and also creates the vector representation even for rare words. For deep learning task, we can not use words as such to the model. One way is to turn it into some representation to capture some essence of the word. The fast text also uses the CBOW and Skip-gram model in order to represent word vectors. The fast text is very fast and sentence vectors can be easily computed.

4.2 Recurrent Networks

Recurrent Neural Networks are the networks which address a special issue in neural architecture by persisting the information within them. The ideology of recurrent networks is that to connect previous information with the present state. In the theory of RNN, it is capable of long-term dependencies but when it comes to practice these networks cant be able to learn them. This is where we need to overcome the long-term dependency problem, for which the LSTMs works better on these issues. Long short-term memory (LSTM) networks are the special kind of neural networks which are designed explicitly to avoid dependency problem. LSTMs is having the four interaction layers instead of one from recurrent networks. LSTMs are having cell state, which is a main feature for them. Cell state resembles the kind of conveyor belt because it is straight down in the chain system. LSTMs is having the ability to add or remove the information in cell state through a carefully regulated structure called gates.

LSTMs are having mainly 3 gates in order to protect and control cell state information. The first step in LSTM network is to get the decision on what information, we need to leave away from the cell state. This operation is based on the sigmoid layer. This is also known as Forget gate. In this state, it takes the values of previous state and the value of present state input in order to represent whether to keep this information or not. The output of this layer is between 0 and 1 since it is the sigmoid layer. 0 represent completely to avoid this information and '1' represent completely to keep the information. The mathematical representation of the forget state is

$$\textbf{Forget state } (F_t) = \sigma(W_f * [h_{t-1}, x_t] + b_f) \tag{4}$$

where

f_t Forget state of the cell.
W_f Weights of the forget state.
h_{t-1} Information from the Previous state.
x_t Input of Present state.
b_f Bias of the forget gate.

The second step in the LSTMs layer is all about what new information is to store in the cell state. This layer is also known as "Input Gate". which values need to be updated based on sigmoid function. Tanh layer is which creates a vector of new candidate values C_t which is added to the state.

$$\textbf{Input state } (I_{st}) = \sigma(W_{In} * [h_{t-1}, x_t] + b_{In}). \tag{5}$$

$$(C_t) = \tan h(W_c * [h_{t-1}, x_t] + b_c). \tag{6}$$

where

I_{st} Input state of the cell.
W_{In} Weights of the Input state.
b_{In} Bias of the Input gate
C_t present cell state.
W_c Weights of the cell state.
b_c Bias of the cell gate.

Now we need to update old cell state C_{t-1} information with the new cell state C_t information. This updating activity is followed by the two steps. In first step, the previous cell state is needed to be forget-ed by multiplying with the forget cell. In second step, the information of new candidate value is to added in the cell state.

$$C_t = f_t * C_{t-1} + I_t * C_t. \tag{7}$$

The last and final step is to form the output of the layer. This is also known as the "Output Gate". In this sigmoid layer is the decider for which part of the state should go out. The cell state from Tanh layer is to normalize the vector values between -1 to 1 by multiplying with the output of sigmoid gate.

$$\textbf{Output state } (O_{st}) = \sigma(W_{Ot} \cdot [h_{t-1}, x_t] + b_{Ot}). \qquad (8)$$

$$\textbf{Output } (h_{st}) = O_{st} * \tan h(C_t) \qquad (9)$$

where

O_{st} Output state of the cell.
W_{Ot} Weights of the Output state.
b_{Ot} Bias of the Output gate.

As it is shown in the methodology, the embedding model is a pre-trained vector model with the different word embeddings such as word2vec, glove and fasttext. These models are incorporated with in the training phase through embedding layer. The training data with embedding layer is passed to the LSTM cells for sequence to sequence learning. This final model is used for the prediction of the test data.

5 Results and Discussion

The following section refers to the results of Telugu sentence pairs to which long short term memory is predicting. In this module the results which are implemented from the methodology is exhibited. In order to validate the training accuracy we separated 0.1% of the data in the training dataset and the same is used for the validation data whether to analyze the training model. The Table 2 states the results of the LSTMs layer with the different word embedding model.

Since long short term memory holds the long term dependencies the results of test data attains 73.10%. In my previous work which is "Paraphrase Identification in

Table 2 Results of LSTMs with various word-embedding

	Word embedding dimension	Training model		Test results
		Training accuracy	Validation accuracy	
LSTM-Word2Vec	100	72.14	70.03	71.34
	200	70.91	68.42	69.59
	300	69.86	66.67	67.21
LSTM-Glove	100	71.42	69.86	71.49
	200	70.81	68.59	69.34
	300	68.21	65.31	67.62
LSTM-Fasttext	100	74.21	70.03	73.10
	200	70.89	68.12	68.93
	300	71.02	67.82	68.71

Table 3 Results of LSTM with combined feature

	Word embedding dimension	Training accuracy	Validation accuracy	Test accuracy
LSTM-combined features	100	76.02	69.60	74.12
	200	71.39	63.26	65.46
	300	68.48	61.17	64.35

Telugu using Machine Learning", is a model in which we considered only the machine learning techniques with word share and tfidf share features given an accuracy of 0.71 in the testing phase. Since in this model there is only single validation set where as in the machine learning model we used tenfold cross validation in the training phase even though the test accuracy is better than the machine learning model.

The word embedding features that are applied in the above results are taken separately for each model. In this model, we applied the word embeddings in combined format. All the three word embedding model such as word2vec, glove and fasttext is combined and taken as a single word embedding model. The Table 3 shows the results of the combined word embedding model with different embedding dimension.

Since the three word embedding models is combined into a single word embedding model and we already know that LSTMs works better for long term dependencies and attained better test accuracy results of 74.12% for the combined word embedding model with 100 dimensional vectors.

6 Conclusion and Future Work

In this proposed work, the main focus is to identify the paraphrases from the Telugu. Deep learning techniques such as Recurrent Neural Network (RNN) are the only model which are different from the neural network model since they capture the long term dependency. The scope of Long Short Term Memory lies in over coming the vanishing gradient problem and achieve good results. In this method also LSTMs is used with different word-embedding model such as word2vec, glove and fasttext. Out of these feature learning model LSTM-fastext is given better accuracy for the testing phase is 73.10%. Since fasttext is free light weight open sourced library, it will generate all the possible combinations of words and characters. As we also used the combination of all the three word embedding features. Out of these LSTM with 100 dimensional word embedding feature attains better accuracy in the testing phase is 74.12%. As a future work, there are some more deep learning techniques such as Recursive Auto-Encoder (RAEs) to which parser is needed and in some cases it is mentioned that Convolutional Neural Network (CNN) based LSTMs will also improve the results further.

References

1. Brockett, C., Dolan, W.B.: Support vector machines for paraphrase identification and corpus construction. In: Proceedings of the 3rd International Workshop on Paraphrasing (IWP2005), pp. 1–8 (2005)
2. Socher, R., Huang, E.H., Pennin, J., Manning, C.D., Ng, A.Y.: Dynamic pooling and unfolding recursive autoencoders for para-phrase detection. In: Advances in Neural Information Processing Systems, pp. 801–809 (2011)
3. He, H., Gimpel, K., Lin, J.: Multi-perspective sentence similarity modelling with CNN. In: International Conference on Emperical Methods in NLP, pp. 1576–1586 (2015)
4. Praveena. R., Anand Kumar, M., Soman, K.P.: Chunking based Malayalam paraphrase identification using unfolding recursive autoencoders, pp. 922–928. https://doi.org/10.1109/ICACCI.2017.8125959
5. Mahalaksmi, S., Anand Kumar, M., Soman, K.P.: Paraphrase detection for Tamil language using deep learning algorithms. Int. J. Appl. Eng. Res. **10**(17), 13929–13934 (2015)
6. Abraham, S.S., Idicula, S.M.: Comparison of statistical and semantic similarity techniques for paraphrase identification, pp. 209–213. IEEE (2012)
7. He, H., Gimpel, K., Lin, J.: Emperical Methods in NLP, pp. 1576–1586. (2015)
8. Chitra, A., Rajkumar, A.: Paraphrase extraction using fuzzy hierarchical clustering. Appl. Soft Comput. **34**, 426–437 (2015)
9. Fernando, S., Stevenson, M.: A semantic similarity approach to paraphrase detection. In: Proceedings of the 11th Annual Research Colloquium of the UK Special Interest Group for Computational Linguistics, pp. 45–52 (2008)
10. Mahalaksmi, S., Anand Kumar, M., Soman, K.P.: Paraphrase detection for Tamil language using deep learning algorithms. Int. J. Appl. Eng. Res. **10**(17), 13929–13934 (2015)
11. Chitra, A., Rajkumar, A.: Paraphrase extraction using fuzzy hierarchical clustering. Appl. Soft Comput. **34**, 426–437 (2015)
12. Aravinda Reddy, D., Anand Kumar, M., Soman, K.P.: Paraphrase identification in Telugu using machine learning. In: Advances in Intelligent Systems and Computing. Springer (2018)

A Distributed Support Vector Machine Using Apache Spark for Semi-supervised Classification with Data Augmentation

S. S. Blessy Trencia Lincy and Suresh Kumar Nagarajan

Abstract One of the popular and extensively used classification algorithms in the data mining and the machine learning technique is the support vector machine (SVM). Yet, conversely they have been traditionally applied to a small dataset or to an extent medium dataset. The current requirement and demand to scale up with the evolving size of the datasets have fascinated the research notice and attention such that new techniques and implementations can be carried out for the SVM, and as a result can scale well with large datasets and tasks. Recently, the distributed SVM is studied by the researchers, but the data augmentation with semi-supervised classification using the distributed SVM is not yet implemented. In this paper, a distributed implementation of support vector machine along with the data augmentation upon the SparkR, which is a recent and effective platform for performing distributed computation, is introduced and analyzed. This framework—A Distributed Support Vector Machine under Apache Spark for Semi-supervised Classification with Smart Data Augmentation—is implemented with a large-scale dataset with more than million data points. The results and analysis show that the proposed approach greatly enhances the predictive performance of the method in terms of execution time and faster processing.

Keywords Big data · SVM · Data augmentation · Apache Spark
Semi-supervised classification

1 Introduction

In this evolving era of big data, various machine learning algorithms and techniques have to be extended to meet the challenges raised by the big data. Algorithms applied for small and medium applications and datasets have to be remodeled or have to be re-

S. S. Blessy Trencia Lincy (✉) · S. K. Nagarajan
School of Computer Science and Engineering, Vellore Institute of Technology, Vellore, India
e-mail: blessylincy@gmail.com

S. K. Nagarajan
e-mail: sureshkumar.n@vit.ac.in

© Springer Nature Singapore Pte Ltd. 2019
J. Wang et al. (eds.), *Soft Computing and Signal Processing* ,
Advances in Intelligent Systems and Computing 898,
https://doi.org/10.1007/978-981-13-3393-4_41

formulated to cope with the evolving need. While dealing with data, beginning from pre-processing, transformation, classification, or regression analysis to the prediction and interpretation processes various measures have to be taken in terms of making use of the data appropriately for the corresponding applications or tasks [1–3]. Various traditional and evolutionary machine learning techniques exist for performing the classification. But the problem lies in the scalability of the algorithm to meet the requirement. Thus, scalability is one of the major problems while dealing with the big data. Hence, the traditional algorithms can be re-studied and formulated to work in a distributed environment coping up with the limitations caused by the scalability issues. The data can be augmented during training, and then, the classification algorithm can be applied to the augmented data. The results obtained after performing the required processing on the augmented data can be compared with the results obtained by applying the processing on the normal data without augmentation.

In this paper, the distributed support vector machines for semi-supervised classification with the data augmentation are proposed which utilizes the above-discussed framework and platforms [4–6]. Section 2 discusses the related work for the proposed system describing various distributed traditional algorithms and approaches. It also explains the issues with the existing system and limitations. Section 3 explains the proposed system with the data augmentation concepts with the framework of the processing involved. Section 5 discusses the experimental setup made here for the proposed work. Finally, in Sect. 6, the outcome of the proposed system is analyzed and interpreted to be used effectively.

2 Related Works

The literature here explains two important and particular concepts of data augmentation and semi-supervised classification. When training a machine learning algorithm for classification, the data can be augmented with artificially produced samples. It is doable to carry out generic augmentation in the feature space, provided if some plausible transformations for the data are recognized and further augmentation in data space endows with a larger benefit for improving the performance and in reducing overfitting.

Polson and Scott in their paper discussed the data augmentation for the binary classification problem [7]. Their aim was to improve the complexity involving the computation of the SVMs. Touloupou et al. [8] discussed the data augmentation done using the Bayesian inference calculating the posterior probability using the sampling techniques. Consoli et al. [9] described the method of automatically tuning the SVM parameters for processing. This is based on a heuristic iterative approach and can be referred to as a modification to the hill climbing technique [10]. This approach was tested with several heterogeneous datasets for performance evaluation. Cheng et al. [11] discussed the data augmentation with MCMC inference, and the computation burden is addressed by the parallel computing [12]. The speeding up of the regression process with the Bayesian regression was the main objective of the approach described in this paper. Velasco et al. [13] described performing the data augmentation with the diabetes mellitus dataset to produce enough data and make the model robust. This handles and problem of the lacking of significant data required for processing. Based on the literature done, there is no approach for the distributed implementation of the support vector machine for the classification with the data augmentation. With the big data which involves variety of data, several challenges arise to deal with the classification. Thus, the data augmentation techniques studied in the literature can be employed for dealing with the semi-supervised classification problems.

3 Proposed Methodology

In the proposed system, the support vector machines are re-devised or re-formulated under a probabilistic background. This is done making the original optimization issue to be produced as a maximum a posteriori (MAP) estimation. With the semi-supervised classification problem, the data augmentation using the MCMC techniques is done and its effectiveness can be proved in comparison with other approaches in terms of training the SVM. To be precise, Polson et al., the data augmentation techniques were employed in the binary classification problem with the aim of improving the computation complexity of the SVMs. This can be extended with the distributed support vector machine upon the Spark framework.

3.1 Data Augmentation—Iterative Bayesian Inference

Technically, the Bayesian inference involves the computation of the posterior probability and the iterative Bayesian inference involves the computation of the posterior probability, and considering the same as the prior probability, compute the posterior probability again. In the proposed approach, the marginal likelihood of the data is computed by using the iterative Bayesian inference technique. The logic behind iterative Bayesian inference is that after the computation of the posterior probability again the result estimate can be considered as the prior probability and the posterior probability are computed again. The number of iterations can be set according to the purpose and the dataset. The idea of iterative Bayesian inference is used in this work for computing the probability estimates, and the result analysis shows the effectiveness of the strategy.

3.2 Semi-supervised Classification and Data Augmentation

The semi-supervised classification involves both the labeled data and unlabeled data for processing. The following part discusses how the data augmentation can be applied to the available data for processing. Provided with the labeled data for processing, represented as LD, and unlabeled data, represented as ULD; need to create an expanded set LD+ by extracting hidden relations with the ULD to LD. For this purpose, there is a need to sort the elements of ULD depending upon their likelihood with the LD. The iterative Bayesian inference approach can be used for this scenario. The input data which contains both labeled and unlabeled data is split up into training data (T) and the test data (TS). TR represents the sampling of the training data, and this can be used to identify the correlation for the unlabeled data. The data augmentation technique is applied in this phase, and the augmented data is obtained which is represented as (TR+). The testing data contains the data which is not processed by using the data augmentation procedure. Thus, the comparison can be made between the data with data augmentation and the data without data augmentation. This is the procedure involved in the evaluation of the data augmentation framework.

3.3 SVM Model

SVM aims to find an optimal hyperplane that maximizes the margin between different labeled sets of data samples.

More formally, let

$$D = \{f(x_n, y_n)\}_{n=1}^{N}$$

denote the dataset wherein
$x_n \in R^D$ is the D-dimensional vector of the data sample and
$y_n \in f\{1, -1\}$ is the data label.

The learning of SVM with l_α-norm regularization is to minimize the objective function as follows:

$$L(w; C, \alpha, \sigma) = \sum_{n=1}^{N} \max\{1 - y_n w^T x_n, 0\} - \frac{1}{C^\alpha} \sum_{d=1}^{D} \left| \frac{w_d}{\sigma_d} \right|^\alpha \qquad (1)$$

where,

w   is the weight of coefficient parameters
σ_d is the standard deviation of the dth feature of x
$C > 0$ is the penalty of the hyperparameter that can be tuned for the best performance using cross-validation.

The second term is a regularization penalty corresponding to a prior distribution $P(w_d \mid C, \alpha, \sigma_d)$.

A pseudo-likelihood of the label y has been introduced to represent SVMs as latent variable models, so that Bayesian inference techniques can be employed to perform parameter estimation.

The pseudo-likelihood is:

$$p(y|x, w) = \exp\{-2\max(1 - yw^Tx, 0)\}$$

Minimizing the loss function in Eq. (1) now turns into estimating the maximum a posterior (MAP) of the following pseudo-posterior distribution:

$$\hat{w}_{MAP} \alpha \, argmax \, \exp[-L(w; C, \alpha, \sigma)]$$
$$\alpha \, argmax \, Z_\alpha(C, \alpha) p(y|x, wp(w|C, \alpha, \sigma)$$

$Z_\alpha(C, \alpha)$ is a pseudo-posterior normalization constant.

For the purpose of model simplicity, we consider a special case with Gaussian prior $(\alpha - 2)$ and a fixed penalty constant $C = 1$.

Data augmentation approach further introduces an auxiliary variable $\lambda > 0$ for each observation label y in such a way that p (y | x, w) becomes the marginal of the joint distribution p (y, λ| x, w).

More importantly, the auxiliary variable λ can be efficiently sampled from an inverse Gaussian (IG) distribution:

$$p(\lambda^{-1}|x, y, w) \sim IG(|1 - ywx^T|^{-1}, 1)$$

Assuming a Gaussian prior for

$$w : p(w) \sim N(\mu_0; \Sigma_0)$$

the data pseudo-likelihood and the posterior conditional distribution of w can be shown to have the following forms.

$$p(y|x, w, \lambda) = \int_0^\infty \frac{1}{\sqrt{2\pi\lambda}} \exp\left\{\frac{[\lambda + (1 - yw^Tx)]^2}{2\lambda}\right\} d\lambda \qquad (2)$$

$$p(w|x, y, \lambda) = N(\mu, \Sigma) \qquad (3)$$

where

$$\Sigma^{-1} = X^T diag(\lambda^{-1})X + \Sigma_0^{-1}$$
$$\mu = \Sigma[X^T(1 + \lambda^{-1}) + \Sigma_0^{-1}]$$

Here, the data matrix is denoted by $X = [x_1, x_2, \ldots, x_N]^T$.

Suppose that there are K classes, the label y_n now takes values on the set$\{1, 2, ..., K\}$.

We consider a set of parameters $\{w_1, w_2, ..., w_k\}$ wherein the parameter for the kth class is w_k. These parameters can be initially set to 1.

Then, the auxiliary variable λ n for the nth data point is independently sampled as follows:

$$\lambda_n^{-1} \sim IG(|1 - w_{y_n} x_n^T|^{-1}, 1) \tag{4}$$

Consider the case where the prior distribution for w_k is a normal distribution with zero mean ($\mu_0 = 0$) and unit variance ($\Sigma_0 = I$): p ($w_k | \mu_0, \Sigma_0$) ~ N (0; I).

We can then derive the MAP for the posterior distribution for each w_k as:

$$w_k = \mu_k = \Sigma_k X^T [I + diag(\lambda^{-1})]Z + \mu_0 \Sigma_0$$

where $Z \in \{-1, 1\}^{NxK}$ denotes the indicator matrix for the labels:

$$z_{nk} = 1 \text{ if } y_n = k,$$
$$\text{otherwise } z_{nk} = -1$$

Let:

$$S \in R^{DxD} = X^T diag(\lambda^{-1})X + I \tag{5}$$
$$T \in R^{DxK} = X^T [I + diag(\lambda^{-1})]Z \tag{6}$$

These equations represent the distributed SVM approach for multiclass classification which is explained in the following section.

4 Distributed SVM on Spark

The distributed SVM is the distributed implementation of the support vector machine upon the Spark framework. The algorithm explains the tasks being separated for worker nodes. The initial vector of coefficient parameters is set to one for all the nodes. The data augmentation is applied with the iterative Bayesian inference upon each partition [14–16]. This step is repeated according to the purpose of the user. Then, the classification of the data is done as specified in the steps 4, 5, and 6. The output is then received by the reducer nodes and sent to the driver program. The algorithm is represented as shown below.

Distributed SVM algorithm:
   Input : $y \in \{1, 2, ..., k\}^{Nx1}$, $X \in R^{NxD}$

1:   Initialize the vector of coefficient parameters for all worker nodes to be one.

$$\text{i.e. } w_k = 1, \forall k = 1, 2, \ldots, K$$

2: The driver then ships this $w_{1:K}$ to each and every worker.
3: The worker node then performs the succeeding steps for the mth partition:

$$\lambda_n^{-1} \sim IG(|1 - w_{y_n} x_n^T|^{-1}, 1) \forall n \in \text{mth partition}$$

$$p(\theta|E) = \frac{p(e|\theta)p(\theta)}{p(e)}$$

// Data augmentation applied in iterations which can be set by the user. Hence, this step can be repeated according to the purpose.

4: $S^{(m)} = X^{(m)^T} diag\left(\lambda^{(m)^{-1}}\right) X^{(m)}$

5: $Z^{(m)} = [Z_{nk}^{(m)}]_{N \times K} : Z_{nk}^{(m)} = 1 \Big| y_n = k, Z_{nk}^{(m)} = -1 \Big|$
   $y \neq k$

6: $T^{(m)} = X^{(m)^T}\left[I + diag\left(\lambda^{(m)^{-1}}\right)\right] Z^{(m)}$

7: The workers then enter the reduce phase which then send $S^{(m)}$ and $T^{(m)}$ back to the driver to obtain S and T.

8: Finally, the driver computes: $w_k = S\backslash T_k : \forall k = 1, 2, \ldots, K$

Output: $\{w_1, w_2, \ldots, w_K\}$

Where X—data matrix, Y—labels, w_k—vector of coefficient parameters, m—partitions, λ—auxiliary variable, IG—inverse Gaussian distribution, S, T—matrix to compute w, and Z—new indicator matrices for the label. The above algorithm is used to perform the classification using the distributed support vector machine with the semi-supervised data. The representation of the attributes and their usage is also described above.

Figure 1 describes the framework of the processing of the data augmentation. The data augmentation can be done by using the iterative Bayesian inference. This evolves the unlabeled data for which the label has to be determined based on the inference obtained from the labeled data. This implies the semi-supervised classification (i.e., the iterative Bayesian inference) technique will be applied to the unlabeled data to determine the label based on the posterior probability distribution obtained. Thus, the unlabeled data can be studied and moved as the training data. The output can be obtained for two different cases here. The first case is the distributed SVM without data augmentation, and the other one is the distributed SVM with the data augmentation. The evaluation and comparison of the two cases can be studied and analyzed for further processing.

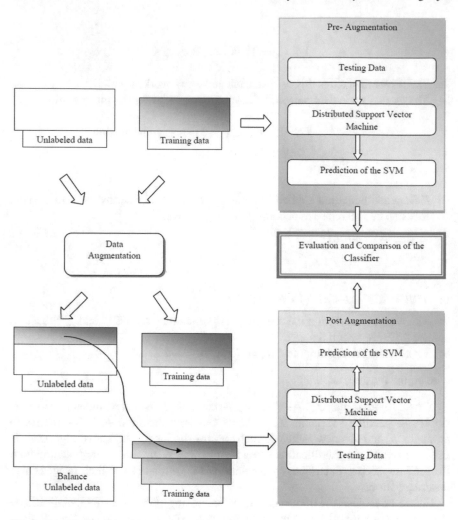

Fig. 1 Framework of the data augmentation process

5 Experimental Setup

For the experiment, the number of clusters we used is three with one master node. The software configuration details include: the open-source Apache Hadoop, HDFS replication factor set to 2, default block size of HDFS 128 MB, the version of Apache Spark, and MLlib is 1.2.0. The HDFS and the Spark master processes, i.e., the name node of the HDFS and the Spark master hosted in the main node. The task of the name node is to control the HDFS and to coordinate the slave machines for the tasks to be carried out according to their respective data node. The task of the Spark master is to control all the executors involving the worker node. The Spark framework makes use of the HDFS file system in order to load the data and to save the data in a similar way as the Hadoop framework.

6 Results and Analysis

The datasets taken for the evaluation of the proposed distributed framework include the ECBDL14, epsilon, and the Susy dataset of sizes 2.7 GB, 2.6 GB, and 2.3 GB, respectively. The details about the datasets are shown in the table below. Figure 2a, b shows the results obtained for the distributed support vector machine in comparison with the different algorithms like the Spark-LR_SGD, Spark-RF, Spark-LSVM, and Spark-DT upon the epsilon dataset.

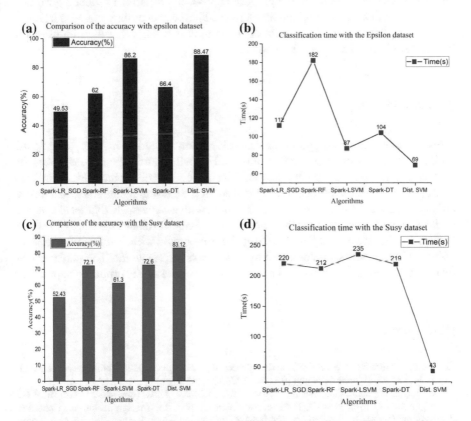

Fig. 2 **a** Classification accuracy of distributed SVM with epsilon dataset, **b** execution time of distributed SVM with epsilon data, **c** classification accuracy of distributed SVM with Susy dataset, and **d** execution time of distributed SVM with epsilon dataset

The performance measures are analyzed in terms of the accuracy and the time taken for the processing. The accuracy obtained is seen as 49.53%, 62%, 86.2%, 66.4%, and 88.47% for the algorithms Spark Linear Regression Stochastic Gradient Descent, Spark Random Forest, Spark Linear SVM, Spark Decision Tress, and the proposed distributed support vector machine, respectively. Figure 2c describes the results obtained for the distributed support vector machine in comparison with the different algorithms Spark-LR_SGD, Spark-RF, Spark-LSVM, and Spark-DT upon the Susy dataset. The performance measures are analyzed in terms of the accuracy and the time taken for the processing. The accuracy obtained is seen as 52.43%, 72.1%, 61.3%, 72.6%, and 83.12% for the algorithms Spark-LR_SGD, Spark-RF, Spark-LSVM, and Spark-DT, respectively. It is seen that the proposed approach comparatively performs well when compared with other approaches in both performance metrics. Similarly, the graph shown in Fig. 2d describes the time taken for the algorithms in comparison with the proposed approach which is represented in seconds. On observation, it is seen that the proposed approach outperforms other algorithms in terms of both accuracy and the time involved in the processing. It is also inferred that the proposed algorithm is more effective in comparing other algorithms when implemented upon the Spark framework.

7   Conclusion

A distributed version of support vector machine is introduced upon the Apache Spark with the smart data augmentation techniques. The main contribution of the paper is to provide an effective semi-supervised classification and the distributed algorithm. The concept of iterative Bayesian inference is used for the data augmentation to handle the unlabeled data to be classified. The experiment with a large-scale dataset shows the scalability of the approach and the increase with the performance measures. This can be seen through the evaluation made from the comparison with the other baseline and traditional approaches implemented upon the Spark framework. The performance metrics are analyzed in terms of the accuracy and speed of the proposed approach.

References

1. Dean, J., Ghemawat, S.: Simplified data processing on large clusters. Commun. ACM **51**(1), 107–113 (2008). https://doi.org/10.1145/1327452.1327492
2. Chang, F., Dean, J., Ghemawat, S., Hsieh, W.C., Wallach, D.A., Burrows, M., Gruber, R.E.: Bigtable: a distributed storage system for structured data. In: 7th Symposium on Operating Systems Design and Implementation (OSDI '06), pp. 205–218, Seattle, WA, USA, 6–8 Nov 2006. https://doi.org/10.1145/1365815.1365816
3. Le Guennec, A., Malinowski, S., Tavenard, R.: Data Augmentation for time series classification using convolutional neural networks. In: ECML/PKDD Workshop on Advanced Analytics and Learning on Temporal Data (2016)

4. Gelman, A.: Parameterization and Bayesian modeling. J. Am. Stat. Assoc. **99**(466), 537–545 (2004). https://doi.org/10.1198/016214504000000458
5. Tanner, M.A.: Data Augmentation, pp. 105–126 (2004). https://doi.org/10.1002/0471667196.ess0283
6. Van Dyk, D.A., Meng, X.L.: The art of data augmentation. J. Comput. Graph. Stat. **10**(1), 1–50 (2001)
7. Polson, N.G., Scott, S.L.: Data augmentation for support vector machines. Bayesian Anal. **6**(1), 1–24 (2011). https://doi.org/10.1214/11-BA601
8. Touloupou, P., Alzahrani, N., Neal, P., Spencer, S.E., McKinley, T.J.: Efficient model comparison techniques for models requiring large scale data augmentation. Bayesian Anal. (2017)
9. Consoli, S., Kustra, J., Vos, P., Hendriks, M., Mavroeidis, D.: Towards an automated method based on Iterated Local Search optimization for tuning the parameters of Support Vector Machines, pp. 1–3 (2017)
10. Triguero, I., Peralta, D., Bacardit, J., Garca, S., Herrera, F.: MRPR: a MapReduce solution for prototype reduction in big data classification. Neuro Comput. **150**, 331–345 (2015)
11. Cheng, H., Fernando, R., Garrick, D.: Parallel computing to speed up whole-genome bayesian regression analyses using orthogonal data augmentation. bioRxiv, 148965 (2017)
12. Nandimath, J., Banerjee, E., Patil, A., Kakade, P., Vaidya, S., Chaturvedi, D.: Big data analysis using Apache Hadoop. In: 2013 IEEE 14th International Conference on Information Reuse and Integration (IRI), pp. 700–703. IEEE, Aug 2013
13. Velasco, J.M., Garnica, O., Contador, S., Lanchares, J., Maqueda, E., Botella, M., Hidalgo, J.I.: Data augmentation and evolutionary algorithms to improve the prediction of blood glucose levels in scarcity of training data. In: Evolutionary Computation (CEC), 2017 IEEE Congress on, pp. 2193–2200. IEEE, June 2017
14. Piza, D.L., Schulze-Bonhage, A., Stieglitz, T., Jacobs, J., Dümpelmann, M.: Depuration, augmentation and balancing of training data for supervised learning based detectors of EEG patterns. In: 2017 8th International IEEE/EMBS Conference on Neural Engineering (NER), pp. 497–500. IEEE, May 2017
15. Zhong, Z., Zheng, L., Kang, G., Li, S., Yang, Y.: Random Erasing Data Augmentation (2017). arXiv:1708.04896
16. van Doorn, J., Ly, A., Marsman, M., Wagenmakers, E.J.: Bayesian Estimation of Kendall's tau Using a Latent Normal Approach (2017). arXiv:1703.01805

An Extensive Review on Various Fundus Databases Use for Development of Computer-Aided Diabetic Retinopathy Screening Tool

Kalyan Acharjya, Girija Shankar Sahoo and Sudhir Kr. Sharma

Abstract Present days, the development of computer-aided diagnosis of diabetic retinopathy (DR) is an active research in the field of medical image processing. The true quality performance of medical image understanding classifier depends on tested fundus database by some means. Therefore, a researcher must have the knowledge of all types of database available to facilitate the development of DR screening tool and this paper is intended to provide an idea of various fundus databases. Trends are seen that a number of researchers use different databases to evaluate their code, in that way the purpose to attain universally accepted diabetic retinopathy screening tool will not be achieved at earliest. However, there are no unique databases available to evaluate the program for all types of DR symptoms on fundus image. In this paper, present the detail comparative discussion on various publicly available databases of fundus images, which established to facilitate to do research work for the development of computer-aided diabetic retinopathy screening tool (CAD-DRST). The selection of test database or multiple databases depends on the researcher and the quality of the database; along with test results signify the acceptability level of the proposed algorithms. Besides that, through this paper, we urged to all scientific communities to establish a common database for the finest development of CAD-DRST.

Keywords Diabetic retinopathy · Fundus database · Retinopathy screening tool
DR symptoms

K. Acharjya (✉) · G. S. Sahoo · S. Kr. Sharma
Jaipur National University, Jaipur 302017, India
e-mail: kalyan.acharjya@gmail.com

G. S. Sahoo
e-mail: gssahoo07@gmail.com

S. Kr. Sharma
e-mail: sudhir.732000@gmail.com

© Springer Nature Singapore Pte Ltd. 2019   407
J. Wang et al. (eds.), *Soft Computing and Signal Processing*,
Advances in Intelligent Systems and Computing 898,
https://doi.org/10.1007/978-981-13-3393-4_42

1 Introduction

Diabetic is a chronic disease due to uncontrolled insulin hormone in the human body; besides other major consequences, diabetic retinopathy (DR) [1] is one among most severe disease. For diabetic patients, it is certain that doctors recommend for frequently go to eye checkup and advise to consider serious note for the abnormalities appear (lesion) on the retina due to diabetic retinopathy. Medical sciences classified it in two categories—initial stage of DR is called non-proliferative diabetic retinopathy (NPDR) and a more advanced stage is proliferative diabetic retinopathy (PDR). Regular eye examination to detect it at an early stage and take preventive action is the only way to avoid partial or entire vision loss from such severe symptoms. When the patient suffers from diabetic retinopathy, then various abnormalities appearing on the retina and doctors are observed those lesions and grade the risk factor of disease on retina based on the total lesion accounted, a healthy and pathological retinal image (fundus) is shown in Fig. 1.

When retina suffers from diabetic retinopathy, there are various symptoms appear on the retina, likewise exudates, microaneurysms and dot hemorrhages. Based on the percentage of severity appearance, doctors suggest various treatments to avoid loss of vision power or in the more severe case complete vision loss. Early detection and take preventive measure is the only way to prevent it from mid-level severity, in most severe cases (last stage), patients have to go for laser therapy (photocoagulation) to protect the vision; however, treatment does not have the guarantee to save the vision in most severe cases. In the time being of the twenty-first century, health emergency due to diabetics ignited for persuading active research to develop an automatic computer-aided diagnosis (CAD) tool [2] for self-monitoring and early detection of diabetic retinopathy without doctor's interference.

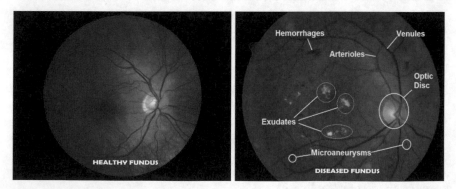

Fig. 1 Healthy and diseased fundus

2 Ongoing Research Work

As per World Health Organization data, there are approximately 10% people are suffering from diabetic around the world [3], and another agency reported 425 million are suffered from diabetics [4]. These statistic figures indicate the one common sign, which is twenty-first century will be health emergency due to diabetic and modern lifestyle is considered one major reason, the scenario already shows the alarming situation and numbers of patients are increasing day by day around the globe. Developed countries having sufficient numbers of ophthalmologists are managing the situation till yet, but if the numbers of patients increase with similar fashion, and then it will be unrestrained situation very soon, whereas in the developing countries, the situation is already worsening, because there are no sufficient eye specialists to regular monitoring of diabetic patients, besides that health infrastructure is very poor in rural and semi-urban areas. The ground rule for diabetic patients has to regular monitoring the retina to detect it at the early stage and take preventive actions to avoid the vision power. Developing countries are facing the situation very hard due to multiple causes; therefore, such alarming situation urges to find an alternative way to monitor the retina without interference by medical experts for screening purposes.

The situation is so divesting that scientific communities are looking for an alternate computer-based solution to monitoring fundus and grading its risk factor based on abnormalities manifestations. In medical image analysis field, researcher is considering affirmative to develop a universal acceptance computer-aided DR screening software tool for clinical practices, where doctor's presence is not needed for examining retina but severe or crucial cases monitoring by a doctor are always sought for. The aforesaid tool will immensely help to all communities across the globe, besides that such system will help to the unsupervised community or people from remote villages in context to get the eye care benefit at the local community center with negligible cost. Therefore, in image understanding research field, scientist are looking for recognition or identification of some features from fundus image based on predefined characteristics, in all cases program developer train and test their code and evaluate the performance by using various image databases, where selection of the database is very crucial role for truly recognize the performance and acceptability of the code.

3 Role of Databases to Develop Universal Acceptance Code

The role of a database which having quality images is very crucial for the development of high-performance retina screening tools. The basic steps involved in the development of CAD-based diabetic retinopathy tool [5] is shown in Fig. 2. The evaluation of classifier with one specific database may give the different results in other databases. The results and testing environment of any algorithm must have global acceptance prior to considering it for clinical practices; therefore, the clas-

Fig. 2 CAD-based
development of DR
screening tool

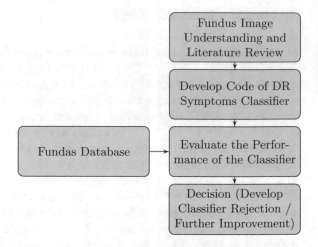

sifier must have tested with different and multiple databases, besides that it should show the satisfactory result. From this motivation, it would be better to have one large database of retinal fundus capture in different image acquisition background across different ethnic subjects having all levels of deceased. Nowadays it has been seen that few researchers [6, 7] presented their result based on private datasets, as per our limited knowledge such work cannot be followed by predecessors and the authenticity of such results might have doubtful until check the work with other databases. Thus presented results based on the different private databases are not a welcome step to develop a universal diabetic retinopathy screening tool for medical practices. Furthermore, it has been seen that there are numerous fundus databases, which has very fewer numbers of images and there are no ground truth data to verify the results for unsupervised methods (image segmentation). Besides, there may be slight color variation in retinal images with different ethnic backgrounds, so such factor might consider before developing a universal acceptance DR screening tool. From this study, we would like share, the truthful selection of fundus database is an imperative task before evaluation process. The point should be noted, considering the particular single database for getting better quality results is different to develop the universal acceptance computer-aided diabetic retinopathy screening systems for medical practices.

4   Introduction to Various Databases

There are numerous fundus databases are publicly available, which are created by various scientific communities to facilitate researchers to develop an automatic diabetic retinopathy screening tool, and some sample fundus images are shown in Fig. 3. The variation of fundus databases depends on various factors, such as fundus camera

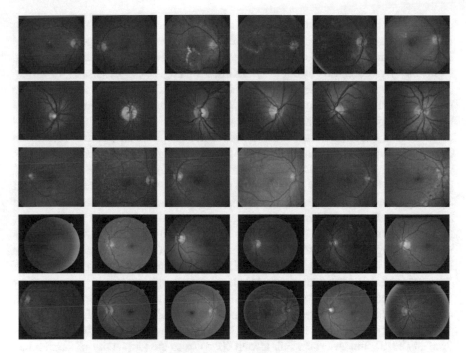

Fig. 3 Samples of fundus from different database

use to acquire images, Field of View (FOV), light illumination, resolution, compression format, image sizes, total number of images in the particular database, bit depth in per pixels, whether raw data or any image enhancement algorithms applied before allow to public access, details of diagnoses code (disease information per subjects) and ground truth (gold data) results for comparison. In 1996–2004, at California University, National Institute of Health (US) was funded a project named as structure analysis of retina (STARE), under this project STARE fundus database [8] was established, which has 400 images to facilitate the ongoing research in development of DR screening tool to detect the retinal eye disease like diabetic retinopathy, macular edema and glaucoma. In between 2003–2006, a joint project between Liverpool University and Liverpool University Hospital U.K. successfully carried out to develop an image analysis tool, additionally created a retinal image database automated retinal image analysis (ARIA) [9, 10], where Zeiss FF450+ fundus camera had been used to acquire the images at St. Paul's Eye Unit, U.K. Same year Image Sciences Institute, University Medical Center, Netherland, established a Digital Retinal Images for Vessel Extraction (DRIVE) database [11] comprising 40 images (20 training and 20 testing), this fundus database especially for evaluation of retinal vessels segmentation classifiers. In same time frame, at University of Lincoln, researcher had been trying to develop algorithm for optic disk extraction, to do the evaluation for optic disk localization algorithms, optic nerve head database (ONHSD) [12] dataset was created, having 99 images acquired by Canon CR6 45MNf fundus camera from different ethnic background subjects. The largest database MESSIDOR [13] having 1200 high-quality fundus images developed under a research program supported by French Ministry Defense under Techno-Vision program 2004, and images were captured by three ophthalmologist departments using 3CCD camera from different variations in subjects. In the progressive development of fundus databases, the machine vision and pattern recognition group from the Lappeenranta University of Technology, Finland consecutively developed three standard diabetic retinopathy databases underneath the project IMAGERET [14]. The aim of the project was to develop a medical decision-making tool for automatic diabetic retinopathy risk grading, those databases named as DIARETDB0 (2006) [15], DIARETDB1 (2007) [16] and DIARETDB2 (2009) [17] were developed with calibration level 0, 1 and 2 having different objectives or improvement in successive development of fundus databases. In the year 2008, a research group from Spain, Universidad Complutense, Hospital Miguel Servet and Universidad Nacional de Educacion a Distancia (UNED) jointly developed the Digital Retinal Images for Optic Nerve Segmentation Database (DRIONS-DB) [18] to assist for optic disk segmentation in ongoing development of computer-aided diabetic retinopathy monitoring system. Furthermore, a new dataset was developed at University of Lincoln, U.K., for a new retinal vessel reference dataset Retinal Vessel Image set for Estimation of Widths (REVIEW) [19] used to calculate the performance of retinal vessel segmentation algorithms. In latest development (2013), Image Processing Group (IPG) from the University of Zagreb has been established new database Diabetic Retinopathy Image Dataset (DRiDB) [20], here we requested to access the dataset, but the response is not received till the final manuscript is prepared. The latest addition in a new database named as High-Resolution Fundus (HRF) database [21] using Canon CR-1 fundus camera under collaborative research between pattern recognition laboratory, Friedrich-Alexander University Germany and Brno University of Technology, Czech Republic. Another latest development and mostly used E-optha [22], a new addition from MESSIDOR database [13] group, primarily focuses on hard exudates and microaneurysms for the finest development of universal DR screening program.

5   Various State of Arts

Make a decision to select the dataset to evaluate the performance of any developed classifier is very important for truly recognizing the acceptability of developed program. The most datasets are created to identify specific regions or single diabetic retinopathy symptom, for instance, optical nerve extraction, vessels segmentation or any retinal vasculature segmentation and diabetic retinopathy symptoms detection. The ideal database should consist of all level of severity of disease (preliminary, mid-level and highly severe condition), sufficient number of images, having all types of DR symptoms (both non-proliferative and proliferative DR), images acquire by a variety of fundus camera with varying different Field of views (FOVs), image acquiring environment should have highly similarity of light artifacts as real clinical applications and it would be better if the fundus is acquired from all ethnic backgrounds. STARE datasets [8] consisting of 400 raw images along with diagnoses of each fundus in a text data file. As the dataset was created as a part to development of self-assessment and monitoring DR tool, besides the images, database package also includes the source code of classifier in MS Visual C and C, expert's annotations with features also available in each image and hand labels ground truth images for result verification purposes. The ground truth includes 40 blood vessel segmentations, vein labeling of 10 images by two experts, 80 images with a localized optic disk. Though it was one prominent dataset used in old days, presently advise to use in unsupervised classifiers for evaluation of optic disk and blood vessels extraction algorithms. In ARIA database [9, 10], images are acquired by Zeiss FF450+ fundus camera with 50° field of view and stored in uncompressed TIFF. Datasets not only consist of 50 images of pathological and healthy images, but also images embraced with other eye-related disease too, like age-related macular degeneration. Additionally, database comprises of the ground truth datasets of blood vessels traced by experts. The DRIVE [11] database was created under retinopathy screening program by govt. of the Netherlands, which consists of 40 uncompressed JPEG images, where 20 each for training and testing set, the color fundus images acquired by Canon CR5 non-mydriatic 3CCD camera with a 45° FOV. All images were captured from 400 randomly selected diabetic subjects with different age group, mostly 25–90. Out of 40 images, 33 having no DR lesion and rest 7 has a mid-level severity of diabetic retinopathy. The database also provides the expert-traced gold datasets for comparison purpose with extracted vessels segmentation results.

ONHSD [12] dataset having 99 images from 50 randomly selected diabetic retinopathy subjects with diverse ethnics backgrounds. The images were captured using Canon CR6 45MNf fundus camera with 45° FOV and each image has 640×480 resolutions. The images are converted to gray level from HSI representation; datasets package also provides the optical head center location in a text file and manually traced boundary edge of the optic head by a trained clinician. The largest datasets MESSIDOR [13] comprising 1200 color fundus images acquired by three ophthalmologic centers using color video 3CCD camera with 45 FOV. The image with 8-bit depth having three categories of image resolution 1440×960, 2240×1488 and 2304×1536 pixels. The acquired 800 images subjects are pupil dilated (a drop of Tropicamide at 0.5%) and the remaining 400 images are a normal acquisition. The datasets also provide data file of medical diagnoses details by clinic experts with retinopathy risk grade associated and severity of macular edema in the individual image.

IMAGERET [14] (2006–2009) was a collaborative project between various universities and industry experts to develop a new imaging hardware to acquire the fundus images. Under this research program, three databases were established—DIARETDB0 [15], DIARETDB1 [16] and DIARETDB2 [17] to providing the support in research for the progressive development of diabetic retinopathy screening tool. In DIARETDB0 (Standard Diabetic Retinopathy Database-Calibration Level 0), images were captured with unknown camera settings and 50 FOV. It contains 130 colors fundus images, where 100 having all levels of DR severity, like hard or soft exudates, hemorrhages microaneurysms and neovascularization. In DIARETDB1 (Calibration Level 1) contains the 89 images, where 84 showing least mild non-proliferative signs (Microaneurysms) and rest 5 are normal healthy images. Image captured through varying camera settings with 50 FOV, as much as similar to real-time clinical applications. DIARETDB2 (Calibration Level 2) is an up gradation of DIARETDB1, with a separate ddb2 package; in addition to fundus which includes the ddb2 MATLAB-customized functions to perform training and evaluate the develop classifier. All these dataset packages include images, gold datasets, detail documentation and MATLAB script use for performance evaluation. In DRIONS-DB [18] database, which abbreviation is digital retinal images for optic nerve segmentation consists of 110 colors fundus with variations of retinopathy or macular edema, like cataract, light artifacts, rim blur, concentric peripapillary atrophy and strong pallor distracter. The dataset also incorporated with ground truth images traced by two experts using drawing software tool. REVIEW [19] is a new fundus dataset benchmarking for retinal vessels extractions algorithms; it contains 16 images with 139 vessel segments, demonstrating in a variety of vessels with multiple pathologies and provides the ground truth data (vessels traced) done by three experts using drawing tool. This dataset is good for evaluation of code, which is intended to measure the vessel profile or measurements of vessel width. The result of a collaborative research on retinal image analysis, High-Resolution Fundus (HRF) datasets [20], has been developed to support the comparative study and performance evaluation of various diabetic retinopathy screening algorithms. The datasets contain 45 images, where 15 healthy subjects, 15 images with DR symptoms and 15 images of glaucomatous patients. The dataset includes masks and ground truth images evaluated by a group of ophthalmologists; it is ideal for macula detection, vessels extraction (arteries/veins) and optic disk segmentation algorithms. E-optha [22] dataset is developed under a shared network ANR-TECSAN-TELEOPHTA [23] project between ADCIS and the French government for automatic diabetic retinopathy evaluation mechanism. Dataset has two subdatabases, MicroAneurysms (MA) and Exudates (EX), where e-optha EX contain 47 fundus with exudate and 35 without lesions, whereas, e-optha-MA comprise of 148 with microaneurysms or small hemorrhages and 233 images with no lesions. As per our limited knowledge, it has been second datasets (first in DIARETDB1) to provide ground truth results of microaneurysms and exudates, furthermore some sample images from different databases are shown in Fig. 3 and brief details of each database are shown in Table 1.

Table 1 Brief summary of retinal fundus databases

Sl no.	Database (year) country	Total images	FOV	Camera	Remarks
1	STARE (1996–2004) USA	397	–	–	Diagnosis codes and diagnoses for each image, PPM format. Includes text file with doctors annotations (visible features).Ground truth datasets for blood vessels and OD
2	ARIA (2003–2006) UK	59	50	Zeiss FF450+	59 diabetics subjects, 61 healthy images and 23 AMD subject. Includes ground truth images for vessels segmentation, uncompressed TIFF format
3	DRIVE (2004) Netherlands	40	45	Canon CR5	Ground truth of vessels segmentation. Each image having 768 by 584 pixels with 8 bit in each image, 20 images each for training and testing
4	ONHSD (2004) UK	99	45	Canon CR6	99 Fundus images taken from randomly selected 50 patients of all ethnic backgrounds. Resolution is 640×480. Includes the gold data which is edge of optic nerve mark by clinicians
5	MESSIDOR (2004) struct France	1200	45	Topcon TRC-NW6	File with mentioning the retinopathy grade and risk of macular edema, color images having 1440×960, 2240×1488 or 2304×1536 pixels resolutions with 8 bit depth
6	DIARETDB0 (2006) Finland	130	50	–	Out of 130 images, 20 healthy subjects and 11 images with diseased appearance. Includes the ground truth images, but in different format
7	DIARETDB1 (2007) Finland	89	50	–	Out of 89 images (PPM Format), 84 contain at least mild non-proliferative signs (MA) of DR, and 5 are healthy fundus images. Database package includes ground truth of Hemorrhages, Soft exudates and Small Dots, but not in binary format
8	DIARETDB2 (2009) Finland	89	–	–	Includes Matlab-files for operating with raw images and ground truth (read data sets using XML files). Ground truth images in EPS format, also available classifier MATLAB code
9	DRIONS-DB (2008) Spain	110	–	–	Ideal for optic nerve segmentation algorithms, database includes expert annotation and Matlab file. Images with 600×400 pixels resolution with and 8 bits depth. Includes ground truth images for OD contour localization
10	REVIEW (2008) UK	16	50	Zeiss and JVC 3CCD	Total 16 images with 193 vessel segments, this datasets can use for performance evaluation of vessels segmentation algorithms. The dimension of images with high-resolution 3300×2600 pixels
11	DRiDB (2013) Croatia	–	–	–	We have requested to access the database, but did not receive any response till the paper is final drafted
12	HRF (2013) Germany	45	45	Canon CR	15 Images with DR, 15 of healthy subjects and 15 images with glaucomatous sign, datasets includes ground truth of vessels and OD
13	E-Ophtha (2013) France	34 (EX), 68 (MA)	–	–	Consists two sub datasets one for MA, which contains 148 images with microaneurysms or small hemorrhages and 233 images with no lesion. Other exudates datasets having 47 images with EX and 35 images with no lesion. Datasets available with ground truth images of EX and MAs

6 Articulation Between CAD-DRST and Database

In the development phase of integrated computer-aided diabetic retinopathy screening system, to identify the various anatomical regions of fundus anatomy and extract the lesions separately in one-by-one basis, the precise selection of database is very important to pursue the development of universal acceptance CAD-DRST system. Each step system involves multiple individual DR lesion detections, starting from first step of noise elimination, image enhancement, optic disk segmentation, followed by vessel tree tracing, DR symptoms detection likewise exudates extraction (both hard and soft exudates detected by different algorithms), microaneurysms detection, finding the dot blot hemorrhages and many more depending on depth level of develop program. Separate algorithms have been developed to target one specific task, in each step of the evaluation process of developed code, a database is required. Therefore the perfect selection of database is very vital to find the true sense of result matrices like sensitivity, specificity and ROC curve [24]. From this study, it has been seen that there are some problems exist in the present databases, firstly all those databases were created for specific DR symptoms evaluation, like vessels segmentation and exudate detection. But in development of complete integrated DR screening system, the developer has to select the another dataset in different phases depending on suitability of database, tested with multiple databases is again tedious and time-consuming process, also same code may give the different results tested with multiple databases. Secondly, most databases do not provide the complete ground truth datasets of input fundus images, it is very important to evaluate the quality metrics based on image segmentation approach. Thirdly, the number of images should be large so that its result validation is truly accepted. In addition, it has been seen that the number of work [6, 7, 25] use the private datasets to evaluate the performance of DR monitoring codes, which is an unwelcome trend for the development of universal acceptance CAD-DRST system. Because without accessed those private datasets, the method cannot be followed by predecessors to further improvement on that work. The author would like to inform all researchers, those used the private datasets to evaluate their program, after the evaluation is done, and datasets should available for public access so that people can work on it and eventually progressive development of the work to be followed. Nowadays, the retinal image analysis is very active research topic in the field of medical image processing, author would like to urge to research communities to develop a single fundus database, which would include ground truth images of all lesions, having large number of images, acquired from various ethnic subjects and detail diagnoses of individual image or any other relevant information, so that as earliest finest classifier to be developed for the purpose of inclusive DR monitoring system for clinical practices, which will be the great contribution toward human mankind, specially for diabetic suffering people.

7 Discussion

As people adopting modern lifestyle, the numbers of diabetic patients are also increasing in alarming fashion, to protect the vision of diabetic patients from the severe disease diabetic retinopathy, and it is utmost important to test the eye in regular interval. As drastically increasing numbers of retinopathy patients, ophthalmologists are not sufficient to tackle this twenty-first-century health emergency due to the diabetic. Hence this technological era demands to develop an automatic DR detection and monitoring system. There are numerous research communities across the globe are trying hard to develop such universal acceptance screening tools. In this development phase of such system, the role of fundus databases is essential to correctly evaluate the performance of developing classifiers, so that it necessary to have the finest dataset, which can allow evaluating all DR symptoms consecutively. At present most publicly available datasets are created to evaluate the specific symptom (lesions) having less number of images, in addition, a common database should be there for evaluation all types of algorithms for DR detection. Therefore author would like to urge all medical image research groups, whose having world's best infrastructure to create one single common database with a large number of images (like MESSIDOR [13]), which may be sufficient to evaluate the all steps of DR-related diagnoses screening algorithms, so that the medical sciences will have ultimate, high-performance and globally accepted integrated diabetic retinopathy screening system; as a result, it will help to save the visions of millions.

References

1. Wild, S., Roglic, G., Green, A.: Global Prevalence of Diabetes: Estimates for the Year 2000 and Projection for 2030, Diabetes Care, pp. 1047–1053 (2004)
2. Amin, J., Sharif, M., Yasmin, M.: A review on recent developments for detection of diabetic retinopathy. Scientficavol (Hindawi Publishing Corporation) (2016). Article ID: 6838976
3. World Health Organization (WHO): Action Plan for the Prevention of Avoidable Blindness and Visual Impairment 2009–2013 (2010)
4. International Diabetes Federation. https://www.idf.org/
5. Agarwal, S., Acharjya, K., Sharma, S.K., Pandita, S.: Automatic computer aided diagnosis for early diabetic retinopathy detection and monitoring: a comprehensive review. In: International Conference on Green Engineering and Technologies (IC-GET) IEEE Xplore
6. Osareh, A., Shadgar, B., Markham, R.: A computational-intelligence-based approach for detection of exudates in diabetic retinopathy images. IEEE Trans. Inf. Technol. Biomed. **13**(4) (2009)
7. Jaya, T., Dheeba, N., Singh, A.: Detection of hard exudates in colour fundus images using fuzzy support vector machine-based expert system. J. Digit. Imag. Soc. Imag. Inform Med. (2015)
8. Hoover, A., Kouznetsova, V., Goldbaum, M.: Locating blood vessels in retinal images by piecewise threshold probing of a matched filter response. IEEE Trans. Med. Imag. **19**(3), 203–210 (2000)
9. Zheng, Y., Hijazi, M.H.A., Coenen, F.: Automated disease or no disease grading of age-related macular degeneration by an image mining approach. Invest. Ophthalmol. Vis. Sci. (2012)

10. Farnell, D.J.J., Hatfield, F.N., Knox, P., Reakes, M., Spencer, S., Parry, D.: Enhancement of blood vessels in digital fundus photographs via the application of multiscale line operators. J. Franklin Inst. (2008)
11. Staal, J.J., Abramoff, M.D., Niemeijer, M., Viergever, M.A., van Ginneken, B.: Ridge based vessel segmentation in color images of the retina. IEEE Trans. Med. Imag. 23, 501–509 (2014)
12. Lowell, J., Hunter, A., Steel, D., Ryder, B., Fletcher, E.: Optic nerve head segmentation. IEEE Trans. Med. Imag. 23(2) (2004)
13. Decenciere, E., Zhang, X., Cazuguel, G., Lay, B., Cochener, B., Trone, C., Gain, P., Ordonez, R., Massin, P., Erginay, A., Charton, B., Claude Klein, J.: Feedback on a publicly distributed database: the messidor database. Image Anal. Stereol. 33(3), 231–234 (2014)
14. Forsstrom, J., Kalesnykiene, V., Kuivalainen, M., Sorri, I., Uusitalo, H., Kamarainen, J.: Automated pattern recognition for the detection of diabetic changes in digital fundus images. In: Abstract for ARVO 2005 Annual Meeting Fort Lauderdale, Florida (2005)
15. Kauppi, T., Kalesnykiene, V., Kamarainen, J.K., Lensu, L., Sorri, I., Uusitalo, H., Kalviainen, H., Pietila, J.: DIARETDB0: evaluation database and methodology for diabetic retinopathy algorithms. Technical report
16. Kauppi, T., Kalesnykiene, V., Kamarainen, J.K., Lensu, L., Sorri, I., Raninen, A., Voutilainen, R., Uusitalo, H., Kilviinen, H., Pietili, J.: DIARETDB1 diabetic retinopathy database and evaluation protocol. Technical report
17. Kauppi, T., Kalesnykiene, V., Kamarainen, J.K., Lensu, L., Sorri, I., Raninen, A., Voutilainen, R., Uusitalo, H., Kalviainen, H., Pietila, J.: The DIARETDB1 diabetic retinopathy database and evaluation protocol. In: Proceedings of the British Machine Vision Conference BMVA Press, pp. 15.1–15.10 (2007)
18. Carmona, E.J., Rincon, M., Garcia-Feijoo, J., Martinez-de-la-Casa, J.M.: Identification of the optic nerve head with genetic algorithms. J. Artif. Intell. Med. 43(3), 243–259 (2008)
19. Al-Diri, B., Hunter, A., Steel, D., Habib, M., Hudaib, T., Berry, S.: Review-a reference data set for retinal vessel profiles. In: 30th Annual International IEEE EMBS Conference Vancouver British Columbia, Canada, 20–24 Aug 2008
20. Pavle, P., Sven, L., Zoran, V., Goran, B., Marko, S., Tomislav, P., Lana, D., Maja, M.R., Nikolina, B., Raseljka, T.: Diabetic retinopathy image database (DRiDB): a new database for diabetic retinopathy screening programs research. In: Proceedings of 8th International Symposium on Image and Signal Processing and Analysis-ISPA Trieste, pp. 704–709 (2013)
21. Attila, B., Ridiger, B., Andreas, M., Joachim, H., Georg, M.: Robust vessel segmentation in fundus images. Int. J. Biomed. Imag. 2013 (2013)
22. Decenciere, E., Cazuguel, G., Zhang, X., Thibault, G., Klein, J.C., Meyer, F., Marcotegui, B., Quellec, G., Lamard, M., Danno, R., Elie, D., Massin, P., Viktor, Z., Erginay, A., Lay, B., Chabouis, A.: TeleOphta-machine learning and image processing methods for teleophthalmology. IRBM, Elsevier Masson SAS IRBM 34, 196–203 (2013)
23. Quellec, G., Lee, K., Dolejsi, M., Garvin, M.K., Abramoff, M.D., Sonka, M.: Three-dimensional analysis of retinal layer texture: identification of fluid-filled regions in SD-OCT of the macula. IEEE Trans. Med. Imag. 29(6), 1321–1330 (2010)
24. Fawcett, T.: An introduction to ROC analysis. Pattern Recogn. Lett. 27, 861–874 (2006). https://doi.org/10.1016/j.patrec.2005.10.010
25. Kaur, J., Mittal, D.: Segmentation and measurement of exudates in fundus images of the retina for detection of retinal disease. J. Biomed. Eng. Med. Imag. 2(1) (2014)

Estimation of Skew Angle from Trilingual Handwritten Documents at Word Level: An Approach Based on Region Props

M. Ravikumar, B. J. Shivaprasad, G. Shivakumar and P. G. Rachana

Abstract In this work, an efficient technique for segmenting the words from multilingual unconstrained handwritten documents at word level is proposed. In the proposed model, morphological operations and connected component analysis are used for word identification. Based on that, the bounding box for each word is drawn and then words are segmented. The proposed algorithm also works on documents with any orientation. We conducted experimentation on our own 300 unconstrained multilingual handwritten documents. The result shows the performance of the proposed algorithm.

Keywords Connected component analysis · Region props · Skew angle

1 Introduction

Nowadays, the world is moving toward digitization. Hence, the transformation from paper to paperless office is very much essential. In this process, analysis of any handwritten document plays an important role. In today's life, processing of office document has become major research area in the field of document image analysis. Skew must be estimated and corrected or else analysis of handwritten document becomes difficult and it drastically reduces the recognition rate. Estimation of skew angle can be at block level, line level, and also at word level [1]. Most of the work has been done on monolingual documents, but very few works have been done on

M. Ravikumar (✉) · B. J. Shivaprasad · G. Shivakumar · P. G. Rachana
Department of Computer Science, Kuvempu University, Shimoga 577451, Karnataka, India
e-mail: ravi2142@yahoo.co.in

B. J. Shivaprasad
e-mail: shivaprasad1607@gmail.com

G. Shivakumar
e-mail: g.shivakumarclk@gmail.com

P. G. Rachana
e-mail: pgrachana@gmail.com

© Springer Nature Singapore Pte Ltd. 2019
J. Wang et al. (eds.), *Soft Computing and Signal Processing* ,
Advances in Intelligent Systems and Computing 898,
https://doi.org/10.1007/978-981-13-3393-4_43

419

skew estimation and correction for multilingual documents. Reason is difficulty faced while finding the skew angle of multiple words, which are of multiple languages and are at different orientations. Besides its complexities, estimation and correction of multiple skews from multilingual handwritten documents has wide range of applications which include analysis of real-time office documents, analysis of ancient script, and analysis of documents prescribed by the doctors.

Remaining part of the paper is divided into four sections. Section 2 briefly explains related work, Sect. 3 focuses on proposed methodology in detail, experimentation with results are discussed in Sect. 4, and finally, conclusion is given in Sect. 5.

2 Related Work

In this section, we brief the related work on skew estimation and correction of handwritten documents. In most of the cases, documents will be multilingual in nature comprising different languages with multiple skews. A multiple skew estimation technique is proposed [1–3], where skew is estimated by fitting a minimum circumscribing ellipse and k-means clustering is used to estimate skew of multiple blocks. A method for estimating document image skew angle is presented [4–7], where it depends on objects with rectangular shape such as paragraphs, texts, and figures. The angle of that rectangle represents the angle of document skew. Skew detection and correction using linear regression technique is proposed [8, 9], and this method uses the Hough transform to detect skew with large angles. A method for skew estimation in binary images is proposed. This method is based on binary moments, where moment-based method to each binary object evaluates their local text skews [10]. A geometrical technique for line and word segmentation is presented in [11], which also estimates multiple skews if present in the document and corrects it by natural method which helps in finding top and bottom border points of shirorekha, and accuracy of 94% is recorded for Indian government office documents. Some of the related works of skew estimation can also be found in the papers [4, 12–15]. Next section discusses proposed methodology in detail.

3 Proposed Methodology

In this section, we discuss proposed method for skew estimation and correction of multilingual handwritten documents. The block diagram of proposed method is shown in Fig. 1.

At the beginning, a multilingual handwritten document image is taken from the dataset and is given as an input to preprocessing stage. In preprocessing, first banalization operation is done in order to perform document layout analysis using Otsu's approach [16]. After this, noise is removed using median filters and it preserves edges. Unwanted marginal borders are eliminated using clear border method, and

Fig. 1 Block diagram of the proposed method

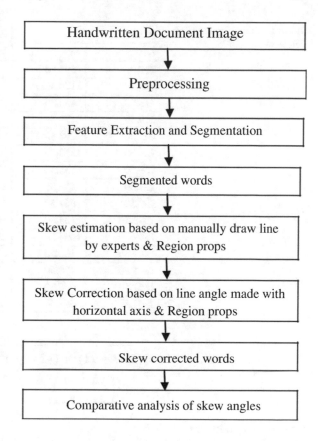

special characters with small area are removed by setting threshold value. When all the noises are removed, the document image is given as input to feature extraction and segmentation method, where we consider the spaces between words are more compared to spaces between characters and by using CCA and measuring properties of words are segmented.

After this, next step is to estimate and correct the skew angle, which can be done using two approaches, and finally, the two approaches are compared for calculating error rates.

In first approach, the skew angle is estimated based on manually drawn lines by human experts, where human experts considered are of different age-group and they are unaware of languages present in the multilingual handwritten documents. They were asked to click manually at two endpoints on the ellipse constructed over word. In order to estimate skew angle, a line is drawn using the clicked points. These clicked points are considered as coordinate values (x_1, y_1) and (x_2, y_2) which are stored in the form of an array of two dimensions for further calculations, which are shown in Table 1. Then the angle of deviation from the horizontal axis is calculated by the formula which is given below:

Table 1 Points on coordinate values of the words

Point 1		Point 2	
X_1	Y_1	X_2	Y_2
21.50	594.50	269.50	458.50
213.50	1170.50	633.50	1166.50
313.50	450.50	545.50	310.50
349.50	2122.50	637.50	1910.50
605.50	298.50	777.50	166.50
721.50	1850.50	929.50	1714.50
709.50	1138.50	1205.50	1126.50
821.50	154.50	1061.50	−1.50
993.50	1662.50	1453.50	1306.50
1273.50	1134.50	1437.50	1110.50
1517.50	1094.50	2133.50	1090.50
1569.50	186.50	1737.50	390.50
1765.50	470.50	1825.50	574.50
1857.50	630.50	1973.50	810.50
2037.50	858.50	2081.50	926.50

$$h = \sqrt{(x2 - x1)^2 + (y2 - y1)^2}$$
$$\theta = \tan^{-1}\left(\frac{y2 - y1}{x2 - x1}\right) \tag{1}$$

where h = hypotenuse.

Finally, the estimated skew must be corrected, where skew correction is based on angles of lines drawn manually by experts, which is given in Eq. 2.

$$x2 = x1 + h \cos(\theta)$$
$$y2 = y1 + h \sin(\theta) \tag{2}$$

In second approach, the skew angle is estimated using the measuring properties of an image region, which is corrected by using geometric transformation function with the help of centroid, major axis, and minor axis. The below formula is used for rotation by an angle θ:

$$x = u \cos(\theta) - v \sin(\theta)$$
$$y = u \sin(\theta) + v \cos(\theta) \tag{3}$$

where u and v represent lengths of the major and minor axes.

The below formula is also used for rotation by an angle θ in matrix format:

$$R = \begin{bmatrix} \cos(\theta) & \sin(\theta) \\ -\sin(\theta) & \cos(\theta) \end{bmatrix} \tag{4}$$

Finally, the comparative analysis of skew correction among two approaches has been done.

4 Experimentation and Results

In order to compare the error rates of two approaches explained above, experimentation is carried out on our own dataset with 300 unconstrained handwritten documents. Once the document is obtained, preprocessing step is applied on the document. The words are labeled in order to avoid the confusion, which also helps in reconstruction after the skew is corrected. We considered five human experts, and they are asked

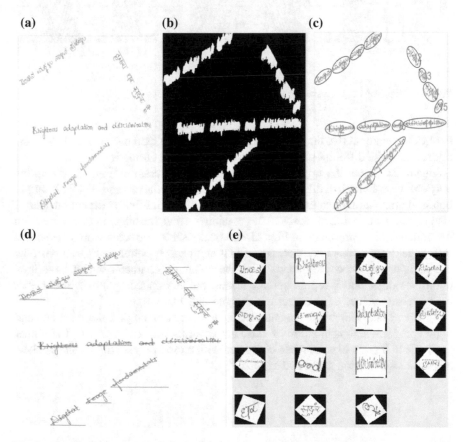

Fig. 2 **a** Original input image, **b** image after CCA, **c** labeling and line dilation drawn based on expert clicks, and **d** angle found using line drawn by (**e**) word segmentation and skew correction experts

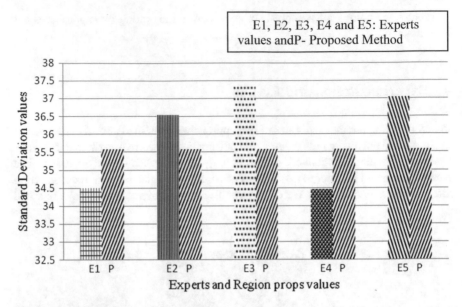

Fig. 3 Comparison between skew angle of experts and region props

to click manually at two endpoints on the ellipse constructed over word. The below figures show the different stages of skew estimation and correction.

Figure 2a shows the original image before any processing, Fig. 2b shows the output of morphological dilation and CCA after noise removal, and Fig. 2c is output obtained after human experts clicking two endpoints and line is drawn on that; it also represents labeling of words. After obtaining a line, the angle of deviation from the horizontal is represented in Fig. 2d, and finally skew-corrected words are shown in Fig. 2e. Sometimes the error rate of first approach is better compared with the second approach, because experts click the points on character which have low-intensity values, and it represents minor axis, but automated system will consider high-intensity values of character which represents major axis.

Figures 3 and 4 show the comparative analysis of skew angle among experts and region props, where estimation of skew angle varies and result obtained is higher error rate in human experts when compared with region props. Table 2 shows skew angle values of region props and experts.

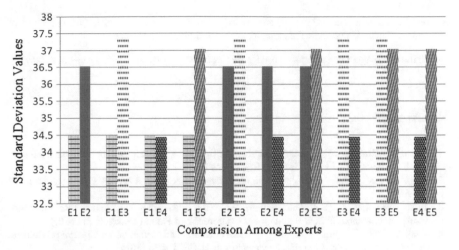

Fig. 4 Comparison between skew angles among experts

Table 2 Angle of experts and region props angle

Expert 1 angle	Expert 2 angle	Expert 3 angle	Expert 4 angle	Expert 5 angle	Region props angle
27.29	33.43	28.96	29.62	30.96	29.81
2.14	5.71	5.09	1.09	7.74	1.74
29.53	37.20	33.23	38.15	37.72	31.31
38.36	31.73	35.31	34.77	29.89	36.45
34.24	32.85	33.40	37.69	37.97	35.97
37.40	37.79	34.18	43.60	29.85	27.98
3.751	1.43	2.46	3.25	2.70	3.50
34.18	37.36	30.06	37.72	30.51	33.28
37.24	41.32	37.24	38.77	37.33	39.38
4.085	7.69	11.30	11.00	18.82	13.67
0.73	1.10	1.53	−0.38	−0.75	1.35
−48.90	−50.90	−54.56	−49.39	−48.99	−48.03
−53.13	−65.55	−66.57	−59.30	−60.46	−68.87
−47.07	−51.66	−53.32	−40.60	−53.97	−45.93
−48.36	−38.65	−53.13	−29.74	−57.99	55.97

5 Conclusion

In this paper, an efficient method is proposed for estimation and correction of multiple skews from unconstrained multilingual handwritten documents at the word level, based on the region props. The proposed method is compared with the skew angle, which is obtained from five human experts, and finally, the result is obtained. The result shows that proposed method is efficient than human experts.

Acknowledgements The authors will acknowledge Dr. D. S. Guru and HPC Laboratory, Dept. of Computer Science, University of Mysore, Mysore, for their encouragement.

Declaration Images and the datasets used in this work are our own and not from any other's work.

References

1. Guru, D.S., Ravikumar, M., Manjunath, S.: Multiple skew estimation in multilingual handwritten documents. IJCSI Int. J. Comput. Sci. Issues **10**(5), 2 (2013)
2. Tang, Y., Lee, S., Suen, C.: Automatic document processing: a survey. Pattern Recogn. **29** (1996)
3. Kasiviswanathan, H., Ball, G.R., Srihari, S.N.: Top down analysis of line structure in handwritten documents. In: 20th International Conference on Pattern Recognition (2010)
4. Shah, L., Patel, R., Patel, S., Maniar, J.: Skew detection and correction for Gujarati printed and handwritten character using linear regression. IJARCSSE **4**(1) (2014)
5. Babu, D.R., Kumat, P.M., Dhannawat, M.D.: Skew angle estimation and correction of handwritten, textual and large areas of non-textual document images: a novel approach. In: International Conference on Image Processing, Computer Vision and Pattern Recognition (2006)
6. Kavallieratou, E., Fakotakis, N., Kokkinakis, G.: Skew angle estimation for printed and handwritten documents using the Wigner-Ville distribution. ELSEVIER Image Vis. Comput. **20**, 813–824 (2002)
7. Singha, C., Bhatiab, N., Amandeep Kaur, C.: Hough transform based fast skew detection and accurate skew correction methods. ELSEVIER-Pattern Recogn. **41**(2008), 3528–3546 (2008)
8. Wagdy, M., Faye, I., Rohaya, D.: Document image skew detection and correction Method based on extreme points. Centre Intell. Signal Imaging Res. (IEEE) **6**(1) (2014)
9. Boukharouba, A.: A new algorithm for skew correction and baseline detection based on the randomized Hough Transform. J. King Saud Univ. Comput. Inf. Sci. **29**, 29–38 (2017)
10. Brodić, D., Milivojević, Z.N.: Estimation of the handwritten text skew based on binary moments. Radio Eng. **21** (2012)
11. NarasimhaReddy, S., Deshpandande, P.S.: A novel local skew correction and segmentation approach for printed multilingual Indian documents. Alex. Eng. J. (2017)
12. Ravikumar, M., Rachana, P.G., Shivaprasad, B.J., Shivakumar, G.: Segmentation of words from unconstrained multilingual handwritten documents. J. Innov. Comput. Sci. Eng. **6** (2017)
13. Ghosh, R., Mandal, G.: Skew detection and correction of online Bangla handwritten word. IJCSI Int. J. Comput. Sci. Issues **9**(4), 2 (2012)
14. Rani, A., Singh, H.: A review on various techniques for skew detection and correction in handwritten text documents. IJREAT Int. J. Res. Eng. Adv. Technol. **3**(3) (2015)
15. Ramakrishna Murty, M., Murthy, J.V.R., Prasad Reddy, P.V.G.D.: Text document classification based on a least square support vector machines with singular value decomposition. Int. J. Comput. Appl. (IJCA) 21–26 (2011)
16. Sun, C., Si, D.: Skew and slant correction for document images using gradient direction. In: Fourth International Conference on Document Analysis and Recognition (1997)

FPGA Implementation of Coding for Minimizing Delay and Power in SOC Interconnects

N. Chintaiah and G. Umamaheswara Reddy

Abstract In this paper, the bus encoding technique is introduced to reduce the number of transitions. The data word transmission on an on-chip bus causes the switching of data bits on the bus wires. This switching charges and discharges the capacitance associated with the wires and consequently causes dynamic power dissipation and increase in delay. A substantial amount of power is dissipated from buses compared with the power dissipation of the remaining circuit. The data bits sent through these buses should be encoded to decrease the switching activity, thereby reducing the power consumption and delay of the buses. In this approach, encoding is widely used to reduce the number of transitions. The encoder comprises four subdivisions, and the average power conserved was 10.17%. Overall, the average delay is decreased by 4.5%. Therefore, more amount of average power is conserved, and the delay is shortened.

Keywords Crosstalk · Encoder · Path delay · Even–odd inversion
Upper–lower inversion

1 Introduction

The size of the transistor in very-large-scale integration (VLSI) has been consistently decreasing. Currently, the semiconductor technology with a transistor size of more than a few tens of nanometers is under development. Buses distribute data and clock signals to various circuits on the chip. Moreover, the performance of integrated circuits depends on the power consumption and delay of the interconnections (buses). In deep submicron (DSM), the capacitance effect is crucial [1] for characterizing on-

N. Chintaiah (✉) · G. Umamaheswara Reddy
Department of Electronics and Communication Engineering, Sri Venkateswara University,
Tirupati, India
e-mail: chintu.svu@gmail.com

G. Umamaheswara Reddy
e-mail: umaskit@gmail.com

© Springer Nature Singapore Pte Ltd. 2019   427
J. Wang et al. (eds.), *Soft Computing and Signal Processing* ,
Advances in Intelligent Systems and Computing 898,
https://doi.org/10.1007/978-981-13-3393-4_44

chip buses, which experience capacitance crosstalk and delay in signal propagation. These factors (power and delay) reduce performance, reduce the reliability of data transmission, and increase the power consumption. These effects occur in embedded processors, digital signal processors, and microprocessors.

In DSM (<130 nm), the coupling capacitance between wires on silicon becomes more dominant than the interlayer capacitance. This coupling capacitance causes logic failures and timing degradation in VLSI circuits, that is, crosstalk. Crosstalk is a phenomenon that occurs when a logic transmitted in VLSI circuits generates undesired effects on neighboring circuits due to coupling. Major objectives of DSM technologies include reducing interconnect delay, reducing power consumption, and increasing design productivity [2].

The total power consumed by buses includes static power, dynamic power, and leakage power. Static power dissipation is due to the parasitic diode leakage current, which is negligible. Dynamic power dissipation occurs due to the transition of input (0–1 or 1–0).

Various techniques have been used for reducing the coupling effect in bus interconnects. The even–odd bit encoding technique is the base of the proposed technique. In this study, even–odd encoding is modified to the upper–lower bit encoding. Some encoding techniques focus on reducing the power only, some are discusses minimizing the delay and some other designers are concentrate reducing on both delay as well as power. The literature review of various techniques and the introduction of the proposed technique are provided in Sect. 1. A detailed explanation of the proposed technique with a logic diagram is provided in Sect. 2. Section 3 provides the experimental analysis and results. Finally, Sect. 4 concludes the study.

2   Overview

This study is focused on reducing the dynamic power for decreasing the coupling transitions of buses. Moreover, the following equation demonstrates the dependence of power consumption on input transitions (α) in an on-chip bus:

$$P_{TOTAL} = P_{DYNAMIC} + P_{STATIC} + P_{LEAKAGE} \tag{1}$$

$$P_{DYNAMIC} = \sum \alpha \cdot V_{DD}^2 \cdot f \cdot C_L \tag{2}$$

where α is the transition activity factor, V_{DD} is the power supply, f is the frequency, C_L is the load capacitor, α is equal to $\alpha_s c_s + \alpha_c c_c$, α_s is the self-transition activity, c_s is the self-capacitance, α_c is the coupling transition activity, and c_c is the coupling capacitance.

Equations (1, 2) suggest that the capacitance is controlled by the transition activity factor (α), which depends on input transitions. Bus coding [3] is the most commonly used technique to reduce dynamic switching on buses. The bus encoding technique prevents crosstalk by manipulating the input data before transmitting them on buses. Bus encoding can reduce certain unwanted data patterns, thereby reducing crosstalk. Various encoding techniques have been proposed to reduce the self-capacitance and coupling capacitance. Moreover, encoding techniques are classified into address bus and data bus encoding techniques.

T0 coding is the basic address encoding technique for a bus. This technique requires an extra line, T0, [4] to code the source word when it is transmitted on a bus. T0-XOR coding [5] is an extended capability of T0 coding, which combines the XOR operation with the T0 code. The T0 code is used to obtain successive values, that is, an XOR of the present and previous value, when the data on the bus lines are highly correlated by using an offset code. In T0-offset coding, the T0 code is combined with an offset operation. Thus, the T0 code is used for in-sequence values, whereas the difference between the previous and current address values (offset) is used for out-of-sequence values. In Address Level Bus Power Optimization (ALBORZ) [6] coding, the transition activity is performed in the address bus by using the offset and limited weight code. T0-C coding is conducted for sequential addresses. However, in this method, the adaptive codebook (ACB) method [7] employs JUMP and BRANCH operations for non-sequential addresses on a bus. Bus invert [1] coding is one of the encoding techniques of the basic data buses that employ the ACB method. Bus invert coding can effectively reduce crosstalk.

TS [8] is another approach in which logic 0 represents no transition and logic 1 represents a change from low to high or high to low. The ACB method utilizes the following two steps to generate code words: (i) storing model pattern index within a codebook, and (ii) determining the XOR of a source word by using the related model pattern in the codebook. An even–odd encoding technique [9] examines activity changes in neighboring lines. All wires are divided into odd and even lines. The low energy set scheme (LESS) [10] uses an XOR–XNOR or XNOR–XOR operation to transmit data.

3   Proposed Encoding–Decoding Techniques

Zhang et al. proposed an even–odd [9] method, which encodes the multiplexed address and data bus, that is, it encodes both in-sequence and out-sequence data. The internal architecture of the encoder comprises subdivisions, such as only even bit inversion, only odd bit inversion, all bit inversion, and no inversions. Among the previous and subdivision inversion data, data with the shortest Hamming distance is transmitted using control signals.

The proposed encoding technique utilizes upper and lower bit inverts. Let data on an n-bit wide bus at time t be denoted as $X^t = \{x_{n-1}^t, x_{n-2}^t, \ldots x_1^t x_0^t\}$. Then, data transmitted on the bus is denoted as $X^{t(enc)}$. The function *Calculate st_n(data1, data2)* determines the number of self-transitions between (data1, data2). Here, data1 and data2 should be n-bit wide.

3.1 Encoding Scheme

(i)   Let $X^{(t-1)\,enc}$ be the previously coded data that is transmitted on a bus, and X^t be the present data that has to be encoded and transmitted.

(ii) Invert upper-side bits in X^t, and affix them with 00 for decoding. Let this new data be denoted as $X^{t(Lower)}$. Evaluate $st\_Upper = Calculate\ ST\_n(X^{t(Lower)}, X^{(t-1)\,enc})$.

(iii) Similarly, invert the lower-side lines in X^t, and affix them with 01. Let this new line be denoted as $X^{t(Upper)}$. Evaluate $st\_Lower = Calculate\ ST\_n(X^{t(Upper)}, X^{(t-1)\,enc})$.

(iv) Invert all lines in X^t, and affix them with 10. Let these new lines be denoted as $X^{t(inv)}$. Evaluate $st\_inv = Calculate\ ST\_n(X^{t(inv)}, X^{(t-1)\,enc})$.

(v)   Consider the present data as no change data, and affix X^t with 11. Let this new data be denoted as $X^{t(same)}$. Evaluate $st\_same = Calculate\ ST\_n(X^{t(same)}, X^{(t-1)\,enc})$.

(vi) Determine $min(st\_Upper, st\_Lower, st\_inv, st\_same)$.

(vii) The code pattern corresponding to the minimum value in step (vi) is transmitted on the bus.

3.1.1   Encoder Architecture

Bit Changer

The encoder comprises the following two parts: (i) bit change (upper, lower, all inverse, no change) and (ii) comparator. The bit change part exhibits the following four subparts (i) st_upper, (ii) st_lower, (iii) st_inv, and (iv) st_same. Each part comprises two inputs, present input $X^{t(enc)}$ and previous input $X^{(t-1)\,enc}$, and an output with two internal signals and one external output count, as displayed in Fig. 1.

Figure 1 displays that each part exhibits two inputs, the present input $x^t = x_1x_2x_3x_4x_5x_6x_7x_8$ and the previous input $x^{(t-1)enc}$. The first part $M1$ is the upper bit changer, that is, st_upper. The output of st_upper inverts the upper four bits,

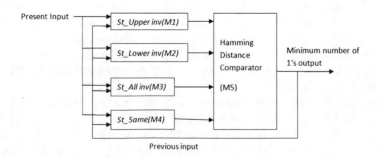

Fig. 1 Bit changer and Hamming distance calculation

Fig. 2 Upper-bit changes

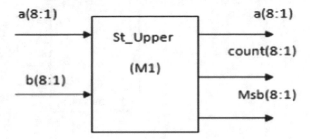

that is, $x^{t(Upper)} = \bar{x}_1\bar{x}_2\bar{x}_3\bar{x}_4x_5x_6x_7x_8$, and compare with second input $x^{(t-1)enc}$ and the number of bits calculated is different. As shown in Fig. 2 i.e the sub module of the encoder it invert the upper four bits present input data compare it and count the number of data changes of present and previous input.

Similarly, the lower-side bit change block $M2$ and the related different bit count *count*2, all inverse bit change block $M3$ and the related different bit count *count*3, no bit change block $M4$ and the related different bit count *count*4 must be calculated. Lower, inverse, and no change bit changes are given as follows:

(1) The output of st_Lower : $x^{t(Lower)} = x_1x_2x_3x_4\bar{x}_5\bar{x}_6\bar{x}_7\bar{x}_8$
(2) The output of st_inv : $x^{t(inv)} = \bar{x}_1\bar{x}_2\bar{x}_3\bar{x}_4\bar{x}_5\bar{x}_6\bar{x}_7\bar{x}_8$
(3) The output of st_same : $x^{t(same)} = x_1x_2x_3x_4x_5x_6x_7x_8$.

3.1.2 Hamming Distance Calculator

The Hamming distance calculator block transmits minimum number of 1's to the data line from the given input lines. Hamming distance is determined by counting the number of 1's present in the given data. The output of the bit changes are *count*1, *count*2, *count*3, and *count*4 for corresponding data lines. The calculator block compares all input lines and outputs the data signal with the minimum number of one's (lowest Hamming distance) and appends the control signal to the output for decoding purposes.

If $min(st\_Upper, st\_Lower \; st\_inv, st\_same) = st\_Upper$, $X^{t(Upper)}$ is transmitted. Hence, $X^{(t)enc} = X^{t(Upper)}$, and the number of transitions is reduced from five to one. Equation (2) indicates that the power dissipation depends on the transition activity factor (α). The reduction in the number of transitions reduces the power dissipation.

3.2 Decoder Architecture

The original data was retrieved from the encoded data by using the two control signals, as shown in Fig. 3.

The decoder had a 10-bit input (encoded input and controls signals were of 8 and 2 bits, respectively) with internal signals and an 8-bit output. The two control signals are the two LSB bits that are 00—invert only upper bits, 01—invert only lower bits, 10—invert all bits, 11—no inversion.

Fig. 3 Decoder

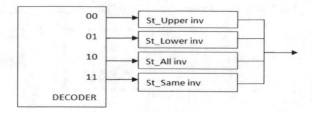

4 Results and Discussion

The simulation work is carried by giving the different data inputs to the proposed design and then calculate the power and delay of the different FPGA devices using the Xilinx Vivado tool. The delay obtained by using various coding schemes was tested with random vectors for 8, 16, and 32-bit wide buses, and the observations are tabulated in Table 1.

The proposed upper–lower inversion technique is compared with the existing even–odd inversion technique [9]. The results are applicable to data, address, and multiplexed lines. In the even–odd inversion technique, all the 8, 16, and 32 bits are compared. However, in the proposed technique, only upper–lower half bits are compared, which reduces the encoder–decoder circuit complexity. Furthermore, the overall circuit delay is reduced.

In the proposed technique, the overall average delay decreased by 4.5% compared with that of the existing technique. In the proposed technique, the delay in 8, 16, and 32-bit encoders decreased by 20%, 3.4%, and 2.2%, respectively. The encoder delay is increased with an increase in the design complexity. Table 1 shows delays in the

Table 1 Total delay in various field-programmable gate array (FPGA) devices

Device	Delay					
	Total delay (ns)					
	8 bit		16 bit		32 bit	
	Even–odd (existing)	Upper–lower (proposed)	Even–odd (existing)	Upper–lower (proposed)	Even–odd (existing)	Upper–lower (proposed)
Artix	3.51	2.74	21.51	18.81	46.60	46.49
Artix7 low voltage	4.01	3.14	21.51	21.39	46.60	46.49
Automatic Spartan	9.57	8.47	15.48	16.50	47.55	47.48
Automatic Zynq	3.37	2.68	15.48	15.40	33.17	33.10
Kintex7	2.73	2.15	13.62	13.56	29.38	29.31
Spartan 3	18.37	12.75	46.74	43.33	90.59	83.40
Spartan 6	9.57	8.47	24.57	24.50	47.55	47.48

Fig. 4 Average percentage
total delay

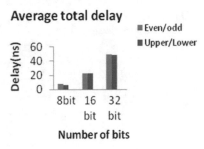

Table 2 Comparison of
average power and delay

Technique	Power (mw)	Delay (ns)
Even–odd technique	16.5	3.514
Upper–lower technique	13.57	2.746

Fig. 5 Dynamic power
conservation

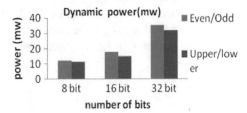

encoders with 8, 16, and 32-bit wide buses for various FPGA family devices, such as Spartan, Artix, and Zynq. All devices exhibit a reduced delay for the proposed method. The 8-bit encoder exhibited lesser delay than the 16 and 32-bit encoders (Fig. 4).

Power consumption is a crucial factor in digital design. In this study, a simulation was performed for the aforementioned different random patterns with various frequencies, number of bits, and FPGA devices. The proposed technique exhibited a decrease in the power consumption. Table 2 indicates that the power consumption of the proposed techniques when the number of bits changes. The average dynamic power conservation is 10.17%, as shown in Fig. 5.

The frequency range of the proposed design is 2–20 MHz. Communication equipment, such as routers, TV, radio, aviation, and navigation devices, operates in this frequency range. Compared with previous encoding–decoding techniques, the average total power change is 5.5%. Table 3 displays the comparison between the even–odd and the proposed method.

Table 3 Power conserved through the upper/lower techniques	Dynamic power (mw)			Power conserved (%)
	Bit (bit)	Even/odd	Upper/lower	
	8	12.2	11.47	6.67
	16	17.7	15.18	14.43
	32	35.4	32.18	9.28

5 Conclusion

The upper–lower bus encoding technique was compared with the existing even–odd encoding technique. The comparison indicated that the proposed technique reduces the total delay up to 5%. When the number of bits increased to 16 and 32 bits, the complexity of the design increased and the delay reduction decreased. By using this encoder–decoder technique, the number of transitions was reduced. Moreover, the average power requirement decreased to 10.17% compared with that of the existing method. The limitation of the encoder–decoder design is the additional overhead of the design. However, in the long term, the power and the delay are reduced by using the design.

References

1. Stan, M.R., Burleson, W.P.: Bus invert coding for low power I/O. IEEE Trans. Very Large Scale Int. Syst. **3**, 49–58 (1995)
2. Kim, K.W., Beck, K.H., Shanbhag, N., Liu, C.L., Kang, S.M.: Coupling-driven signal encoding scheme for low power interface design. In: IEEE/ACM International Conference on Computer-aided Design. 318–321 (2000)
3. Ramprasad, S., Shanbhag, N.R., Hajj, I.N.: A coding framework for low-power address and data busses. IEEE Trans. Very Large Scale Integr. Syst. **7**, 212–221 (1999)
4. Benini, L., Micheli, G.De., Macii, E., Sciuto, D., Silvano, C.: Asymptotic zero-transition activity encoding for address busses in low-power microprocessor-based systems. In: Proceedings of IEEE/ACM Great Lakes Symposium on VLSI (GLS-VLSI), Urbana, IL. pp. 77–82, Mar 1997
5. Fornaciari, W., Polentarutti, M., Sciuto, D., Silvano, C.: Power optimization of system-level address buses based on software profiling, In: Proceedings of International Conference on Hardware/Software Codesign (CODES), pp. 29–33 (2000)
6. Aghaghiri, Y., Fallah, F., Pedram, M.: ALBORZ: address level bus power optimization. In: Proceedings of International Symposium of Quality Electronic Design (ISQED), pp. 470–475 (2002)
7. Komatsu, S., Fujita, M.: Irredundant address bus encoding techniques based on adaptive code-books for low power, In: Proceedings of the Conference on Asia South Pacific Design Automation (ASPDAC), pp. 9–14 (2003)
8. Stan, M.R., Burleson, W.P.: Limited-weight codes for low-power I/O, In: Proceedings of International Workshop on Low Power Design, Napa, CA, pp. 209–214, Apr 1994

9. Zhang, Y., Lach, J., Skadron, K., Stan, M.R.: Odd/even bus invert with two-phase transfer for buses with coupling. In: Proceedings of the International Symposium on Low Power Electronics and Design (ISLPED), pp. 80–83 (2002)

10. Baek, K.H., Kim, K.W., Kang, S.M.: A low energy encoding technique for reduction of coupling effects in SOC interconnects. In: Proceedings of 43rd IEEE Midwest Symposium Circuits and Systems, Lansing, MI, pp. 80–83, Aug 2000

Instrumentation for Measurement of Geotechnical Parameters for Landslide Prediction Using Wireless Sensor Networks

Mohammed Moyed Ahmed, Gorre Narsimhulu and D. Sreenivasa Rao

Abstract Landslides are the most serious geological disasters in our country, because of its short time of occurrence, causing heavy casualties, and huge economic losses. Due to the complexity of geographical conditions and some of the influence of technical factors such as ground displacement, groundwater conditions, and pore water pressure, a real-time dynamic monitoring of soil parameters is necessary to understand landslide dynamics and to provide an early warning mechanism. In this paper, we discuss an instrumentation model developed based on the digital geotechnical sensors and NI wireless sensor network platform with LabVIEW software, for monitoring and prediction of landslides.

Keywords Wireless sensor networks · Clustering · Geo-sensors
Early warning system

1 Introduction

Natural hazards, such as landslides, rock falls, and snow slides, are significant threat to the habitat in the highest mountain chain of the Himalayas, northeastern hill ranges, the Western Ghats and Konkan Hills [1]. In the areas where the presence of a large number of mountain landscapes, numerous urban population and roads are located near the mountains, during the rainy season, due to heavy rainfall, these areas are prone to landslides and pose a great threat to the lives and property of residents. Therefore, a flexible and stable system for monitoring and early warn-

M. M. Ahmed (✉) · G. Narsimhulu · D. Sreenivasa Rao
Department of ECE, College of Engineering, Jawaharlal Nehru Technological University
Hyderabad (JNTUH), Hyderabad, India
e-mail: mmoyed@gmail.com

G. Narsimhulu
e-mail: narsiroopa@gmail.com

D. Sreenivasa Rao
e-mail: dsraoece@gmail.com

© Springer Nature Singapore Pte Ltd. 2019
J. Wang et al. (eds.), *Soft Computing and Signal Processing* ,
Advances in Intelligent Systems and Computing 898,
https://doi.org/10.1007/978-981-13-3393-4_45

437

ing of landslides is very much necessary. However, because of monitoring areas are often inaccessible mountains and lack of roads, field wiring and power supply are all restricted to deploy cable system. In addition, the cable method is often used to record collected data by the data logger connected to nearest site, and it requires a dedicated person to regularly go to the monitoring point to download data. Therefore, the system cannot obtain real-time data and has poor flexibility. Wireless sensor networks (WSN) as relatively new technology with its unique ability to solve this problem with improved monitoring capability can be used for a flexible and stable system for monitoring and early warning of landslides. The WSN, in a broad sense, refers to any system that has sensors that sense the external environment variables and that are interconnected through wireless transmission. In a narrow sense, WSN generally refers to a sensor module using the Zigbee protocol (IEEE 802.15.4 standard). Zigbee is simply a short distance (approximately 50–150 m) and a low transmission rate (20–250 kbps). It sends and receives small amounts of data with low-power wireless transmission technology. Each sensor module can be regarded as a very small computer with three functions, including a microprocessor, micro-power sensors, and a low-voltage operation transmitting/receiving antenna, so that each sensor module can sense micro-sensors. The device can sense ambient environment information or monitor specific targets. It is customary to refer to each sensing module as a node. WSN combines wireless network technology, sensors, data loggers, and information technology for a variety of applications such as disaster (landslides) monitoring.

1.1 Design and Functioning of Wireless Sensor Networks

WSN [2] shown in Fig. 1 can be understood as computer networks of small intelligent devices, so-called sensor nodes, which deal with cooperation, a common task and wirelessly communicate with each other. To a sensor node, a variety of sensors can be connected. Sensor nodes receive the data from the sensors, caches, and if necessary further processes the information and finally passes to another node of the network.

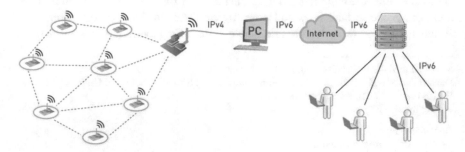

Fig. 1 WSN topology

The core of a sensor node is its processor with associated memory that can detect physical parameters of environment over long periods of time from the connected sensors. The sensor node has its own power source, usually a battery. But it may also be equipped with a charger provided by photovoltaic cells. It has also equipped with communication unit (transceiver) which can share information with other sensor nodes. In addition to the normal sensor nodes, each wireless sensor network has a base station, which serves as a data sink. More complex data processing is needed at some level, which cannot be performed on simple sensor nodes, so this job is done by the base station. Finally, the base station is responsible for forwarding of the data to stations outside the network. This can be done via General Packet Radio Service (GPRS) or similar technologies (GSM, UMTS, EDGE, landline, and radio link). On the other hand, the base station is also used for control and regulation of the entire network, for example, sometimes changes in the measurement cycle are necessary. To meet these responsibilities, the base station is usually equipped with more powerful hardware and software than the normal sensor nodes and has a stronger, long-lasting energy source.

2 Related Work

Active, slow-running landslides are to be monitored for determining the displacement direction of the ground surface as well as displacement rate [3]. This may be done for example by its geometric shape, its orientation in space, and its gravity field—as well as how they change over time. In many cases, however, this alone is not sufficient, in addition to the geometry of the sliding body and the kinematic observable properties at the ground surface including information on the hydro-geological and geotechnical properties of the substrate required for risk analysis of landslides [4] (e.g., pore water pressures and the depth of slip surface, etc.). This requires the installation of additional sensors in the field using geotechnical methods by which the collected spot data are forwarded to the skilled person who performs each of the assessment. Wireless sensor networks offer a solution for narrower measuring grid at a lower cost for the monitoring of landslides, which allows you to capture both geodetic and geotechnical data.

2.1 Landslide Monitoring System

The landslide monitoring system should be able to measure geotechnical parameters in addition to the field displacement. For this purpose, it is necessary that geotechnical sensors can be connected to the sensor nodes. For the designed alert system, parameters to be measured are inclination change, pore water pressure, soil temperature, and rainfall measurement.

Vertical In-place Inclinometer System For the purpose of long-term monitoring of lateral movement using in-place inclinometer system in soil of the slopes, an array of inclination measurement probes (tilt sensors) is installed inside a standard grooved inclinometer casing for real-time lateral movement monitoring soil layers. Inclinometer system measures the significant displacement of soil mass at different layers of soil. Real-time monitoring of inclination with in-place inclinometer system helps in observing behavior of ground movement that may adversely affect stability of the slope. Each hole will deploy a string of level sensor at the bottom and deploy several tilt sensors at different depths. Since the landslide phenomenon in the area is mainly caused by rain erosion, the depth of the groundwater table is the first indicator of the risk of landslides. These data are collected by a liquid-level depth sensor deployed at the bottom of the hole and sent by the wireless network, inside the inclinometer casing, there are a number of tilt-measuring sensors, and the spacing between each sensor is connected with a spacer rod over which inclination is measured. Spacer tubing length (mm) = gauge length (mm) −381 mm. Outside diameter of spacer tubing is 19 mm. The gauge length and the angle of tilt θ decide the lateral movement which is $L \, Sin \, \theta$, when there is a movement of soil mass, casing of the inclinometer shifts position, resulting in the change in the position of the sensors inside the inclinometer. Sensor output gives the tilt angle from the vertical, which can be converted into the lateral displacement using the formula $L \, Sin \, \theta$. The output from the sensors is acquired by the data acquisition system for real-time monitoring of the displacement of the soil. From an in-place inclinometer string assembly, the lateral displacement of the soil can be obtained by subtracting initial displacement from the current displacement, and since one end of the casing is fixed at the bottom end, complete profile can be easily calculated by adding the reading from preceding sensors.

Vibrating Wire Piezometer Pore pressure plays an important role in prediction of landslides, and it is the pressure experienced due to the water contained in the pores of the soil, rocks, or some concrete built structures. The study of pore pressure has following main purposes: Effect of water in pores of soil or rock is to reduce load-bearing capacity of soil or rock [5]. Effect is more pronounced with higher pore water pressure leading eventually in some cases to total failure of load-bearing capacity of the soil. One has to determine the ground water level and its flow pattern for the purpose of mass movement. Also we need to determine flow pattern of water in earth/rock fill and concrete dams and their foundations and to delineate the phreatic line. Calibration coefficients individually provided with each transducer. The gauge factor for polynomial linearity correction of pressure is calculated by following equation:

$$P = \left[A(R_1)^2 + B(R_1)^2 + C \right] mPa$$

where P is the pressure experienced in mPa units, R_1 is the reading observed during observation, and A, B, C are the polynomial coefficients. For transducers with a built-in interchangeable thermistor is used to record and monitor the temperature of

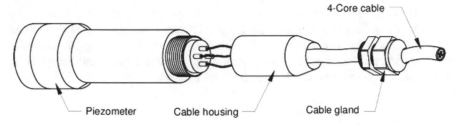

Fig. 2 Pore pressure sensor with thermistor

Fig. 3 Rain gauge (tipping bucket type)

the soil in degree Celsius. Vibrating wire sensor with the exception of the temperature sensor has a thermistor incorporated (Fig. 2) in it for temperature measurement.

Rain Gauge As shown in the Fig. 3 Rain gauge is used for monitoring the amount of rainfall in the field of landslide prone area. Tipping bucket mechanism is usually used in the rain gauge which also consists of a magnet and switch assembly, for the measurement of daily rainfall. Rain water is collected through a funnel-type arrangement with a debris filter. Water tips into the buckets. Tipping bucket mechanism activates magnetic switch that produces a pulse for each 0.2 mm of rainfall. Calibration of buckets can be done using two screws.

2.2 Connecting Sensor to Data Acquisition System

In sensor used in IPI chain tilt meter having SDI-12 interface, a power supply of 12 Vdc is required to interface the sensor chain which is provided by the data logger. Since the SDI-12 network is connected to the data logger by using a 3/6 pair cable

22 AWG size, the resonant frequency for measuring pore pressure can be read out by connecting it to second channel of data logger, which is connected to data acquisition system. Rain gauge is connected to the third channel of the data logger where the rainfall data can be stored into the memory of the data logger, and later can be transmitted to the remote server through WSN network.

2.3   Landslide Prediction Using WSN

Landslides' monitoring and early warning network architecture have to be designed keeping the following characteristics into consideration:

- Should support long-term monitoring of information from multiple sensors;
- Scalable, collaborative self-organized hierarchical network architecture;
- Support rapidly deploying multimedia broadband network for post-disaster emergency communications;
- Features of landslide monitoring sensor are adaptable to deployment densities [6], transfer location, and the deployment characteristics;
- Layered network architecture that supports long-term monitoring;
- Easy integration with the mobile communication network.

Wireless sensor network-based network management platform has been developed using NI nodes [7], gateway, and digital sensors with Encardio data logger. Network management, monitoring, and early warning system application are developed using LabVIEW on Windows server. Figure 4 shows the complete network architecture for the landslide monitoring. This network consists of three modules—first one is the WSN module, second one a gateway module, and the third one is the remote monitoring and control center where the data are stored and analyzed.

Wireless Sensor Network Module This module uses the Zigbee protocol (IEEE 802.15.4 standard), which is self-organizing property in a wireless sensor network that can automatically detect neighboring nodes and form a meshed multi-hop route that eventually converges to a gateway node. The entire process does not need to be manual intervention. At the same time, the entire network has dynamic flexibility. When any node is damaged or a new node is added, the network can automatically adjust the route and adapt to changes in the physical network at any time. This is the so-called self-healing property. **Gateway Module** The function of this module is to gather the information from the landslide-prone area collected by WSN and to transmit the information to the cloud for storage and analysis, and it uses the GPRS/Internet network for transmitting the information to the cloud. Remote data center has a real-time access to the cloud, and it can download the data at any point of time for the monitoring and analysis purpose. **Remote Monitoring Center Module** The function of the third module is located in the disaster monitoring center. For the analysis and prediction of landslides, LabVIEW platform has been used, data from

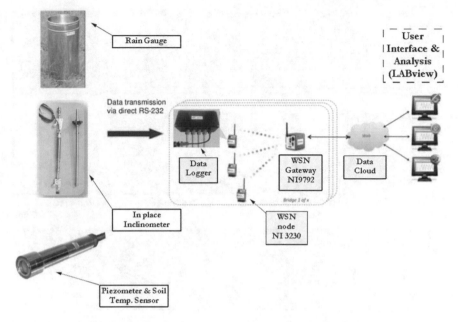

Fig. 4 Complete WSN setup for monitoring and prediction of geological disasters

the cloud are downloaded periodically at the remote monitoring center and analyzed, and in case of abnormality prevails, early warning can be sent to the disaster-prone areas [8].

3 Results and Analysis

Figure 5 shows the data collected for a period of time, and data from the geological sensors are pore pressure, soil temperature, amount of rainfall, and displacement parameters. These values are crucial for prediction of landslides and monitoring of the slip surface. For the development of early warning system, collected data will be compared with the threshold values and the warning will be generated.

4 Conclusion

An effective geological disaster monitoring and early warning system are currently being in great need to be developed in our country. By using WSN and geological digital sensors with integration of network management system, one can automati-

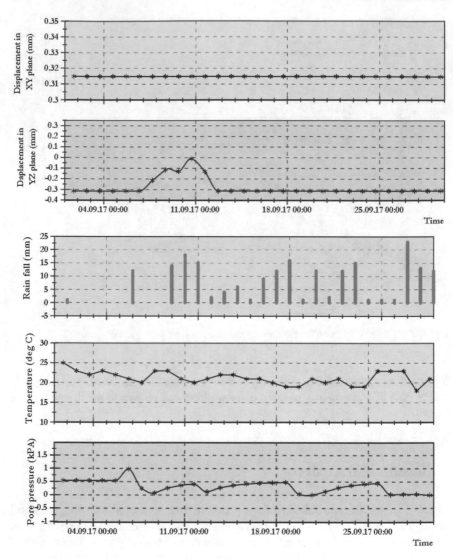

Fig. 5 Geotechnical data acquired from the field

cally collect and timely monitor information for detecting the slope failures. A comprehensive disaster monitoring system, which can produce early warning to avoid loss of lives and property, can be built using this technique. Instrumentation for the measurement of geotechnical parameters for the prediction of landslides has been successfully developed and tested. Using pore pressure, soil temperature, rain gauge, and in-place-inclinometer sensors, interfaced to Encardio data logger and by using National Instrument's WSN hardware (NI 3230 RS-232 node, NI 9792 gateway), a WSN has been developed. Disaster monitoring, data storage, and landslide prediction application have been developed using LabVIEW software. The system can be installed at the landslide-prone areas for real-time implementation in the future, and shortcomings will have to be addressed in future work.

Acknowledgements This research is supported in part by AICTE project under Research Promotion Scheme titled "Development of Early Warning System for Landslide Prediction"—DEWSLP-2017, JNTUHCEH, ECE Dept. No. 8–4/RFID/RPS/Policy-1/2016-17.

References

1. Sekhar, L., Kuriakose, Æ.G., Sankar, Æ., Muraleedharan, C.: History of landslide susceptibility and a chorology of landslide-prone areas in the Western Ghats of Kerala, India. Environ. Geol. **57**, 1553–1568 (2009). https://doi.org/10.1007/s00254-008-1431-9
2. Akyildiz, I.F., et al.: Wireless sensor networks: a survey, Comput. Netw. **38** (4), 393–422 (2002)
3. Arnhardt, C, Fernández-Steeger, T.M, Azzam, R.: Sensor fusion of position- and micro-sensors (MEMS) integrated in a wireless sensor network for movement detection in landslide areas. Geophys. Res. Abstr. **12**, EGU2010-8828 (2010)
4. Di Maio, C., Vassallo, R.: Geotechnical characterization of a landslide in a Blue Clay slope. Landslides **8**(1), 17–32 (2011)
5. Ramesh, M.V., Raj, R., Freeman, J., Kumar, S., Rangan, P.V.: Factors and approaches towards energy optimized wireless sensor networks to detect rainfall induced landslides. In: Arabnia, H.R., Clincy, V.A., Yang, L.T. (eds.) Proceedings of the 2007 International Conference on Wireless Networks (ICWN 2007), Las Vegas, NV, USA, pp. 435–438, 25–28 June 2007
6. Lambot, S., Weihermuller, L., Huisman, J.A., et al.: Mapping spatial variation in surface soil water content: comparison of ground penetrating radar J. J. Hydrol. **42**, W11403 (2006)
7. National Instruments. LabVIEW manual. http://www.ni.com/labview//
8. Peak, Y.L., Wen, Z.: Crop water based wireless sensor network condition monitoring system research and design. Agric. Eng. **25**(2), 60–67 (2009)

Improving Diagnostic Test Coverage from Detection Test Set for Logic Circuits

Bommidi Madhan and J. P. Anita

Abstract The proposed work aims at generating a diagnostic test set which is a compact test set derived from a large set of test vectors generated from any automatic test pattern generator (ATPG). This diagnostic test set is required to find out the exact location of the faults. The patterns generated from the ATPG may be sufficient to find out whether the circuit is fault free or not, but will not give the location of the fault. Hence, the proposed method aims at identifying the exact location of the faults. The experiment has been carried out on several ISCAS'85 and ISCAS'89 benchmark circuits.

Keywords Fault location · Fault diagnosis · Test coverage · Test set

1 Introduction

Testing is an integral part of the VLSI design cycle. With the advancement in IC technology, designs are becoming more and more complex and the feature size of transistors keeps decreasing making testing a challenging task. Testing occupies 60–80% time of the design process. Hence, the goal of the semiconductor industry is to produce quality devices. Fault diagnosis plays an important role in testing. Fault diagnosis [1] is used to understand the cause of failure in the defective chip. This is done by modeling the faults to be diagnosed, and the circuit is then simulated for the detection of these faults.

Vector inputs or patterns for checking semiconductor device faults are automatically generated using ATPG process. The test devices' response to known vector inputs is compared against the expected response from true circuit [2]. A device is

B. Madhan · J. P. Anita (✉)
Department of Electronics and Communication Engineering, Amrita School of Engineering,
Amrita Vishwa Vidyapeetham, Coimbatore, India
e-mail: jp_anita@cb.amrita.edu

B. Madhan
e-mail: madhanbommidi@gmail.com

© Springer Nature Singapore Pte Ltd. 2019
J. Wang et al. (eds.), *Soft Computing and Signal Processing* ,
Advances in Intelligent Systems and Computing 898,
https://doi.org/10.1007/978-981-13-3393-4_46

faulty, if an error occurs in this response. Fault coverage and cost of performance are the parameters which decide the effectiveness of ATPG process.

The most important factor of the VLSI process is the fault analysis, which is the method of identifying the reasons of failure in a defective chip. Fault coverage is the ratio of the detected faults to the total number of faults in the list. The huge set of fault list is reduced by collapsing the set of faults in the list by using the concept of fault equivalence and fault dominance [3]. Hence, there should be at least one pattern to detect each fault.

The proposed work aims at minimizing the test set and maximizing the test coverage. In the proposed work, the stuck-at faults [4] are only considered. Here some lines in the circuit have stuck-at faults which is either at logic 0 or at logic 1. The motivation behind the diagnosis is to find the origin of cause of failure of the chip and also to find out the exact location of the fault. By this process, we can find the diagnostic coverage [5, 6]. The proposed work aims at finding the fault coverage and improving the diagnostic coverage.

Stuck-at faults are among one of the most naturally occurring faults [7]. Hence, in the proposed work only stuck-at faults are considered. The initial step in the testing is the identification of the fault. A fault can occur at any gate input or at the gate output or at any intermediate node in the circuit. Next, this fault has to be sensitized so that the cause of the fault propagates to the output. This makes the detection of the fault easier.

2 Literature Survey

Tang et al. [8] have used a new technique called statistical learning algorithm to analyze the diagnostic results. Using Monte Carlo simulation, they showed achievability and effect of different factors.

Pomeranz [9] recommended that the effects which support the expulsion of tests likewise support the removal of noticeable yields from consideration during defect diagnosis. In particular, a test may make a yield reaction that an imperfection conclusion system will not have the capacity to translate effectively. This may influence some observable yields more strongly than others. In this manner, the evacuation of perceptible yields from thought can enhance the precision of analysis. This paper portrays a summed up, expanded diagnosis finding system that removes tests and perceptible yields from consideration. It presents test results to show the impacts of removing observable outputs on the accuracy of diagnosis.

Ye et al. [2] have proposed the volume diagnosis assuming an imperative part in the testing process. To get good result, extreme noticeability design is fundamental. In any case, the test designs utilized by volume diagnosis generally have low noticeability to particular flaws. In the tests, it was found that all things considered, under programmed produced test designs, faults in the same fan-out-free locale (FFR) represent just small percentage of conceivable fault sets, however their offer altogether indistinct deficiencies are 70%; faults in various FFRs yet with a similar perception focuses account for 4% of all fault sets, yet their offer altogether indistinct issues are 22%. Declining the statement that issues in the same FFR are harder to recognize, we give a technique called SFPAT-ADPG. By using different circuits, another fault was rundown to a current ATPG device [10]. We produce the compacted test designs and likewise the symptomatic examples with extreme noticeability for the first circuit.

Investigations on ISCAS'89 circuit demonstrate the applicability of the proposed method strategy.

McCluskey and Clegg [3] have proposed the study of the effects of faults on combinational logic circuits. Specifically, the conditions whereby two different faults can create a similar modification in the circuit behavior were investigated. This connection between two faults [10] appeared to be an equivalence relation, and three different types of equivalence relations are indicated. Important and adequate conditions for the presence of these equivalence relations are demonstrated. An algorithm for deciding the equivalence classes for one of the kinds of equivalence was discussed.

Veneris et al. [11] have proposed an algorithm which compares whether two faults are same or not. If they are not the same, a test vector distinguishing the two faults is returned by the algorithm. The effectiveness of this proposed method is experimented on the benchmark circuits.

3 Proposed Work

The proposed work aims at obtaining a reduced diagnostic test set. Figure 1 shows the block diagram of the proposed work.

The netlist of the circuit under test (CUT) for which the test set has to be obtained is given to the ATPG. The ATPG generates the test vectors for all the faults in the fault list. The complete test vectors constitute the test set. Fault list is the set of all faults that can occur in any CUT. When any circuit is tested by applying the test vectors, the output responses are compared to the golden response. When there is a

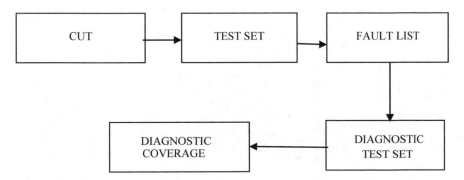

Fig. 1 Block diagram of proposed work

mismatch between the output response and the golden response, a fault is identified. The set of all such faults constitute the fault list.

The set of test vectors [12] from the test set which detect the faults are the pass vectors. The set of all such pass vectors constitute the diagnostic test set. The overall coverage of the circuit with a total number of faults is known as diagnostic coverage. The diagnostic test set gives the location of the fault when the fault list and the test set are given. The diagnostic coverage is determined from the diagnostic test set. The experiment has been carried out on several ISCAS'85 and ISCAS'89 benchmark circuits. The results were obtained using Synopsys TetraMAX DC Compiler tool.

4   Results and Discussion

The circuit was simulated, and the collapsed set of faults were obtained for the benchmark circuits. Table 1 gives the results obtained for ISCAS'85 benchmark circuits. In c17 circuit, all the faults have been detected as it is a very small circuit. However, in other large circuits shown, not all the faults are detected, and hence, the fault coverage is reduced. CPU time is the CPU execution time in seconds which again depends on the complexity of the circuit. Similarly, Table 2 gives the results obtained for ISCAS'89 benchmark circuits. Here, again the results depend on the circuit complexity, location of the fault, and the nature of the fault.

Table 3 gives the diagnostic coverage obtained for ISCAS'85 benchmark circuits. Here, a measure of the diagnostic coverage is taken. Also, the number of collapsed faults based on equivalence and dominance is obtained. Table 4 gives the diagnostic coverage obtained for ISCAS'89 benchmark circuits.

Table 1 Number of faults and the coverage obtained for ISCAS'85 benchmark circuits

Circuit	Number of test patterns	Number of faults	Number of faults detected	Number of faults unde-tected	Fault coverage (%)	Fault efficiency (%)	CPU time (s)
c17	8	36	36	0	100.0	100.00	0.02
c432	25	1110	149	961	13.42	100.00	0.07
c499	79	2390	2318	72	96.99	96.99	0.72
c880	73	2104	1254	850	59.60	100.00	0.85
c1908	53	3816	2196	1620	57.55	100.00	1.37
c1355	75	2726	1702	1024	62.44	100.00	1.02
c2670	167	6520	6278	242	96.29	98.29	38.23
c3540	79	7904	2419	5485	30.60	99.75	35.95
c5315	200	10094	10038	56	99.45	99.97	2.86

Table 2 Number of faults and the coverage obtained for ISCAS'89 benchmark circuits

Circuit	Number of test patterns	Number of faults	Number of faults detected	Number of faults unde-tected	Fault coverage (%)	Fault efficiency (%)	CPU time (s)
s27	9	64	64	0	100.00	100.00	0.00
s298	31	528	528	0	100.00	100.00	0.01
s344	44	652	652	0	100.00	100.00	0.02
s713	73	1426	1353	73	94.88	100.00	0.36
s1423	92	2846	2820	26	99.09	99.61	1.82
s5378	329	10538	10417	121	98.85	99.49	2.51
s13207	583	26358	26060	298	98.87	100.00	6.24
s35932	77	71224	63842	7382	89.64	100.00	659.1
s38417	1373	76678	76433	245	99.68	99.95	71.80
s385841	856	76864	73457	3407	95.57	100.00	46.95

Table 3 Diagnostic coverage obtained for ISCAS'85 benchmark circuits

Circuit	Collapsed faults	Diagnostic coverage (%)	Total time (s)
c17	22	100	026
c432	520	97.86	1316
c499	750	98.76	1.15
c880	942	95.26	2.49
c1355	1566	59.87	14.89
c1908	1872	86.89	16.45
c2670	2632	94.86	68.48
c3540	3297	96.89	31.54

Table 4 Detected and undetected faults for the ISCAS'89 benchmark circuits

Circuit	Detected	Undetected	Total fault	Diagnostic coverage (%)
S24	9	4	16	75.00
S298	66	2	68	100
S344	147	15	182	88.02
S510	24	19	44	54.55
S526	32	2	36	94.12
S1238	140	10	180	100

5 Conclusion

The proposed method has obtained a diagnostic test set for stuck-at faults from a group of detection vectors generated from an ATPG tool and obtained the collapsed faults and its diagnostic coverage for both ISCAS'85 and ISCAS'89 benchmark circuits. The proposed work can be extended to other faults like transition faults and delay faults.

References

1. Anita, J.P., Sudheesh, P.: Test power reduction and test pattern generation for multiple faults using zero suppressed decision diagrams. Int. J. High Perform. Syst. Arch. **6**(1), 51–60 (2016)
2. Ye, J., Zhang, X., Hu, Y., Li, X.: Substantial fault pair at-a-time (SFPAT): an automatic diagnostic pattern generation method. In: Proceedings of ATS, pp. 192–197, Dec 2010
3. McCluskey, E.G., Clegg, F.W.: Fault equivalence in combinational logic networks. IEEE Trans. Comput. **C-20**(11), 1286–1293 (1971)
4. Mohan, N., Anita, J.P.: A zero suppressed binary decision diagram based test set relaxation for single and multiple stuck-at faults. Int. J. Math. Model. Numer. Optim. **7**(1), 83–96 (2016)
5. Lin, Y.-C., Lu, F., Cheng, K.T.: Multiple-fault diagnosis based on adaptive diagnostic test pattern generation. IEEE Trans. Comput. Aided Des. Integr. Circuits Syst. **26**(5), 932–942 (2007)
6. Tang, H., Reddy, S.M.: Diagnosis of multiple faults based on fault-tuple equivalence tree. In: Proceedings of DFTS, pp. 217–225, Oct 2011
7. Pomeranz, I.: A test selection procedure for improving the accuracy of defect diagnosis. IEEE Trans. Very Large Scale Integr. (VLSI) Syst. **24**(8), 2759–2767 (2016)
8. Tang, H., Manish, S., Rajski, J., Keim, M., Benware, B.: Analyzing volume diagnosis results with statistical learning for yield improvement. In: Proceedings of ETS, pp. 145–150, May 2007
9. Pomeranz, I.: Improving the accuracy of defect diagnosis by considering reduced diagnostic information. In: Proceedings of VTS, p. 16, Apr 2015
10. Kundu, S., Jha, A., Chattopadhyay, S., Sengupta, I., Kapur, R.: Framework for multiple-fault diagnosis based on multiple fault simulation using particle swarm optimization. IEEE Trans. Very Large Scale Integr. (VLSI) Syst. **22**(3), 696–700 (2014)
11. Veneris, A., Chang, R., Abadir, M., Amiri, M.: Fault equivalence and diagnostic test generation using ATPG. In: Proceedings of ISCAS, pp. V-221–V-224, May 2004
12. Lee, H., Ha, D.: ATALANTA: an efficient ATPG for combinational circuits. Technical Report 93-12, Virginia Polytech. Inst. State Univ., Blacksburg, VA, USA, pp. 12–93

Design and Modelling of Different Types of SRAMs for Low-Power Applications

V. Vijayalakshmi and B. Mohan Kumar Naik

Abstract In this work, different types of SRAMs were designed and characterized these cells in terms of temperature, voltage, static noise margin, leakage power, and total power. These SRAMs were designed and stimulated in Cadence using 45-nm technology.

Keywords SRAM cell · Leakage power · Static noise margin · Cadence

1 Introduction

Many devices were scaled down to nanometre regime to achieve lower power consumption and lower power dissipation. But many problems had occurred in these devices as they are scaled down such as more static power dissipation, more noise and reduced battery life in cell phones [1]. Therefore, efficient SRAM designs are required to improve battery life in cell phones as well as to reduce noise and reduced static power dissipation.

This work focuses on the design and modelling of different types of SRAMs for low-power applications. These different types of SRAM cells were characterized in terms of temperature, voltage, static noise margin, and total power consumption. These different types of proposed circuits were stimulated using Cadence in 45-nm technology.

Figure 1 below shows the BASIC 6T SRAM CELL. In this circuit, it has the combination of two inverters (one inverter consisting of N0 transistor and P0 transistor and another inverter consisting of P1 transistor and N1 transistor) and also has two access transistors (N2 transistor and N3 transistor).

V. Vijayalakshmi · B. Mohan Kumar Naik (✉)
New Horizon College of Engineering, VTU, Bengaluru, India
e-mail: mohan72003@gmail.com

V. Vijayalakshmi
e-mail: vijupushpendra@gmail.com

© Springer Nature Singapore Pte Ltd. 2019
J. Wang et al. (eds.), *Soft Computing and Signal Processing* ,
Advances in Intelligent Systems and Computing 898,
https://doi.org/10.1007/978-981-13-3393-4_47

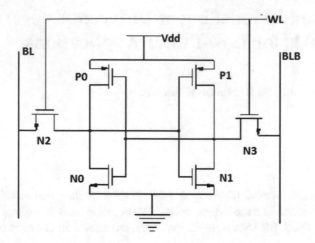

Fig. 1 BASIC 6T SRAM CELL

Here these two inverters are used to store zero bit or one bit on N0 transistor, P0 transistor, N1 transistor and P1 transistor. Access transistors N2 and N3 are made ON and OFF by using word line (WL). The bit lines that are used in this circuit are BL and BLB.

The three different modes or states of operation for BASIC 6T SRAM CELL are as follows: write mode (data are written), read mode (data are read) and standby mode (data cannot be read or written) [2].

2 Proposed Work

2.1 Circuit 1

A PMOS transistor was added between BASIC 6T SRAM CELL and supply which acts as a power-gating transistor as shown in Fig. 2, i.e. proposed circuit 1. This circuit reduces the supply voltage to SRAM cell as it acts a resistance between power supply and SRAM cell and that produces a virtual V_{dd} at SRAM cell. The MP3 transistor is self-biased, and its width is made four times SRAM cell inverting transistors' width. Resistance produced by MP3 transistor decreases by increasing its width so that leakage power is reduced in this proposed circuit 1 when compared to BASIC 6T SRAM CELL.

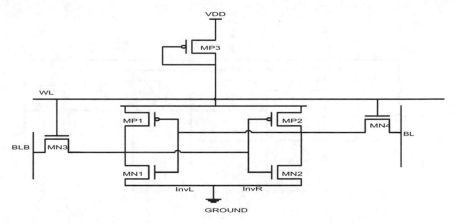

Fig. 2 Proposed circuit 1

2.2 Circuit 2

A NMOS transistor was added in the ground path of a BASIC 6T SRAM CELL as
shown in Fig. 3, i.e. proposed circuit 2. This circuit reduces the subthreshold leakage
currents by increasing ground level during read mode or write mode or standby mode
as it acts as a resistance between ground and SRAM cell. The power-gating transistor
is self-biased, and its width is made four times SRAM cell inverting transistors' width.
Resistance produced by power-gating transistor decreases by increasing its width so
that leakage power is reduced in this proposed circuit 2 when compared to BASIC
6T SRAM CELL.

2.3 Circuit 3

In this proposed circuit, power-gating transistors are used in both at source voltage
node and at ground path so this may be called a combination of above two proposed
circuits 1 and 2 as shown in Fig. 4, i.e. proposed circuit 3. Here MP3 transistor reduces
the supply voltage to SRAM cell as it acts a resistance between power supply and
SRAM cell and that produces a virtual V_{dd} at SRAM cell. The MP3 transistor is
self-biased, and its width is made four times SRAM cell inverting transistors' width.
Similarly, MN5 transistor reduces the subthreshold leakage currents by increasing
ground level during read mode or write mode or standby mode as it acts a resistance
between ground and SRAM cell. The power-gating transistor is self-biased and its
width is made four times SRAM cell inverting transistors width. Resistance produced
by power-gating transistors decreases by increasing its width so that leakage power
is reduced in this proposed circuit 3 when compared to BASIC 6T SRAM CELL [3].

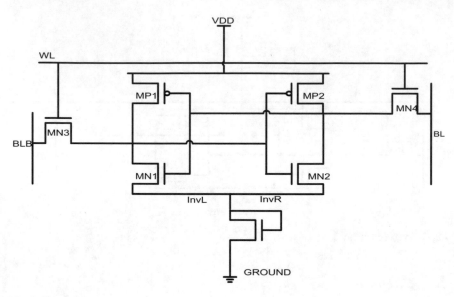

Fig. 3 Proposed circuit 2

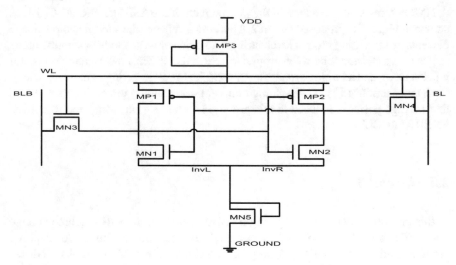

Fig. 4 Proposed circuit 3

2.4 Circuit 4

An NMOS and a PMOS transistor are added between SRAM cells as shown in Fig. 5, i.e. proposed circuit 4 and supply voltage having different supplies to the sources of both the transistors. Whenever WL = 1, then SRAM cell would get supply from both the sources while when WL = 0 (standby state) only would provide supply to cell which is half of actual V_{dd} so less power would be required to maintain the bit stored in cell.

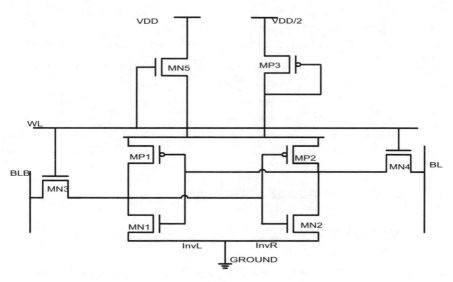

Fig. 5 Proposed circuit 4

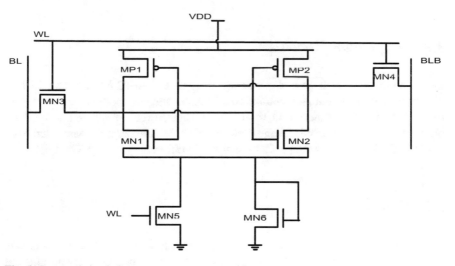

Fig. 6 Proposed circuit 5

2.5 *Circuit 5*

A couple of NMOS transistors is added between SRAM cell and ground node as shown in Fig. 6, i.e. proposed circuit 5. One of this having gate connected to WL while other is self-biased. Whenever WL = 1 (active mode), it provides 0 V to SRAM cell while when WL = 0 (standby mode), it provides voltage somehow greater than 0 V.

Table 1 Temperature analysis in different types of SRAM cells

Temperature (°C)	BASIC 6T CELL	Circuit 1	Circuit 2	Circuit 3	Circuit 4	Circuit 5
0	699	0.96	1.25	0.0998	0.00289	1.09
10	46.4	1.15	1.52	0.128	0.0041	1.34
20	50.1	1.35	1.88	0.162	0.00568	1.7
30	53.7	1.41	2.41	0.203	0.00769	2.22
40	981	1.52	3.16	0.251	0.0102	2.99

3 Simulation Results and Waveforms

3.1 EDA Tool Used

Cadence tool is provided by Cadence Design Systems. The various tools used were the Virtuoso Schematic Editor and Spectre Simulator.

3.2 Temperature Analysis

Leakage power of each proposed circuits along with BASIC 6T SRAM CELL is observed under different temperature conditions in standby mode operation. When the circuit is in standby mode (WL = 0), then leakage power is observed through each proposed circuits. Observed leakage power is in E^{-8} W. Temperature was varied from 0 to 40 °C. Table 1 gives temperature analysis in different types of SRAM cells.

3.3 Voltage Analysis

Performance of each proposed circuits along with BASIC 6T SRAM CELL was observed under different source voltages given to the circuits. When the circuits are in standby mode (WL = 0), then leakage power is observed through each circuits. Observed leakage power is in nW. Voltages varied from 0.6 to 1.1 V. Table 2 gives the voltage analysis in different types of SRAM cells.

3.4 Static Noise Margin Analysis

While reducing leakage power in circuit stability of circuit is also considered which is observed using DC analysis to obtain a butterfly curve which is used to calculate Static Noise Margin (SNM). Here static noise margin analysis is done for standby

Table 2 Voltage analysis in different types of SRAM cells

Voltage (V)	BASIC 6T CELL	Circuit 1	Circuit 2	Circuit 3	Circuit 4	Circuit 5
0.6	15.3	0.507	0.63	0.119	0.008	0.546
0.7	188	1.67	2.07	0.308	0.0182	1.8
0.8	2450	5.18	6.67	0.774	0.036	5.8
0.9	526	29.18	22.3	1.9	0.0704	20.4
1	1450	45.3	80.06	4.56	0.135	77.3
1.1	3630	221	297	10.8	0.254	292

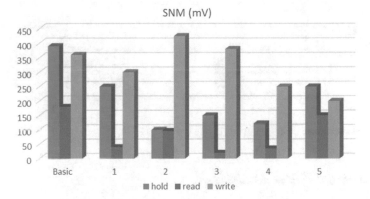

Fig. 7 Static noise margin in different types of SRAM cells for hold mode, read mode and write mode operations

Table 3 Static noise margin in different types of SRAM cells

Different types of SRAM cells	Static noise margin		
	(Standby) Hold mode operation	Read mode operation	Write mode operation
BASIC 6T CELL	390	180	360
Circuit 1	250	40	300
Circuit 2	100	96	425
Circuit 3	150	20	380
Circuit 4	122	35	250
Circuit 5	250	150	200

mode, read mode and write mode operations for different types of SRAM cells as shown in Fig. 7. Static noise margin is measured in milli volts. Table 3 gives the static noise margin in different types of SRAM cells.

Table 4 Leakage power in different types of SRAM cells

Circuit	Leakage power (nW)
BASIC 6T CELL	526
Circuit 1	29.18
Circuit 2	22.3
Circuit 3	1.9
Circuit 4	0.07
Circuit 5	20.4

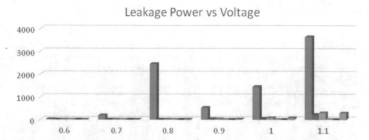

Fig. 8 Leakage versus voltage in different types of SRAM cells

3.5 Leakage Power Analysis

Leakage power analyses are done for BASIC 6T SRAM CELL and also for the different types of proposed circuits under normal room temperature condition, and source voltage is set at 0.9 V as shown in the Table 4. Leakage power is measured in nW [4]. Figure 8 shows leakage power versus voltage in different types of SRAM cells.

3.6 Total Power

Total power drawn from source in active mode of cell is observed for all the proposed circuits along with BASIC 6T CELL. The source voltage is set at 0.9 V. Table 5 gives total power in different types of SRAM cells. Figure 9 shows the total power consumption in different types of SRAM cells.

3.7 Leakage Power Versus Temperature Variation

Leakage power for any cell increases with increase in temperature, and in the graph shown below it is observed that leakage power for all proposed circuits is less than BASIC 6T SRAM CELL at any temperature. Figure 10 shows leakage power versus temperature variation in different types of SRAM cells.

Table 5 Total power in different types of SRAM cells

Circuit	Total power (nW)
BASIC 6T CELL	8.8
Circuit 1	83.21
Circuit 2	3.2
Circuit 3	174
Circuit 4	2.08
Circuit 5	5.6

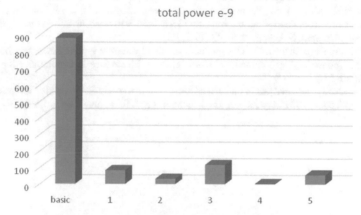

Fig. 9 Total power consumption in different types of SRAM cells

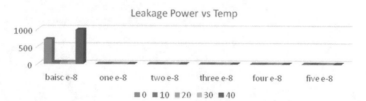

Fig. 10 Leakage power versus temperature variation for different types of SRAM cells

4 Conclusion

In this paper, different types of SRAM cells were designed and modelled using Cadence tool at 45-nm technology. From simulation results, it was found that leakage power is reduced in all proposed circuits with respect to BASIC 6T SRAM CELL. Leakage power increases on increasing source supply voltages as well as on increasing temperatures at a particular source voltage. Total power consumption in all proposed circuits is also reduced to a significant amount. However, stability of proposed circuits is affected as SNM for each state (hold, read and write) is reduced in almost each proposed circuits. Hence, SRAMs can be used in low-power applications.

Acknowledgements The authors wish to thank New Horizon College of Engineering for supporting this work.

References

1. Gupta, D.C., Raman, A.: Analysis of leakage current reduction techniques in SRAM cell in 90 nm CMOS technology. Int. J. Comput. Appl. **50**(19), 0975–8887 (2012)
2. Shukla, N.K., Singh, R.K., Pattanaik, M.: Design and analysis of a novel low-power SRAM bit—cell structure at deep—sub-micron CMOS technology for mobile multimedia applications. Int. J. Adv. Comput. Sci. Appl. **2**(5), 43–49 (2011)
3. Vamsi Kiran, P.N., Saxena, N.: Design and analysis of different types SRAM cell topologies. In: IEEE Sponsored 2nd International Conference on Electronics & Communication System (ICECS 2015)
4. Mitra, A: Design and analysis of an 8T read decoupled dual port SRAM cell for low power high speed applications. World Acad. Sci. Eng. Technol. Int. J. Electron. Commun. Eng. **8**(4) (2014)

Analysis of Coding Unit Size and Coding Depth for Class A Sequence in HEVC

Shruti D. Sawant and Shilpa P. Metkar

Abstract High Efficiency Video Coding (HEVC) standard is the most recent video coding standard which outperforms its predecessors in terms of Bit Rate saving while maintaining the video quality. Flexible block partitioning structure is one of the prominent features of HEVC encoder. It overcomes limitations of H.264 such as limited block size or depth of partitioning. Even though it enhances performance of the encoder, it does add substantial computational load. In order to accelerate performance, we address the complexity reduction of block partitioning in HEVC in this paper. Thorough analysis of block partitioning structure with detailed results is presented for Class A test sequence.

Keywords HEVC · CU · Block partitioning

1 Introduction

HEVC is jointly finalized by the ITU-T and the ISO/IEC working together as the Joint Collaborative Team on Video Coding [1]. HEVC standard aims to double coding efficiency relative to the H.264, with the same video quality at an average Bit Rate saving of 50% [2].

Major source of improvement from one generation of video coding standard to next is increasing number of possibilities of coding a frame. These include supporting larger block sizes, increased precision of motion vectors, and increased number of intra-prediction as well as inter-prediction modes. These features while improving coding efficiency of the standard impose a tremendous computational burden on the encoder. HEVC encoder fully exploiting the capabilities is expected to be several

S. D. Sawant (✉) · S. P. Metkar
College of Engineering, Pune, Pune, India
e-mail: sawantsd16.extc@coep.ac.in

S. P. Metkar
e-mail: metkars.extc@coep.ac.in

© Springer Nature Singapore Pte Ltd. 2019
J. Wang et al. (eds.), *Soft Computing and Signal Processing*,
Advances in Intelligent Systems and Computing 898,
https://doi.org/10.1007/978-981-13-3393-4_48

times more complex than an H.264/AVC encoder [3]. Hence, reducing complexity of HEVC encoder has been a topic for research in recent years.

A fast intra-mode decision algorithm proposed in [4] uses two-step approach to reduce unlikely modes in rate-distortion optimization (RDO). Lee et al. [5] used by-products of encoding process of HEVC to propose fast algorithm for inter-coding. A number of efforts have been made to investigate an early coding unit (CU) termination, early detection of skip mode, and coded block flag algorithms [6].

Although higher extent of adaptability is achieved by quadtree-based block partitioning in HEVC, it may cause unnecessary transmission of redundant sets of motion information. These redundancies can be avoided by merging leaves of quadtree. In [7], such algorithm for block merging in HEVC is proposed.

In this paper, focus is on complexity of block portioning in HEVC. An extensive analysis of improvement in coding efficiency with respect to various coding depths and block sizes is carried out. The paper contains brief description of block partitioning structure of HEVC in Sect. 2. Methodology is explained in Sect. 3. Section 4 presents CU size analysis for Class A test sequence. Detailed experimental results are shown in Sect. 5. Section 6 concludes the paper.

2 Block Partitioning Structure in HEVC

HEVC adopts a flexible quadtree-based block partitioning structure along with motion-compensated prediction which distinguishes the standard from its predecessors and contributes to substantial Bit Rate saving offered by HEVC relative to H.264 [8].

Each frame in HEVC is divided into equally sized distinct square blocks, which act as the roots of quadtree structure and thus are called coding tree blocks (CTBs). Coding tree unit (CTU) contains a luma CTB, two chroma CTBs along with syntax information. They can have sizes from 8×8 up to 64×64. The CTU can be a single CU or further split into four smaller CUs of equal sizes which are nodes of quadtree [8]. Figure 1a shows division of largest CU (LCU) into CUs of various sizes. The coding depth for LCU is 0, and it increases by one every time CU is split into four equal CUs. Prediction decision (intra-coding or inter-coding) is made at CU level.

A CU can be split into one or more PUs as per the PU splitting type. There are three types of PUs based on mode of prediction, namely skipped CU, inter-coded CU and intra-coded CU. Inter-coded PU with zero motion vectors and residual energy is called skipped PU [8].

HEVC supports square and symmetric partition modes as well as newly introduced asymmetric motion partition (AMP) modes [8].

The TU is a basic unit to apply the integer transformation and the quantization. CUs can be split into one or more TUs. It forms another quadtree, called residual quadtree with CU at its root. TU splitting types in HEVC are based on whether corresponding PU uses square symmetric partitioning or AMP [8].

(a)

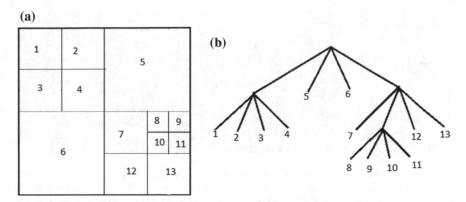

Fig. 1 **a** Partitioning of LCU (64 × 64) into CUs of various ranging from *8 × 8*, *16 × 16*, and *32 × 32* and **b** corresponding quadtree. The numbers indicate coding order of CUs

3 Methodology

For analysis of block partitioning, HEVC test model (HM) reference software has been used [9]. To acquire optimum block partitioning structure for a frame, HM checks rate-distortion (RD) costs for all the possible prediction modes by increasing coding depth at each subdivision until maximum coding depth is reached. RD cost for skip mode is checked first followed by intra-modes and inter-modes. Out of three units, CU size determinations play a major role in RD performance. PU and TU size determinations further add large computational costs. Also, this process of exhaustive search is highly time consuming. Thus, there is a need to address the complexity of block partitioning in HEVC encoder for its practical implementation. Thorough knowledge about its structure is essential, to come up with fast block partitioning algorithm.

4 Analysis of CU Size

While working with block partitioning structure of HEVC, we observed that certain parameters like CU size, depth of partitioning play a major role in deciding the best-suited partitioning structure for a particular frame. As a test case, Class A test sequence called "Traffic" is tested for different combinations of CU size and coding depths. The major objective is to analyze the effect of these parameters on encoder performance as well as their impact on computational complexity of encoder.

Consider Fig. 2, where two combinations of CU size and depths are tested. In first case, higher CU sizes are available which are suitable for homogeneous regions. On the other hand in second case, smaller CU sizes cover regions of complex motion more effectively. In high-resolution videos, a frame contains larger smooth regions,

Fig. 2 Motion-compensated frame no. 3 from "Traffic" sequence corresponding to c64d2 and c32d2 (*left* and *right*). 'c' indicates CU size, and 'd' indicates coding depth. The differences due to use of larger CU size can be seen around homogeneous regions corresponding to road. Use of smaller CU size is observed in regions of moving vehicles

and thus larger block sizes prove beneficial whereas smaller block sizes are suitable for adapting local properties of frame. Thus, it is vital to carry out extensive analysis of block partitioning structure for various depths and CU sizes.

5 Experimental Results

HEVC test model [9] referred to HM16.14 is used in Visual Studio 12.0 and Ubuntu environment using common test conditions [10]. HEVC main profile and low delay B configuration are used. Class A test sequence, "Traffic" of resolution 2560×1600 with frame rate of 30 fps, 5 s duration (150 frames), is used. It has stationary camera motion, no background change, and slow motion of vehicles. For thorough analysis of the encoder performance, we evaluated the following two cases.

5.1 Variable CU Size

In this case we have kept minimum CU size constant to 8×8 and maximum CU size is varied from 64×64, 32×32 and 16×16. Quantization parameter (QP) values of 22, 27, 32, and 37 are used. These are the standard values used to plot RD curve as well as for computation of BD-Bit Rate and BD-PSNR.

For thorough analysis of block partitioning structure, Y-PSNR, Bit Rate, BD-Bit Rate, BD-PSNR are recorded as performance parameters.

Peak signal-to-noise ratio (PSNR) is calculated for each YCbCr component. As human visual system is more sensitive to luminance, Y-PSNR values are considered. Higher PSNR values for larger CU sizes indicate that larger CU size results in better visual quality. Table 1 lists the Y-PSNR values for variable CU sizes.

Table 1 Y-PSNR values for variable CU size

Test sequences	QP	Y-PSNR		
		16×16	32×32	64×64
Class A	22	41.7702	41.8513	41.8669
	27	39.0683	39.1638	39.2108
	32	36.5860	36.7288	36.7951
	37	34.0309	34.2556	34.3257

Table 2 Bit Rate values for variable CU size

Test sequence	QP	Bit Rate		
		16×16	32×32	64×64
Class A	22	13915.4948	13623.9318	13523.8663
	27	4778.0489	4507.9587	4465.9345
	32	2104.1009	1902.8923	1865.9630
	37	1072.5632	910.8548	866.7737

Table 3 BD-PSNR and BD-Bit Rate values for variable CU size

Test sequence	BD-PSNR		BD-PSNR	
	16×16	32×32	16×16	32×32
Class A	−0.4882	−0.1022	17.4369	3.6675

An efficient encoder is said to have low Bit Rate. With improvements in computation capabilities, next-generation video coding standards provide a significant reduction in Bit Rate. Table 2 shows Bit Rate values for various CU sizes, and higher CU sizes have lower Bit Rate as when we encode larger blocks, lesser bits are required.

BD-PSNR and BD-Bit Rate metrics are an ITU-T approved metric for computing the average gain in PSNR and the average percent Bit Rate saving [11–13]. Table 3 shows average BD-PSNR difference and percentage BD-Bit Rate saving. These values are calculated for CU sizes 32×32 and 16×16 considering 64×64 as reference. Ideally, BD-PSNR values should be positive as PSNR values are higher for 64×64 than those for lower CU sizes. BD-Bit Rate values should ideally be negative as higher CU sizes require lesser Bit Rate.

As shown in Fig. 3, a curve corresponding to size 64×64 and 32×32 has negligible gap. This test sequence contains frames where background is present at very less extent and most part of a frame contains vehicles moving. BD-PSNR corresponding to 32×32 size is negligible which confirms that the performance of 32×32 is comparable to that of 64×64.

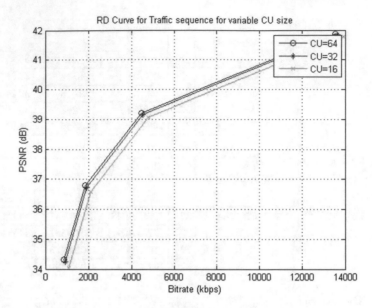

Fig. 3 RD plot for Class A sequence for variable CU size. It contains three RD plots corresponding to three CU sizes

Table 4 Available CU sizes at various CU sizes and coding depths	Coding size (c)	Coding depth (d)	Available CU sizes
c64d2	64×64	2	$64 \times 64, 32 \times 32$
c32d2	32×32	2	$32 \times 32, 16 \times 16$

5.2 Variable CU Size and Coding Depths

In this case, two combinations of CU size and coding depths are tested. Table 4 lists these combinations and available CU sizes in each combination. Here, 'c' and 'd' indicate the depth of coding, respectively. The support to various sizes of CU is more beneficially compared to fixed size of macroblock in preceding video coding standard. This ability proves useful, especially for adapting to variety of applications and to support lower resolution videos which are still widely used.

Figure 4 shows RD plot corresponding to c64d2 and c32d2. It shows that c32d2 outperforms c64d2 at higher Bit Rate. The reason behind this behavior is that smaller CU sizes available in c32d2 combination provide better coverage of regions corresponding to complex motion of vehicles involved in "Traffic" sequence.

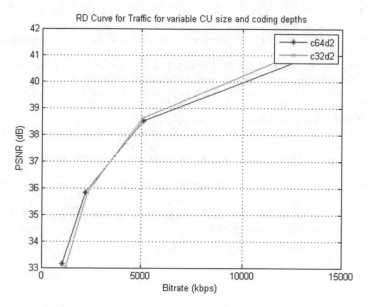

Fig. 4 RD plot for Class A sequence for variable CU sizes and coding depths

6 Conclusion

Experimental results reveal that testing RD costs for all possible CU sizes is computationally an expensive task, since some of these CU sizes do not result in a significant performance improvement. There is a possibility to accelerate the HEVC encoder by early termination of the CU determination process. Future work of the project is to develop an adaptive algorithm for block partitioning of HEVC where the suitable partitioning structure as in the CU size and coding depth is chosen according to the resolution of the test sequence as well as the motion of the objects involved.

References

1. Gary, J.S., Woo-Jin, H., Thomas, W.: Overview of the high efficiency video coding (HEVC) standard. IEEE Trans. Circuits Syst. Video Technol. **22**(12), 1649–1668 (2012)
2. Ohm, J.-R., Gary, J.S., Schwarz, H., Tan, T.K., Wiegand, T.: Comparison of the coding efficiency of video coding standards—including high efficiency video coding (HEVC). IEEE Trans. Circuits Syst. Video Technol. **22**(12), 1669–1684 (2012)
3. Frank, B., Benjamin, B., Karsten, S., David, F.: HEVC complexity and implementation analysis. IEEE Trans. Circuits Syst. Video Technol. **22**(12), 1720–1731 (2012)
4. Zhang, H., Ma, Z.: Fast intra mode decision for high efficiency video coding (HEVC). IEEE Trans. Circuits Syst. Video Technol. **24**(4), 660–668 (2014)
5. Ahn, S., Lee, B., K., M.: A novel fast CU encoding scheme based on spatiotemporal encoding parameters for HEVC inter coding. IEEE Trans. Circuits Syst. Video Technol. **25**(3), 422–435 (2015)

6. Belghith, F., Kibeya, H., Loukil, H., Ayed, M.A.M.: A new fast motion estimation algorithm using fast mode decision for high efficiency video coding standard. J. Real Time Image Process. **11**, 675–691 (2016)

7. H., Philipp, O., Simon, B., Benjamin: Block merging for quadtree-based partitioning in HEVC. IEEE Trans. Circuits Syst. Video Technol. **22**(12), 1649–1668 (2012)

8. Kim, K., Min, J., Lee, T., Han, W.-J., Park, J.: Block partition structure in the HEVC standard. IEEE Trans. Circuits Syst. Video Technol. **22**(12), 1697–1706 (2012)

9. HEVC Test Model: https://hevc.hhi.fraunhofer.de/

10. Bossen, F.: JCT-VC-L1100: common test conditions and software reference configurations. In: Proceedings of the 12th JCTVC Meeting, Geneva (2013)

11. Bjontegaard, G.: Calculation of average PSNR differences between RD-curves. In: Doc. VCEG-M33, Austin, TX (2001)

12. Bjontegaard, G.: Improvement of BD-PSNR model. In: Doc. VCEG-AI11, Berlin, Germany (2008)

13. MATLAB code for BD-PSNR and BD-bitrate: https://www.mathworks.com/matlabcentral/fileexchange/27798-bjontegaard-metric

A Novel Approach of Low Power, Less Area, and Economic Integrated Circuits

Sanjeeda Syed, R. Thriveni and Pattan Vaseem Ali Khan

Abstract Foremost leading-edge integrated circuit applications like microprocessors, digital signal processors, RF processors, and mixed-signal processors need low power and less area of die utilization. This paper presents entirely a different method of designing and analyzing the primary modules of four-bit ripple carry adders, multiplexers of various sizes, full adders, and full subtractors based on decomposite Shannon's expansions and MGDI techniques. The proposed systems are modules in nature indicating that it can easily be upgraded to various leading-edge ICs. The core benefit is less transistors and less power consumption. The implementation and analysis of these digital systems are done by using Microwind and Digital Schematic at $0.12\,\mu m$ technology and analyzed area and total power consumption with different technologies and power supply voltages.

Keywords CMOS · MSD · MSDFA · MGDI · RCA

1 Introduction

The requisite for low-power designs in today's technology is very high as the VLSI technology is advancing, so, from last few decades, much attention has been given to low-power integrated circuit design with the reduction in supply voltage because the power consumption is an important aspect in VLSI circuits in every corner of

S. Syed (✉) · R. Thriveni
Department of Electronics & Communication Engineering, Aditya College of Engineering, Valasapalle (Post), Punganur Road, Madanapalle 517325, India
e-mail: sanjusyed786@gmail.com

R. Thriveni
e-mail: thriveni.phd@gmail.com

P. V. A. Khan
Department of Information & Technology, Honda Motor Cycles & Scooters India PVT., LTD-3F, Bangalore, India
e-mail: pvaseem80@gmail.com

© Springer Nature Singapore Pte Ltd. 2019   471
J. Wang et al. (eds.), *Soft Computing and Signal Processing* ,
Advances in Intelligent Systems and Computing 898,
https://doi.org/10.1007/978-981-13-3393-4_49

electronics and communications in the field of wireless communication, biomedical application, computers, processors, etc. An important factor pertaining to CMOS digital circuit is the threshold voltage which does not guarantee to decrease much below than the present threshold voltage. In accumulation to cost, the reliability problem is a major aspect in CMOS circuits. With the doubling of on-chip transistors in every two years, minimizing the power consumption has become currently a challenging area of research. This power deliberation has become very great as it required a position as a design parameter along with the other two parameters like speed and chip area. In this study, we have designed, simulated, and analyzed four-bit ripple carry adders, multiplexers of various sizes, full adders, and full subtractors based on Shannon's expansions or Shannon's Decomposite and MGDI techniques and compared its power utilization with respective CMOS designs [1, 2].

2 Propose Methodology

2.1 Shannon's Decomposition (SD)/Shannon's Expansion Theorem

The proposed Shannon's decomposition circuits are implemented by pass transistors. Pass transistors logic reduces the number of transistors used to make different logic, by eliminating redundant transistors [3]. These pass transistors are used as switches to pass logic levels between nodes of a circuit. This reduces the number of active devices. Pass transistors are simple and fast as they conduct current in either direction; complex gates are implemented with less number of transistors. Shannon's decomposition theorem states that it is a method by which any Boolean function can be represented as a sum of two sub-functions [4].

2.2 Modified Shannon's Decomposed Logic Designs

In this paper, we have five proposed modified Shannon's decomposed logic designs they are MSD full adder, MSD full subtractor, MSD ripple carry adder, MSD 4:1 mux, and MSD 2:1 mux.

2.2.1 Modified Shannon's Decomposed Full Adder

Full adder is a digital circuit which performs summation of three bits and produces two outputs. The first two inputs are x and y, and third input is an input carry as z_{in}. The outputs of full adder are sum and carry [5].

SD Boolean equations of sum and carry are shown in Eqs. (1) and (2) [6, 7].

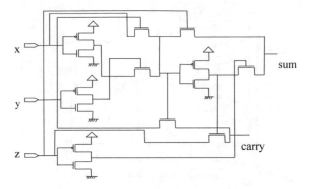

Fig. 1 MSD full adder

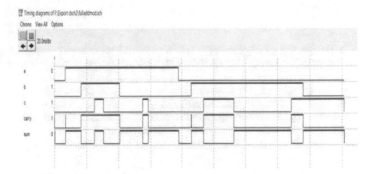

Fig. 2 Timing waveforms of modified Shannon's decomposed full adder

$$\text{Sum} = (x \text{ xor } y)z' + (x \text{ xnor } y)z \tag{1}$$

$$\text{Carry} = (x \text{ xor } y)z + (x \text{ xnor } y)y \tag{2}$$

This full adder is shown in Fig. 1 and is constructed using pass transistor logic. In PTL, inputs are directly applied to both drain or source and gate of pass transistor, and the two inputs are multiplied, and output is taken at source or drain [8]. These full adders are mainly used in ALU's, to construct higher adders, multipliers, comparators, and for DSP processors, etc. Timing waveforms of modified Shannon's decomposed full adder are shown in Fig. 2.

2.2.2 Modified Shannon's Decomposed Full Subtractor

Full subtractor is a digital circuit that performs subtraction of three bits. One is minuend, second is subtrahend and the final is borrow of LSB. Hence, there are three inputs for full subtractor and it produces two outputs as difference (D) and borrow (B).

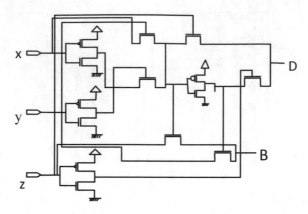

Fig. 3 MSD full subtractor

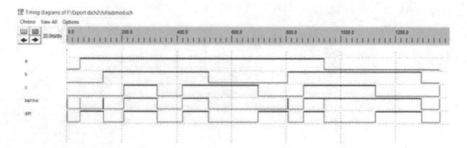

Fig. 4 Timing waveforms of full subtractor

Full subtractor is shown above in Fig. 3 and it is constructed by pass transistors using Shannon's decomposition boolean expression. In PTL, inputs are directly applied to both drain and gate of pass transistor, and the two inputs are multiplied, and output is taken at source [9]. Full subtractors are mainly used in ALU's and used to calculate addresses in processor. These are also utilized in many other parts of processor, table indices, and also in decrement operators. Timing waveforms of full subtractor are shown in Fig. 4.

2.2.3 Modified Shannon's Decomposed Ripple Carry Adder

RCA shown in Fig. 5 is a circuit that is used to add two n-bit numbers. It is constructed using one-bit MSD full adders connected in cascade, the first MSD full adder output C0 is given as carry input to another MSD full adder. Generally, as it is a four-bit adder, it consists of inputs as A and B (four-bit numbers) [10]. Carry from one full adder is given to another preceding full adder, similarly for all the four full adders, and finally this ripple carry adder produces sum and carry of total sum as four bits and one-bit carry. Small ripple carry adder is also used in many high-speed adders

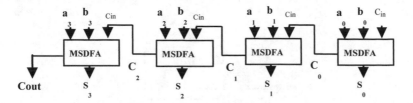

Fig. 5 Modified Shannon's decomposed ripple carry adder

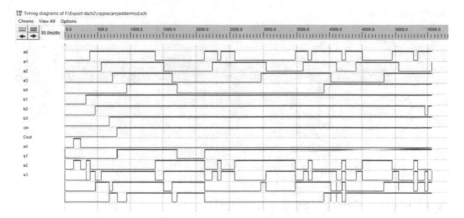

Fig. 6 Timing waveforms of ripple carry adder

like carry select adder [11] and high-speed multipliers like Vedic multipliers. Timing waveforms of MSD ripple carry adder are shown in Fig. 6.

2.2.4 Modified Shannon's Decomposed Multiplexer

Multiplexer or mux is a circuit that selects one of the input signals and forwards the selected input into a single line. For this selection of inputs, we have selection lines based on number of inputs; if there are 2^n number of inputs, then there are n selection lines. These multiplexers are used in communication systems for data transmission. For high-performance speed switching, we go for mux which is constructed by basic electronic elements, and these are accomplished by handling both analog and digital applications; this increases the efficiency of the communication system by allowing the transmission data. Multiplexers are used in computer memory to retain large amount of memory and also to decrease the quantity of copper lines required to connect memory to other parts of the computer [12]. The multiplexers simply decide which input is to be connected to the output, also called as data selector, and also it

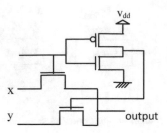

Fig. 7 2 to 1 mux

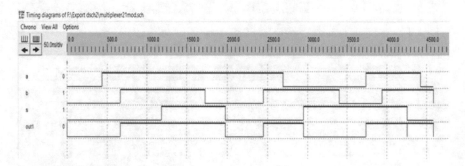

Fig. 8 Timing waveforms of 2:1 mux

Fig. 9 4 to 1 MSD mux

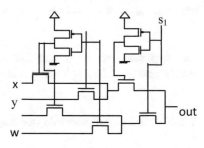

is used in complex combinational circuits like barrel shifters and high-speed adder like carry select adder.

2:1 MSD Mux: 2:1 MSD multiplexer selects any one input from given two inputs. Now, these 2:1 multiplexers are implemented using pass transistor logic which is shown in Fig. 7, and timing waveforms of 2:1 MSD mux are shown in Fig. 8.

4:1 MSD Mux: This 4:1 multiplexer selects any one of the input signals from given four input signals through selection lines. As it is a 4:1 mux, it has four inputs and two selection lines as shown in Fig. 9, and timing waveforms of 4:1 MSD mux are shown in Fig. 10.

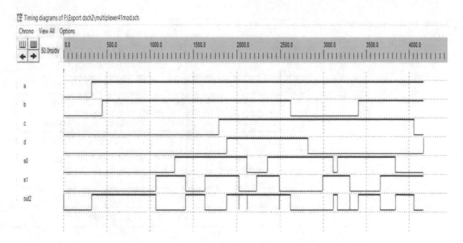

Fig. 10 Timing waveforms of 4:1 MSD mux

2.3 MGDI Technique

MGDI technique refers to modified gate diffusion. The design of MGDI cell consists of both pMOS and nMOS transistors [8]. MGDI design structure uses a four-terminal MOSFET; we have four terminals as diffusion nodes of both p-transistor and n-transistor as P and N and common gate terminal as G and common diffusion node as output. In this MGDI technique, inputs are given to terminals P, N, G and output is taken at out terminal. The output equation of MGDI technique [8] is given as in Eq. (3).

$$Out = G'P + GN \tag{3}$$

The output equation is given based on pass transistor logic, i.e., both using pMOS pass transistor and nMOS pass transistor in Table 1. Here the substrate of pMOS is connected to Vdd, and the substrate of nMOS transistor is given to GND [13]. This makes the GDI for constant body biasing of MGDI cell which increases the loading effect and circuit stability. Through this MGDI technique, we can develop many logic functions such as AND, OR, NOT, XOR, XNOR, MUX, NAND, NOR.

2.3.1　MGDI Full Adder

The same Shannon's decomposition-based full adder Boolean expression [14, 15] is also implemented using MGDI technique which makes the use of less number of transistors and is shown in Fig. 11, providing less area and less power consumption by making use of MGDI output equation [12, 16].

Table 1 Truth table of MGDI

N	P	G	Out	Function
0	y	x	$z'y$	f1
y	1	x	$x'+y$	f2
x	y	x	$x+y$	or
y	x	x	xy	and
y	y'	x	$(xy)'$	nand
x'	y	x	$(x+y)'$	nor
z	y	x	$x'y+xz$	mux
0	1	x	x'	not
b	b'	a	$x'y+xy'$	xor
b'	b	a	$x'y'+xy$	xnor

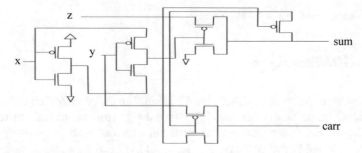

Fig. 11 MGDI full adder

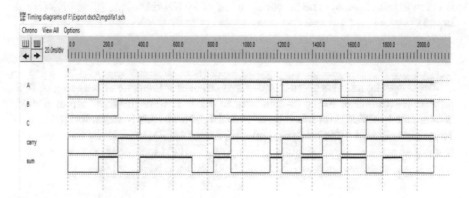

Fig. 12 Timing waveforms MGDI full adder

The timing waveforms of MGDI full adder are shown in Fig. 12 as these timing waveforms are generated by Digital Schematic tool.

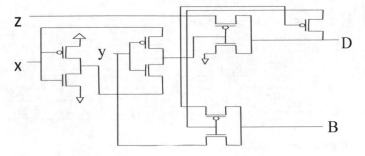

Fig. 13 MGDI full subtractor

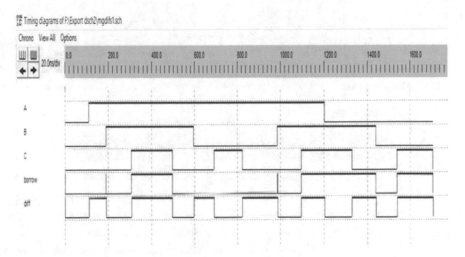

Fig. 14 Timing waveforms of MGDI full subtractor

2.3.2 MGDI Full Subtractor

The same Shannon's decomposition-based full subtractor Boolean expression is also implemented using MGDI technique [17]; the circuit diagram of full subtractor using MGDI is shown in Fig. 13 with less transistor count. Timing waveform of MGDI full subtractor is shown in Fig. 14.

2.3.3 MGDI Ripple Carry Adder

Four-bit MGDI ripple carry adder needs four MGDI full adders and must connect them in series to obtain a design of MGDI four-bit ripple carry adder. The timing waveforms of MGDI ripple carry adder are shown in Fig. 15.

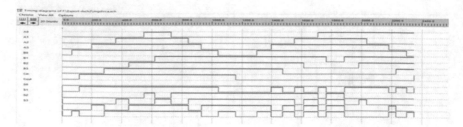

Fig. 15 Timing waveforms of RCA

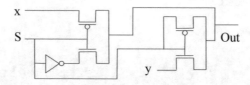

Fig. 16 MGDI 2:1 mux

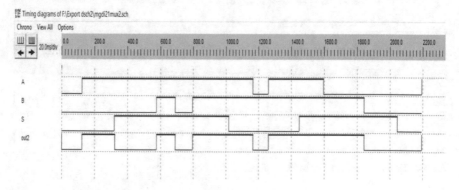

Fig. 17 Timing waveforms of MGDI 2:1 mux

2.3.4 MGDI 2:1 Mux

A MGDI 2:1 mux is designed with less transistors. By using selection lines, we can select any one input from given two inputs as shown in Fig. 16, and also timing waveforms are shown in Fig. 17.

2.3.5 MGDI 4:1 Mux

A MGDI 4:1 mux is designed with less transistors. By using selection lines, we can select any one input from given four inputs as shown in Fig. 18, and also timing waveforms are shown in Fig. 19.

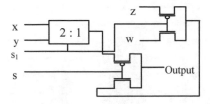

Fig. 18 MGDI 4:1 mux

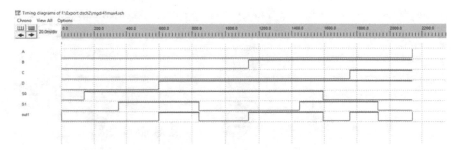

Fig. 19 Timing waveforms of 4:1 mux

3 Power Analysis

All the power analysis and area analysis of various logic families are shown in Table 2 and Table 3, respectively, and also it is shown in Fig. 20 and Fig. 21, respectively.

Table 2 Performance comparison of conventional CMOS versus proposed MSD, MGDI designs

Technique	Full adder (μw)	Full subtracter (μw)	RCA (μw)	2:1 mux (μw)	4:1 mux (μw)
CMOS	18.473	17.445	25.057	12.420	10.453
MSD	14.031	14.031	22.206	1.234	0.380
MGDI	3.623	3.623	4.008	0.604	0.251

Table 3 Performance comparison of conventional CMOS versus proposed MSD, MGDI designs

Technique	Full adder (μm * μm)	Full subtracter (μm * μm)	RCA (m * μm)	2:1 mux (μm * μm)	4:1 mux (μm * μm)
CMOS	62 * 12	90 * 13	163 * 14	28 * 13	54 * 14
MSD	28 * 13	28 * 13	130 * 15	31 * 7	81 * 9
MGDI	19 * 12	19 * 12	54 * 13	11 * 10	17 * 11

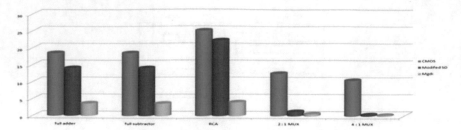

Fig. 20 Average **power** consumption of CMOS versus proposed MSD, MGDI designs

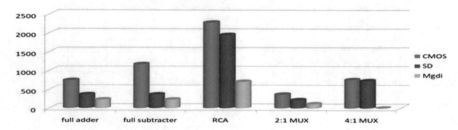

Fig. 21 Area required of conventional CMOS versus proposed MSD, MGDI designs

4 Conclusion

MGDI- and MSD-based multiplexer (2:1 and 4:1), four-bit ripple carry adder, one-bit full subtractors are designed and analyzed in this paper. MGDI and MSD designs of this paper are successfully operated at low power as compared with CMOS designs [8]. The result shows that among all MGDI, MSD, and CMOS designs, MGDI is best suited for low-power applications. All the designs are designed with minimum count of transistors. Reduced number of transistors results in great reduction of dynamic power and area which finally improves the speed of design. Performance analysis clearly specify that proposed MSD and MGDI designs are best compared to CMOS techniques. Hence, MGDI technique and above designs can be used in low-power applications like processors, smart cards.

References

1. Saravanan, S., Madheswaran, M.: Design of low power, high performance area efficient Shannon based adder cell for neural network training. In: Control, Automation, Communication and Energy Conversation, INCACEC (2009)
2. Lee, J.D., Yoony, Y.J., Leez, K.H., Park, B.-G.: Application of dynamic pass-transistor logic to an 8-bit multiplier. J. Kor. Phys. Soc. **38**(3), 220–223 (2001)
3. Purnima, K., AdiLakshmi, S., Sahithi, M., Rani, A.J., Poornima, J.; Design of modified Shannon based full adder cell using PTL logic for low power applications

4. Rawat, S., Sah, A., Pundir S.: Implementation of Boolean functions through multiplexers with the help of Shannon expansion theorem
5. Jiang, Y., Al-Sheraidah, A., Wang, Y., Sha, E., Chung, J.-G.: A novel multiplexer-based low-power full adder. IEEE Trans. Circuits Syst. II: Express Briefs **51**(7), 345 (2004)
6. Singh, M., Pandit, M.K., Jana, A.K.: Economic full adder circuit in VLSI using Shannon expansion
7. Aliotoa, M., Di Cataldob, G., Palumbob, G.: Mixed full adder topologies for high-performance low-power arithmetic circuits. Microelectron. J. **38** (2007). Received in revised form 4 Sept 2006; Accepted 11 Sept 2007
8. Vishalatchi, S., Dhanam, B., Ramasamy, K.: Design, analysis and implementation of various full adder using GDI and MGDI technique
9. Bhattachatyya, P., Kundu, B., Ghosh, S., Kumar, V., Dandapat, A.: Performance analysis of a low-power high-speed hybrid 1-bit full adder circuit. IEEE Trans. VLSI Syst. **19**(4) (2015)
10. Mohanty, B.K., Patel, S.K.: Area–delay–power efficient carry-select adder. IEEE Trans. Circuits Syst. II: Express Briefs **61**(6) (2014)
11. Ruiz, G.A., Granda, M.: An area-efficient static CMOS carry-select adder based on a compact carry look-ahead unit. Microelectron. J. Received in revised form 23 Aug 2004; Accepted 2 Sept 2004
12. Ramkumar, B., Kittur, H.M.: Low-power and area-efficient carry select adder. IEEE Trans. Very Large Scale Integr. (VLSI) Syst. **20**(2) (2012)
13. Chang, C.H., Gu, J.M., Zhang, M.: A review of 0.18 μm full adder performances for tree structured arithmetic circuits. IEEE Trans. Very Large Scale Integr. (VLSI) Syst. **13**(6) (2005)
14. Hassoune, I., Flandre, D., O'Connor, I.: ULPFA: a new efficient design of a power-aware full adder. IEEE Trans. Circuits Syst. I: Regul. Pap. **57**(8) (2010)
15. Navia, K., Moaiyeri, M.H., Mirzaeem, R.F.: Two new low-power full adders based on majority-not gates. Microelectron. J. (2009)
16. Goel, S., Kumar, A., Bayoumi, M.A.: Design of robust, energy efficient full adders for deep submicrometer design using hybrid CMOS logic style. IEEE Trans. Very Large Scale Integr. (VLSI) Syst. **14**(12) (2006)
17. Aguirre-Hernandez, M., Linares-Aranda, M.: CMOS full adders for energy-efficient arithmetic applications. IEEE Trans. Very Large Scale Integr. (VLSI) Syst. **19**(4) (2011)

Design and Implementation of Car for Smart Cities—Intelligent Car Prototype

Anita Chaudhari, Dhvani Shah, Kiran Mungekar and Vidhan Wani

Abstract Transportation is the most essential need of human beings. Making cars smart will be a breakthrough, where they automatically learn to drive on streets. We will look forward to map the road by itself and self-learn the survival on the roads. Self-driving vehicles detect and avoid obstacles and objects. It may happen that while driving a person suffers from a heart attack or severe headache, then based on his facial expressions, our system will automatically send SMS to his family members. Also, if the user is feeling sleepy, using mobile phones, or looking outside for long time mistakenly, then our system will raise an alarm in such situations. Autonomous car is making sure of reaching its last stop safely and cleverly, thus escaping the risk of human mistakes and taking the necessary decisions related to the real world. Lane- and obstacle-detection algorithms are used to provide the required control to the car.

Keywords Self-driving · Road and obstacle finding algorithms · Autonomous car Intelligent

1 Introduction

Vehicles have been the most powerful breakthrough in technology. Whoever has ever anyone thought about making these vehicles smart. Over the decades, man has been working on the perfection of vehicles to make them extremely fast. Making cars smart

A. Chaudhari (✉) · D. Shah · K. Mungekar · V. Wani
St. John College of Engineering and Management, Palghar, India
e-mail: anitac@sjcet.co.in

D. Shah
e-mail: dhvanis@sjcet.co.in

K. Mungekar
e-mail: mungekarkiran05@gmail.com

V. Wani
e-mail: vpwani_bhayandar@yahoo.co.in

© Springer Nature Singapore Pte Ltd. 2019
J. Wang et al. (eds.), *Soft Computing and Signal Processing* ,
Advances in Intelligent Systems and Computing 898,
https://doi.org/10.1007/978-981-13-3393-4_50

will be a breakthrough, where they automatically learn to drive on streets. We will look forward to map the road by itself and self-learn the survival on the roads. This is one of the prominent implications of artificial intelligence. A persuasive part of an eventual autonomous car's development is communication over cars and structure. Linked vehicle machinery provides a great chance to implement an efficient and smart map-reading system [1].

We can monitor our vehicle from anywhere by using the concept of Internet of things. After an accident, alert message should be sent to the nearest hospital and also to the family members. By using machine learning, we improve decision-making system of the car. From that, car can decide if an obstacle is detected, then the car has to stop or overtake the obstacle. In case of security, if the car gets into accident with another car, a quick notification is sent to the driver's family members and the nearest hospital to get immediate medical help. In another case, when the driver gets heart attack or is having headache, then just by giving voice command, the car shall drive you to the nearest hospital. If the car is stolen, it also tracks and gives the location of the car to police. The car must follow all traffic rules. This car consists of in-built alcohol level detection system which measures the level of alcohol in the atmosphere and if it is high then the car won't be allowed to start for the safety of life. Not only for human's safety and security, but also car helps in making a global eco-friendly city by giving air pollution and noise pollution values to municipal corporations.

2 Related Work

This section details the literature survey of different algorithm and machine learning techniques used for training of autonomous cars. We did the study on different hardware devices that support our system.

A new technique to select the mark and unmark lane edges is explained in specifics depending on OpenCV. The algorithm has been successfully implemented on a small self-directed car [2, 3]. The outcomes from the Canny Edge detection algorithm are used to differentiate various objects and for image analysis [4]. Higher performance is inhibited by slowly bringing current inputs to the steering control system [5]. Car tracing is done using Kalman [6]. Stereo vision method is used for monitoring the obstacle and calculating the final shortest path [7]. Car owner assistance is provided by noticing the sideway lane by video streaming through the camera attached to smart car. It is possible to show the distance apart from obstacle on the display [8]. In [9], the authors presented the methodology for tracking the location of smart car, fuel level, current temperature and humidity. The best part of the idea is to improve a remote-controlled vehicle which can be implemented using IoT above a protected server. So that it can drive itself safely in case of connectivity loss. The main goal here is to minimize the risk of human life and guarantee that highest safety during driving. A small car including the above features has been developed which shows optimum performance in a simulated environment [10, 11]. Robots can take decision like in which direction they have to take right or left [12]. Collision avoidance system is implemented which detects the presence of an obstacle coming toward the vehicle as well as on the blind spot of a vehicle and alerts the driver accordingly. This system used ultrasonic sensor for detection of real-time moving or stationary obstacle under all-weather environment [13].

In this connection, an attempt is made to integrate obstacle detection, vehicle-to-car communication, and voice control module to provide the required controls to the car using the Raspberry Pi. A unique and novel approach in the above work is controlling the vehicle by using the idea of IoT [14]. In the proposed work, Raspberry Pi Camera module is used for detection of objects and collection of images. Analysis is made on an image in order to confirm the algorithm suitable for edge detection of images [15].

The main focus is to make the use of face recognition technique for the vehicle. The face recognition methodology enables face recognition of the authorized users of the vehicle to be enrolled in the records. Before any user can access the vehicle, the image of his face is checked with the faces in the database. The driver with no match in the database is not allowed to access the vehicle [16]. Capturing real-world images using camera and then masking and contour methods are used to identify the red signals of the transportation. So car can take own decision and avoid accidents [17].

For authentication, different authors proposed different schemes; two-factor authentication can be used by using unique features of devices [18]. Authentication plays an important role in the verification of documents, digital certificate, biometric identifiers [19] ascertained the approach involved in the authentication of the activities performed by the users and we defined as inherence factors. Subsequently, these factors were generalized as static and dynamic biometric parameters which are considered. Furthermore, biometric identifiers were categorized as distinctive and measurable. These parameters were considered to label and describe individuals. Li [20] developed a fingerprint multibiometric system cryptosystems based on decision level fusion whose accuracy was higher than cryptosystems with single biometric [21].

2.1 Convolutional Neural Network

We mainly have the images recorded from human driving in different settings like highways, cities. After having the dataset, we preprocess the information to make our algorithm work. We design a useful function in Python that allows us to dynamically load a small batch of data, reprocess it, and then output it directly into a neural network. We used Keras for that, making it completely readable. The main model is the classical CNN, which eliminates the need for feature extraction which is very important and time-consuming. This model can work in real time on a laptop or on a minicomputer like Raspberry Pi, and we predict the steering angle based on the current frame.

3 System Methodology

Making cars smart will be a breakthrough, where they automatically learn to drive on streets. We are proposing prototype for the intelligent car by using machine learning and IoTs. We will look forward to map the road by itself and self-learn the survival on the roads. This is one of the prominent implications of artificial intelligence. Autonomous car senses dynamic objects using sensors like cameras, laser scanner. The car can use map from many data streams, combine it with sensors, and take decision. It may happen that while driving a person suffers from a heart attack or severe headache, then based on his facial expressions and his facial images, our system will automatically send an SMS to his family members. Also, if the user is feeling sleepy, using mobile phones while driving, or looking outside the window for long time mistakenly, then our system will raise an alarm in such situations. This helps in minimizing the accidents and may be life-saving for a lot of people. The increasing pollution has started producing trouble for humans. So by attaching humidity and noise sensor for autonomous car, it will give a reading of noise and toxic gases present in a particular area. The moving autonomous car sends the data report after every one hour to environmental management team so that necessary action can be taken. From that, government can come to know the pollution rate of a particular area at a particular time and make a plane for managing rate. Figure 1 shows the system architecture for this research.

To incorporate security, we will use fingerprint scanners to unlock the door of the car and send notification to Android phone for confirmation. On the other hand, in case of any accident happened to autonomous car, the emergency notification will be sent with the current location to the nearest hospital along with the owner's family members. Similarly, in case of any medical emergency happened with the driver, the autonomous car drives him safely to the nearest hospital along with informing to the family. Autonomous car has not just followed the traffic rules but also make sure the human interaction with the machine is in the proper way. We use many exciting algorithms to control the car.

Hardware technology used for the smartness of the car. The Raspberry Pi is a low-cost, small-sized computer used for the classification procedure. The camera is connected to Raspberry Pi. The camera is used to capture video as well as photographs to identify different states of the traffic light (red, yellow and green), different types of sign boards, and road directions. The camera also identifies human body, animals, and vehicles moving on the road for safety. The camera also detects drowsiness of driver. An ultrasonic sensor measures the distance to an object and detects obstacles by using sound waves. Gas sensor is used for detecting smoke and harmful gases (e.g.,

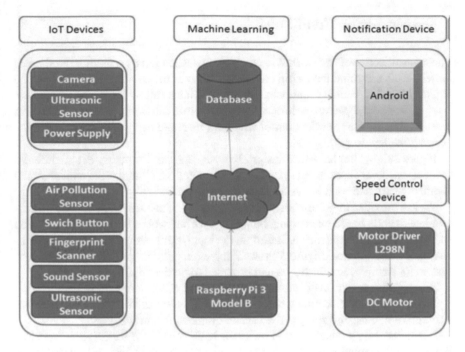

Fig. 1 Architecture of the proposed system

portable air pollution detector). The sound sensor can detect the sound frequency and thus helps to monitor the sound pollution level in a particular area. Fingerprint scanner offers high levels of security and enhanced authentication experiences to car users; these biometric fingerprint scanner technologies make cars more secure. Switch button (push button) is used for in case of any emergency situation like heart attack, headache, or accidental condition; just by pushing button, the emergency message with geolocation will be sent to the hospital center and a notification to the family members.

The prototype is said to be smart for the following reasons. The research aims to build an autonomous car prototype for smart cities using Raspberry Pi. Camera is used to give information from the actual world to the car. The car is used to escape the risk of human mistakes. Many live algorithms like road detection, obstacle detection are used to control the car. In addition to handle some healthcare problems or any accidental situation, the car is capable of contacting the helpline center and sending message to family members. It also provides security from car robbery by adding fingerprint authentication. To reduce air and noise pollution, the car's additional sensors are used to get and send the readings of environment to municipal database.

4 Experimental Conditions

The main objective of our work is to reach given destination by selecting the shortest route, keeping in mind the traffic conditions. The car must avoid and overcome the objects like other vehicles, animals, humans and must follow the traffic rules while driving. Driver and passenger interaction with (semi-)autonomous vehicles: The car must be able to take over the control whenever there are danger indications like the driver is feeling drowsy.

Figure 2 shows our hardware design. To design the car, Raspberry Pi and different types of sensors are used in our system. We had trained our system using convolutional neural network and used Keras for training our system.

We had provided large number of images for training to our system. It selects direction intelligently based on training. Same way, we had worked on obstacle detection and traffic signal identification. Based on the input, our system takes decision without human interference. Figure 3 shows sample images that we had provided to our system for training. Similarly, we had provided more than 6000 images.

Figure 4 shows our image count for training and validation process.

Figure 5 shows the accuracy chart of our system after training, and the results of left turn, right turn, and straight road accuracy are shown in the graph.

In similar way, we had worked on obstacle detection and traffic sign detection. We had implemented authentication mechanism using biometric system that helps the car to be accessed by legal users only. In our research work, we are detecting facial expression of drivers to avoid accidents. As of late, driver drowsiness/laziness

Fig. 2 Smart car

Fig. 3 Sample images for training the machines about left, right, and straight turn

Fig. 4 Validation versus
training

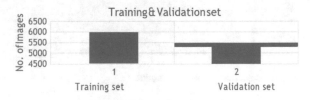

Fig. 5 Accuracy of the car
while driving on the road
(1-left turn, 2-right turn, and
3-straight)

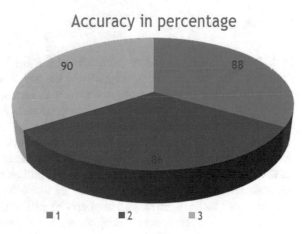

has been one of the real reasons for street mischances. We had defined the face and
eye function which is used to figure out the distance between vertical eye and the
distance between the horizontal eyes. Using the above-mentioned function, it figures
out that whether the eye is open or not during the drive. If the eye is not open during
driving, the system will give an alarm to the (drowsy) driver. Figure 6 shows the
working of drowsiness detection module.

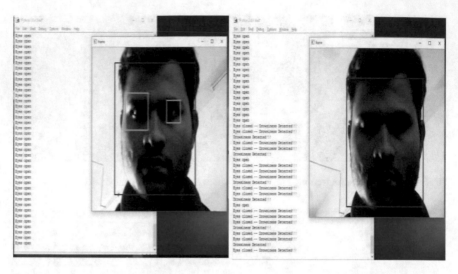

Fig. 6 Drowsiness detection module

5 Conclusion

In this research work, we have implemented Car for Smart Cities—Intelligent Car Prototype. This work will be beneficial to the society to save lives and avoid accidents. It aims to build a prototype using Raspberry Pi interfaced with intelligent sensors that overall will make a revolution in the automobile field. Though there are many intelligent cars currently in existence developed by Google and Tesla, this study aimed to corporate certain features like lane detection, object detection, alarm system during crucial conditions like drowsiness, heartache etc. All these ideas will be built on deep learning techniques which will bring intelligence to the car. Also, at regular intervals, data for air pollution will be collected that will further help to maintain green environment. At the initial stage, Raspberry Pi will be used to build the prototype of the smart car. After its successful deployment, we will replace it with more intelligent and high-performance IoT chips of Intel Corporation. Smart cars will not only help to save life, but it will also be helpful in medical emergency situations when a person is severely facing issues and there is no assistance around him. The car will be able to drive him to the nearby hospital, provided he is able to give commands to the car. This project will incorporate all the leading technologies like machine learning, deep learning, IoT, and cloud computing. This study will definitely make progress in the technical field and more importantly in India. India's vision today is 'Digital India,' and through technologies, we can promote inclusive growth of India both in terms of research and transforming India into a digitally empowered society and knowledge economy.

Declaration We declare that required permission is taken from the concerned authority for the use of image or dataset in the work (Fig. 6), and we take responsibility if any issues arise later.

References

1. Bagloee, S.A., Tavana, M., Asadi, M., Oliver, T.: Autonomous vehicles: challenges, opportunities, and future implications for transportation policies. J. Modern Trans. **24**(4), 284–303 (2016)
2. Pannu, G., Ansari, M., Gupta, P.: Design and implementation of autonomous car using Raspberry Pi. Int. J. Comput. Appl. **113**, 22–29 (2015). https://doi.org/10.5120/19854-1789
3. Bhangale, M., Dabhade, G., Khairnar, A., Bhagat, M.: Self-driving car to demonstrate real time obstacles and object detection. IRJET **03**(11) (2016). e-ISSN 2395-0056
4. Ujjainiya, L., Chakravarthi, M.K.: Raspberry-Pi based cost effective vehicle collision avoidance system using image processing. ARPN J. Eng. Appl. Sci. **10**(7) (2015). ISSN 1819-6608
5. McFall, K.: Using visual lane detection to control steering in a self-driving vehicle. https://www.researchgate.net/publication/289272518. Accessed Oct 2015
6. Anandhalli, M., Baligar, V.P.: A novel approach in real-time vehicle detection and tracking using Raspberry Pi. Alex. Eng. J. (2017). ISSN 1110-0168
7. Tembhurkar, M., Dakhane, A.: An autonomous aquatic vehicle routing using Raspberry PI. Int. J. Adv. Res. Comput. Sci. Softw. Eng. (2277 128X) (2015)
8. Pawar, N.P., Patil, M.M.: Driver assistance system based on Raspberry Pi. Int. J. Comput. Appl. **95**(16), 36–39 (2014)
9. Gupta, V., Mane, V., Pradhan, M.R., Kotangale, K.B.: IOT based car automation using Raspberry Pi. Imp. J. Interdiscip. Res. (IJIR) **3**(4) (2017). ISSN 2454-1362
10. Shahjalal, M.A., Ahmad, Z., Arefin, M.S., Hossain, M.R.T.: A user rating based collaborative filtering approach to predict movie preferences. In: 2017 3rd International Conference on Electrical Information and Communication Technology (EICT), pp. 1–5 (2017)
11. Patel, B., Pandya, S., Patel, S.: IoT based automated car. Int. J. Recent Innov. Trends Comput. Commun. **5**(5), 420–422 (2017). ISSN 2321-8169
12. Karthikeyan, M., Kudalingam, M., Natrajan, P., Palaniappan, K., Prabhu, A.M.: Object tracking robot by using Raspberry PI with open computer vision (CV). IJTRD **3**(3) (2016). ISSN 2394-9333
13. Garethiya, S., Ujjainiya, L., Dudhwadkar, V.: Predictive vehicle collision avoidance system using Raspberry-Pi. ARPN J. Eng. Appl. Sci. **10**, 3655–3659 (2015)
14. Kharade, P., Laxmi, M., Pooja, A.S., Savadatti, P., Marali, K.: Prototype implementation of IoT based autonomous vehicle on Raspberry Pi. Bonfring Int. J. Res. Commun. Eng. **6**, 38–43 (2016). https://doi.org/10.9756/bijrce.8197
15. Ujjainiya, L., Chakravarthi, Kalyan: Raspberry-Pi based cost effective vehicle collision avoidance system using image processing. ARPN J. Eng. Appl. Sci. **10**, 3001–3005 (2015)
16. Khan, T.J., Bhadange, M.R., Pagar, P.S., Salve, V.: Smart vehicle monitoring system using Raspberry Pi. SEST **3**(2). ISSN 2394-0905
17. Tiwari, R., Singh, D.K.: Vehicle control using Raspberry Pi and image processing. Innov. Syst. Des. Eng. **8**(2) (2017). ISSN 2222-2871
18. Rodrigues, B., Chaudhari, A., More, S.: Two factor verification using QR-code: a unique authentication system for Android smartphone users. In: 2016 2nd International Conference on Contemporary Computing and Informatics (IC3I), pp. 457–462, Noida (2016)
19. Kisku, D.R., Gupta, P., Sing, J.K. (eds.).: Advances in Biometrics for Secure Human Authentication and Recognition. CRC Press

20. Li, C., Hu, J., Pieprzyk, J., Susilo, W.: A new biocryptosystem-oriented security analysis framework and implementation of multibiometric cryptosystems based on decision level fusion. IEEE Trans. Inf. Forensics Secur. **10**(6), 1193–1206 (2015)
21. Dey, S., Samanta, D.: Unimodal and Multimodal Biometric Data Indexing. Walter de Gruyter GmbH & Co KG (2014)

Security Systems in Design for Testing

M. I. Shiny, R. Reshma, Shine Ross, Siji John, R. Sreevidya
and Sumana Moothedeth

Abstract It has become a great problem nowadays that the information contained in an IC chip is not provided with adequate security against the cunning attackers. The modern chip designs inbuilt with a design for testing (DfT) that enable testing process to become more easier, but it also acts as backdoor tool to the hackers to retrieve the sensitive data from the chip. So, in order for providing security for the information contained in the chip, we introduce a reconfigurable PUF design technique with the DfT. This security is achieved by providing a barrier against the hamming distance-based attack. In this paper, we compare the analysis of modified linear feedback shift registers (LFSRs), pseudorandom sequence generator, and physical unclonable function (PUF) in terms of randomness, hamming distance, and security. The result of comparison provides the PUF design have maximum security without effecting the testability of the circuits.

Keywords Linear feedback shift register · Physical unclonable function Design for testing

M. I. Shiny (✉) · R. Reshma · S. Ross · S. John · R. Sreevidya · S. Moothedeth
Department of Electronics and Communication Engineering, Jyothi Engineering College,
Cheruthuruthy, Thrissur, India
e-mail: shinyissac122001@gmail.com; shiny@jecc.ac.in

R. Reshma
e-mail: reshmavattoli00@gmail.com

S. Ross
e-mail: shinedavis1008@gmail.com

S. John
e-mail: sijijohn1997@gmail.com

R. Sreevidya
e-mail: athirajose6@gmail.com

S. Moothedeth
e-mail: sumanasneha123@gmail.com

© Springer Nature Singapore Pte Ltd. 2019
J. Wang et al. (eds.), *Soft Computing and Signal Processing* ,
Advances in Intelligent Systems and Computing 898,
https://doi.org/10.1007/978-981-13-3393-4_51

1 Introduction

In very large-scale integration (VLSI) technology, it is very necessary to build a mechanism in order to test the functionality of the integrated circuits (IC). The technique used for achieving the functionality verification is called design for testability (DfT). The two basic properties that determine the testability of node are observability and controllability. Here data in the IC is converted into certain codes by combining it with small piece of information during these operations. As a consequence, the chips are encrypted so that hackers have serious difficulties in hacking the contents in chip. Thus, designers are introducing more and more tamper-resistant hardware in order to prevent security failure. In order to add testability, both scan-based testing and built-in self-test (BIST) methods are used. DfT facilitates a design for postproduction testability of an IC. It is an additional circuitry (extra logic) which is inserted in the normal design during the process. These newly features make it easier to develop and apply manufacturing tests to the designed hardware. Controllability and observability are the two key features that must be considered as the characteristics of the DfT.

1.1 Basics of Scan Chain

Scan design is one of the simplest and widely accepted testing methods in DfT due to its high fault coverage and low area overhead. Here in Fig. 1 shows that the scan chain extracts the contents from the circuit under test when it is in test mode. That is, the original data can be shifted through the chain, whereas in normal mode the test vector and responce are secured.

Many solutions are proposed to protect the core against scan chain-based attack. In [1], the scan chain is divided into small segments, and each segment is connected through the combinational circuit which manages order in the chain. To improve the testability, the scan chain needs to satisfy the controllability and observability of the internal nodes. But this method provides encrypted scan out that violate the observability of the circuit and with area overhead is due to additional gates.

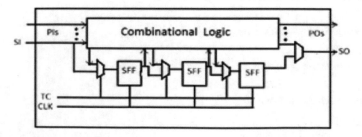

Fig. 1 Basic block diagram of a scan chain

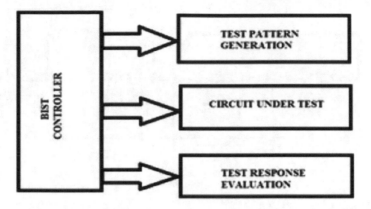

Fig. 2 Block diagram of BIST

Built-in self-test (BIST) is an important testability technique. In BIST, a part of circuit itself is used to test the system. It is fast, efficient and hierarchical design in which some hardware is capable of testing boards, chips, and systems [2]. Testing can be done during operations and maintenance. BIST architecture includes test pattern generation, test response evaluation, and circuit under test (CUT) as mentioned in Fig. 2. This technique is secure than the scan chain-based test but more expensive.

1.2 Review of Random Number Generators

The main purpose of the random number generator in the DfT is to protect the sensitive data enclosed in a chip. It provides first level of security due to the random nature of the generator.

1.2.1 Linear Feedback Shift Register (LFSR)

An example of hardware pseudorandom number generator is a linear feedback shift register (LFSR). A set of flip-flops are connected in series and the feedback is provided with Exclusive-OR configuration. Seed is the initial value feed to the LFSR. It acts as the second level of security. When the inputs of the registers are fed with a seed value and the LFSR is clocked, it generates a pseudorandom pattern of 1's and 0's. Figure 3 shows the four-bit LFSR using D flip-flops and XOR gate. A maximal length LFSR is the one that generate a stream of random numbers of maximum length $[(2^n) - 1]$ before it starts to repeat, where n is number of register element in it. LFSRs require less combinational logic and can work at high frequencies.

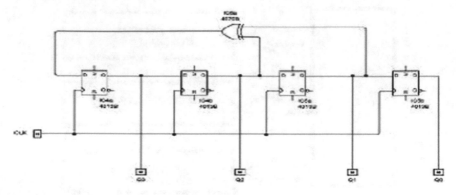

Fig. 3 Circuit diagram of LFSR

1.2.2 Modified Linear Feedback Shift Register

Modified LFSR is the extended version of LFSR which generates patterns of higher randomness. Certain modifications are done by adding additional logic gates in between the flip-flops for pseudorandom pattern generation as shown in Fig. 3. In paper [3], built-in self-test (BIST) is used in which the DfT is embedded along with the IC. Here the LFSR is used for producing highly randomized test patterns. This is accomplished by the selection of characteristic polynomial and seed to obtain high fault coverage along with minimized test patterns, but with more area overhead and time taken. The LFSR can be more randomized by adding multiplexers with select lines as the select lines in it are more variable. Also here a low power designing techniques are used to achieve energy savings [4].

1.2.3 Pseudorandom Number Generator (PRNG)

PRNG is a random number generator also called deterministic random bit generator. These are not really random as it is completely determined by an initial value called PRNG's Seed. It is used in wide variety of applications such as simulations, electronic game, and cryptography. A modified random generator is used in [5] for protecting secret data from the scan chain. The proposed technique presents to prevent unauthorized access of the scan-out data.

1.2.4 Physical Unclonable Functions (PUFs)

In chip testing, physical unclonable functions (PUFs) are emerging as a powerful security primitive that can potentially solve several security problems. For a physical unclonable functions device, it is a physical entity which is easy to evaluate but hard to predict. PUFs depend on the uniqueness of their physical microstructure.

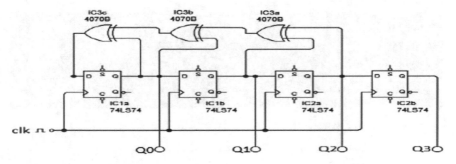

Fig. 4 Circuit diagram of modified LFSR

The random physical factors introduced during manufacturing are unpredictable and uncontrollable, which makes it virtually impossible to duplicate or clone the structure.

The first PUF design used in FPGA is introduced in [6]. It proposed Arbiter PUF design which consists of two paths created by connecting a number of switch delay elements in series. The next design is discussed in [7] relies on the SRAM that is present on most modern FPGA. Since a SRAM, assumes a random value of 0 or 1 when it is powered. Most efficient method of PUF is mentioned [8] in which 256 ring oscillators are used consisting of five inverting stages. A decoder is used to generate bits for the oscillator and a MUX is used to choose which ring oscillator is to be fed in the counter based on the input bits. The counters are compared and randomized and sequence of 0's and 1's are generated

1.3 Proposed Methodology

The proposed random generator mainly focused on how to improve the randomness of the circuits. This paper proposed modified random number generator which generates maximum randomness data with more hamming distance. The proposed solutions with desired security level.

1.3.1 Modified LFSR

The existing LFSR can be modified by adding basic gates within the circuit. It is necessary that a maximal length of $[(2^n) - 1]$ random patterns must be generated. Here modified LFSR as shown in Fig. 4 is designed to change the position of the tape point to obtain the maximum randomness pattern. In this structural modification, we can get more hamming distance than the conventional LFSR.

Here we have used four-bit modified LFSR; hence, fifteen random patterns are obtained. The simulated result of the above circuit is shown in Fig. 5. In this circuit, a key is use for protecting the random numbers. It generates more randomness than the existing LFSR.

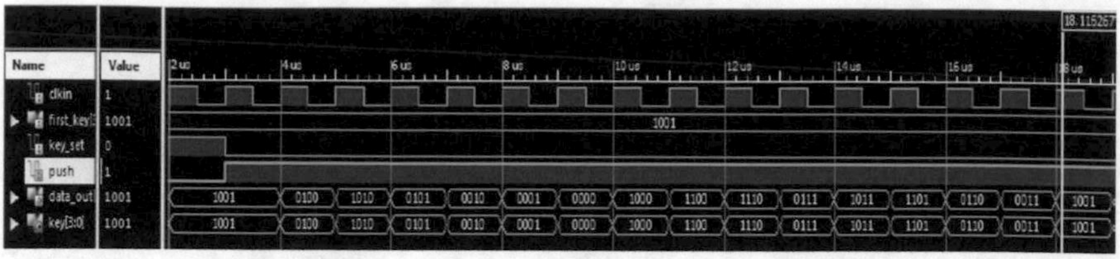

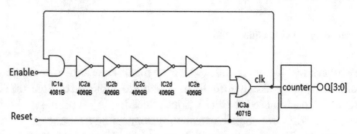

Fig. 5 Waveform of modified LFSR

Fig. 6 Proposed PUF circuit diagram

1.3.2 Reconfigurable Physical Unclonable Functions (PUFs)

The proposed PUF is a ring oscillator-based design. This circuit includes five inverter stages and single four-bit counter. The output of the fifth inverter is fed as one of the inputs of the OR gate. And the output of the OR gate is given as the input to counter and also fed back as the input to the overall system. The output from the counter is considered as the generated random pattern. The proposed circuit diagram of the ring oscillator PUF is as shown in Fig. 6. The main advantages of this circuit always generate random number. Nobody can predict the next value the security level PUF design is very suitable.

1.3.3 Pseudorandom Number Generator

Pseudorandom number generator is kind of linear feedback shift register. The shift registers are made up of T flip-flops. The feedback connection is made by Exclusive-OR gate as shown in Fig. 7. The registers are fed with a seed value. When the flip-flops are clocked, the register values are shifted. As a result, random patterns are generated at the output of each shift register and its simulated waveform as shown in Fig. 8.

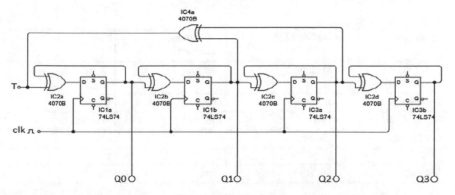

Fig. 7 Pseudorandom number generator

Fig. 8 Pseudorandom number generator waveform

1.4 Parametric Analysis

Implementation result of each modified structure is shown in Table 1. Here the parameters are the area dependent on the number of lookup table (LUT) utilization each circuit and cost is directly proportional to the number of components and design complexity. And the security depends on the hamming distance between the original data with the reconfigurable structure. Modified structure with more hamming distance means the structure can access only the granted authority. More hamming distance between accessed key with generated key that serves confuse to the attackers. Moreover it provides higher degree of security. All the parameters of modified LFSR, PN sequence, and PUF design are mentioned below in Table 1.

Table 2 compares the comparative study of hamming distance parameter of modified LFSR, PN sequence generator, and PUF design. In terms of randomness and hamming distance, PUF design is more secure. It also applicable to secure scan chain-based testing without compromise the testability.

Table 1 Parametric analysis

Parameters	Modified LFSR	PN sequence generator	PUF
Cost (no. of components)	Medium	Medium	High
Complexity	Medium	High	Less
Hamming distance	High	Medium	Medium
Randomness	Medium	Medium	High
LUT	6	7	9
Security	Medium	High	Highest

Table 2 Hamming distance comparison of sequence generator

SI. no.	Input	Modified LFSR	Hamming distance	PN sequence generator	Hamming distance	PUF	Hamming distance
1	0000	0100	1	0001	1	0001	1
2	0001	1010	3	0011	1	0010	2
3	0010	0101	3	0101	3	1110	2
4	0011	0010	1	1110	3	1011	1
5	0100	0001	2	0010	2	1010	3
6	0101	0000	2	0110	2	0000	2
7	0110	1000	3	1011	3	1101	3
8	0111	1100	3	1100	3	1010	3
9	1000	1110	2	0100	2	1111	3
10	1001	0111	3	1101	2	0011	2
11	1010	1011	1	0111	3	0110	2
12	1011	1101	2	1000	2	1101	2
13	1100	0110	2	1001	2	1001	2
14	1101	0011	3	1010	3	1011	2
15	1110	1001	3	1111	1	1100	2
16	1111	0100	3	0001	3	1000	3

1.5 Conclusion

In this paper, we presented different methods of random pattern generation for designing for testing such as modified LFSR, pseudorandom number generator, and PUF design. Compared to all these test pattern generation PUF design produce maximum hamming distance with more randomness. The proposed methodologies modified LFSR and PUF design can be implemented in BIST in order to secure the device from malicious attackers. Parameters such as hamming distance and the area of each pattern generator are obtained from the simulation results in a Xilinx ISE Design suit and comparison done in Tables 1 and 2. Comparative study of these patterns are listed on the tables and finally concluded that PUF design is the most suitable method for random pattern generation because the pattern generation is always randomness. The comparative result also evaluates the area, cost, and randomness for the different proposed.

References

1. Hely, D., Flottes, M.-L., Frederic, B., Rouzeyre, B., Berard, N., Renovell, M.: Scan design and secure chip. In: IOLTS, vol. 4, pp. 219–224 (2004)
2. John, P.K.: BIST architecture for multiple RAMs in SoC. Proc. Comput. Sci. **115**, 159–165 (2017)
3. Haridas, N., Nirmala Devi, M.: Efficient linear feedback shift register design for pseudo exhaustive test generation. In: BIST 3rd IEEE International Conference Electronics Computer Technology (ICECT), vol. 1 (2011)
4. Shaer, L., et al.: A low power reconfigurable LFSR. In: 18th Mediterranean IEEE Electrotechnical Conference (MELECON) (2016)
5. Shiny, M.I., Nirmala, D.M.: LFSR based secured scan design testability techniques. Proc. Comput. Sci. **115**, 174–181 (2017)
6. Suh, G.E., Clarke, D., Gassend, B., van Dijk, M., Devadas, S.: Efficient memory integrity verification and encryption for secure processors. In: Proceedings of the 36th annual IEEE/ACM International Symposium on Microarchitecture, p. 339. IEEE Computer Society (2003)
7. Guajardo, J., Kumar, S.S., Schrijen, G.-J., Tuyls, P.: Physical unclonable functions and public-key crypto for FPGA IP protection. In: FPL 2007 International Conference on Field Programmable Logic and Applications, 2007, pp. 189–195. IEEE (2007)
8. Suh, G.E., Devadas, S.: Physical unclonable functions for device authentication and secret key generation. In: Proceedings of the 44th annual Design Automation Conference, DAC '07 (2007)

Design of an Area-Efficient FinFET-Based Approximate Multiplier in 32-nm Technology for Low-Power Application

V. M. Senthil Kumar and S. Ravindrakumar

Abstract For applications, where the performance and accuracy are less important inexact computing can be used. The power efficiency increases as the bit-based multiplication is not done for full word length. Applications where speed and power are dominant compared to accuracy inexact multipliers are the first choice. In this approach, the partial products of the approximate multiplier are changed. Two variants of multipliers are designed. One is with CMOS and other with FinFET. The proposed FinFET-based multiplier circuit reduces the leakage current which ultimately results in the power consumption reduction. The proposed compressor-based multiplier reduces the number of operands and partial products. The inexact computing reduces the number of interconnects and components. The proposed multiplier circuits are compared with the existing counterparts and found that the performance improvement is 40.61%. In future, a FinFET-based Multiply-Accumulate unit (MAC) for biomedical application will be proposed.

Keywords FinFET · CMOS · Approximate computing · Power saving
Array multiplier · Multipliers · Low area

1 Introduction

Digital logics form the basic building blocks of computer arithmetic architecture and work with high reliability, precision, and speed. The errors happening in circuits for multimedia and image processing application are adjustable. Due to the large volume of data and high resolution, exact computation in the stipulated time is a problematic one [1]. The inexact computation is suitable for multimedia applications,

V. M. Senthil Kumar
Malla Reddy College of Engineering and Technology, Hyderabad, Telangana, India
e-mail: vmspraneeth@gmail.com

S. Ravindrakumar (✉)
IRRD Automatons, Karur, Tamil Nadu, India
e-mail: gsravindrakumar7@gmail.com

© Springer Nature Singapore Pte Ltd. 2019
J. Wang et al. (eds.), *Soft Computing and Signal Processing* ,
Advances in Intelligent Systems and Computing 898,
https://doi.org/10.1007/978-981-13-3393-4_52

where the approximate results are acceptable. Designing energy efficient systems using inexact computing are done in literature [2]. The design of approximate or inexact computing circuits is based on circuits working with low-power and higher-performance operation. Compared with precise or exact logic circuits, the inexact computing saves time and power. These inexact computing methods are widely used to design adders and multipliers [3]. Addition operation is performed using full-adder cells. Compressors are used to perform arithmetic operations. Several other types of multipliers like vedic based are also been implemented using compressors in literature [4]. The objective is to implement the multiplier using compressor-based and FinFET-based circuits. The FinFET-based multipliers have less delay and low power [5]. In this chapter, a FinFET-based multiplier is presented.

1.1 Low-Power Arithmetic Circuit Design

The signal and image processing algorithms are based on addition, multiplication, and accumulation. The most of the chip area is occupied by the multipliers. The multipliers consume more power, and the delay due to multipliers is critical. So, speed is an important factor in real-time signal and image processing application. Nowadays, systems are designed with wireless units embedded. So the data rate and speed are important. However, improvements in speed result in larger areas. So, area and speed are usually conflicting constraints. In conventional multiplication [6], repeated addition is performed based on the multiplicand. Since conventional multiplication uses repeated sum of partial products the area occupied is more and power consumed is more. But the approximate multipliers are designed by an approximate adder. The errors occurring due to reduction and rounding are found during the consecutive steps. To reduce the error distance, the correction constant should be closer to the estimated sum value. Several multipliers are designed using multiplier arrays [7]. Compressor-based multiplier is mostly used in high-speed applications. Usage of compressor reduces the partial product and optimizes the power dissipation [8]. However, there are only few researches carried out in compressor for inexact computations. This type of multipliers is more effective to approximate computation or multiplication.

1.2 Inexact Computing/Compression

In digital processing where inexact computing becoming an attractive paradigm and VLSI Implementation becomes much easier [8], then complex arithmetic operations like Fourier analysis or discrete wavelet transform (DWT) are to be performed the inexact computing becomes more efficient. Compressor circuits were initially used for generating partial products in BiCMOS technology [9], and BiCMOS circuit provided better power consumption but failed in the reduction of leakage current.

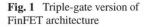

Fig. 1 Triple-gate version of FinFET architecture

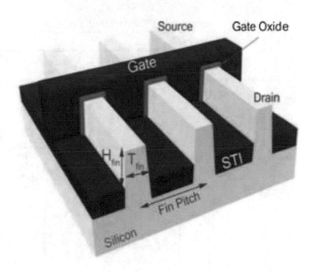

A pass transistor-based CMOS multiplier using compressor was reported in the literature [10].

1.3 FINFET Fundamentals

The advancement in nanometer technologies decreased the power usage and increased the density [11]. The operating frequency can be increased with a small amount of increase in power. The power supplies for non-portable devices have different challenges. The design of cooling system is complex, and alternate power options are required [12]. This is due to the problems occurring in the increase of package density [13]. The power budget puts a limitation on the design goals to be met with the required performance [14]. Even though CMOS devices are working with low-power and high-operating frequencies, new devices are required when the device size decreases. The Fin-type Field Effect Transistors (FinFET), an emerging transistor technology is an optional one during its entry. But nowadays, FinFET devices have replaced most of the CMOS circuits. This is due to the fact that below 32 nm the CMOS circuits suffer from second-order effects. On the other hand, continued transistor scaling limits the performance. The fundamental material and process technology limit the design [15]. In literature, FinFET designs are presented [16] and the performance was compared with the CMOS circuits [17]. The fabrication of FinFET is compatible with the CMOS which helps in rapid deployment. A triple-gate version of FinFET is shown in Fig. 1. Using FinFET, several circuits are designed in literature [18, 19].

The primary parameters used for the FinFET device are shown in Table 1.

Table 1 Primary parameters in PTM for FinFET 32-nm technology

Primary parameters in PTM					
n-type FinFET	$L_{gate} = 32$ nm	$H_{fin} = 40$ nm	$W_{fin} = 8.6$ nm	$T_{ox} = 1.4$ nm	$V_{DD} = 1$ V
p-type FinFET	$L_{gate} = 32$ nm	$H_{fin} = 50$ nm	$W_{fin} = 8.6$ nm	$T_{ox} = 1.4$ nm	$V_{DD} = 1$ V

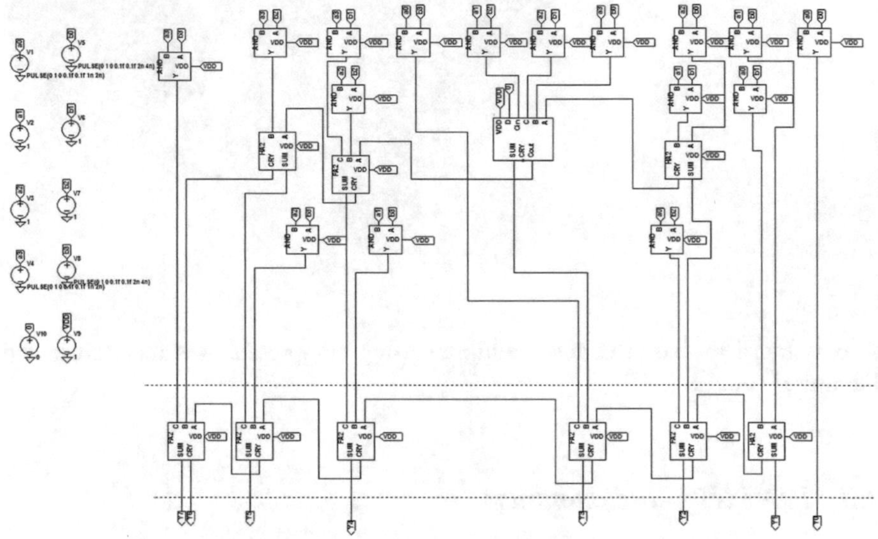

Fig. 2 Schematic diagram of DADDA multiplier

2   Existing Method

All approximate multipliers are designed using CMOS technology with supply voltage 1 V for 32 nm. In the existing methods, the partial product reduction was carried out in few least significant columns or all columns [8, 20, 21]. The approximation does not affect the accuracy. The partial products generation and approximation were done using adders and compressors [22]. The schematic of an existing CMOS multiplier is shown in Fig. 2. The schematic of array multiplier is shown in Fig. 3. For N input bits, N − 1 rows of N full adders are used.

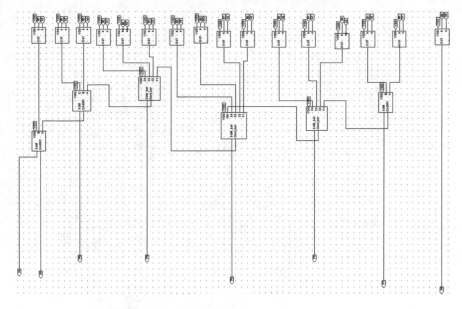

Fig. 3 Schematic for 4 × 4 array multiplier

3 Proposed Method

In proposed method, the CMOS circuits are replaced and implemented using shorted gate FinFET-based full adder as shown in Fig. 4. A double-gate MOS (DGPMOS and DGNMOS) is designed for the full adder and multiplier. This proposed FinFET-based full adder contains 10 transistors. Here at each point, two transistors with their source and drain terminals tied together.

Based upon the multi-bit full adder, the power dissipation and delay parameters will vary with a great extent. So to reduce leakage current loss and to increase the control over channel charges, the shorted gate FinFET-based full adder and multiplier are used. Even though the number of transistors is doubled compared to 10-transistor logic, it will work efficiently with FINFET. The short channel effects will be reduced. The array multiplier worked well for this shorted gate FinFET-based full adder. The simulation results proved that the power and delay decrease as the feature size decreases.

4 Results and Discussion

The efficiency of the proposed multipliers is better when compared with the existing approximate multipliers. The performance comparison with respect to average power, delay, and power-delay-product is shown in Table 2.

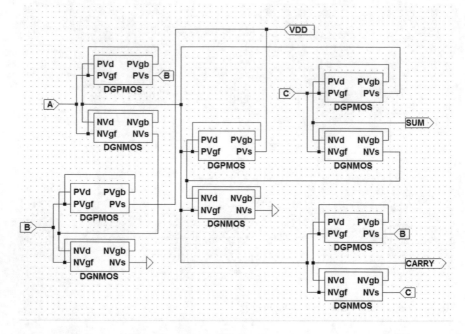

Fig. 4 Schematic of FinFET-based 1_bit full adder

Table 2 Performance of CMOS and FinFET-based multiplier

Component	Average power in mW		Delay in ns		PDP in W ns	
	CMOS	FinFET	CMOS	FinFET	CMOS	FinFET
Array multiplier	9.50	9.11	0.06	0.15	5.9E−04	1.3E−03
Proposed array multiplier with 10 transistor adder	5.59	5.41	0.19	0.20	1.0E−03	1.1E−03

Table 3 Performance of CMOS for different technologies

Feature size (nm)	Operating voltage (V)	Power (μW)	Delay (ps)	PDP
180	5	227.84	2074.4	472631.29
90	3.5	384.53	759.14	391912.10
32	2.5	604.04	604.04	354897.66
20	1.5	762.20	762.20	278287.20

It can be observed that as the feature size of CMOS reduces below 45 nm and smaller, short channel effects will rise. Due to this, power dissipation increases. This is depicted in Table 3.

The above results shows that as the feature size decreases, the delay is decreasing at the cost of increasing the power dissipation, but overall PDP is good at smaller

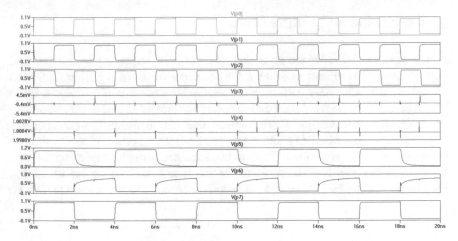

Fig. 5 Output waveform of 4 × 4 array multiplier

Table 4 Performance of FinFET-based proposed multiplier for different technologies

Technology (nm)	Power (W)	Average current (A)	Average power (W)	Average energy (J)
180	92.94×10^{-6}	98.05×10^{-6}	46.05×10^{-6}	126.57×10^{-15}
90	27.68×10^{-6}	28.57×10^{-6}	13.67×10^{-6}	39.89×10^{-15}
65	16.95×10^{-6}	16.97×10^{-6}	7.69×10^{-6}	22.81×10^{-15}
20	3.93×10^{-6}	4.12×10^{-6}	1.86×10^{-6}	5.61×10^{-15}

technologies (20 nm). With increase in device sizes, the power dissipation becomes approximately four times greater, as we use four basic array multipliers. Like that delay also increases. In practical, both the power and delay are exceeded the expected ones because of the multi-bit adder that we used to add the partial product terms. The usage of FinFET reduces this problem by increasing the control over channel charges. Both triple-gate and double-gate transistors can be used. In this work, the double gate is used since it is flexible in design. The shorted gate FinFET mode technique is compatible with planar CMOS circuit, and the replacement for the multiplier is done. In this paper, the array multiplier is designed with the shorted gate FinFET-based full adder, which is having less number of transistors compared to conventional one. So that we can achieve less area, lower power and delay compared to previous designs. The performance of 4 × 4 array multiplier using shorted gate FinFET-based full adder has been evaluated. The power and delay results of array multiplier are shown in Table 2. The simulation results are shown in Fig. 5.

The performance of 4 × 4 array multiplier using shorted gate FinFET-based full adder for different technologies with variable power supply voltage has been evaluated. The power, current, and energy measured are shown in Table 4.

5 Conclusion and Future Work

In this paper, a FinFET-based multiplier using inexact computing is designed and implemented. The problem with the CMOS was experimented using different designs by varying the technology parameter and multiplier size. From the analysis, it has been found that the CMOS-based device performance is inferior at lower technology or smaller size devices below 45 nm. To overcome this small area, low-power design is proposed using FinFET in shorted gate mode. Further optimization was done at the adder level. The array multiplier was implemented using inexact computing for both CMOS and FinFET. The performance was reported. From the analysis, it has been found that the FinFET design seems to be efficient when compared to CMOS.

References

1. Momeni., A, Han, J., Montuschi, P., Gu, J., Chung, C.-H.: Ultra low voltage, low power 4-2 compressor for high speed multiplications. In: Proceedings of the International Symposium on Circuits and Systems, pp. 321–324 (2003)
2. Chang, C.H., Jiangmin, G., Zhang, M.: Ultra low voltage low power CMOS 4-2 and 5-2 compressors for fast arithmetic circuits. IEEE Trans. Circuits Syst. I Regul. Pap. **51**, 1985–1997 (2004)
3. Margala, M., Durdle, N.G.: Low power low voltage 4-2 compressors for VLSI applications. In: Proceedings of conference on IEEE Alessandro Volta Memorial Workshop on Low Power Design, pp. 1–7 (1999)
4. Kaur, H., Prakash, N.R.: Area efficient low PDP 8-bit vedic multiplier design using compressors. In: Proceedings of 2nd International Conference on Recent Advances in Engineering and Computational Sciences (RAECS), pp. 1–4 (2015)
5. Whitehouse, J., John, E: Leakage and Delay analysis in FinFET array multiplier circuits. In: Proceedings of Conference on 57th International Midwest Symposium on Circuits and Systems (MWSCAS), pp. 909–912 (2014)
6. Radhakrishnan, D., Preethy, A.P.: Low power CMOS pass logic 42 compressor for high speed multiplication. In: Proceedings of IEEE Conference on ROC, pp. 1296–1298 (2000)
7. Maheshwari, N., Yang, Z., Han, J., Lombardi, F.: A design approach for compressor based approximate multipliers. In: Proceedings of 28th International Conference on VLSI Design and Embedded Systems, pp. 209–214 (2015)
8. Hsiao, S.F., Jiang, M., Yeh, J.: Design of high speed low power 3-2 counter and 4-2 compressor for fast multipliers. Electron. Lett. **34**, 341–343 (1998)
9. Law, C.F., Rofail, S.S., Yeo, K.S.: Low power circuit implementation for partial product addition using pass transistor logic. In: IEEE Proceedings Circuits, Devices and Systems, pp. 124–129 (1999)
10. Ohkubo, N., Suzuki, M., Shinbo, T., Yamanaka, T., Shimizu, A., Sasalu, K., Nakagome, Y.: A 4.4 ns CMOS 54 × 54b multiplier using pass transistor multiplexer. IEEE J. Solid State Circuits **30**, 251–257 (1995)
11. Hadia, S.K., Patel, R.R., Kosta, Y.P.: FinFET architecture analysis and fabrication mechanism. IJCSI Int. J. Comput. Sci. Issues **8**, 235–240 (2011)
12. Rudenko, T., Kilchytska, V., Collaert, N., Nazarov, A., Jurczak, M., Flandre, D.: Electrical characterization and special properties of FinFET structures. In: Proceedings of Conference on Nanoscaled Semiconductor on Insulator Structures and Devices, pp. 199–220 (2007)

13. Rasouli, S.H., Endo, K., Banerjee, K.: Variability analysis of FinFET based devices and circuits considering electrical confinement and width quantization. In: Proceedings of IEEE/ACM International Conference on Computer Aided Design, pp. 505–512 (2009)
14. Sharma, S.: Performance analysis of inverter gate using FinFET and planar bulk MOSFET technologies. Int. J. Electr. Electron. Eng. (IJEEE) 7, 493–506 (2015)
15. Joshi, P., Khandelwal, S., Akashe, S.: High performance FinFET based D flipflop including parameter variation. In: Springer Proceedings in Physics—Advances in Optical Science and Engineering, pp. 239–243 (2015)
16. Mishra, P., Muttreja, A., Jha, N.K.: Gate sizing: FinFET's vs. 32 nm bulk MOSFETs. Nano Electron. Circuit Des. 2, 23–54 (2011)
17. Rasouli, S.H., Dadgour, H.F., Endo, K., Koike, H., Banerjee, K.: Design optimization of FinFET domino logic considering the width quantization property. IEEE Trans. Electron Devices 57, 2934–2943 (2010)
18. Tawfik, S.A., Kursun, V.: Low power and compact sequential circuits with independent gate FinFETs. IEEE Trans. Electron Devices 55, 60–70 (2008)
19. Bhushan, S., Khandelwal, S., Raj, B.: Analyzing different mode FinFET based memory cell at different power supply for leakage reduction. In: Proceedings of Seventh International Conference on Bio Inspired Computing: Theories and Applications (BICTA), pp. 89–100 (2012)
20. Lombardi, F.: Design and analysis of approximate compressors for multiplication. IEEE Trans. Comput. 64, 984–994 (2014)
21. Menon, R., Radhakrishnan, D.: High performance 5-2 compressor architectures. IEEE Proc. Circuits Devices Syst. 153, 447–452 (2006)
22. Saravanan, S., Senthil Kumar, V.M.: Design of a reduced carry chain propagation adder using FinFET. Asian J. Inf. Technol. 15, 1670–1677 (2016)

Adaptive Noise Cancellation Using NLMS Algorithm

R. Rashmi and Shweta Jagtap

Abstract This paper studies the behaviour of normalized least mean square (NLMS) adaptive filter algorithm-based noise canceller to eliminate intense background noise of high and low frequency from a desired signal. Noise signal filtration requires a filter which automatically adapts with the variation of input signal and noise signal. Performance is measured by optimizing rate of convergence and mean square error (MSE) using MATLAB. The experimental results indicate that adaptive noise canceller can remove low- and high-frequency noise of signals conveniently, and for small values of step size MSE decreases and for larger value of step size the rate of convergence increases. The computation time increases with the increase of filter length.

Keywords Adaptive filters · LMS algorithm · NLMS algorithm
Active noise control

1 Introduction

Many applications such as radar [1], sonar, video, audio signal processing, noise reduction [2], signal processing [3], communication systems [4], control systems [5] encounter contamination of signal of interest by strong interference of noisy environment. Adaptive techniques have been used in these applications successfully during the last two decades. If signal and noise are stationary and their characteristics are known, then the fixed coefficient-based conventional digital filter shows a satisfactory performance. But in dynamic environment, desired signal extraction requires a continuous updated weights for the optimum performance [3]. Since conventional filters cannot change the filter weight to cancel the noise, the adaptive filter with the

R. Rashmi (✉) · S. Jagtap
Department of Instrumentation Science, Savitribai Phule Pune University, Pune, India
e-mail: ruchirashmi@gmail.com

S. Jagtap
e-mail: shweta.jagtap@gmail.com

© Springer Nature Singapore Pte Ltd. 2019
J. Wang et al. (eds.), *Soft Computing and Signal Processing* ,
Advances in Intelligent Systems and Computing 898,
https://doi.org/10.1007/978-981-13-3393-4_53

characteristic of iterative altering weights are used. The analog adaptive filter shows fast response with the lower power consumption but their adaptive algorithm does not show the optimum result due to offset problem [6]. Digital filters work without offset problem, highly immune to noise, better operation in wide range of frequencies and so used in most digital signal processing to extract the desired signal. The adaptive filter works to reduce noise by altering the filter coefficient. Coefficient of algorithm changes to minimize the cost function. The most common adaptive algorithm for design and implementation is least mean square (LMS) due to its computational simplicity.

LMS algorithm is less complicated than normalized least mean square (NLMS) as its step size does not vary with time but LMS requires advance information of characteristics of signal, which is difficult to achieve. In this paper, NLMS is used to overcome this problem by calculating maximum step size with number of iterations [7]. The main application of the adaptive filter is in prediction, system identification [8], inverse modelling and noise cancellation. The major contribution of this paper is to design and implementation of noise canceller using filter finite impulse response (FIR) based on adaptive algorithm NLMS for noise elimination. Different sections of the paper are organized as follows. In Sect. 2, operational principal of LMS and NLMS is presented. In Sect. 3, system design of the noise canceller is proposed. In Sect. 4, simulation results are discussed, and conclusion is provided in Sect. 5.

2 Adaptive Filter

Design technique and algorithm of adaptation decide efficiency of adaptive filter. Adaptive filters are consisting of two processes: (1) filtering process and (2) adaption process. In filtering process, the input signal is processed and filtered signal is produced, while in adaptation process, the coefficients of the filter adjust to minimize cost function. Adaptive filter can be combination of different kinds of filter structures and filtering algorithms [9]. Filter structure is classified into two categories based on their impulse response as finite impulse response (FIR) filter and infinite impulse response (IIR) filter. FIR filter's impulse response goes to zero after a finite time so it has a finite duration while IIR adaptive filter responses indefinitely due to its internal feedback mechanism. The most popular adaptive algorithms used for the adaptation process are LMS and NLMS.

2.1 Least Mean Square (LMS)

LMS algorithm is based on steepest descent method. Filter tap weight w(n) is updated with each iteration to minimize the error e(n) [10]. Here output of the adaptive filter is defined by (1)

$$y(n) = \sum_{i=0}^{N-1} w(n)\, x(n-i) \qquad (1)$$

$$w(n+1) = w(n) + 2\mu e(n)\, x(n) \qquad (2)$$

where in (2) $w(n)$ is coefficient of adaptive FIR filter tap weight vector at time N, $x(n)$ is input vector, and μ is a small positive constant called as step size parameter which governs both rate of convergence and stability of the system. With lower step size, convergence speed increases and mean square error increases. For each iteration, 2N additions and 2N + 1 multiplication are required in LMS algorithm.

2.2 Normalized Least Mean Square (NLMS)

When the signals are not known, NLMS shows greater stability than LMS as step size parameter is based on instantaneous values of coefficient of input vector $x(n)$ [11]. Here output of the adaptive filter is defined by (3)

$$y(n) = \sum_{i=0}^{N-1} w(n)\, x(n-i) \qquad (3)$$

The filter tap weights for the next iteration are given by (4)

$$w(n+1) = w(n) + \mu(n)e(n)\, x(n) \qquad (4)$$

For each iteration, 2N additions and 3N + 1 multiplications are required in NLMS algorithm.

3 Adaptive Filter System Design

In this paper, the MATLAB Simulink toolbox is used for simulation of standard NLMS algorithm in noise cancellation configurations. Two filters, high-pass filter (HPF) and low-pass filter (LPF), are designed for 2000 frames which are Gaussian pseudorandom distributed as shown in Fig. 1. The random source generates wide range of frequency with Gaussian random values by using the Ziggurat method as shown in Fig. 2. High LPF and HPF are designed with the normalized frequency, and its magnitude response is shown in Fig. 3 and Fig. 4, respectively. LPF and HPF filter the low-frequency and high-frequency signals. Filtered and unfiltered signals at different points are analysed by the spectrum analyser. Here the discrete sine wave is generated by the trigonometric function by sampling the continuous function $y_i = A_i \sin(2\pi f_i t + \phi_i)$ with a sample time of 0.05 s, frequency $f = 0.5$ Hz and amplitude

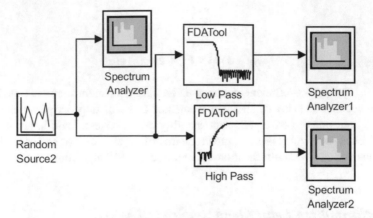

Fig. 1 High- and low-frequency noise generation

Fig. 2 Noise signal

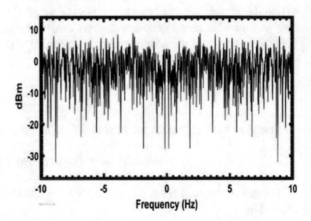

$A = 1$ as shown in Fig. 5. Then, a low-frequency noise signal and a high-frequency noise signal as shown in Figs. 6 and 7 are added individually to the input signal. The separately contaminated noisy sine wave signal $(d(n) = s(n) + N(n))$ with high- and low-frequency noise as shown in Figs. 8 and 9. Hence in this structure, NLMS is used for fast adaption of the filter coefficient with time to converge and to minimize MSE. The noise canceller system model where NLMS algorithm is applied to remove the low-frequency noise and high-frequency noise from sine wave is as shown in Figs. 10 and 11. Here the desired wave is discrete sine wave which is contaminated with the low-frequency noise which is generated using random source and low-pass filter (LPF). The ambient noise which is generated from random source is processed by adaptive filter to generate a signal of same frequency as the frequency of the noise contaminating the desired discrete sine wave, and then noise is cancelled out from the desired signal. Here the signal from the random source is highly correlated with the signals of the noise component that is the low-frequency signals.

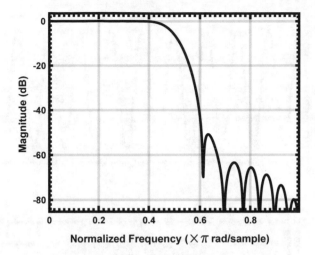

Fig. 3 Magnitude response of LPF

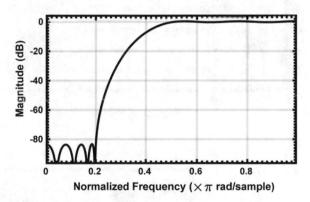

Fig. 4 Magnitude response of HPF

4 Result and Discussion

Simulation-based performance of NLMS algorithms based on additive white Gaussian noise (AWGN) with different filter length and step size is compared. The time analysis of the recovered sine wave with the filter lengths of 32 taps and 42 taps with step size of 0.1 is shown in Fig. 12 and with step size of 0.01 is shown in Fig. 13. As the step size decreases, the rate of convergence decreases but smaller value of step size requires more iteration to eliminate the noise signal. As the tap size of filter weight increases, it takes more time to recover the signal. Here as the filter length increases, the computation time increases, and the maximum convergence rate and stability decreases. Hence to overcome this issue, poles and zeros are added to maintain the stability in increased filter length system. The time response of the error

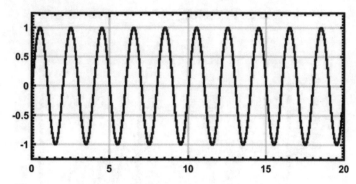

Fig. 5 Desired signal

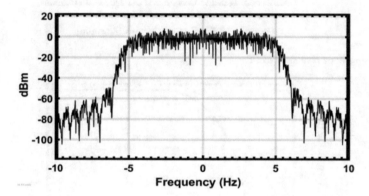

Fig. 6 Low-frequency noise

Fig. 7 High-frequency noise

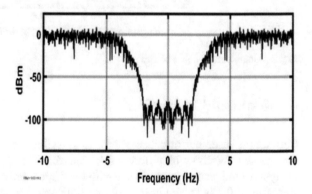

signal with the step size of 0.1 and with filter lengths of 32 taps and 42 taps is shown in Fig. 14 and with the step size of 0.01 is shown in Fig. 15. The results show that for smaller values of step size, mean square of error (MSE) decreases, but smaller value of step size requires more iteration to decrease the MSE. As the filter length is increased, it takes more time to recover the signal due to more computation. The

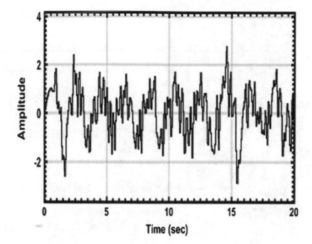

Fig. 8 Low-frequency noise and desired signal

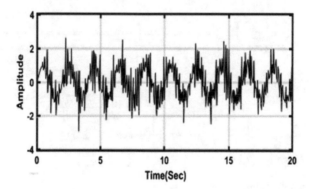

Fig. 9 High-frequency noise and desired signal

MSE will converge to nonzero constant if the system does not have required number of poles and zeroes. The recovered sine wave with the filter lengths of 32 taps and 42 taps with step size of 0.1 is shown in Fig. 16 and with step size of 0.01 is shown in Fig. 17 for high-frequency noise. The error with the filter lengths of 32 taps and 42 taps with step size of 0.1 is shown in Fig. 18 and with step size of 0.01 is shown in Fig. 19 for high frequency noise. Timing diagram of signal shows that noise canceller eliminates all undesired frequency components successfully and MSE converses to zero.

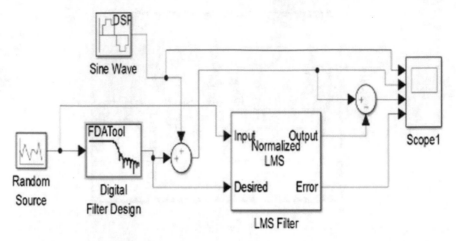

Fig. 10 Noise canceller for low-frequency noise

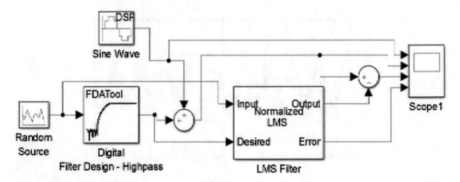

Fig. 11 Noise canceller high-frequency noise

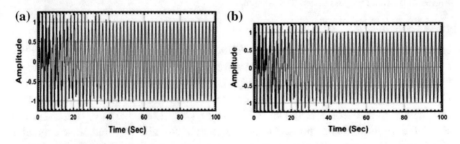

Fig. 12 Time waveform of recovered signal from low-frequency noise with 0.01 filter step size and **a** 32 and **b** 42 filter lengths

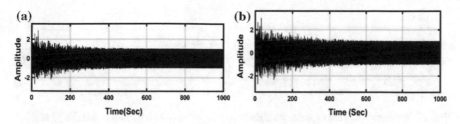

Fig. 13 Time waveform of recovered desired signal with 0.01 filter step size and **a** 32 and **b** 42 filter lengths

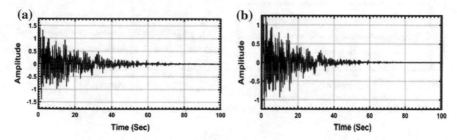

Fig. 14 Time response of the error signal with 0.1 filter step size and **a** 32 and **b** 42 filter lengths

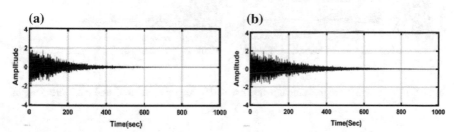

Fig. 15 Time response of the error signal with 0.01 step size 0.01 and **a** 32 and **b** 42 filter lengths

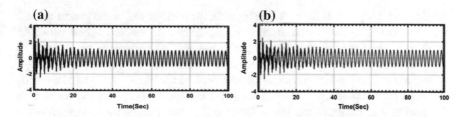

Fig. 16 Recovered desired signal from high-frequency noise with 0.1 step size and **a** 32 and **b** 42 filter lengths

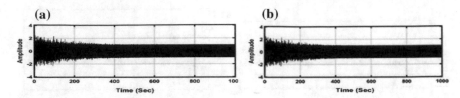

Fig. 17 Recovered desired signal from high-frequency noise with 0.01 step size and **a** 32 and **b** 42 filter lengths

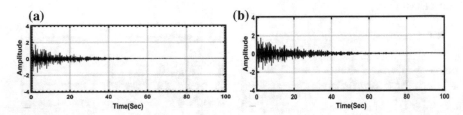

Fig. 18 Error signal for high-frequency noise with 0.1 step size and **a** 32 and **b** 42 filter lengths

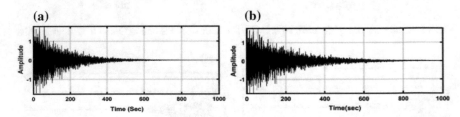

Fig. 19 Error signal for high-frequency noise with 0.01 step size and **a** 32 and **b** 42 filter lengths

5   Conclusion

The detailed design of noise canceller system for background noise cancellation is described and analysed. Variation of error signal and the desired signal with filter length and step size is shown. Fundamental issues like conversion, time computational difficulties, stability of the system with variation in step size and filter length are discussed. The mentioned noise canceller can be used in aircraft inside the vehicle cabins for the passenger comfort, in active headphone for pilots to cancel noise of propeller and in active mufflers. The result allows continuing future works in adaptive algorithm-based system for fast adaptation with stability.

Acknowledgements This work has been supported by DST-PURSE, Savitribai Phule Pune University. One of the authors R. Rashmi is thankful to the University Grant Commission (UGC) for SRF.

References

1. Huleihel, W., Tabrikian, J., Shavit, R.: Optimal adaptive waveform design for cognitive MIMO radar. IEEE Trans. Signal Process. **61**(20) (2013)
2. Schneider, M., Kellermann, W.: Multichannel acoustic echo cancellation in the wave domain with increased robustness to nonuniqueness. IEEE/ACM Trans. Audio Speech Lang. Process. **24**(3) (2016)
3. Haykin, S.: Adaptive Filter Theory, 4th edn. Prentice Hall
4. Mahbub, U., Shahnaz, C., Fattah, S.A.: An adaptive noise cancellation scheme using particle swarm optimization algorithm. In: 2010 International Conference on Communication Control and Computing Technologies (2010)
5. Althahab, A.Q.J.: A new robust adaptive algorithm based adaptive filtering for noise cancellation. In: Analog Integrated Circuits and Signal Processing (2017)
6. Qiuting, H.: Offset compensation scheme for analogue LMS adaptive fir filters. Electron. Lett. **28**(13) (1992)
7. Sultana, N., Kamatham, Y., Kinnara, B.: Performance analysis of adaptive filtering algorithms for denoising of ECG signals. In: 2015 International Conference on Advances in Computing Communications and Informatics (ICACCI) (2015)
8. Gowri, T., Rajesh Kumar, P., Rama Koti Reddy, D.V.: An efficient variable step size least mean square adaptive algorithm used to enhance the quality of electrocardiogram signal. In: Advances in Intelligent Systems and Computing (2014)
9. Shaik, B.S., Naganjaneyulu, G.V.S.S.K.R., Chandrasheker, T., Narasimhadhan, A.V.: A method for QRS delineation based on STFT using adaptive threshold. Procedia Comput. Sci. (2015)
10. Gorriz, J.M., Ramirez, J., Cruces-Alvarez, S., Puntonet, C.G., Lang, E.W., Erdogmus, D.: A novel LMS algorithm applied to adaptive noise cancellation. IEEE Signal Process. Lett. **16**(1), 34–37 (2009)
11. Dewasthale, M M., Kharadkar, R.D.: Improved NLMS algorithm with fixed step size and filter length using adaptive weight updation for acoustic noise cancellation. In: 2014 Annual IEEE India Conference (INDICON) (2014)

Optimizing MPLS Tunnel Creation Performance by Using SDN

Snehal Patil and Mansi S. Subhedar

Abstract In today's world, many high-speed enterprise links are running on MPLS. For enterprises, it is not possible to migrate to SDN technology directly and smooth transition of MPLS networks onto SDN needs to be ensured. This paper aims at optimizing the MPLS performance by coupling it with SDN. SDN controller uses some features of MPLS-TE to read network statistics. Based on the input of OSPF extension headers, SDN will reroute the traffic whenever there is congestion. The controller is preprogrammed with flows written from OpenFlow Manager. Whenever there is some change in topology or network statistics, packet header will be modified as per the flows and it will be rerouted.

Keywords MPLS · VPN · Traffic engineering · Software-defined networking
OpenFlow controller

1 Introduction

Multiprotocol label switching (MPLS) can be implemented as per the network requirements, within Internet service provider (ISP) network. This will lead to the formation of MPLS labels for each network and creation of Label Forwarding Information Base (LFIB) on each provider router as well as and Provider Edge Router to enable multiprotocol label switching. For MPLS traffic engineering, tunnels need to be created through which particular customer data can be routed from source to destination [1]. The Resource Reservation Protocol (RSVP) let reserve resources for a particular customer. However, a dynamic allotment of resources can be set just to avoid wastage of resources in idle or low traffic hours. It may lead to a problem

S. Patil · M. S. Subhedar (✉)
Department of Electronics and Telecommunication Engineering, Pillai HOC
College of Engineering and Technology, Rasayani, Raigad, Maharashtra, India
e-mail: mansi_subhedar@gmail.com

S. Patil
e-mail: snehalspatil12@gmail.com

© Springer Nature Singapore Pte Ltd. 2019   527
J. Wang et al. (eds.), *Soft Computing and Signal Processing* ,
Advances in Intelligent Systems and Computing 898,
https://doi.org/10.1007/978-981-13-3393-4_54

when the reserved resources are used by some other clients and at the same time the owning network of the reserved path has peak traffic hours. This would ultimately lead to congestion. It can be avoided by either dropping existing traffic or making the reserved resources free for the owning network or by providing an alternate path for the existing traffic and set the reserved path clear for new traffic of the owning network. To provide alternate paths to existing traffic and/or incoming traffic, we need to create alternate tunnels from source to destination. The alternate tunnels can be created statically or dynamically. In static method, ISP's network administrator has to learn the entire ISP network and configure alternate tunnels for each network (client). In dynamic method, alternate tunnels can be created automatically and can be brought into up state whenever required. However, this dynamic learning may take up some time to learn the current network status and pass it to other provider routers to form new tunnel [2].

In this work, practical implementation of software-defined networks (SDNs) in real-world networks and its ability to increase the efficiency of a network if working alongside MPLS or as a standalone application is demonstrated. High network availability and no congestion are most important criteria for ISPs, in order to meet service-level agreement (SLA) terms agreed by ISP and the client. We have primarily focused on avoiding congestion or link breakdown by providing alternate paths in case of congestion or link failure. Multiple paths must be made available to increase network availability, convergence and to maintain a proper flow. These paths can be learnt dynamically or statically [3].

2 SDN Architecture

The SDN framework is illustrated in Fig. 1 which is divided into three parts: southbound interface, northbound interface, and the controller. The southbound interface contains all the forwarding devices. The northbound interface consists of all the applications used to write flows. The controller is used to actually manipulate traffic or change the routing policies on the forwarding devices. The forwarding devices should run an agent to communicate with controller [3]. Here, the role of the agent is performed by Open Version Switch (OVS). The flows are instructions stating how the traffic should be forwarded or which modifications need to be done if required before forwarding. The flows are written into the OVS. It performs forwarding and path manipulation task on the basis of the written flows. OpenFlow Manager (OFM) which uses OpenFlow protocol is employed to write the flows. Southbound interface provides a way for SDN controller to communicate with network forwarding devices that includes packet handling instructions, loads, notification of links going up or down, and providing statistics like flow counters. Examples are OpenFlow, OVSDB, NETCONF, and SNMP [4, 5].

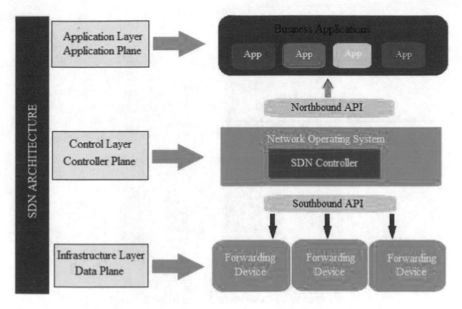

Fig. 1 SDN architecture

SDN controller provides services like:

1. Topology: It finds out connectivity between the devices. It basically describes the structure of the topology.
2. Inventory: It keeps record of devices which are enabled with SDN and itemization-related data of the same. This information includes rendition for the protocols that have been used and their capabilities.
3. Statistics: It reads counter information to monitor traffic on flows, interfaces, and flow table.
4. Host Tracking: It is used to determine where IP address or MAC address is located in a network.
5. Application interfaces include java API which forms the northbound interface (mostly RESTConf is used) [6]. It allows usage for https announcement approaching controller to govern system actions along with a collection of data. The network application lets us to write and/or edit network policies.

3 OpenFlow Protocol

The OpenFlow protocol module is depicted in Fig. 2 OpenFlow channel is used to form a connection between the switch and the controller. At a time, switch can be connected to one or more controllers. The OpenFlow channel carries basic Packet-IN, Packet-OUT, and Flow-MOD packets. There can be single otherwise additional flow

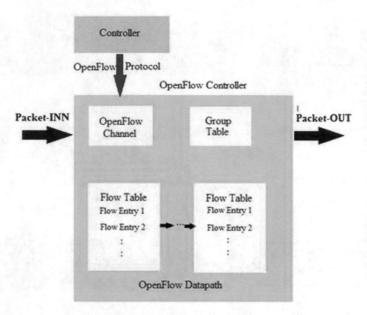

Fig. 2 OpenFlow protocol module and components of OpenFlow

tables within switch. Every flow table has a unique ID. By default, a flow table with 0 ID is created [7]. The flow tables are scanned in ascending sequence. Each group table has a unique identifier and an action bucket. The identifier is used to uniquely identify a group table, and action bucket contains a set of actions. Group tables are meant for same flow actions with different match criteria. This is similar to multicast. It allows you to change Flow-MOD for multiple flows in just one instance. Figure 3 exhibits working of OpenFlow protocol with controller. Whenever an OpenFlow switch receives a packet, it first checks its local flow table. If an entry is found, it takes the following action, else it is a table-miss and it passes packets to the controller which is the default action.

The switch forwards a Packet-IN report to the controller. This message might consist of either whole packet or a part of it along with a Buffer ID which is used as a reference to the packet. The controller now can send either:

- Packet-OUT: It is a normal forwarding information message.
- Flow-MOD: It instructs the switch to create a new entry and perform one or more actions on the packet.

Timers are basically used to restrict a count for entries of flow table which avoid flooding. Each timer can be disabled by setting it to 0. The switch might have multiple flow entries for the same packet. The choice is based on the priority of the flow [8, 9]. The flow having a numerically highest priority gets the first preference. Timers associated with the OpenFlow entries are:

Fig. 3 OpenFlow protocol
working with controller

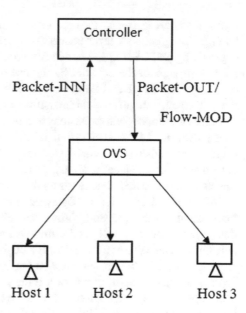

1. Idle Timer: If no match is found for an entry, it will be removed from the table. Default timer value is 20 s.
2. Hard Timer: When this timer is over, entry is discarded irrespective of whether any match exists or not.

4 RSVP Operation and MPLS Traffic Engineering

RSVP is a Resource Reservation Protocol which retains routers (resources) beside the paths in network. It operates over an IPv4 and IPv6 Internet layer and gives recipient-originated setup for reservations of resources of multicast or unicast data flows with scaling and robustness. Application data is never transported by RSVP. However, it resembles protocols, like Internet Control Message Protocol (ICMP) or Internet Group Management Protocol (IGMP). Host or routers use RSVP to demand or hand over required level of quality of service (QoS) to application data streams or flows. RSVP specifies how applications keep reservations and how they can discontinue the reserved resources until the necessity of it has ended. RSVP actions commonly outcome in resources being reserved in each node along a path [10].

Traffic engineering is extremely important for service provider and Internet service provider (ISP) backbones. As high bandwidth availability and no latency are important criteria for ISP, they should take care of bandwidth availability. Also, a network should be highly flexible, therefore such backbones be able to bear up failures of link or node. MPLS is combination of data link layer and network layer technologies. By combining data link features with network layer, MPLS performs traffic engineering. Hence, the one-tier system of connections should be obtained just by combining network layer connections with data link connections. MPLS traffic engineering [MPLS-TE] spontaneously builds and also retains label switched paths (LSPs) throughout the network by the use of Resource Reservation Protocol. The route utilized by a given LSP at indiscriminate bit of period is determined based on the LSP resource requirements and network resources, such as bandwidth. Available resources are flooded via extensions to a link-state-based Interior Gateway Protocol (IGP). Paths for LSP are calculated at LSP head based on a fit between required and available resources (constraint-based routing). IGP automatically routes the traffic onto these LSPs. Typically, a packet crossing MPLS traffic engineering backbone travels on a single LSP that connects the ingress point to the egress point.

MPLS-TE uses Interior Gateway Protocols like Intermediate System-to-Intermediate System, Open Shortest Path First, to inescapably map packets toward the proper traffic flows. It send traffic flows over the system by the use of MPLS expedition. The paths of packet flows throughout the backbone are determined based on required traffic flow resources also feasibility of resources. MPLS-TE also uses "constraint-based routing," where a route of a packet forwarding is the shortened route which appropriates resource needs (constraints) of the packet flow. Traffic flow has different requirements like bandwidth, media, and priority versus other flows. It repairs link or node breakdowns which alters connections of network by modifying to advanced set of constraints [10, 11].

5   Our Contribution

This paper aimed to optimize the procedure for dynamic learning of alternate paths or tunnels. The technic of optimizing the performance is explained as follows. Each P and PE router will be now connected to a central controller. The routers will inform their traffic statistics to the controller. For communication between controller and routers, an interface is required. Open Version Switch (OVS) is used for this function. The OVS instance will run for each network, i.e., for each interface of a router. Whenever a packet is going through the network which is running an instance of OVS, the packet is first sent to the controller and controller replies with the flows which are then populated in the flow table. These flows will decide the forwarding path for traffic.

Consider a topology consisting of nine routers. There are three customer sites, namely Customer-A, Customer-B, Customer-C. For customer-A to reach it's another site, there are nine possible paths. For customer-B to reach it's another site, there are nine possible paths. For customer-C to reach it's another site, there are 11 possible paths. These all paths are preprogrammed in SDN controller [12] (Fig. 4).

OpenFlow is about forwarding data to appropriate destination. In this approach, path reservation protocol like RSVP used in literature is not employed. Here, SDN controller dynamically provides the path to appropriate router obtained using OSPF extensions and by considering congestions. The role of router inside the network is

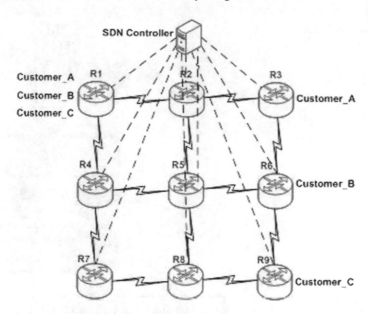

Fig. 4 Network with MPLS+SDN

to just keep records of link congestion based on available bandwidth and bandwidth utilization of the concerned outgoing interfaces and report the same to SDN controller. The controller is already programmed on how to calculate the best path based on different bandwidth requirements and associated SLAs. If a particular link gets crowded which is part of the best path, controller will sense the congestion and will notify to the penultimate congested router to use less congested path. SDN controller will be the only decision maker. It will notify the routers about the path to be used.

The northbound interface contains OpenFlow Manager (OFM) which uses Open-Flow protocol (OFP) as shown in Fig. 5.

The OVS is connected to the OFM through port 8080 which is a standard port for connectivity amid the controller and the switch. OFM's user interface can be obtained at port 9000. The OFM can be used to change network policies or write new flows on to the forwarding devices. The integration of OFM is done using C programming and Python and is an open-source platform. Figure 6 shows congestion control using SDN. The OFM together with OVS reads the network statistics from the forwarding device remotely. The new flows according to the changing traffic conditions through the OFM can be written. Every packet consists of the Explicit Congestion Notification (ECN) bit, which adopts the two rightmost bits of the Diffserv field in the IPv4 or IPv6 header for encoding the following code points:

1. Non-ECN Capable Transport, Non-ECT—00
2. ECN Capable Transport, ECT(0)—10
3. ECN Capable Transport, ECT(1)—01
4. Congestion Encountered, CE—11.

Fig. 5 SDN and OFM
interaction

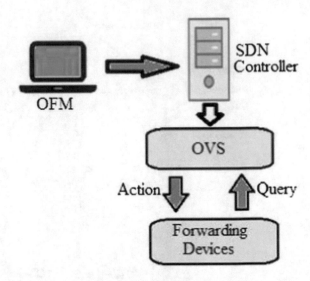

Fig. 6 Congestion control
using SDN

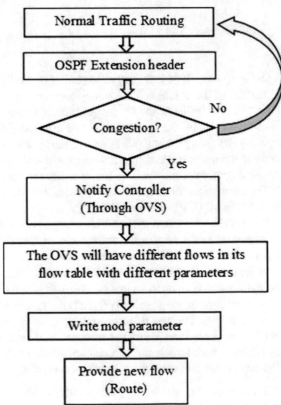

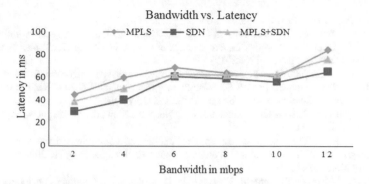

Fig. 7 Comparison of bandwidth versus latency of MPLS, MPLS+SDN, and SDN

If one and other extreme ends support ECN, their packets are stamped as ECT(0) or ECT(1) as congestion encountered intimation may only be managed efficiently by an upper-layer protocol that supports it. ECN is used in alliance with upper-layer protocols, such as TCP. Figure 7 displays a comparison of latency and bandwidth for MPLS, SDN, and MPLS+SDN. It can be observed that coupling with SDN yields better results and latency can be reduced considerably [13].

6 Conclusion

MPLS coupled with SDN responses to sudden topology changes at a faster rate as compared to conventional MPLS-TE performance. When SDN controller is used to calculate statistics collected from OSPF extension headers, latency for end-to-end packet delivery is reduced and response time of network to tackle congestion has also been reduced. Programming enables the administrator to control congestion at first time and write the flows based on the varying network statistics to maintain the requirements of service-level agreement.

References

1. Cisco Systems. MPLS traffic engineering, constrained-base routing and operation in MPLS-TE. Cisco Press. http://www.cisco.com/articles/article.asp-p=426640&seqNumber=3
2. Kar, K., Kodialam, M., Lakshman, T.V.: Minimum interference routing of band-width guaranteed tunnels with MPLS traffic engineering applications. IEEE J. Sel. Areas Commun. **18**(12), 2566–2579 (2000)
3. Goransson, P., Black, C.: Software De ned Networks-A Comprehensive Ap-proach. ISBN: 9780128045794
4. Sezer, S., Scott Hayward, S., Kaur Chouhan, P.K., Fraser, B.B., Lake, D., Finnegan, J., Viljoen, N., Miller, N.M., Rao, N.: Are we ready for SDN? Implementation challenges for software-

defined networks. IEEE Communications Magazine (July 2013)

5. Porras, P., Shin, S., Yegneswaran, V., Fong, M., Tyson, M., Gu, G.: A security enforcement kernel for Openflow networks. In: Proceedings of the 1st Workshop Hot Topics in Software De ned Networks, pp. 121–126 (2012)

6. Elsayed, K.M.F.: HCASP: a hop-constrained adaptive shortest path algorithm for routing bandwidth guaranteed tunnels in MPLS networks. In: Proceedings ISCC 2004 Ninth International Symposium on Computers and Communications, 2 (July 2004)

7. ONF.: Software-De ned Networking: The New Norm for Networks. White paper. https://www.opennetworking.org

8. Tu, X., Li, X., Zhou, J., Chen, S.: Splicing MPLS and open-flow tunnels based on SDN paradigm. In: IEEE International Conference on Cloud Engineering (IC2E), pp. 489–493 (11–14 March 2014)

9. Michael, J., Zinner, T., Hobfeld, T., Tran-Gia, P., Kellerer, W.: Interfaces, attributes, and use cases: a compass for SDN In: IEEE Communications Magazine, June (2014)

10. Cisco Systems. MPLS Traffic Engineering, Cisco Press. https://www.cisco.com/c/en/us/td/docs/ios/12_0s/feature/guide/TE_1208S.html

11. Hasan, H., Cosmas, J., Zaharis, Z., Lazaridis, P., Khwan-dah, S.: Creating and managing dynamic MPLS tunnel by using SDN notion. In: International Conference on Telecommunications and Multimedia (TEMU) (2016). https://doi.org/10.1109/temu.2016.7551923

12. Hasan, H., Cosmas, J., Zaharis, Z., Lazaridis, P., Khwan-dah, S.: Development of FRR mechanism by adopting SDN notion. In: 24th International Conference on Software, Telecommunications and Computer Networks (SoftCOM) (2016). https://doi.org/10.1109/softcom.2016.7772133

13. RFC 7560: Problem Statement and Requirements for Increased Accuracy in Explicit Congestion Notification Feedback, tools.ietf.org. IETF. August 26, 2015. https://tools.ietf.org/html/rfc7560

Dictionary Learning-Based MR Image Reconstruction in the Presence of Speckle Noise: Greedy Versus Convex

M. V. R. Manimala, C. Dhanunjaya Naidu and M. N. Giri Prasad

Abstract The trend of acquiring samples at a rate far below Nyquist rate is termed as compressive sensing (CS). CS enables the accurate recovery of signals/images by exploiting the underlying sparsity in either transform or signal domain. CS is useful in reducing acquisition time in MR imaging. In this paper, CS based on dictionary learning has been applied for MR imaging in the presence of speckle noise. Performance of two classes of sparse recovery techniques, namely convex optimization and greedy iterative algorithms, has been investigated and compared, when employed for the sparse coding stage of an adaptive patch-based dictionary. Two greedy algorithms, orthogonal matching pursuit (OMP) and compressive sampling matching pursuit (CoSaMP), are considered and contrasted with convex techniques: basis pursuit (BP) and least absolute shrinkage and selection operator (LASSO). Experimentation has been done on spine, knee, and brain image by varying the speckle noise variance, sparsity threshold per patch for various sampling schemes. Results show that convex techniques achieve higher peak signal-to-noise ratio (PSNR) than greedy algorithms in the presence of high noise.

Keywords Compressive sensing · Orthogonal matching pursuit
Compressive sampling matching pursuit · Basis pursuit
Least absolute shrinkage and selection operator

M. V. R. Manimala (✉) · M. N. Giri Prasad
JNTUA, Ananthapuramu, India
e-mail: svrmanimala_ece@mvsrec.edu.in

M. N. Giri Prasad
e-mail: mahendragiri1960@gmail.com

C. D. Naidu
VNRVJIET, Hyderabad, India
e-mail: cdnaidu@vnrvjiet.in

© Springer Nature Singapore Pte Ltd. 2019
J. Wang et al. (eds.), *Soft Computing and Signal Processing* ,
Advances in Intelligent Systems and Computing 898,
https://doi.org/10.1007/978-981-13-3393-4_55

1   Introduction

Sparse signal processing finds numerous applications in the area of image compression, analysis, and medical imaging. Compressive sensing (CS) exploits the sparsity inherent in the signal in the transform domain or image domain and allows the accurate image/signal recovery from measurements that are far below the Nyquist rate. A nonlinear reconstruction procedure should be employed to recover CS signal, which is computationally expensive.

Adaptive patch-based dictionaries provide higher degree of undersampling as patch-based sparsity is exploited rather than the global sparsity. Higher undersampling rates reduce the number of samples acquired and hence reduce the acquisition time. This makes CS useful in many applications such as medical imaging, wireless sensor networks, seismic imaging. In the recent past, CS has been applied to MR imaging by Lustig et al., is termed as LDP [1], and displays a high quality of reconstruction from a reduced set of measurements. In [2], dictionary learning (DL) has been applied to MR imaging by Ravishankar et al. denoted as DLMRI in which K-SVD [3] was used to train the dictionary and orthogonal matching pursuit (OMP) was employed for sparse coding stage. Although analysis and comparison of sparse codes exist, however its analysis in the dictionary learning framework in the presence of speckle noise is less studied. In the present work, DL paradigm has been applied to three MR images in the presence of speckle noise [4] and comprehensive analysis has been presented on the sparse recovery stage of the dictionary learning. Two classes of sparse recovery techniques, namely greedy algorithms (orthogonal matching pursuit, compressive sampling matching pursuit) [5, 6] and convex solutions (basis pursuit and least absolute shrinkage and selection operator) [7, 8], have been compared with respect to speckle noise variance, sampling scheme, computation time, and sparsity threshold.

2   Compressive Sensing

Compressive sensing can be achieved by designing a good sparse model for the data. Sparsity can be achieved via a synthesis dictionary model [2] or a sparsifying transform model [9]. In a synthesis dictionary model, a signal $y \in C^m$ can be sparsely represented as a linear combination of a small number of atoms or columns from a synthesis dictionary $D \in C^{m \times n}$, given by

$$y = Dx \tag{1}$$

with $x \in C^n$ being a sparse vector, i.e., $\|x\|_0 \ll n$. Solution to the above problem involves finding l_0 norm of x (NP-hard) by counting a number of nonzero elements of x. Alternative solution to the above equation can be found by employing convex optimization (l_1 norm) or greedy algorithms.

Greedy pursuits are the algorithms that iteratively build up an estimate of x. These algorithms start with an empty set and estimates x by iteratively adding new components such as OMP [5], CoSaMP [6].

Basis pursuit (BP) finds signal representation in overcomplete dictionaries by minimizing the l_1 norm of the coefficients occurring in the representation. For noiseless measurements, BP is given by

$$\min_x \|x\|_1 \quad \text{subject to} \quad Dx = y \tag{2}$$

Unlike OMP, BP starts from a full model and iteratively prunes the model by swapping the relatively useless terms with the useful terms. BP may not lead to sparse solution every time, but under right conditions it can provide a sparse solution or even sparsest solution [7].

Least absolute shrinkage and selection operator (LASSO) [8] is a variant of BP, also known as basis pursuit with denoising (BPDN) in certain areas. LASSO shrinks or sets some coefficients to zero by pacing restriction to the value of l_1 norm unlike BP which minimizes the l_1 norm. Mathematically, LASSO is represented by

$$\min \|Dx - y\|_2 \quad \text{subject to} \quad \|x\|_1 < \lambda \tag{3}$$

where λ represents the threshold value. LASSO overcomes the drawbacks of prediction accuracy and interpretation that is prominent in ordinary least squares (OLS) and is particularly useful when the measurements are noisy.

3 Speckle Noise

Speckle noise is prominent in coherent imaging techniques like ultrasound imaging. Main source of speckle noise is due to the random variations of the backscattered waves from the objects and transmission errors [4]. Speckle noise degrades the image quality in ultrasound imaging and to some extent in MRI. It reduces the contrast resolution. Speckle pattern depends on scatterer number density, spatial distribution of scatterers, and characteristics of an imaging system. Speckle noise is also known as a texture. Speckle noise has both multiplicative component and additive component and is given by

$$f(x, y) = g(x, y)_{\eta_m}(x, y) + {}_{\eta_a}(x, y) \tag{4}$$

where $g(x, y)$ is an unknown function representing the original noise-free image, $f(x, y)$ is the noisy observation of $g(x, y)$, and η_m and η_a are multiplicative and additive components of noise, respectively. x and y are variables to represent spatial location.

4 Dictionary Learning

Adaptive dictionaries are useful in many image processing applications such as image compression, denoising, segmentation, and classification. These dictionaries offer a high degree of undersampling. An adaptive patch-based synthesis dictionary has been developed in [2] termed as DLMRI. Patch size in DLMRI has been selected to be equal to the number of atoms or columns of the dictionary, termed as a complete dictionary. K-SVD (singular value decomposition) [3] has been employed to train the dictionary. Sparse coding is an important stage in DL. To reconstruct MR images, DLMRI iteratively trains the dictionary first and then employs the dictionary to sparse code all the patches of the image. Sparse coding stage for DLMRI has been analyzed in [10], by employing OMP, CoSaMP, and BP with respect to undersampling limit, number of iterations, operational complexity for various sampling schemes.

Although numerous sparse coding algorithms exist, in the present work two representative greedy algorithms (OMP and CoSaMP) are selected for comprehensive analysis and comparison with representative convex solutions (BP and LASSO) for sparse coding of patches in DLMRI. Performance of these algorithms is evaluated for brain, knee, and spine image with respect to varied noise levels and ratio m/k (number of measurements per patch to sparsity threshold per patch) for 7-fold pseudoradial, 4-fold random and Cartesian sampling schemes. BP has been implemented by using l_1-magic toolbox [11], and LASSO has been implemented with SparseLab package [12].

5 Results

5.1 Parameters

Peak signal-to-noise ratio (PSNR) in dB signifies the reconstruction quality and is the ratio of peak intensity value of reference image to the root-mean-square (RMS) reconstruction error relative to the reference image. Reconstruction of edges and fine edges is measured by high-frequency error norm (HFEN). For a good reconstruction quality, high PSNR and low HFEN are desired. PSNR0 represents zero-filled reconstruction. Sparsity is investigated by varying the m/k ratio, where m is the number of measurements per patch and k is the sparsity threshold for the patch. Runtime of the algorithm is evaluated with respect to m/k.

An Intel Core i5 CPU at 2.3 GHz and 4 GB memory with 64-bit Windows 8 operating system has been employed to carry out all the computations. All implementations have been coded in MATLAB v7.13 (R2011b). DLMRI algorithm depends on various parameters, which include the patch size $\sqrt{m} \times \sqrt{m}$, sparsity threshold per patch k, and data consistency in the presence of noise is determined by the weighting factor λ. Total number of overlapping patches is denoted by N. To train the dictionary, only a fraction of patches δ are employed and J represents the number of iterations

(a)   **(b)**   **(c)**

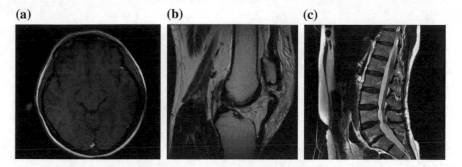

Fig. 1 Reference images of **a** *brain* [2], **b** *knee* [13], **c** *spine* [14]

in learning. Experimentation has been done on brain [2], knee [13], and spine [14] image shown in Fig. 1. DLMRI iterations have been fixed to one, and DL stage of DLMRI employs 5 iterations of K-SVD for all the results in this article.

5.2 Noise Analysis

Figure 2 presents the plots of PSNR versus noise for the spine image [14] with m/k = 5 and various sampling schemes, by increasing the noise variance from 0.05 to 0.5. Results with pseudoradial sampling (Fig. 2d) demonstrate that PSNR obtained is almost same for all the algorithms; however, for LASSO, it is noted to be comparatively lower. At noise variance of 0.3, zero-filled PSNR is almost equal to the PSNR obtained with LASSO, CoSaMP, and OMP. Hence, for the noise variance of 0.3 and above, DLMRI reconstruction of spine image with 7.11-fold pseudoradial undersampling does not show any improvement over zero-filled reconstruction. Figure 2e illustrates the PSNR plot with 4-fold Cartesian undersampling. For low noise of 0.05, greedy algorithms obtain comparatively higher PSNR, but as the noise variance increases above 0.1, convex solutions provide higher PSNR. BP provides better recovery compared to other algorithms but takes high time to converge. Similar results have been noted with reconstruction using 4-fold random sampling (Fig. 2f).

Figure 3 displays the reconstruction of spine image with 4-fold random sampling of the k-space, speckle noise variance of 0.2 m/k has been fixed to 5, and the reconstructions with four sparse coding techniques have been displayed in this figure. PSNR of the noisy image (Fig. 3b) is about 16.65 dB with respect to the reference image. Although speckle noise was reduced to a great extent, however it could not be eliminated totally. PSNR of BP (21.41 dB) is about 0.3 dB higher than OMP, CoSaMP, and LASSO at a noise variance of 0.2. Reconstruction error has some visible spine structure.

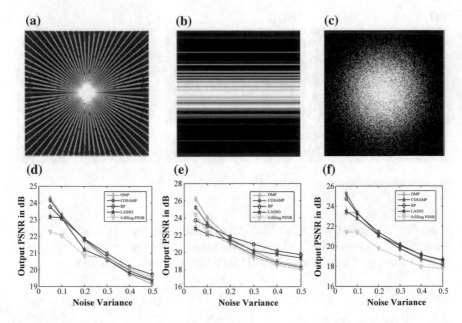

Fig. 2 PSNR plot for spine image [14]: **a** 7.11-fold pseudoradial sampling mask in k-space [2], **b** 4-fold variable density Cartesian sampling [2], **c** 4-fold random undersampling [2], **d** *PSNR versus noise variance* with 7.11-fold pseudoradial sampling, **e** *PSNR* versus *noise variance* with 4-fold Cartesian sampling, **f** *PSNR* versus *noise variance* with 4-fold random undersampling

Figure 4 depicts the effect of noise on the reconstruction of knee image with DLMRI algorithm. Experiments were done with m/k = 5 and by varying the noise level. Pseudoradial mask at 7.11-fold undersampling has been employed for the simulations in Fig. 4a. At a noise variance of 0.1, PSNR for all the sparse coding techniques considered has been observed to be around 26 dB. As the noise variance increases to 0.5, PSNR decreases to 21 dB for LASSO, 20.72 dB for BP, 19.84 dB for CoSaMP, and 19.63 dB for OMP. Similar trends were observed with 4-fold random undersampling (Fig. 4b) and Cartesian undersampling (Fig. 4c). PSNR was observed to be somewhat higher for Cartesian sampling.

5.3 m/k Effect

Variations in PSNR as m/k is increased from 2 to 7 have been analyzed for the brain image (Fig. 1a) acquired with 4-fold random undersampling of k-space in Fig. 5. Experiments were performed by increasing the noise variance N from 0 to 0.5, in steps of 0.1. PSNR increases by about 0.2–0.6 dB as m/k increases (Fig. 5a), when OMP is employed for sparse coding stage. For noiseless measurements, PSNR is seen to be almost constant when CoSaMP is employed for sparse coding (Fig. 5b). A small

(a)   (b)   (c)

(d)   (e)   (f)

(g)   (h)   (i)

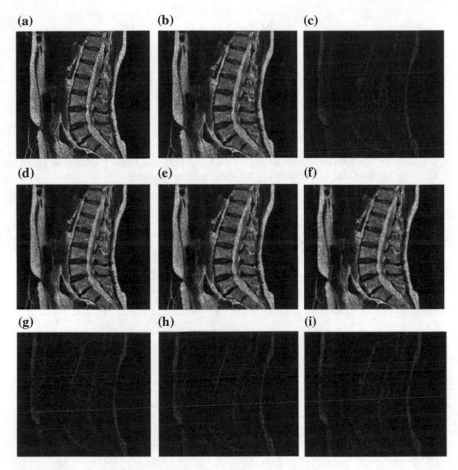

Fig. 3 **a** Noisy spine image [14] with speckle noise variance of 0.2. **b** Reconstruction of spine image with 4-fold random undersampling and m/k = 5 for: OMP (PSNR = 21.11 dB), **d** CoSaMP (PSNR = 21.14 dB), **e** BP (PSNR = 21.41 dB), **f** LASSO (PSNR = 21.07 dB). Reconstruction error with **c** OMP, **g** CoSaMP, **h** BP, **i** LASSO

increase in PSNR has been noted for CoSaMP in the case of noisy measurements. A 3 dB drop in PSNR was noted for reconstructions with LASSO under noiseless condition. Corresponding noisy measurements show about 0.3 dB drop in PSNR, as m/k is increased from 2 to 7 (Fig. 5c). Similar result has been noted for BP (Fig. 5d).

The above experimental results demonstrate that the PSNR improves marginally for greedy algorithms as m/k ratio increases, whereas for convex solutions PSNR decreases with m/k. Greedy algorithms perform better than convex solutions under noiseless or low noise conditions, but as the speckle noise variance is increased to 0.5, convex solutions obtain comparatively higher PSNR.

Figure 6 displays the plot of PSNR versus m/k variation for the brain image acquired by random sampling of k-space as in Fig. 5, by considering all the four

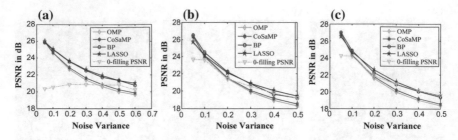

Fig. 4 Plots of PSNR versus speckle noise for knee image [13] with m/k = 5: **a** 7.11-fold pseudo-radial sampling, **b** 4-fold random sampling, **c** 4-fold Cartesian sampling

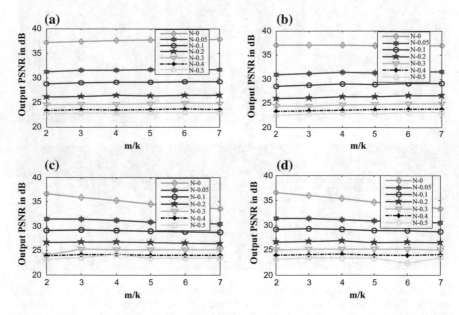

Fig. 5 PSNR versus m/k plots for brain image [2] with speckle noise variance N = 0 to 0.5 for **a** OMP, **b** CoSaMP, **c** LASSO, **d** BP

algorithms with a fixed noise variance of 0.5. PSNR has been observed to increase marginally with m/k for OMP, CoSaMP, and BP, whereas for LASSO it decreases by 0.23 dB. From Fig. 6b, it has been observed that the runtime increases with the increase in m/k value, particularly for the values of m/k greater than 4. Highest runtime of 1595 s (for m/k = 7) has been noted for BP and least runtime of 174 s has been observed for CoSaMP. Corresponding runtime for LASSO and OMP is 189 s and 267 s, respectively. Greedy algorithms need less time for computations compared to convex solutions. Experimental results demonstrate that computation time for OMP is higher than CoSaMP for m/k greater than 5.

Table 1 demonstrates the computation time, HFEN, PSNR, and PSNR0 for a 7 × 7 patch with sparsity threshold of 7 (m/k = 7). Computation time for OMP and LASSO

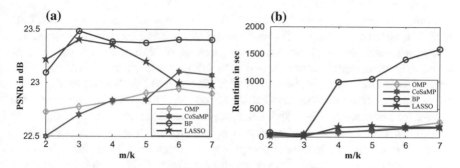

Fig. 6 Brain image (noise variance 0.5): **a** *PSNR versus m/k* and **b** *runtime versus m/k*

Table 1 Computation time, HFEN, PSNR, and PSNR0 for knee [13], brain [2], and spine image [14] with m/k = 7, noise variance = 0.1, random sampling—4-fold

Algorithm	Computation time (s)	HFEN	PSNR (dB)
Knee image			
OMP	338	6.32	24.03
CoSaMP	**165.9**	6.27	24.27
BP	1588	**6.12**	**24.44**
LASSO	236.9	6.82	23.81
Brain image			
OMP	**85.72**	**3.45**	**29.23**
CoSaMP	168.8	**3.45**	29.21
BP	1640	3.49	28.76
LASSO	154	4.1	27.12
Spine image			
OMP	257.8	6.23	23.4
CoSaMP	**170.1**	**6.13**	**23.55**
BP	1610	6.4	23.16
LASSO	183	6.88	22.83

varies by a great extent, contrary to BP and CoSaMP. Although BP takes more time to converge, it provides better guarantees and higher PSNR in the presence of speckle noise. As noise increases, convex solutions like BP and LASSO attain higher PSNR and better recovery ability, which is evident from the results in Figs. 3 and 4.

6   Conclusion and Future Scope

In this work, a comprehensive analysis of two classes of sparse coding techniques, namely greedy algorithms and convex solutions, has been presented for reconstruction of images affected by the speckle noise. Various images have been considered to evaluate and compare the performance of the four sparse coding algorithms (OMP, CoSaMP, BP, and LASSO) in DL with respect to the sampling scheme, m/k ratio.

Experimental results show that in the presence of high speckle noise, convex solution provides better reconstruction when compared to greedy algorithms. High computation time is a major limitation with BP, whereas LASSO has an advantage of low computation time with high PSNR. Greedy algorithms are faster compared to convex techniques; however, PSNR decreases as the noise level increases. For greedy algorithms, PSNR increases with m/k, whereas converse is true for convex solutions. As m/k increases, computational complexity increases for all the algorithms and hence runtime also increases. Convex solutions are more stable even in the presence of high noise compared to greedy algorithms.

Reconstruction with DLMRI proves to be better than LDP employed in [1] even at higher undersampling rates. Although patch-based adaptive dictionary can reduce the speckle noise, to enhance the perceptual quality of the image some preprocessing/post-processing is required. Similar analysis can be carried out on ultrasound images.

Declaration The brain image and various masks used in this work are made available online by Dr. Saiprasad Ravishankar and Dr. Yoram Bresler for research purpose, and the authors would like to thank them for the same.

References

1. Lustig, M., Donoho, D.L., Santos, J.M., Pauly, J.M.: Compressed sensing MRI. IEEE Signal Process. Mag. **25**(2), 72–82 (2008)
2. Ravishankar, S., Bresler, Y.: MR Image reconstruction from highly undersampled k-space data by dictionary learning. IEEE Trans. Med. Imaging **30**(5), 1028–1041 (2011)
3. Aharon, M., Elad, M., Bruckstein, A.: K-SVD: an algorithm for designing overcomplete dictionaries for sparse representation. IEEE Trans. Signal Process. **54**(11), 4311–4322 (2006)
4. Khaled, Z., et al.: Real-time speckle reduction and coherence enhancement in ultrasound imaging via nonlinear anisotropic diffusion. IEEE Trans. Biomed. Eng. **49**(9), 997–1014 (2002)
5. Tropp, J., Gilbert, A.C.: Signal recovery from random measurements via orthogonal matching pursuit. IEEE Trans. Inf. Theory **53**(12), 4655–4666 (2007)
6. Needell, D., Tropp, J.A.: CoSaMP: iterative signal recovery from incomplete and inaccurate samples. Commun. ACM **53**(12), 93–100 (2010)
7. Cheny, S.S., et al.: Atomic decomposition by basis pursuit. Soc. Ind. Appl. Math. **20**(1), 33–61 (1998)
8. Tibshirani, R.: Regression shrinkage and selection via the lasso. J. R. Statist. Soc. B **58**, 267–288 (1996)
9. Ravishankar, S., Bresler, Y.: Learning sparsifying transforms. IEEE Trans. Signal Process. **61**(5), 1072–1086 (2013)
10. Manimala, M.V.R., et al.: Sparse recovery algorithms based on dictionary learning for MR image reconstruction. In: IEEE 2016 International Conference on Wireless Networks, Signal Processing and Networking, pp. 1354–1360, March (2016)
11. Candès, E., Romberg, J.: l_1-magic: recovery of sparse signals via convex programming. Caltech, October (2005)
12. Donoho, D., et al.: About SparseLab, Stanford University, Version.100, May (2006)
13. http://www.philosophyinaction.com/blog/?p=3187
14. http://si-instability.com/wp-content/uploads/2013/03/MRI5-16-2012.jpg

Towards Improving the Intelligibility of Dysarthric Speech

Arpan Roy, Lakshya Thakur, Garima Vyas and Gaurav Raj

Abstract Humans utilize many muscles to produce intelligible speech, including lips, face and throat. Dysarthria is a speech disorder that surfaces when one has weal muscles due to brain damage. Primary characteristics of a dysarthric patient are slurred and slow speech that can be difficult to understand based on the severity of the condition. This paper proposes an approach to improve the intelligibility of the dysarthric speech using a simple yet effective speech-transformation technique such as warping the frequency of LPC poles and mapping coefficients of linear predictive coding. This technique was applied to dysarthric audio from the UA-speech database to obtain the desired results. Both objective and subjective measures are used to evaluate the transformed speech. The obtained results pointed towards a significant improvement in the dysarthric speech's intelligibility. This method can be used to develop special voice-enabled search platforms for dysarthric patients and in helping rehabilitation of the patients by developing a speech therapy based on auditory feedback.

Keywords Dysarthria · LPC · Speech intelligibility · Frequency warping
Speech therapy · Speech enhancement · Auditory feedback

A. Roy (✉) · L. Thakur
Department of Electronics and Communication Engineering, ASET, Amity University Uttar
Pradesh, Noida, UP, India
e-mail: royarpan09@gmail.com

L. Thakur
e-mail: lapstjup@gmail.com

G. Vyas · G. Raj
Computer Science Engineering, ASET, Amity University Uttar Pradesh, Noida, UP, India
e-mail: gvyas@amity.edu

G. Raj
e-mail: graj@amity.edu

© Springer Nature Singapore Pte Ltd. 2019
J. Wang et al. (eds.), *Soft Computing and Signal Processing* ,
Advances in Intelligent Systems and Computing 898,
https://doi.org/10.1007/978-981-13-3393-4_56

1 Introduction

A damage to the central or peripheral nervous system (brain damage) can often hinder with muscular control over the speech-producing organs leading to a group of speech disorders referred to as dysarthria. The severity and the type of dysarthria in a patient depend majorly on the area severity and area of nervous system that is affected. It is an impairment in speech production processes such as articulation of building blocks of speech like consonants and vowels. The neurological damage responsible for dysarthria can also affect various other physical activities in an individual restricting mobility and computer interaction. For example, patients with severe dysarthria are found to be about 150–300 times slower than normal users in keyboard interaction [1, 2]. In contrast to this, dysarthric speech is only about 10–17 times slower as compared to the typical speakers [3, 4].

Numerous studies have been done on enhancing the dysarthric speech. Vijay-lakshmi et al. [5] employed a recognition system based on HMM and thereafter synthesized the speech adaptively in order to preserve the naturalness and intelligibility of the speech [6, 7]. Proposed a method that transforms the vowels of the dysarthric speaker and matches it to the vowels of a healthy speaker, this method claimed to have resulted in improvement in intelligibility from about 48 to 54% baseline. Morales et al. (2009) built weighted transducers into an ASR system according to the observed confusion matrices of the phonemes, which improved word-error rates by approximately 5% on severely dysarthric speech and approximately 3% on moderately dysarthric speech. Selouni et al. [8] used a concatenating algorithm combined with a grafting technique to correct the poorly uttered phonemes and further managed to improve the PESQ score value by more than 20%. Acoustic transformations were used by [9] in order to correct the inserted and dropped phoneme errors.

This paper successfully utilizes speech and audio transformation techniques such as LPC coefficient mapping and frequency warping to increase the speech intelligibility of a dysarthric speaker while preserving the naturalness of the speech. This improved audio can then be played back to the speaker after a delay. Doing so helps the patient recover faster as they listen to an improved quality of their speech as an auditory feedback, which is crucial for stimulating their nervous system to produce intelligible speech. This results in a faster rehabilitation of dysarthric speakers. The transformed speech is then evaluated using both subjective and objective methodologies for male and female speakers, respectively. Subjective measure of intelligibility and naturalness of speech are calculated using the degradation mean opinion score (DMOS). Objective measure of intelligibility is evaluated on the basis of perceptual evaluation of speech quality (PESQ) score.

The paper is organized as follows: Sect. 2 describes the database used; Sect. 3 discusses the methodology employed; Sect. 4 evaluates the enhanced speech quality using objective and subjective measures; and Sect. 5 concludes the paper.

2 Speech Corpus

The UA-speech database has been used for the purpose of this project, as the data is available for free via ftp and a secure http.

The subjects used for this database are diagnosed with spastic dysarthria, which was informally confirmed by a certified speech-language pathologist listing to these recordings.

Subjects read isolated words from a computer monitor. Prompt words included:

- Digits (DGT): 10 digits, e.g., "one, two, three ..."
- International Radio Alphabet (IRA): the 26 letters of the International Radio Alphabet, e.g., "alpha, bravo, charlie..."
- Computer Words (CPW): the 19 word processing commands, e.g., "command, line, paragraph, enter..."
- Common Words (CMW): the most common 100 words in the Brown corpus, e.g., "the, of, and..."
- Uncommon Words (UCW): A total of 300 words selected from Project Gutenberg novels using an algorithm that sought to maximize biphone diversity, e.g., "naturalization, faithfulness, frugality...".

3 Methodology

3.1 System Architecture

In order to improve the intelligibility of dysarthric speech, the architecture has been proposed in Fig. 1. Speech signal transformations have been performed on both the control and the dysarthric speech after clipping the useful audio using a voice activity detector and dynamic time warping the two signals to perform time alignment. Linear predictive analysis along with frequency warping transformation has been done on the speech signals of dysarthric and control speech. The result of the final resynthesized speech after the transformation is then measured according to their intelligibility scores using PESQ objective measure and DMOS subjective measure.

3.2 Speech Analysis

In most of the studies done previously to improve the speech intelligibility of dysarthric speakers, variations in frequency, temporal and prosodic characteristics of their speech were made [10, 11].

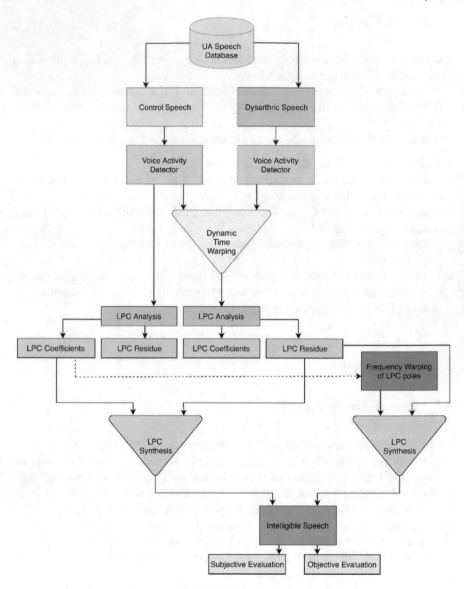

Fig. 1 System architecture to improve the intelligibility of dysarthric speech

In order to differentiate the dysarthric and the control audio, a speech analysis was performed on spectral energy and the formant features of both the audios. The features that were perceptually more important were then selected to improve the intelligibility of speech in this study.

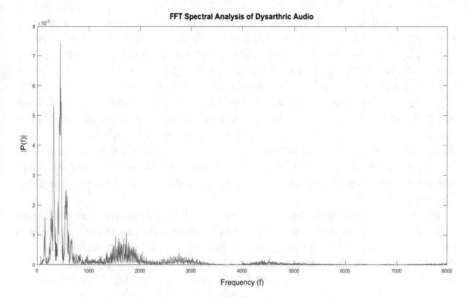

Fig. 2 FFT spectral analysis of dysarthric audio

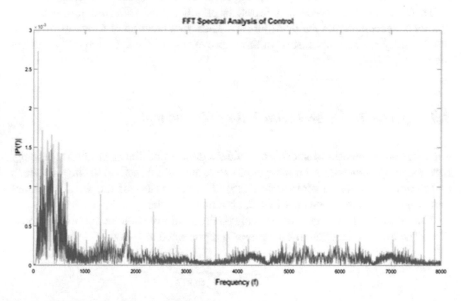

Fig. 3 FFT spectral analysis of control audio

Dysarthric speech signals do not show the necessary spectral changes that should take place when compared to their control counterparts [12]. Taking the fast Fourier transform of both control and dysarthric signals for a word "Yes" has been shown in the following Figs. 2 and 3.

These dysarthric patients also show high levels of a condition known as a vocal fry which results in the voice coming down to its lowest natural register. This changes the way a person's vocal cords vibrate. These changes result in subsequent inconsistencies in the vibrations and thus make the voice of the speaker less intelligible. A vocal fry may not necessarily result in reduced intelligibility, but the reduced quality of speech certainly makes it harder for the listener to understand the garbled voice of the dysarthric patient. This condition is more observed in male, about 25%, as compared to 10% in female. This is due to the fact that male patients have a lower overall pitch as compared to women. As a result, their voices descend into vocal fry much more easily. This condition can be digitally eliminated in their speech signals by smoothening the energy trajectory by using median filtering and zero-phase filtering [13].

Resonant frequency of the vocal tract also varies from patient to patient depending on their length of the tract. These resonant frequencies are called formants, which help us distinguish between two sounds or phonemes in case of speech. These formants also help in understanding acoustic and articulation factors which result in the reduction of intelligibility of the dysarthric speaker's speech [14]. After performing formant modifications on the dysarthric audio, it was observed that the speech quality had improved without any noticeable improvement in the speech's intelligibility.

Therefore, it is observed that formants play a crucial role in determining the quality of speech but not the intelligibility. Thus, modification of formants on a low intelligibility speech only results in the improvement of the speech quality and not the intelligibility.

3.3 Speech Transformation Using LPC Mapping

Analysis and synthesis of speech are widely done using linear predictive coding. LPC mapping allows for a simple, yet effective and robust method to encode speech and perform various transformations [15]. The parameters of the vocal track are represented by the coefficients of LPC, obtained from the speech.

The LPC analysis is done by calculating a linear all-pole infinite impulse response filter. The transfer function of the system is represented as follows:

$$H(z) = \frac{1}{1 - \sum a_k z^{-k}} \tag{1}$$

where a_k is the filter coefficient.

The speech articulators are accurately modelled by this all-pole filter. It results in a uniform mixture of all frequencies, which has a flat frequency spectrum known as the LP residual. This LP residual is essential for the synthesis of the speech signal. It contains all the required information necessary for the synthesis process to give a natural quality.

As the control and dysarthric speech vary greatly in their length. The latter being longer than the former due to the slow speech rate of the speaker. This results in a significant difference in both the speakers. To compensate for these differences in timings and to successfully enable featurewise comparison between the two audio files, dynamic time warping (DTW) [16–18] is used. This maps frame by frame the dysarthric audio with the respective utterance of control audio. The LPC coefficients are then calculated from the mapped normative speech, and the LP residuals are extracted from the mapped dysarthric speech. These are then used to resynthesize the audio to produce a more intelligible speech after performing transformations [19]. The LPC coefficients from the control speech are used to improve the intelligibility while the LP residuals are used to improve the originality and naturalness of the resultant-transformed speech [20].

3.4 Frequency Warping of LPC Poles

All the frequencies have the same weight in the spectrum after LPC analysis. However, it is known that the lower frequencies of the human speech are more important; therefore, it is important to improve the speech signals in that frequency range in order for the technique to be more effective. The techniques such as frequency warping of linear predictive poles, sub-band coding with linear predictive coding and perceptual linear predictive coding can be used to accentuate the lower frequencies in the linear predictive analysis [21]. In our model, we have used the frequency warping technique on the linear predictive poles to improve the perceptual quality of dysarthric audio.

The application of a warping function to the analysis filter causes a shift in the frequencies of the formants in the linear predictive model [22]. Here, (z) is the warping function, which replaces the delay element to achieve frequency warping of the signal.

$$T(z) = \frac{z^{-1} - \alpha}{1 - \alpha z^{-1}} \tag{2}$$

The above transformation is performed to boost the perceptual effect of the low frequencies and thereby enhancing the speech intelligibility. Preserving the lower frequencies is important to classify words and discriminate between them. Depending on the degree of warping, the lower frequencies are accordingly boosted. This warping parameter is known as α which can vary between 0 and 1.

After dynamically time warping, the control and dysarthric speech, the LP residual and coefficients are calculated. Frequency warping is then done on the coefficients of the control speech. The value of α is changed to vary the degree of warping depending on the type of speaker and the words uttered. Varying the α is a trade-off between the word intelligibility and its naturalness. To keep the value of α consistent in our model, the value was varied multiple times to select the best value to preserve both the naturalness and the intelligibility of speech. Warping factor of -0.2 was observed to provide the balance between naturalness and intelligibility and was thus chosen as the value of α in this model.

Finally, the linear predictive residual of the dysarthric speech along with the warped linear predictive coefficients of the control speech is used as the input of the linear predictive synthesis filter to reproduce the synthesized speech. These enhancements are made in the low-frequency component by using frequency warping, resulting in a significant improvement in the transformed speech's intelligibility when compared with the linear predictive method.

4 Experimental Results

Our proposed system has been tested on low-intelligible words of male and female speaker samples collected from the following directories—Common Words, Computer Words, Uncommon Words, International Radio Alphabet and Digits of UA-speech database. Evaluation of the transformed speech in terms of naturalness and intelligibility has been done using both subjective and objective measures. For subjectively measuring the intelligibility, we have used a degradation mean opinion score (DMOS). Perceptual evaluation of speech quality (PESQ) measure is used as an objective measure for evaluating the transformed quality of speech with a control audio as a reference.

4.1 Subjective Measure

For the DMOSs, a group of 10 human participants were used to evaluate the transformed speech in a controlled auditory environment [23]. Each person listens to a set of two words chosen randomly for both male and female speakers from each of the five directories of Common Words, Computer Words, Uncommon Words, International Radio Alphabet and Digits of UA-speech database. Thus, each participant listens to 20 words out of the 455 words contained in the UA-speech database. Each word is synthesized with both frequency warping and without frequency warping for both male and female speakers. The participants rate the naturalness and intelligibility of each of these words to provide us with the DMOS as shown in Table 2 and 3. The intelligibility rating is done on a 5-point grading scale (1–bad; 2–poor; 3–average; 4–good; 5–excellent). The naturalness of the resultant speech is also done on a 5-point rating scale (1–highly artificial; 2–artificial; 3–neutral; 4–natural; 5–highly natural). The cumulative average scores of their intelligibility are shown in Figs. 4 and 5, while the naturalness scores are shown in Figs. 6 and 7.

From Figs. 4 and 5, we can compare the intelligibility scores of the speech before and after applying the transformations and observe that there is a substantial improvement in the intelligibility in both male and female speakers. It also shows that the participants saw a higher improvement in the intelligibility of the male speakers

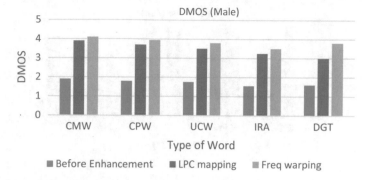

Fig. 4 Subjective evaluation of intelligibility for male speakers

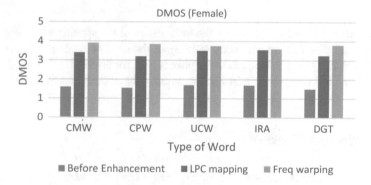

Fig. 5 Subjective evaluation of intelligibility for female speakers

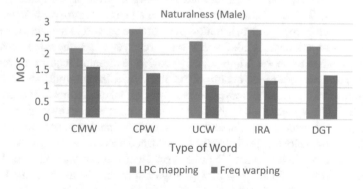

Fig. 6 Subjective evaluation of naturalness for male speakers

than their female counterparts. Although frequency warping clearly improves the intelligibility of the speakers, it reduces the naturalness of the transformed speech as shown in Figs. 6 and 7. This is due to the shift in the formant frequencies when using frequency warping which reduce the speaker's identity. Therefore, residual

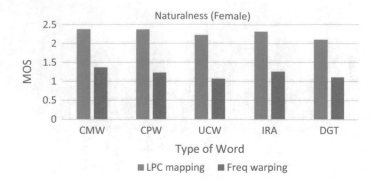

Fig. 7 Subjective evaluation of naturalness for female speakers

transformation of the LPC preserves the naturalness of the speaker to a much higher degree as compared to the frequency warping of the LPC poles. Due to its emphasis on the lower frequencies, frequency warping is best for improving the intelligibility of the while LPC mapping is best for maintaining the naturalness of the speaker.

4.2 Objective Measure

Accurate and reliable estimates of intelligibility can be obtained using subjective tests, but these tests are expensive and time-consuming to conduct. Objective measures provide an easy and straightforward approach to evaluate the quality of audio and speech [24]. This can be done in an intrusive and a non-intrusive manner. When an output speech is compared to a control or reference speech to measure the relative score of the output speech with respect to the reference speech, it is known as an intrusive objective measure. A non-intrusive objective measure requires no reference speech to calculate the speech quality of the output file. In our work, we have used a popular intrusive objective measure known as perceptual evaluation of speech quality (PESQ) [24]. It is widely used in the field telecommunication to evaluate the speech at the output end when compared to the speech of the speaker at the input. We have used PESQ to evaluate the quality of the transformed speech by taking a corresponding control audio as a reference. The PESQ provides a score ranging from 1 to 5, 1 being the worst while 5 being the best quality of speech. The average intelligibility scores are shown in Figs. 8 and 9.

From the above results, we can observe in the improvements in the speech intelligibility of dysarthric speakers, before and after the transformation, using the objective measure.

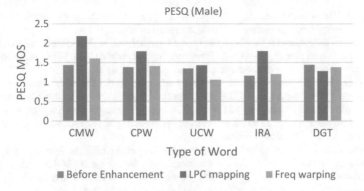

Fig. 8 Objective evaluation of intelligibility for male speakers

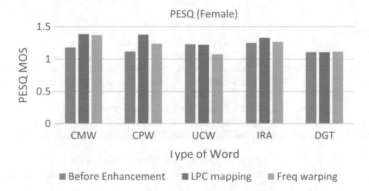

Fig. 9 Objective evaluation of intelligibility for female speakers

5 Conclusion

Our work uses transformation techniques such as linear predictive mapping and frequency warping to increase the intelligibility while maintaining the naturalness of the dysarthric speaker. This improved audio can then be played back to the speaker after a delay and thus creating a simple and effective speech therapy tool for faster rehabilitation of dysarthric speakers. Our transformation system's robustness and simplicity make it a very effective implementation for real-world cheap speech therapy. The delayed auditory feedback of the improved audio builds up a confidence in the speaker to speaker more as they hear a better version the their own voice as compared to their original garbled speech. This confidence in the patients is crucial for their rehabilitation as this makes them more interested in speaking more, which results in the improvement of their natural speech. This algorithm is capable of significantly improving the dysarthric speech intelligibility of the patients suffering from severe speech disability. Therefore, this transformation system is an effective tool for the fast recovery and rehabilitation of dysarthric patients.

Declaration We the authors declare that this manuscript is original, has not been published before and is not currently being considered for publication elsewhere.

We confirm that the manuscript has been read and approved by all named authors and that there are no other persons who satisfied the criteria for authorship but are not listed. We further confirm that the order of authors listed in the manuscript has been approved by all of us.

We further confirm that the UA-speech database used in the work covered in this manuscript has been acquired from the protected ISLE data server with access provided by Mark Hasegawa-Johnson, Professor ECE, University of Illinois. The audio recordings of human patients have been conducted with the ethical approval of all relevant bodies and subjects who refused permission are not represented in the database distribution.

We understand that the Corresponding Author is the sole contact for the Editorial process. The Corresponding Author is responsible for communicating with the other authors about progress, submissions of revisions and final approval of proofs. We confirm that we have provided a current, correct email address which is accessible by the Corresponding Author.

References

1. Rudzicz, F.: Adjusting dysarthric speech signals to be more intelligible. Comput. Speech Lang. **27**(6), 1163–1177 (2013). Special Issue on Speech and Language Processing for Assistive Technology
2. Hosom, J.-P., Kain, A.B., Mishra, T., van Santen, J.P.H., Fried-Oken, M., Staehely, J.: Intelligibility of modifications to dysarthric speech. In: Proceedings of the IEEE International Conference on Acoustics, Speech, and Signal Processing (ICASSP '03), vol. 1, pp. 924–927 (2003 April)
3. Green, P., Carmichael, J., Hatzis, A., Enderby, P., Hawley, M., Parker, M.: Automatic speech recognition with sparse training data for dysarthric speakers. In: Proceedings of Eurospeech 2003, Geneva, pp. 1189–1192 (2003)
4. Tolba, H., El Torgoman, A.S.: Towards the improvement of automatic recognition of dysarthric speech. In: 2009 2nd International Conference on Computer Science and Information Technology. IEEE (2009)
5. Dhanalakshmi, M., Vijayalakshmi, P.: Intelligibility modification of dysarthric speech using HMM-based adaptive synthesis system. In: 2015 2nd International Conference on Biomedical Engineering (ICoBE), IEEE (2015)
6. Yates, A.J.: Delayed auditory feedback. Psychol. Bull., 213 (1963)
7. Blanchet, P.G., Hoffman, P.R.: Factors influencing the effects of delayed auditory feedback on dysarthric speech associated with Parkinsons disease. J. Commun. Disord., Deaf. Stud. Hear. Aids (2014)
8. Downie, A.W., Low, J.M., Lindsay, D.D.: Speech disorder in parkinsonism usefulness of delayed auditory feedback in selected cases. Int. J. Lang. Commun. Disord., 135–139 (1981)
9. Menendez-Pidal, X., Polikoff, J.B., Peters, S.M., Leonzio, J.E.. Bunnel, H.T.: The nemours database of dysarthric speech. In: Fourth International Conference on Spoken Language. ICSLP 96. Proceedings, vol. 3, pp. 1962–1965 (1996)
10. Hosom, J.-P., Kain, A.B. Mishra, T., Van Santen, J.P.H., Fried-Oken, M., Staehely, J.: Intelligibility of modifications to dysarthric speech. In: 2003 IEEE International Conference on Acoustics, Speech, and Signal Processing, 2003. Proceedings. (ICASSP'03), vol. 1, p. I-924. IEEE (2003)
11. Das, D., Santhosh Kumar, C., Reghu Raj, P.C.: Dysarthric speech enhancement using formant trajectory refinement. Int. J. Latest Trends Eng. Technol. (IJLTET) **2**(4) 2013
12. Selouani, S.-A., Yakoub, M.S., O'Shaughnessy, D.: Alternative speech communication system for persons with severe speech disorders. In: EURASIP Journal on Advances in Signal Processing, p. 6 (2009)

13. Tomik, B., Krupinski, J., Glodzik-Sobanska, L., BalaSlodowska, M., Wszolek, W., Kusiak, M., Lechwacka, A.: Acoustic analysis of dysarthria profile in ALS patients. J. Neurol. Sci. **169**(1), 35–42 (1999)
14. Alexander, K., Niu, X., Hosom, J.-P., Miao, Q., van Santen, J.P.H.: Formant Re-synthesis of Dysarthric Speech. In: Fifth ISCA Workshop on Speech Synthesis, pp. 25–30 (2004)
15. Rabiner, L.R., Schafer, R.W.: Digital Processing of Speech Signals. Prentice Hall (1978)
16. Rabiner, L.R., Juang, B.-H.: Fundamentals of Speech Recognition (1993)
17. Ellis, D.: Dynamic time warp (DTW) in Matlab. Web resource http://www.ee.columbia.edu/dpwe/resources/matlab/dtw (2003)
18. Boersma, P.: Praat, a system for doing phonetics by computer. Glot International, pp. 341–345 (2002)
19. Slaney, M.: Auditory Toolbox. Interval Research Corporation, Tech. Rep 10 (1998)
20. van Heuven, V.J., Pols, L.C.: Analysis and synthesis of speech: strategic research towards high-quality text-to-speech generation, vol. 11. Walter de Gruyter (1993)
21. O'shaughnessy, D.: Speech Communication: Human and Machine. Universities Press (1987)
22. Harma A., Laine, U.K.: A comparison of warped and conventional linear predictive coding. In: IEEE Transactions on Speech and Audio Processing, pp. 579–588 (2001)
23. Loizou, P.C.: Speech quality assessment. In: Multimedia Analysis, Processing and Communications, Springer, Berlin Heidelberg, pp. 623–654 (2011)
24. Gu, P.L., Harris, J.G., Shrivastav, R., Sapienza, C.: Disordered speech evaluation using objective quality measures. In: ICASSP, pp. 321–324 (2005)

Modified Adaptive Beamforming Algorithms for 4G-LTE Smart-Phones

Veerendra Dakulagi, Ambika Noubade, Aishwarya Agasgere,
Pradeep Doddi and Kownain Fatima

Abstract Recently, a huge demand for wireless communication, especially for cellular communication, has raised increased expectations. Current and future cellular communications demand wide coverage, high date transmission, very high quality, and spectrum utilization. These expectations will continue to increase because from the last few years use of mobile phones has reached more than one billion worldwide. Hence, effective spectrum utilization is indeed required to meet all these expectations. One of the most promising technologics for future mobile communication is the 'adaptive antenna.' The adaptive antenna is also known as 'smart antenna.' Smart antenna uses adaptive beamforming algorithms to detect and track the mobile user. One of the most commonly used beamforming algorithms is least mean square algorithm. This LMS algorithm is well known for its low complexity, fast tracking, and less prone to numerical errors. It requires only O(L) flops to calculate array weights, where 'L' is the number of antenna elements used. The use of LMS algorithm in wireless communication is widespread. It is used in many fields including cellular communication and surveillances. Standard LMS algorithm requires at least 90 iterations for the satisfactory performance. But this corresponds to almost half cycle of signal of interest (SOI). Due to this, LMS algorithm is not suitable for many wireless communication applications, particularly for 4G LTE, 5G, and beyond. This paper proposes two computationally efficient modified LMS algorithms, namely sign data LMS (SDLMS) and sign error LMS (SELMS). These SDLMS and SELMS

V. Dakulagi (✉) · A. Noubade · A. Agasgere · P. Doddi · K. Fatima
Department of ECE, Guru Nanak Dev Engineering College, Bidar 585403, Karnataka, India
e-mail: veerendra.gndec@gmail.com

A. Noubade
e-mail: ambikanoubade@gmail.com

A. Agasgere
e-mail: aishwaryaagasgere@gmail.com

P. Doddi
e-mail: pradeepdoddi358@gmail.com

K. Fatima
e-mail: konainfatima68@gmail.com

© Springer Nature Singapore Pte Ltd. 2019
J. Wang et al. (eds.), *Soft Computing and Signal Processing* ,
Advances in Intelligent Systems and Computing 898,
https://doi.org/10.1007/978-981-13-3393-4_57

561

algorithms require only 50 and 40 iterations to produce required beamforming in cellular communication. Hence, these algorithms have convergence improvement of 44.45 and 55.56% over standard LMS algorithm. Hence, the proposed SLMS and SELMS beamforming algorithms are most suitable for 4G LTE and beyond mobile communication.

Keywords LMS · SDLMS · SELMS · Smart antenna · 4G LTE · 5G

1 Introduction

Wireless communication is one of the fastest growing fields of communication industry. A huge research in signal processing area for improving the performance of communication systems is being carrying out worldwide [1]. The smart antennas have emerged as frontier in the wireless communication industry to improve the quality, coverage, and capacity. Smart antennas have gained more attention from the last few years due to the impulsive advancement in analog-to-digital converters (ADCs) and digital signal processing (DSP) fields [2, 3]. Smart antennas come under three important categories: single input, multiple output (SIMO); multiple input, multiple output (MIMO); and multiple input, single output (MISO) [4–6]. A single antenna is utilized at the transmitter, and at the target zone, two or more antennas are utilized in case of SIMO. Two or more antennas are utilized at the transmitter, and a single antenna is utilized at the target in case of MISO. And in MIMO, several antennas are used at both the source and the destination [7].

A smart antenna is a collection of elements of antenna coupled with digital signal processor (DSP) [8]. Such a design sarcastically improves the capability of a wireless connection from beginning to end a composition of varieties of gain and intrusion abolition [9].

2 Problem Formulation

In this work, the most popular configuration, ULA, is considered for the implementation of proposed algorithms [10–12]. Figure 1 shows the used ULA configuration. As shown in above ULA representation, 'L' antenna elements are used horizontally with the separation of $d = \lambda/2$, where 'λ' is the wavelength of radio wave impinges with $c = 3 \times 10^8$ m/s and has 'f'Hz frequency. Here, the value of 'd' is strictly considered as $\lambda/2$ only to avoid the problem of grating lobe and mutual coupling effect. When the value of d becomes more than this specified value, the former one will be caused by producing more than one main lobe. More than one main lobe always leads to misjudgment and may make the bad signal estimation. When this value becomes less than specified value, mutual coupling will be resulted and will affect the performance of the system.

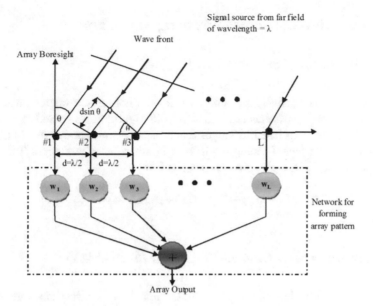

Fig. 1 Geometry of uniform linear array

Consider dimension steering vector for directions in far fields and array observation vector. As the array output of ULA can be expressed as $\theta_1, \theta_2, \ldots, \theta_m$ in the far fields. The array output of ULA can be expressed as:

$$y(n) = w^{H} \cdot x(n) \tag{1}$$

Here, the induced signal in noisy environment can be expressed as

$$x(n) = s(n)a(\theta_o) + \sum_{i=1}^{m} i_i(n)a(\theta_i) + n_o(n) \tag{2}$$

where $x(n)$ = signal induced on ULA, i_n = interference signal, $a(\theta_o)$ = steering vector of desired signal, $a(\theta_i)$ = steering vector of ith interference signal, m = number of jamming (interference) sources. In the above expression, the signal $i(n)$ is not of interest, because it is a jamming source in communication. Let us consider $i(n)_1 = i(n)_2 = i(n)$.

The error signal is the difference between the antenna array output and the desired signal, and it can be expressed as:

$$e(n) = s(n) - y(n) \tag{3}$$

Here, $(.)^T$ is the transpose and w'' is the weight, which is given as: $w = [w_1, w_2, \ldots, w_m]^T$.

3 Proposed Adaptive Beamforming Algorithms

3.1 LMS Algorithm

LMS algorithm is relatively less complex; it does not require matrix inversion as in case of sample matrix inversion (SMI). LMS algorithm uses fixed step size for adaptive beamforming. The LMS algorithm is usually considered as one of the most common adaptive beamforming algorithms in smart antenna design. The design equations of the LMS algorithm are shown below.

$$w(n + 1) = w(n) + \mu e(n)x(n) \tag{4}$$

3.2 Proposed Sign Data LMS Algorithm (SDLMS)

The sign data LMS algorithm is proposed to improve the convergence rate of LMS algorithm. To achieve this, weight upgrading equation of LMS algorithm is modified as:

$$w(n + 1) = w(n) + \frac{\mu}{\alpha_1 e(n) + (1 - \alpha_1)e(n)} e(n)x(n) \tag{5}$$

Here, μ is the fixed step size and α_1 is the shaping parameter.

3.3 Proposed Sign Error LMS Algorithm (SELMS)

The weight upgrading equation of this algorithm to improve the convergence rate is represented as:

$$w(n + 1) = w(n) + \frac{\mu}{1 + \|e(n)\|^2(1 - \alpha_1)} e(n)x(n) \tag{6}$$

4 Results and Discussion

Consider ULA with a number of antenna elements, $L = 8$, inter-element spacing $d = \lambda/2$, AOA of induced signal $= 0°$, interferences at $[-20° \ 0° \ 30°]$, step size $\mu = 0.2$, and noise as white Gaussian with zero mean and 0.01 variance.

Figure 2 shows the comparison of beamforming algorithms in rectangular and polar form.

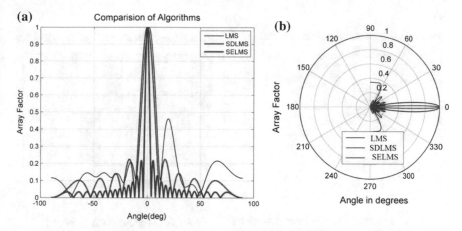

Fig. 2 Comparison of beamforming algorithms: **a** rectangular form, **b** polar form

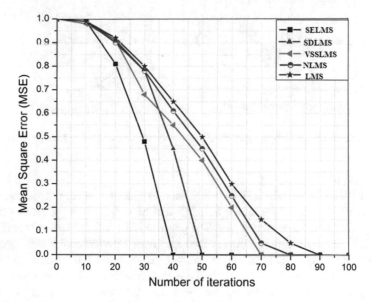

Fig. 3 Convergence analysis of various beamforming algorithms

The above figure shows the comparison of radiation patterns of LMS, SDLMS, and SELMS for eight antenna elements at 0°. The interferer is one with angle 30°. It is clear from the result that the SELMS algorithm produces narrower beam as compared to SDLMS and LMS. Hence, SELMS algorithm has high directivity and accuracy over other two methods. Convergence analysis of various beamforming algorithms is shown in Fig. 3.

The above figure shows the comparison of convergence analysis of LMS, its popular variants, variable step size LMS (VSSLMS), normalized LMS (NLMS),

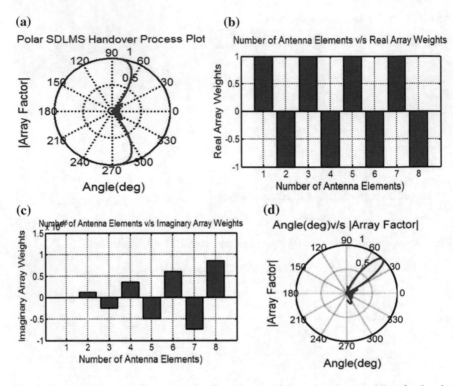

Fig. 4 Handover process: **a** beam scanning, **b** real array weights, **c** imaginary weights, **d** polar plot of SDLMS

proposed SDLMS and SELMS for eight antenna elements. From the above result, it is clear that proposed methods, SDLMS and SELMS, solve the trade-off issue between MSE and convergence rate. The proposed SELMS outperforms and shows about 50% improvement in convergence rate over conventional LMS algorithm. Figures 4 and 5 show the handover process of SDLMS and SELMS, respectively.

The above result is obtained for eight antenna elements. In handover process, the desired user is found by scanning from $+90°$ to $-90°$ that is shown in Fig. 4a and simultaneously the real and imaginary array weights are updated as shown in Fig. 4b and c, respectively. Once the scanning is done, the beam is formed in the desired direction ($45°$) as shown in Fig. 4d. This algorithm is better as compared to LMS algorithm.

In handover process as shown in above figure, the desired signal is found by scanning from $+90°$ to $-90°$ that is shown in Fig. 5a upgrading of real and imaginary array weights are shown in Fig. 5b and c, respectively. Once the scanning is done, the beam is formed in the desired direction ($45°$) as shown in Fig. 5d. This algorithm is better as compared to LMS and SDLMS algorithm. Table 1 compares the various popular beamforming algorithms available in the literature.

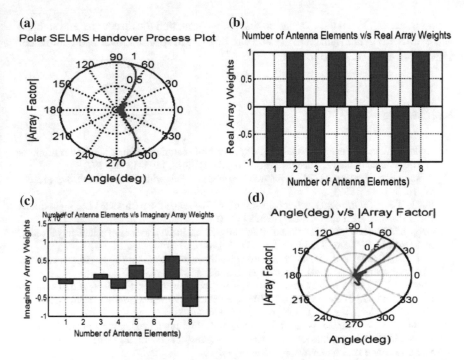

Fig. 5 Handover process: **a** beam scanning, **b** real array weights, **c** imaginary weights, **d** polar plot of SELMS

Table 1 Comparison of LMS and its variants with the proposed methods

Algorithm	Iterations	Performance
LMS [1]	90	Less complex and low convergence
NLMS [2]	80	Less efficient in non-stationary environment
SMI [5]	No iterations	It require large computations
VSSLMS [8]	70	Deep null depth
SDLMS (this work)	50	Less complex and fast convergence
SELMS (this work)	40	Less complex and fast convergence

5 Conclusion

In this paper, new algorithms, namely SDLMS and SELMS, for 4G LTE smartphones have been simulated. Detailed performance analysis of SDLMS and SELMS has been carried out under various conditions. We note that the proposed SDLMS and SELMS produce beamforming for a large number of array elements (even for 100 array elements). From the simulation results, we note that the directivity and stability of SELMS are better than SDLMS and can be used in soft handover using multi-cell MIMO for 4G LTE smartphones.

References

1. Godara, L.C.: Applications of antenna arrays to mobile communications, part II: beamforming and directional of arrival considerations. Proc. IEEE **85**(8), 1195–1245 (1997)
2. Kwong, R.H., Johnston, E.W.: A variable step size LMS algorithm. IEEE Trans. Sig. Process. **40**, 1633–1642 (1992)
3. Slock, T.M.: On the convergence behavior of the LMS and the normalized LMS algorithms. IEEE Trans. Sig. Process. **41**, 2811–2825 (1993)
4. Rupp, M.: The behavior of LMS and NLMS algorithms in the presence of spherically invariant processes. IEEE Trans. Sig. Process. **41**, 1149–1160 (1993)
5. Srar, J.A., Chung, K.S., Mansour, A.: Adaptive array beamforming using a combined LMS-LMS algorithm. IEEE Trans. Antennas Propag. **58**, 3545–3557 (2010)
6. Lopes, P.A.C., Tavares, G., Gerald, J.B.: A new type of normalized LMS algorithm based on the Kalman filter. In: Proceedings of IEEE International Conference on Acoustics, Speech and Signal Processing, Hawaii, U.S.A., pp. 1345–1348 (2007)
7. Dakulagi, V., Bakhar, M., Vani, R.M., Hunagund, P.V.: Implementation and optimization of modified MUSIC algorithm for high resolution DOA estimation. In: IEEE MTT-S International Microwave and RF Conference, pp. 190–193 (2014)
8. Dakulagi, V., Bakhar, M., Vani, R.M.: Smart antennas for next generation cellular mobile communications. i-manager's J. Dig. Sig. Process. **4**(3), 6–11 (2016)
9. Dakulagi, V., Bakhar, M., Vani, R.M.: Robust blind beam formers for smart antenna system using window techniques. Elsevier Proc. Comput. Sci. **93**, 713–720 (2016)
10. Dakulagi, V., Bakhar, M.: Smart antenna system for DOA estimation using 2D-ULA. Indian J. Sci. Res. **13**(1), 241–244 (2017)
11. Dakulagi, V., Bakhar, M.: Efficient blind beamforming algorithms for phased array and MIMO RADAR. IETE J. Res. (2017). https://doi.org/10.1080/03772063.2017.1351319
12. Dakulagi, V., Bakhar, M.: A novel LMS beamformer for adaptive antenna array. Elsevier Proc. Comput. Sci. **115**, 94–100 (2017)

Occluded Ear Recognition Using Block-Based PCA

V. Ratna Kumari, P. Rajesh Kumar and S. Srinivasa Kumar

Abstract In the real world, two non-intrusive biometric methods, namely face recognition and ear recognition, are interesting as they have flexibility in scanning the subject being tested. In particular, ear recognition is quite interesting as the ear pattern is stable in all emotions. This paper presents a novel way using block-based PCA to recognize ear even in case of partial occlusion. The other significance of the proposed method is that it can decide whether an input image is occluded or not. Conducted experiments, used a standard data set and shown that min–min fusion technique with city block distance metric is the apt ear recognition method when block-based PCA is used. The recognition rate with the proposed method is greater than 94%, and the equal error rate (EER) is less than 15% for a 25% occluded ear image.

Keywords Ear recognition · Principal component analysis · Euclidean distance
City block distance · Recognition rate · Equal error rate

1 Introduction

Allowing ear recognition-based access in the real world is always better than face recognition-based access as the ear is always stable as per the words of computer scientist Kevin Bowyer of Notre Dame. It is also established that if you got a profile head shots or video image of someone, then it is very easy to use automatic ear recognition to identify that person. Nowadays, ear recognition has become very

V. Ratna Kumari · P. Rajesh Kumar (✉)
Department of ECE, Prsad V. Potluri Siddhartha Institute of Technology, Kanuru, Andhra Pradesh, India
e-mail: rkpanakala@gmail.com

V. Ratna Kumari
e-mail: vemuriratna2005@gmail.com

S. Srinivasa Kumar
Department of ECE, JNT University, Kakinada, Andhra Pradesh, India
e-mail: samay_ssk@gmail.com

© Springer Nature Singapore Pte Ltd. 2019
J. Wang et al. (eds.), *Soft Computing and Signal Processing*,
Advances in Intelligent Systems and Computing 898,
https://doi.org/10.1007/978-981-13-3393-4_58

popular as it is one of the least intrusive methods in biometrics [1, 2]. In this contest, this paper proposes a method to recognize the occluded ear. Ear recognition also plays an important role in forensic studies as it is possible to identify a person without his cooperation in a non-intrusive way [3]. In particular, it is found that some features of ears of twins will also differ [4].

Chang et al. [5] presented a report to compare ear recognition with face recognition that uses a technique of principal component analysis (PCA) on ear and face images. The PCA uses ear space consisting triangular fossa and antitragus as landmark points, and this paper implementation also considers the same as those points are not occluded in 25% occlusion of upper or lower parts of the ear.

In principle, the PCA is a linear transformation subspace method which is widely known for its performance in dimensionality reduction, visualization, compression and feature extraction. But the PCA computation requires large memory and it is having high computational complexity particularly for large databases. Generally, PCA integrated with Euclidean distance, Mahalanobis distance or City block distance metric is being used to classify ear or face images [6–8]. This method reduces dimensional feature vectors from 'd' to 'h', where h < d. The resultant h—numbers of dimensional feature vectors—is known as basis vectors. The direction of maximum variance generally represents the first basis vector, and the remaining vectors are mutually orthogonal. Each of these basis vectors corresponds to a principal axis, obtained by a dominant eigenvector with the largest eigenvalue of the resultant original data covariance matrix. In this analysis, basis vectors characterize the original feature space and its number is much smaller than the dimensional feature vectors. Golub and van Loan [9] applied Jacobi's method and could able to make computations depend on a number of feature vectors. Liu et al. [10] proposed a merging method in eigenspace which reduces the computational complexity further.

Tharwat et al. [11] proposed a block-based PCA for an ear recognition, which achieved a better recognition rate with four blocks for each ear image. But [11] has not implemented for occluded ear and also not tested for suitability of block-based PCA algorithm for hardware-amenable implementation.

In this paper, an ear recognition system performance is observed by applying PCA and block-based PCA with a minimum distance classifier. It is aimed for accurate recognition, decreasing the complexity and dimensionality by dividing the ear image into four equal-sized blocks (i.e. block-based PCA). City block distance metric provides relatively high contrast compared with Euclidean distance metric in high-dimensional data mining applications. City block minimum distance metric provides less computational complexity compared to Euclidean and Mahalanobis metrics, and hence, it is preferable to be used in hardware implementations [12]. The experimental results confirm that the block-based PCA algorithm followed by score-level fusion with City block distance metric achieves recognition performance superior to other even when occlusion is considered.

2 PCA and Block-Based PCA with Different Distance Measures

An ear image is generally represented as Γ (M × N), where M is the width and N is the height of the image. The feature matrix is calculated for K, the number of training images in the database, and is represented by $\Gamma = [\Gamma_1, \Gamma_2, ..., \Gamma_K]$. The training set average is calculated as $\psi = \frac{1}{K} \sum \Gamma_i$. The data matrix 'A' is created by subtracting the average and is written as $A = [\Phi_1, \Phi_2, ..., \Phi_K]$ ((M × N) × K), where $\Phi_i = \Gamma_i - \psi$. The covariance matrix of A is calculated as the $C = AA^T$.

The eigenvectors and the eigenvalues of the covariance matrix are represented as V_k and λ_k, respectively. Eigenvectors are sorted according to the corresponding eigenvalues. The top h eigenvectors are only retained to achieve dimensionality reduction, and thus, the projection matrix P is yielded [13]. The test image with the size same as training image is first normalized by $\Phi_T = T - \psi$ and then projected into ear space by $\omega = P^T \Phi_T$.

The features of test image are compared with the database images with Euclidean and City block distance. The Euclidean distance between ith and kth feature vectors with length 'h' is given by D_{ik}.

$$D_{ik} = \sqrt{\sum_{j=1}^{h} (w_{ij} - w_{kj})^2} \tag{1}$$

The City block distance between ith and kth feature vectors with length 'h' is given by D_{ik}.

$$D_{ik} = \sum_{j=1}^{h} |w_{ij} - w_{kj}| \tag{2}$$

In case of Euclidean distance, the sum of square of difference of features is taken, and in City block, the absolute sum of difference of features is taken.

In block-based PCA, the ear image of size M × N is first divided into Q-equal-sized non-overlapping sub-images of size (M × N)/Q. The PCA features of the individual blocks are obtained, and each sub-image is classified independently using a minimum distance classifier. Different fusion techniques like abstract, rank and score level are applied to make the decision towards matching by combining the outputs of the classifiers [9].

2.1 Fusion Methods

The accuracy of the biometric system can be increased by combining different resources working independently. Combining independent resources compensates the misclassification of some samples by a single classifier or by a particular method, and the overall performance is also at different levels. The following is the summary of the most common approaches for image fusion.

Multi-sensorial systems use more number of sensors to capture the samples from the same instance, and the robustness of the system is increased by combing these samples using different fusion methods [14].

Comparing fusion at classification-level fusion at the feature level, the second approach leads to improved performance due to the availability of more information. The feature-level combination is generally implemented by concatenating several feature vectors.

The recognition performance is largely improved by classification-level fusion (classifier fusion) when compared with simple individual classifier. Fusion approach is distinguished at three levels as follows.

- Abstract-level fusion: It is a simplest and most commonly used method and also called as a decision-level fusion method. The decision is made by combining the outputs of several classifiers of the same test sample.
- Rank-level fusion: Each class has its own rank, and the rank is obtained by sorting the output of each classifier in decreasing order of confidence. The decision is made towards the highest class rank.
- Score-level fusion: It is also called measurement-level fusion. The distance between the training image and test image is represented by fusion rules on the vectors. Each classifier is represented by its score, and the decision is made towards the class with the minimum score. Classification of an input image by using score-level fusion is explained as follows.

Assume that the input pattern 'Z' is to be classified into one among 'm' possible classes based on distances provided by different classifiers. Let D_{ij} is the distance between the input pattern and ith sample of jth block. The input pattern is then classified into the class 'c' according to the following conditions.

$$c = arg \ \min_i \ \min_j (D_{ij}) \tag{3}$$

$$c = arg \ \min_i \ \max_j (D_{ij}) \tag{4}$$

$$c = arg \ \min_i \ \text{med}_j (D_{ij}) \tag{5}$$

$$c = arg \ \min_i \ \text{mean}_j (D_{ij}) \tag{6}$$

2.2 Computational Complexity of PCA and Block-Based PCA

For a set of K images with each dimensions M × N, the computational complexity of PCA is given by

$$O_{PCA} = O\,(KMN) + O\,(KMN) + O\,(K(MN)^2)$$
$$+ O\,(K(MN)^3) + O\,(K(MN)\log_2(MN)) + O\,(hMN) + O\,(KhMN)$$
$$= O\,(K(MN)^3) \tag{7}$$

where $O\,(KMN)$ is the computational complexity in calculating mean image obtaining by subtracting it from the data matrix. $O\,(K(MN)^2)$ is the computational complexity in calculating covariance matrix. $O\,(K(MN)^3)$ is required for identifying the eigenvectors and eigenvalues of the covariance matrix. $O\,(K(MN)\log_2(MN))$ is to sort the eigenvectors according to corresponding eigenvalues. $O\,(hMN)$ is for the computation of eigenspace. $O\,(KhMN)$ is for projecting the images into eigenspace.

Performing PCA on each sub-image results a computational complexity of $O\,(K(MN/Q)^3)$, and for Q sub-images, the computational complexity is

$$O_{subimagePCA} = O\left(QK(MN/Q)^3\right) = O\left(K(MN)^3/Q^2\right) \qquad . \tag{8}$$

The complexity of performing PCA is thus reduced b a factor of 'Q' when all sub-images are considered, and the recognition rate of the individual sub-image compared to the recognition rate of the entire ear image is also reduced.

3 Experimental Results

The experiments in this work are conducted with the IIT Delhi database. The ear images in this database are acquired using a simple handheld setup from a distance (Touchless) in indoor environments during Oct 2006–Jun 2007 collected from the students and staff at the IIT Delhi campus. The database is acquired from 125 different persons, and for each person, a minimum of three ear images and a maximum of six ear images are collected. All the persons used in the database are in the age group of 14–58 years. The database of 471 processed images has been sequentially numbered for every person with an identification number. The resolution of these images is 180×50 pixels, and all these images are available in bitmap format.

In this work, 3 images, each of 110 different subjects out of the total available 125 subjects in the database, are considered for training. The algorithms are tested with 3 to 6 images of 110 different subjects and ear images of remaining subjects which are considered as unauthorized subjects. For conducting the experiments, a minimum distance classifier with different measures, Euclidian and City block, is used.

The occlusion is considered in different cases. In the first case, the bottom 25% of the ear image is occluded, and in the second case, the top 25% of the ear image is occluded. The occlusion on test images is considered in this way to accommodate the difficulty in taking the snapshots of ear images. In some cases, particularly with the ladies, the ears are sometimes covered by hair from the top portion and by earrings

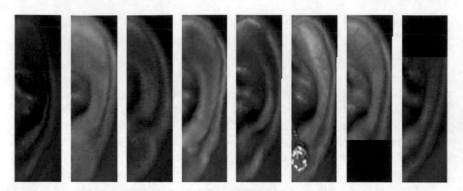

Fig. 1 Sample database ear images (last two ear images are manually occluded)

Table 1 Comparison of recognition rate with PCA algorithm on whole ear image

Test sample	Rank 1 recognition rate (%)	
	Euclidean distance metric	City block distance metric
No occlusion	96.98	98.84
25% occlusion at bottom of ear image	13.46	28.77
25% occlusion at top of ear image	10.67	22.27

Table 2 Comparison of equal error rate with PCA algorithm on whole ear image

Test sample	Equal error rate (%)	
	Euclidean distance metric	City block distance metric
No occlusion	14.48	6.97
25% occlusion at bottom of ear image	89.43	83.78
25% occlusion at top of ear image	92.26	87.10

from the bottom portion. Examples of some images used in the database are shown in Fig. 1.

The testing is done in two phases. In the first phase, classical PCA is applied on whole ear image, and the recognition performance and EER are obtained. The results of the first phase are summarized in Tables 1 and 2.

In the second phase, each ear image is divided into four blocks to reduce the computational complexity and PCA is applied to individual blocks. The feature vectors of the individual blocks are calculated and are applied to a minimum distance classifier with respect to each block. The matching is done at classifier fusion with different fusion techniques. The results obtained in the second phase are tabulated in Tables 3 and 4.

Table 3 Comparison of ear recognition rate with different fusion techniques

Occlusion criteria	Fusion technique	Euclidean distance metric	City block distance metric
No occlusion	Abstract level	95.12	96.75
	Rank level	95.12	96.75
	Score level		
	i. Min–min	94.43	97.22
	ii. Min–max	93.04	95.59
	iii. Min–median	97.68	98.14
	iv. Min–mean	97.45	98.88
25% occlusion at bottom of ear image	Abstract level	79.11	89.79
	Rank level	22.96	38.51
	Score level		
	i. Min–min	89.56	94.66
	ii. Min–max	8.8	23.43
	iii. Min–median	76.57	92.81
	iv. Min–mean	82.6	95.36
25% occlusion at top of ear image	Abstract level	83.06	88.63
	Rank level	24.82	35.73
	Score level		
	i. Min–min	92.81	96.06
	ii. Min–max	8.58	14.85
	iii. Min–median	83.99	93.97
	iv. Min–mean	89.10	95.82

Table 4 Comparison of equal error rate with different fusion techniques

Occlusion criteria	Fusion technique	Euclidean distance metric	City block distance metric
No occlusion	Score level		
	i. Min–min	15.64	10.39
	ii. Min–max	19.76	13.03
	iii. Min–median	15.19	9.77
	iv. Min–mean	14.06	9.18
25% occlusion at bottom of ear image	Score level		
	i. Min–min	24.05	14.22
	ii. Min–max	94.02	83.31
	iii. Min–median	60.21	35.86
	iv. Min–mean	59.33	37.76
25% occlusion at top of ear image	Score level		
	i. Min–min	23.17	12.09
	ii. Min–max	93.44	91.40
	iii. Min–median	62.58	35.42
	iv. Min–mean	58.74	33.82

Lower recognition performance is observed with occluded test images, and it is improved in the block-based principal component analysis. The recognition performance using the block-based PCA algorithm with different fusion techniques is shown in Tables 3 and 4.

4 Conclusions

It is observed that City block distance metric gives a better performance in ear recognition than Euclidean distance metric. If there is no occlusion, then the performance of block-based PCA is same as that of PCA. However, the block-based PCA has shown superior performance, i.e. >90% recognition compared to that of PCA when there is 25% occlusion in the whole ear image. It is also observed that score-level fusion with min–min and min–mean techniques gives the better ear recognition than any other fusion techniques.

Acknowledgements This work is supported by University Grants Commission of India under the scheme—Minor Research Project No. MRP6023 Dated: 31. 10. 2016.

References

1. Abaza, A., Ross, A., Herbert, C., Harrison, M.A.F., Nixon, M.S.: A survey on ear biometrics. ACM Compu. Surv. **45**(2) (2013)
2. Emersic, Z., Struc, V., Peer, P.: Ear recognition: more than a survey. Neurocomputing **255**, 26–39 (2017)
3. Iannarelli, A.: Ear Identification, Forensic Identification Series. Paramount Publishing Company, Fremont, CA (1989)
4. Nejati, H., Zhang, L., Sim, T., Martinez-Marroquin, E., Dong, G.: Wonder ears: identification of identical twins from ear images. In: 21st International Conference on Pattern Recognition, pp. 1201–1204 (2012)
5. Chang, K., Bowyer, K.W., Sarkar, S., Victor, B.: Comparison and combination of ear and face images in appearance-based biometrics. IEEE Trans. Pattern Anal. Mach. Intell. **25**(9), 1160–1165 (2003)
6. Turk, M., Pentland, A.: Face recognition using eigenfaces. In: Proceedings of IEEE Conference on Computer Vision and Pattern Recognition, Maui, Hawaii, USA, pp. 586–591 (1991)
7. Fan, Z., Ni, M., Sheng, M., Wu, Z., Xu, B.: Principal component analysis integrating Mahalanobis distance for face recognition. In: Second International Conference on Robot, Vision and Signal Processing, pp. 89–92 (2013)
8. Cha, S.-H.: Comprehensive survey on distance/similarity measures between probability density functions. Int. J. Math. Models Methods Appl. Sci. **1**(4), 300–307 (2007)
9. Golub, G.H., van Loan, C.F.: Matrix Computations, 3rd edn. John Hopkins University Press, Baltimore (1996)
10. Liu, L., Wang, Y., Wang, Q., Tan, T.: Fast principal component analysis using eigen space merging. IEEE Int. Conf. Image Process. (ICIP) **6**, 457–460 (2007)
11. Tharwat, A., Ibrahim, A., Ali, H.A.: Personal identification using ear images based on fast and accurate principal component analysis. In: INFOS2012, pp. 56–59, May 2012

12. Aggarwal, C.C., Hinneburg, A., Keim, D.A.: On the surprising behavior of distance metrics in high dimensional space, pp. 420–434 (2001)
13. Querencias-uceta, D., Carmen, S.: Principal component analysis for ear-based biometric verification. IEEE conference, pp. 1–6 (2017)
14. Jain, A., Nandakumar, K., Ross, A.: Score normalization in multimodal biometric systems. J. Pattern Recogn. **38**(12), 2270–2285 (2005)

Fractal-Shaped DGS and Its Sensitivity Analysis with Microstrip Patch Antenna for 2.4 GHz WLAN Applications

Smeeta Hota, Guru Prasad Mishra and Biswa Binayak Mangaraj

Abstract This paper offers the impact of DGS on the performance of aperture feed patch antenna which is operating at 2.4 GHz. Many antenna structures are designed, each having DGS of different configurations which includes plain DGS, plain wire DGS, fractal DGS, and fractal wire DGS. On studying the performance of these structures, it is found that the best antenna structure is the antenna with MIF Type 2 Iteration 2 DGS. In order to study the effect of design parameters of DGS on it, sensitivity analysis is conducted. Based on the analysis, the value for which maximum size reduction achieved is found out. Finally, it is observed that for the best structure, size reduction is achieved with an excellent gain when compared to that of the simple patch antenna. The whole process of design and simulation is carried out using ANSYS HFSS software, and the results, i.e., the antenna performance parameters, obtained after simulation are graphically displayed using MATLAB.

Keywords Patch antenna · DGS · Size reduction · Aperture feed
WLAN applications

1 Introduction

Presently, there is a growing demand for wireless communication technology, which has increased the need for implementation of a number of communication devices on a single compact structure. The antenna plays an important role in a wireless communication system, so its integration into a single compact-sized structure should be done efficiently. Conventional antennas are generally bigger, due to which effi-

S. Hota (✉) · G. P. Mishra · B. B. Mangaraj
Veer Surendra Sai University of Technology, Burla, Odisha, India
e-mail: smeetahota@gmail.com

G. P. Mishra
e-mail: guruprasadmishra5@gmail.com

B. B. Mangaraj
e-mail: bbmangaraj@gmail.com

© Springer Nature Singapore Pte Ltd. 2019   579
J. Wang et al. (eds.), *Soft Computing and Signal Processing* ,
Advances in Intelligent Systems and Computing 898,
https://doi.org/10.1007/978-981-13-3393-4_59

cient miniaturization techniques are needed. This will lead to smaller antennas with improved performance [1].

Microstrip antennas are extensively used. They are inexpensive, compact, and of low profile. These antenna structures suffer from low impedance bandwidth, low gain, and polarization problem [2]. There are various methods like material loading, stacking, using metamaterials and creating defects in the ground plane that can be used to enhance the radiation characteristics of these antennas.

Defected ground structures (DGSs) are slots/defects with various geometrical configuration etched on the ground plane of microwave planar circuits. The defects can be single or multiple, also periodic or aperiodic structures. By integrating DGS with microstrip antennas, the performance achieved is size reduction, bandwidth, gain enhancement, higher-mode harmonic suppression, and mutual coupling reduction between adjacent elements.

In many articles on DGS, it is found that by applying DGS technique, the size of the microwave circuit can be reduced. For designing various filters, amplifiers, and for improving the radiation characteristics of antennas, this technique is used. In [3–5], it is concluded that fractal concept can be used for improving the performance of antenna. In [6, 7], DGS is applied for size reduction of the amplifier from around 40–50%. In [8], application DGS as bandpass filter is studied. In [9], a band-stop filter with DGS is designed to improve the Q factor. In [10], five-cascaded double U-shaped DGS is applied to achieve a 42% size reduction. In [11], the DGS technique is used to achieve good impedance matching. In [12], the author concluded that application of DGS is an emerging technology in reducing the size of antennas. In [13], it is found that 44.74% size reduction in patch size is achieved, on applying Koch curve fractal DGS on the ground plane. DGS can also be used with patch antennas for improving their radiation characteristics, for impedance matching and to reduce mutual coupling between antenna elements in an antenna array.

In this paper, we design a rectangular microstrip patch antenna at 2.4 GHz with aperture feeding method. Then several structures each having DGS of a different type, i.e., plain, plain wire-shaped, fractal, and fractal wire-shaped DGSs, are designed. After that, simulation is carried out to study the performance of these designs and to find out the best one. This structure is applicable for WLAN applications.

This paper is organized in fivefolds. Design of aperture feed patch antenna is presented in Sect. 2. A description of the application of DGS to the patch antenna is given in Sects. 3 and 4 contains results along with the discussion. Lastly, the article concludes in Sect. 5 which includes the design applications.

2 Design of Aperture Feed Patch Antenna

We design an aperture feed rectangular microstrip patch antenna, working at 2.4 GHz [1, 2]. After designing, optimization is needed to resonate the antenna structure exactly at 2.4 GHz. Here, the substrate, feed line width, and slot size and position can be optimized. The layout of an aperture feed rectangular microstrip patch antenna

Fig. 1 Layout of aperture
feed patch antenna at
2.4 GHz (both top and side
view)

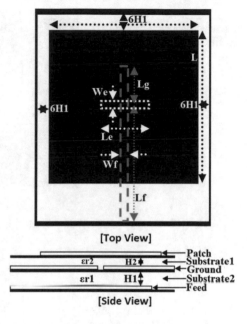

[Top View]

[Side View]

Table 1 Optimized design
parameters of the simple
patch and defected ground
structure

Parameters	Value (mm)	Parameters	Value (mm)
Le	13.38	L1	4.67
We	2.3	W1	4.67
L	38.2	a	4.67
W	62.74	m	0.1
Lf	34.38	εr1, εr2	4.4, 2.2
Wf	3.059	H1	1.6
Lg	11.35	H2	1.5

is shown in Fig. 1. The optimal design parameters of this 2.4 GHz patch antenna are
shown in Table 1. The substrates used here are FR4_epoxy and Rogers RT/duroid
5880 (tm). The performance parameter of this design is shown in the subsequent
section.

3 Application of DGS to the Patch Antenna

Nowadays, for many researchers efficient miniaturization of printed antennas using
DGS has become an area of interest. DGSs can be of any shape like a spiral, V,
U, H, dumbbell, and fractal. When defects are etched on the ground plane, the
effective capacitance and inductance vary by adding slot resistance, capacitance,
and inductance. This leads to disturbances in the current distribution of the ground

Fig. 2 a Layout of
dumbbell-shaped DGS and b
its RLC equivalent circuit

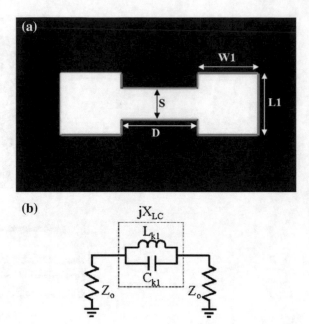

plane. For a constant, physical length application of DGS leads to an increase in the
electrical length. So, there occurs a reduction in the size of the original structure.
When fractal-shaped DGS is applied to printed antennas, it is found that on increasing
the number of iteration of the fractal, the total effective electrical length increases
within the same physical area. This leads to size reduction and improved radiation
characteristics. Figure 2 shows the dumbbell-shaped DGS (which is commonly used)
along with its RLC equivalent circuit.

The inductance (L) and capacitance (C) can be obtained from the following equations.

$$C = \frac{\omega_C}{2Z_0\left(\omega_0^2 - \omega_C^2\right)} \tag{1}$$

$$L = \frac{1}{4\pi^2\omega_0^2} \tag{2}$$

where ω_c is 3-dB cutoff angular frequency, ω_0 is angular resonance frequency, and
Z_0 is characteristic impedance of the strip line.

In Fig. 3, the layout of DGS of various shapes etched on the ground plane of
different antenna structure is shown. In the previous section, it is shown that in
aperture feed patch antenna, a slot on the ground plane below the patch belongs to
the part of the design of aperture feeding technique. Since a slot already exists on
the ground plane because of aperture feeding, we etch two square slots on both sides
of it each at a distance of 4.67 mm from its edge, which is shown in Fig. 3b. The
design parameters for DGS are tabulated in Table 1, L1 and W1 are the length and

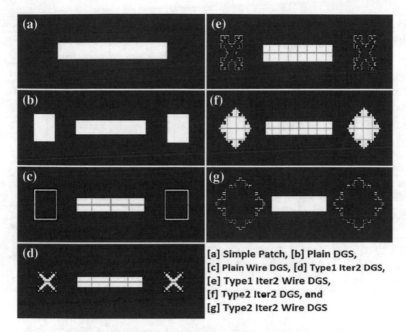

Fig. 3 Layout of DGS of different shapes etched on the ground plane

width of the slot, a is the distance of the square slots from the edge of the slot that exist previously, and m is the thickness of the wire slots as shown in Fig. 3c. We conduct a parametric study on 'a' to get its optimum value which equals 4.67 mm. We apply second iteration of Minkowski Island Fractal (MIF) both of Type 1 and Type 2 to the square slots as shown in Fig. 3d and f, respectively, and also designed its corresponding wire DGSs. Figure 3e and g shows the wire form of the second iteration of MIF Type 1 and Type 2 DGS, respectively. After designing, the performance of these structures is studied, which is described in the following section.

4 Results and Discussion

ANSYS HFSS software is used to conduct the whole design and simulation process. After simulation, all the results obtained are exported to MATLAB and are displayed graphically.

Table 2 shows that on applying DGS to the simple patch, the resonant frequency gets lowered which leads to size reduction of the antenna compared to the simple one. Among all the cases, the Type 2 Iteration 2 gives the best performance. The resonant frequency of this antenna is reduced to 2.344 GHz; the gain is 7.64 dB; impedance is nearly 50 Ω with reactance 0 Ω. Thus, we can achieve size reduction with excellent gain by implementing this structure.

Table 2 Performance parameters for all defected ground patch antenna with fractal structures

Cases	Res. freq. (in GHz)	S(1, 1) (in dB)	Gain (in dB)		Z(1, 1) (in Ω)	
			E-plane	H-plane	Real	Img.
Simple	2.40	−35.73	7.64	7.64	49.98	−1.63
Plain DGS	2.366	−30.49	7.64	7.64	52.97	0.80
Plain wire DGS	2.344	−28.40	7.61	7.61	52.60	2.92
Type 1 Iteration 2	2.366	−20.31	7.71	7.71	48.43	9.41
Type 1 Iteration 2 wire	2.366	−30.91	7.65	7.65	52.85	0.68
Type 2 Iteration 2	2.344	−19.62	7.64	7.64	56.04	−9.34
Type 2 Iteration 2 wire	2.344	−19.14	7.61	7.61	56.08	−10.06

We perform sensitivity analysis on the design parameters of DGS for all the cases that are shown in Table 2. The analysis results are shown in Fig. 4. All the parameters, i.e., L1, W1, a, and m, for each case are plotted in a single graph for comparison.

From the previous section, it is revealed that the best structure is the Type 2 Iteration 2 DGS. We perform sensitivity analysis for this structure and observe that among all the parameters, the parameter W1 plays a very important role in reducing the size of the antenna. At a value of W1 = 14.3 mm, the resonant frequency is 1.88 GHz which is the lowest among all the values of resonant frequencies even when all the other parameters are considered. Thus, a maximum size reduction is achieved at this value of W1 with excellent co-pole gain.

We compare the performance of the simple patch to the antenna with Type 2 Iteration 2 DGS (with W1 = 4.67 mm), which is tabulated in Table 3. The S(1, 1) value is plotted against frequency for both of the 2.4 GHz aperture feed patch antenna and antenna with Type 2 Iteration 2 DGS, which is shown in Fig. 5. The resonant frequency of the best structure is 2.344 GHz which is less than that of our conventional 2.4 GHz simple patch antenna. The Z(1,1) value is plotted against frequency for both of the 2.4 GHz aperture feed patch antenna and antenna with Type 2 Iteration 2 DGS which is shown in Fig. 6, which indicates that good impedance matching is achieved. The co-pole and cross-pole radiation characteristics of the simple patch antenna and antenna with Type 2 Iteration 2 DGS are shown in Fig. 7. The co-pole gain is same as that of the simple patch. Based on the above useful features, this structure is applicable for WLAN applications.

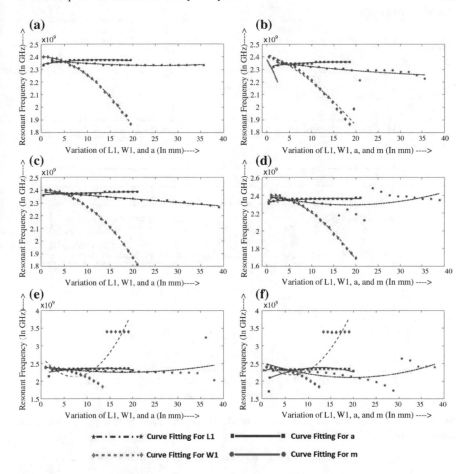

Fig. 4 Results of sensitivity analysis of the parameters L1, W1, a, and m for **a** plain DGS, **b** plain wire DGS, **c** Type 1 Iteration 2 DGS, **d** Type 1 Iteration 2 wire DGS, **e** Type 2 Iteration 2 DGS, and **f** Type 2 Iteration 2 wire DGS

Table 3 Performance comparison of the simple patch antenna with the best case, i.e., Type 2 Iteration 2 DGS

	Simple patch	Type 2 Iteration 2
Res. freq. (in GHz)	2.40	2.344
S(1, 1) (in dB)	−35.73	−19.62
VSWR (abs value)	1.03	1.23
Bandwidth (in MHz) (%)	2.37–2.43 GHz, 60 MHz, 2.5%	2.31–2.36 GHz, 50 MHz, 2.13%

(continued)

Table 3 (continued)

		Simple patch	Type 2 Iteration 2
Co-pole gain (in dB)	E-Plane	7.64	7.64
	H-Plane	7.64	7.64
Cross-pole gain (in dB)	E-Plane	−31.60	−29.99
	H-Plane	−43.75	−44.92
Z(1, 1) (in Ω)	Real	49.48	56.04
	Img.	−1.63	−9.34

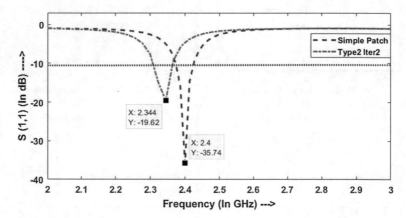

Fig. 5 S(1, 1) versus frequency characteristics of the 2.4 GHz aperture feed patch antenna and antenna with Type 2 Iteration 2 DGS

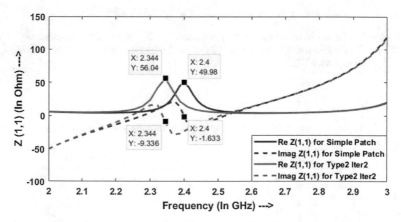

Fig. 6 Z (1, 1) versus frequency characteristics of the 2.4 GHz aperture feed patch antenna and antenna with Type 2 Iteration 2 DGS

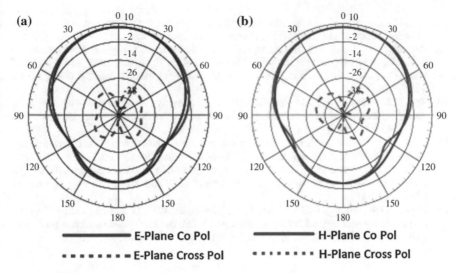

Fig. 7 Simulated radiation patterns **a** simple 2.4 GHz aperture feed antenna and **b** antenna with Type 2 Iteration 2 DGS

5 Conclusion

In this paper, the design and simulation of aperture feed patch antenna structures with DGS of different shapes are presented. DGSs are etched on the ground plane underneath the patch. The best structure is the Type 2 Iteration 2 DGS. The performance achieved by this antenna structure is size reduction with excellent gain. Based on this, we can use this antenna for WLAN applications. However, in future the performance of the antenna can be enhanced efficiently by taking optimal values of the design parameters.

References

1. Balanis, C.A.: Antenna theory: analysis and design, 2nd edn. Wiley, New York, USA (1997)
2. Garg, R., Bhartia, P., Bahl, I., Ittipiboon, A.: Microstrip Antenna Design Handbook. Artech House (2000)
3. Mishra, G.P., Maharana, M.S., Modak, S., Mangaraj, B.B.: Study of Sierpinski fractal antenna and its array with different patch geometries for short wave ka band wireless applications. Proc. Comput. Sci. **115**, 123–134 (2017)
4. Jena, M.R., Mangaraj, B.B., Pathak, R.: A novel Sierpinski carpet fractal antenna with improved performances. Am. J. Electr. Electron. Eng. **2**(3), 62–66 (2014)
5. Jena, M.R., Mangaraj, B.B., Pathak, R.: An improved compact & multiband fractal antenna using the Koch curve geometry. Sci. Educ. Wirel. Mob. Technol. **2**(1), 1–6 (2014)
6. Lim, J.S., Lee, Y.T., Kim, C.S., Ahn, D., Nam, S.: A vertically periodic defected ground structure and its application in reducing the size of microwave circuits. IEEE Microw. Wirel. Compon. Lett. **12**(12), 479–481 (2002)

7. Lim, J.S., Lee, Y.T., Kim, C.S., Ahn, D., Nam, S.: Application of defected ground structure in reducing the size of amplifiers. IEEE Microw. Wirel. Compon. Lett. **12**(7), 261–263 (2002)
8. Rahman, A.A., Verma, A.K., Boutejdar, A., Omar, A.S.: Compact stub type microstrip bandpass filter using defected ground plane. IEEE Microw. Wirel. Compon. Lett. **14**(4), 136–138 (2002)
9. Wang, C.J., Lin, C.S.: Compact DGS resonator with improvement of Q-factor. Electron. Lett. **44**(15) (2008)
10. Tam, K.W., Martins, R.P.: Miniaturized microstrip lowpass filter with wide stopband using double equilateral u-shaped defected ground structure. IEEE Microw. Wirel. Compon. Lett. **16**(5), 240–242 (2006)
11. Pei, J., Wang, A.G., Gao, S., Leng, W.: Miniaturized triple-band antenna with a defected ground plane for WLAN/WiMAX applications. IEEE Antenna Wirel. Propag. Lett. **10**, 298–301 (2011)
12. Reddy, B.R.S., Vakula, D.: Compact zigzag-shaped-slit microstrip antenna with circular defected ground structure for wireless applications. IEEE Antenna Wirel. Propag. Lett. **14**, 678–681 (2015)
13. Prajapati, P.R., Murthy, G.G.K., Patnaik, A., Kartikeyan, M.V.: Design and testing of a compact circularly polarised microstrip antenna with fractal defected ground structure for L-band applications. IET Microw. Antennas Propag. **9**(11), 1179–1185 (2015)

A Novel Adaptive Beamforming Algorithm for Smart Antennas

Ramakrishna Yarlagadda, V. Ratna Kumari and Venkata Subbaiah Potluri

Abstract A new class of adaptive algorithms to process complex signals is used for the adaptive beamforming in smart antennas employed for cellular and mobile communications. Adaptive nonlinear gradient descent (ANGD) algorithm and augmented complex least mean squares (ACLMS) algorithm are proven to be useful to process complex signals of large dynamics. A hybrid system is proposed by employing the convex combination of ACLMS and ANGD algorithms. The new algorithm is tested on smart antennas for mobile communications through MATLAB simulations. From the results, it is shown that the hybrid algorithm outperforms both the individual algorithms in respect of the important array characteristics like side-lobe level (SLL), half power beamwidth (HPBW), desired signal tracking, and mean squared error (MSE) convergence.

Keywords Smart antenna · Adaptive beamforming · ACLMS algorithm
ANGD algorithm · Convex hybridization

1 Introduction

Approximation procedures with cooperative data handling are required in contemporary scientific applications like target localization, sensor networks, life sciences, biomedical engineering, and environmental monitoring. Adaptive procedures are a new selection of approximation procedures that avail gradient-based approach of

R. Yarlagadda (✉)
Gudlavalleru Engineering College, Gudlavalleru, Andhra Pradesh, India
e-mail: ramakrishna.yarlagadda@gmail.com

V. Ratna Kumari
Prsad V. Potluri Siddhartha Institute of Technology, Kanuru, Andhra Pradesh, India
e-mail: vemuriratna2005@gmail.com

V. S. Potluri
Velagapudi Ramakrishna Siddhartha Engineering College, Kanuru, Andhra Pradesh, India
e-mail: pvsubbaiah@vrsiddhartha.ac.in

© Springer Nature Singapore Pte Ltd. 2019   589
J. Wang et al. (eds.), *Soft Computing and Signal Processing* ,
Advances in Intelligent Systems and Computing 898,
https://doi.org/10.1007/978-981-13-3393-4_60

adaptive schemes. Convergence of algorithms can be improved by establishing the cooperation between individual nodes using adaptive techniques. When the communication resources are plenty, the optimizing techniques are particularly useful to compute the local estimate of the desired constraints. The degree of cooperation among network resources and maximum connectivity with the network can be ensured with adaptive algorithms.

The impairments like block nature of the solution, requirement of stationary input, computation of large correlation matrix do not make the traditional adaptive algorithms best suitable for real-world real-time applications. One possible way to overcome large correlation matrix and slow convergence problems is to make use of complex adaptive algorithms which can process different types of data, ranging from stationary to nonstationary, linear to nonlinear, etc., [1].

The application of ACLMS and ANGD algorithms benefits the smart arrays in important features such as directionality characteristics and high gain.

Deployment of ANGD algorithm on smart antennas proved to be appropriate for nonlinear input signals, whereas the ACLMS algorithm is appropriate for nonstationary signals [2, 3].

2 Adaptive Algorithms

2.1 ACLMS Algorithm

The ACLMS algorithm improves the tracking ability when nonstationary condition prevails in the smart array. The error signal is calculated from the array input and the desired signal. The weights are updated such that the error in the next iteration is minimized. Weights are converged at an instant where the error is minimum or zero.

x_0 is assumed as complex random vector in complex domain [4], and its augmented vector is defined as

$$x^a = \begin{bmatrix} x_0 \\ x_0^* \end{bmatrix}. \tag{1}$$

The augmented output of the linear process $y_0(n)$ is given by

$$y_0(n) = h^T(n)x_0(n) + g^T(n)x_0^*(n) \tag{2}$$

where $x_0(n)$ is the input to the FIR filter, $h(n)$ and $g(n)$ are adaptive weight vectors used to minimize the cost function

$$E(k) = \frac{1}{2}|e_0(n)|^2 = \frac{1}{2}|d(n) - y_0(n)|^2 = \frac{1}{2}e_0(k)e_0^*(n) \tag{3}$$

where $e_0(n)$ is the complex error signal and $d(n)$ is the desired signal.

The gradient of the cost function in Eq. (3) yields

$$\nabla E(n)_h = -e_0(n)x_0^*(n) \tag{4}$$

$$\nabla E(n)_g = -e_0(n)x_0(n) \tag{5}$$

The weight updates are now given as

$$h(n+1) = h(n) + \mu e_0(n)x_0^*(n) \tag{6}$$

$$g(n+1) = g(n) + \mu e_0(n)x_0(n) \tag{7}$$

Augmented weight vector $w^a(n)$ of Eqs. 6 and 7 is given as

$$w^a(n) = \left[h^T(n) \ g^T(n) \right]^T \tag{8}$$

Weights of the augmented vector are updated as [5]

$$w^a(n+1) = w^a(n) + \mu \, e_0^a(n)\left(x^a\right)^*(n) \tag{9}$$

2.2 ANGD Algorithm

ANGD algorithm activation function is defined as

$$y_0(n) = \Phi\left(x^T(n)w(n)\right) = \lambda \overline{\Phi}\left(x^T(n)w(n)\right) \tag{10}$$

λ is the amplitude of activation function. The performance of the nonlinear adaptive algorithm is improved by adjusting the parameter λ in Eq. (10). The update for λ [6] is given by

$$\lambda(n+1) = \lambda(n) - \rho\nabla_\lambda E. \tag{11}$$

$\nabla_\lambda E$ is the gradient of $E(n)$, and ρ denotes the step size of the algorithm. The step size is generally preferred to be a small constant with positive amplitude. Standard gradient derivation is being followed to obtain [7]

$$\nabla_\lambda E = \frac{\partial E(n)}{\partial \lambda(n)} = \frac{1}{2}\frac{\partial\left[e_0^2(n)\right]}{\partial \lambda(n)} = -e_0(n)\overline{\Phi}\left(x^T(n)w(n)\right). \tag{12}$$

The error of the nonlinear adaptive ANGD algorithm in [8] is computed as

$$e_0(n) = d(n) - \Phi\left(x^T(n)w(n)\right) \tag{13}$$

$$w(n+1) = w(n) + \eta x_0(n)\Phi'\big(x^T(n)w(n)\big)e_0(n) \qquad (14)$$

$$\lambda(n+1) = \lambda(n) + \rho e_0(n)\overline{\Phi}\big(x^T(n)w(n)\big) \qquad (15)$$

Key features like gradient update, weight update, and error calculation values related to ANGD algorithm can be obtained through Eqs. (13)–(15).

3 Proposed Hybrid Structure

In this work, convex hybridization of ACLMS and ANGD algorithms are used to generate hybrid adaptive weights to handle complex nonlinear and non-stable signals in a large dynamical range with improved stability [8]. In case if one subfilter (algorithm) fails, the output of the hybrid algorithm automatically tracks the other subfilter [9].

The two subfilters of the hybridization architecture shown in Fig. 1 are adapted independently with ACLMS algorithm and ANGD algorithm, respectively. A common input signal $x_0(n) = [x_1(n), \ldots, x_N(n)]^T$ is used for both the algorithms. In the hybridization process, the two subfilters, i.e., ACLMS and ANGD algorithms, share the common input and prediction setting tasks.

The outputs of the two individual algorithms $y_{ACLMS}(n)$ and $y_{ANGD}(n)$ are combined to form overall hybrid output in the following way.

$$y_h(n) = \lambda(n)y_{ACLMS}(n) + (1 - \lambda(n))y_{ANGD}(n) \qquad (16)$$

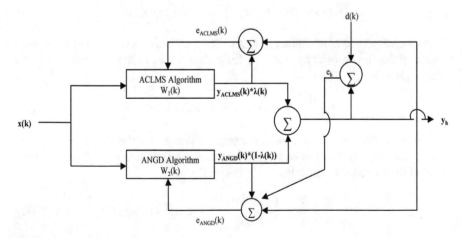

Fig. 1 Convex hybridization of ACLMS and ANGD algorithms

$\lambda(n)$ in Eq. (16) is the adaptive mixing parameter, reorganized by reducing the cost function $E(n)$, derived from the mean squared error [10] of the hybrid system error $e_{0h}(n)$.

$$E(n) = \frac{1}{2}|e_{0h}(n)|^2 = \frac{1}{2}|d(n) - y_{0h}(n)|^2 \tag{17}$$

A stochastic gradient-based method is used to update $\lambda(n)$, and its update is given by

$$\lambda(n+1) = \lambda(n) - \mu_\lambda \nabla_\lambda E(n) \tag{18}$$

where μ_λ is the step size. The complex error of the hybrid system $e_{0h}(n)$ is separated as follows

$$\nabla_\lambda E(n) = \left\{ e_{0h}(n)\frac{\partial e_{h0}^*(n)}{\partial \lambda(n)} + e_{0h}^*(n)\frac{\partial e_{0h}(n)}{\partial \lambda(n)} \right\} \tag{19}$$

The gradients of Eq. (19) are evaluated separately as

$$\frac{\partial e_{0h}(n)}{\partial \lambda(n)} = \frac{\partial e_{0h_r}(n)}{\partial \lambda(n)} + j\frac{\partial e_{0h_i}(n)(n)}{\partial \lambda(n)} \tag{20}$$

$$\frac{\partial e_{0h}^*(n)}{\partial \lambda(n)} = \frac{\partial e_{0h_r}(n)}{\partial \lambda(n)} - j\frac{\partial e_{0h_i}(n)}{\partial \lambda(n)} \tag{21}$$

Mixing parameter λ is updated as.

$$\lambda(n+1) = \lambda(n) + \mu_\lambda Real\left\{ e_{0h}(n)\left[y_{ACLMS}(n) - y_{ANGD}(n) \right]^* \right\} \tag{22}$$

The hybrid error is minimized by adjusting the mixing parameter in convex hybridization and is given as

$$\lambda(n) = sgm(g(n)) = \frac{1}{\left(1 + e^{-g(n)}\right)} \tag{23}$$

$g(n)$ is updated as

$$g(n+1) = g(n) + \mu_a e_0(n)|y_{ACLMS}(n) - y_{ANGD}(n)|\lambda(n)(1 - \lambda(n)) \tag{24}$$

The adaptation of the hybrid algorithm is stable and has fast convergence due to the adaptation step size parameter μ_a. The range of $g(n)$ is constrained to $[-g^+, g^+]$ so that the range of $\lambda(n)$ is limited to $[-1, +1]$.

4   Simulation and Performance Analysis

The input signal $x_0(n) = \cos(2wt)$ is chosen to carry out simulations at a frequency of 1 kHz. In order to simulate real-time antenna working environment, an additive white Gaussian noise (AWGN) is considered along with the input signal. A smart antenna with N number of array elements is used at the input of the receiver. ACLMS, ANGD, and the proposed hybrid algorithm are used in the DSP section of the smart antenna to analyze beamforming characteristics. Beamwidth, SLL, and convergence are the performance indicators considered in the analysis of these algorithms. Beamwidth and SLL are obtained from the array factor of the designed array.

4.1   ACLMS Algorithm Simulation

ACLMS algorithm is adopted in the adaptive beamforming process in the DSP section of smart antenna, and its performance is tested with array length (N) and adaptive step size parameter (μ). The important array characteristic values obtained through various investigations are presented in [2].

4.2   ANGD Algorithm Simulation

ANGD algorithm produces better directional properties with smaller HPBW and smaller SLL at higher values of N [2, 3]. However, to reduce the complexity in array size and physical dimensions, N value is fixed at 8. Similarly, $\mu = 0.02$ case is giving optimum values for HPBW and SLL.

4.3   Hybrid Algorithm Simulation

The characteristic curves of ACLMS, ANGD algorithms, and the hybrid algorithm with low-noise and high-noise environments are shown in Figs. 2, 3 and 4. An array of length N = 8 and $\mu = 0.02$ is considered for simulation. In case of low-noise simulation, the AWGN is considered with SNR = 20 dB, and for high noise, SNR = 8 dB.

The array factors with ACLMS, ANGD, and hybrid algorithms plotted in Fig. 2 reveal that the hybrid algorithm is superior to ACLMS and ANGD algorithms. These characteristic values obtained in simulations are presented in Table 1.

The hybrid algorithm perfectly eliminates the deep fading caused by noisy components present in signal, and it results in early convergence. This fact is presented in Fig. 3.

Fig. 2 Comparison of array factors

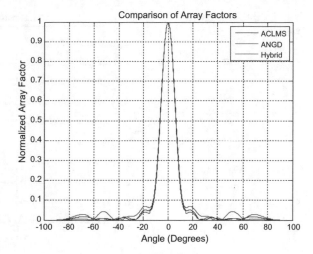

Fig. 3 Assessment of signal tracking

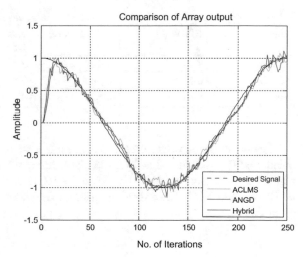

Table 1 Comparison of array characteristic values

Algorithm	HPBW (°)	Side-lobe level	Number of side lobes	CPU time (ms)
ACLMS	9.30	0.0513	4	485
ANGD	8.80	0.0867	3	492
Hybrid	8.60	0.0408	4	485

MSE convergence shown in Fig. 4 reveals that hybrid algorithm converges much faster than other two algorithms. The hybrid filter is converged at 90th iteration, whereas other filters converge only after 120–150 iterations.

Fig. 4 MSE convergence assessment

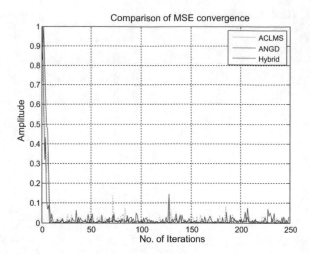

Table 2 Comparison of array characteristic values in noisy environment

Algorithm	HPBW (°)	Side-lobe level	Number of side lobes	CPU time (ms)
ACLMS	7.00	0.4102	3	532
ANGD	6.50	0.2640	7	535
Hybrid	5.50	0.1633	4	530

The characteristic values listed in Table 2 disclose that the ANGD algorithm gives smaller HPBW and SLL than ACLMS algorithm at all noise conditions. The hybrid algorithm outperforms both ACLMS and ANGD algorithms with smaller values of HPBW and SLL. The hybrid algorithm offers enhanced stability over the two algorithms.

5 Conclusion

From the performance characteristics of the hybrid algorithm shown in Figs. 2, 3 and 4 and characteristic values presented in Tables 1 and 2, it is concluded that the hybrid adaptive algorithm provides improved performance over its two subfilters. It provides better performance in a high-noise-signal environment with the lowest possible SLL, which results in the improvement of the quality of services offered. Based on SLL and HPBW values obtained, this algorithm can be preferred for urban cellular communications with improved capacity.

References

1. Ramakrishna, Y., Subbaiah, P.V., Prabhakara Rao, B.: A novel hybrid adaptive algorithm for improvement of mean square error convergence. Int. J. Syst. Control Commun. **6**(1), 59–68 (2014)
2. Ramakrishna, Y., Kumari, V.R., Subbaiah, P.V.: Hybrid adaptive beamforming algorithms for smart antennas. In: 2017 6th International Conference on Computer Applications in Electrical Engineering-Recent Advances (CERA), Roorkee, pp. 117–122 (2017). https://doi.org/10.1109/cera.2017.8343312
3. Jelfs, B.: Collaborative adaptive filtering in complex domain. In: IEEE Workshop on Machine Learning for Signal Processing, pp. 421–425, October 2008
4. Ramakrishna, Y., Subbaiah, P.V., Rao, B.Prabhakara: Adaptive nonlinear gradient decent (ANGD) algorithm for smart antennas. Int. J. Mob. Netw. Commun. Telemat. **2**(6), 11–19 (2012)
5. Goh, S.L., Bozic, M., Mandic, D.P.: A nonlinear neural FIR filter with an adaptive activation function. J. Autom. Control (University of Belgrade) **13**(1), 1–5 (2003)
6. Xia, Y., Took, C.C., Mandic, D.P.: An augmented affine projection algorithm for the filtering of noncircular complex signals. Elsevier Sig. Process. **90**(6), 1788–1799 (2010)
7. Mandic, D.P., Boukis, V.C., Jelfs, B., Goh, S.L., Gautama, T., Rutkowski, T.: Collaborative adaptive learning using hybrid filters. In: Proceedings of IEEE International Conference on Acoustics, Speech and Signal Processing, vol. 3, pp. 921–924 (2007)
8. Xia, Y., Mandic, D.P., Van Hulle, M.M., Principe, J.C.: A complex echo state network for nonlinear adaptive filtering. In: IEEE Workshop on Machine Learning for Signal Processing, pp. 404–408, October 2008
9. Trentin, E.: Networks with trainable amplitude of activation functions. Elsevier J. Neural Netw. **14**(4), 471–493 (2001)
10. Arenas-Garcia, J., Figueiras-Vidal, A.R., Sayed, A.H.: Mean-square performance of a convex combination of two adaptive filters. IEEE Trans. Sig. Process. **54**(3), 1078–1090 (2006)

Fruit Detection from Images and Displaying Its Nutrition Value Using Deep Alex Network

B. Divya Shree, R. Brunda and N. Shobha Rani

Abstract This paper presents a simple and efficient approach to perform fruit detection and predict nutrition information of the fruits using deep Alex networks (DAN). The datasets employed for analysis are acquired from fruit 360 database of image processing challenges. Fruit categories include apples, berries, banana, grape, papaya, peach, avocado, and multiple flavors of apple. And also, the experimentations are carried out on various other fruit samples collected from multiple Web repositories. The network architecture is as usual comprised of to five convolution layers and three fully connected layers including the max pooling, RELU layers. The input images are assumed to be of dimensions $227 \times 227 \times 3$ with number of filters of 96 of size $11 \times 11 \times 3$ with a stride length of 4. The results of experiment prove that fruit detection using DAN is efficient with an accuracy of about 91% to classify the fruits of about 50 different categories on a single machine of configurations 1 GPU, 8 GB RAM, and octa-core processor.

Keywords Deep learning · Fruit image classification · Nutrition prediction
Object recognition · Convolution neural networks

1 Introduction

Consumption of fruits in day-to-day dietary plays an important role in the nutritional supplement intake of humans. Estimation of right nutrient intake through fruits consumption has become increasingly significant in order to maintain the proper health of the well-being. Recommendations on the choice of fruits being consumed have

B. Divya Shree · R. Brunda · N. Shobha Rani (✉)
Amrita School of Arts & Sciences, Amrita Vishwa Vidyapeetham, Mysuru, India
e-mail: n.shoba1985@gmail.com

B. Divya Shree
e-mail: divyabasavaraju21@gmail.com

R. Brunda
e-mail: brundarajanna@gmail.com

© Springer Nature Singapore Pte Ltd. 2019
J. Wang et al. (eds.), *Soft Computing and Signal Processing* ,
Advances in Intelligent Systems and Computing 898,
https://doi.org/10.1007/978-981-13-3393-4_61

been vital in this process which can be accomplished through a specialized system. In this paper, the emphasis is given on the development of the strategies that can perform classification of fruits and suggest recommendations regarding the nutrition details of the fruits. Computer vision has its wide applications in the area automatic object recognition and classification; detection of a particular fruit type from the image of the same is the objective of the computer vision systems today. However, the intensive computational procedures involved in this process are currently being replaced by deep learning procedures due to wide recognition in terms of performance and reliability of the outcomes achieved. Quite a good amount of research is being carried out in this area, few reports depict the fruit detection based on the color, based on the size and shape, based on the geometrical and statistical properties of the fruit, and others include the decomposition of objects in varied levels through wavelets. The recognition and classification of the fruits have several benefits that include agriculture, industrial inspection, automated fruit segregation systems, nutrition prediction based on the images of fruits. Thus, fruit recognition and classification systems are employed in various fields and can be integrated with the latest technology for day to day. Numerous experimentations are carried out on fruit detection and classification; the summary of some of the important contributions are as follows.

Banot and Mahajan [1] had proposed an approach for fruit detection and grading using fuzzy logic technique for feature analysis, and classification is done using artificial neural network. The results are compared with classifiers k-nearest neighbor's, support vector machine, RGB color space method, and color mapping techniques. Morphological features are used to identify the class of fruit using the neuronal network. Suresha et al. [2] had proposed an automatic classification system for apples using the SVM classifier. Database containing 90 images of red and green apples. In addition, apple's RGB image is converted to the HSV format, and the image of the apple is divided from the bottom using a threshold. SVM with a linear core function is used for the classification for which 100% accuracy is achieved. Sagare et al. [3] implement a direct color mapping technique that classifies and grades the fruits based on shape. In this work, fruit images are captured by the camera and processed using shapes, areas, and major and minor axes features. Paulraj et al. [4] had proposed a simple color identification algorithm using a neural network to evaluate the degree of ripeness of a banana. RGB color components and histogram features are used to estimate the ripeness of the banana. An accuracy of 96% is reported using backpropagation neural network classifier. May et al. [5] also used RGB color models and artificial fuzzy logic to study the maturity of oil palm fruits based on similar techniques. Colors based on the RGB color model, average color intensity, and decision-making process using fuzzy logic techniques were trained as data, and the system was able to classify three different categories of oil palm fruit as 86.67% overall accuracy. Yao et al. [6] had proposed a method for generating image text description using image analysis and text generation. The parsing technique displays the correspondence between different shared visual modes within the image and breaks down the image into parts, namely scenes, object, and parts. Finally, the text description is meaningfully generated. Wang et al. [7] had proposed two wheat diseases that were classified according to color, shape, and texture features

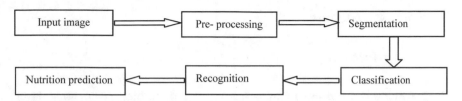

Fig. 1 Block diagram of the proposed system

and classified using a backpropagation neural network. The accuracy of the report exceeded 90%. Donnelly [8] had proposed a technique to automatically generate caption to the images to assist the vision impaired from the Web images. Baseline LSTM model is used to create image caption via generative recurrent neural network. Sumi et al. [9] have used stemming, lookup, the production techniques, and suffix stripping algorithms to generate caption and create a summary by identifying and concatenating important sentence in a document. Vinyals et al. [10] proposed a generative model based on a deep loop structure that combines computer vision and machine translation techniques. A probabilistic neural network is used for generation of descriptions from images. The LSTM-based sentence generator is used to generate description for images. Patel et al. [11] had proposed an algorithm using multiple features by carrying out fruit detection on tree. Features extracted include intensity and color, orientation, edge, feature maps. Arivazhagan et al. [12] used color and texture features, based on statistical and co-occurrence features, to identify fruits using a minimum distance classifier.

The proposed system consists of two stages, first one is the recognition of the fruit type, and the second one is the nutritional value prediction. The above block diagram (Fig. 1) gives an idea of the proposed system. Firstly the image is fed to the system for preprocessing. Preprocessing steps include RGB to grayscale conversion, filtering, and resizing the image to 160×160 pixels. Segmentation is a technique that aims at identification and extraction of four ground object from an image resulting in individual segments. Classification is done with the help of deep Alex network. The network would classify the input image into one of the predefined classes. Fruit type recognition if followed by classification. Based on fruit type, the nutrition values will be displayed.

Most of the works in the literature are devised by using geometrical, statistical, and color features in different RGB color spaces. These techniques are computationally expensive along the comprising level of reliability and throughput and are not ideal for classification of very large-scale image category classification. Also, it is observed that the classification is focused only on particular type of fruits, and its various types like an apple are classified to green, red. Therefore, it is very significant to develop a system that can perform fruit classification over varied types and also predict the nutrition supplements associated with the detected fruit type.

2 Proposed Method

2.1 Deep Learning

Deep learning is a class of machine learning algorithm that uses multiple layers comprising of nonlinear units. Each layer uses the output of the previous layers as its input. Deep learning algorithms use more layers than shallow learning algorithms. Convolution neural network is classified as one of the deep learning algorithms. These networks are composed of multiple convolution layers with a few fully connected layers. They also make use of pooling; its configuration allows convolution networks to take advantage of bidimensional representation of data. Another deep learning algorithm is the recursive neural network. In this kind of architecture, the same set of weights is recursively applied over some data. Recurrent networks have shown good results in natural language processing.

2.2 Alex Net

Alex net is a convolutional neural network for image classification. The task is to categorize the given entry into one of the given classes. Alex's network has eight layers. The first five are convolution layers and the last three are totally connected layers. Convolutional layer is responsible for feature extraction, and fully connected layers are regular neural networks. There are also some layers called grouping and activation. Figure 2 shows the arrangement and configuration of all Alex networks.

Alex Net Architecture: Yellow square indicates the input feature map. Green indicates the convolutional layer, orange indicates the max pooling layer, and blue indicates the fully connected layer. Arrows show the direction of flow of data. Dropout

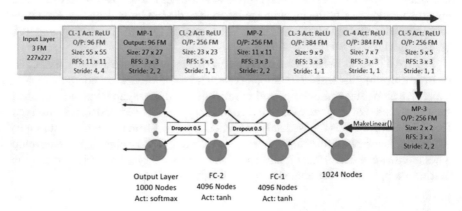

Fig. 2 Architecture of Alex net

Table 1 Experimental statistics of fruit categories

Label	Number of training images	Number of testing images
Apple Braeburn	164	32
Apple golden 1	164	32
Apricot	164	32
Avocado	164	32
Banana	164	32
Cherry	164	32
Grape white	164	32
Kiwi	164	32
Mango	164	32
Orange	164	32
Papaya	164	32
Peach	164	32
Pineapple	164	32
Pomegranate	164	32
Strawberry	164	32

0.5 indicates that a dropout layer exists between two fully connected layers with a retain probability of 0.5. Output indicates the number of feature maps a layer generates, and its dimension is specified by size. RFS indicates receptive field size of the layer. Strides mention the horizontal jump and vertical jump of the receptive field. The complete architecture has 11 processing layers and more than two crore trainable weights.

3 Dataset

The dataset is created synthetically by assuming a white background behind the fruit, accomplished by placing a white sheet of paper as background. However, due to variations in the lighting conditions, the background was not uniform. Fruits were scaled to fit a 160×160 pixels image. In the future, it is planned to work with different textured background and even larger image dimensions, as the practical simulation requires high-end processing units and multiprocessing architecture. The resulted dataset is used for training that consists of 2460 images of fruits spread across 15 labels. Images are obtained by capturing fruits of different orientations. Table 1 lists the details of various fruit types.

4 Performance Analysis

The experimentation in the proposed system is carried out with a synthetically generated datasets of fruits. The dataset includes 164 references for each fruit comprised of
training set as well as test set with different proportions of training and test samples
and total number of sample resulting in 2460. The datasets are collected from various
sources of online, and also few of the images are synthetically gathered by capturing
the fruits images over a plain background. The classification of fruit images using
convolutional neural network is accomplished in training and testing phases.

If Ti1, Ti2, Ti3 … Tin and Ts1, Ts2, Ts3 … Tsn represent the training and
test samples, the accuracy of classification is defined as the number of instances of
sequence Tsi∈Ck, where Ck is the class to which a fruit belongs to and i = 1, 2, 3 …
n, k = 1, 2, 3 … n, and i = k. Here i = k indicates that Ts1∈C1, Ts2∈2… Tsn∈Cn.
The accuracy is given by Eq. (1)

$$Accuracy = \frac{n_c}{\tau} \tag{1}$$

where n_c represents the number of test samples classified correctly by the classifier
and τ is the total of images in the test dataset. The experimental evaluation has also
been analyzed with the help of standard performance measures for testing classification efficiency. The performance measures such as sensitivity/recall rate, specificity,
precision are evaluated.

Let N(TP⁺), N(FN⁻), N(TN⁻), and N(FP⁺) represent the number of true positives,
false positives, true negatives, and false positives, respectively. The performance
measures sensitivity, specificity, precision, F-measure, and accuracy are defined by
Eqs. (2) through (5).

$$sensitivity = \frac{N(TP^+)}{N(TP^+) + N(FN^-)} \tag{2}$$

$$specificity = \frac{N(TN^-)}{N(TN^-) + N(FP^+)} \tag{3}$$

$$precision = \frac{N(TP^+)}{N(TP^+) + N(FP^+)} (2) \text{ to } (4) \tag{4}$$

$$Accuracy = \frac{N(TP^+) + N(TN^-)}{N(TP^+) + N(TN^-) + N(FP^+) + N(FN^-)} \tag{5}$$

In this analysis, N(TP⁺) represents the noisy samples for which the class is correctly predicted, N(FN⁻) represents the noisy samples for which the class is incorrectly predicted, N(FP⁺) represents the non-noisy samples for which the class is
incorrectly predicted, and N(TN⁻) represents the non-noisy samples for which the
class is correctly predicted.

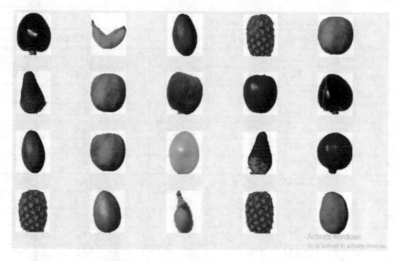

Fig. 3 Few instances of dataset samples in our knowledge base

Fig. 4 Few instances of sample recognized

Above two figures (Figs. 3 and 4) give instances of samples in the knowledge base and test case results. Images of the same fruit type with different orientation were fed to the system as training dataset forming the knowledge base. More than 2000 images build the knowledge base of the deep Alex network. Further based on the knowledge acquired by training, system has recognized the fruit types.

Tables 2 and 3 give a quick glance of the nutritive values of few fruits considered in this experiment. Few major fruit constituent parameters are considered. The energy fruit could give when consumed is given in kilojoules per kilocalories (measured for every 100 g of fruit).

In Table 4 initially, 40:60% is evaluated for training and testing sets from a total of 164 from each category; further, the analysis is carried out by assuming training and testing ratios in terms of 50:50, 60:40, 70:30%. The result of experiments proves that fruit detection using deep Alex network is efficient and accuracy is 91%.

Table 2 Nutrition facts of various fruit categories

Food	Energy	Water	Fiber	Fat	Protein	Sugar
Substance = 100 g	Kj/Kcal	%	G	G	G	G
Apple Braeburn	207/49	84	2.3	0	0.4	11.8
Apple golden	207/95	84	2.3	0	0.4	11.8
Apricot	153/36	87	2.1	0	1.0	8.0
Avocado	523/126	81	0.2	10	2.0	7.0
Banana	375/88	76	2.7	0	1.2	20.4
Cherry	221/52	86	1.2	0	0.0	13.0
Grapes	274/64	83	2.2	0	0.6	15.5
Kiwi fruit	168/40	84	2.1	0	1.1	8.8
Mango	255/60	84	1.0	0	0.0	15.0
Orange	198/47	87	1.8	0	1.0	10.6
Papaya	136/32	91	0.6	0	0.0	8.0
Peach	151/36	89	1.4	0	1.0	7.9
Pineapple	211/50	84	1.2	0	0.4	12.0
Pomegranate	343/86	82	3.4	0	1.0	17.0
Strawberry	99/23	91	2.2	0	0.7	5.1

Table 3 Vitamin contents of various fruit categories

Food	Vit A	Vit C	Vit B1	Vit B2	Vit B6	Vit E
Substance = 100 g	Ug	Mg	Mg	Mg	Mg	Mg
Apple Braeburn	2	15	0.02	0.01	0.05	0.5
Apple golden	2	15	0.02	0.01	0.05	0.5
Apricot	420	5	0.06	0.05	0.06	0.5
Avocado	20	17	0.06	0.12	0.36	3.2
Banana	3	10	0.04	0.03	0.36	0.3
Cherry	40	10	0.02	0.02	0.04	0.1
Grapes	0	3	0.03	0.01	0.08	0.6
Kiwi fruit	5	70	0.01	0.02	0.12	1.9
Mango	210	53	0.05	0.06	0.13	1.0
Orange	2	49	0.07	0.03	0.06	0.1
Papaya	40	46	0.03	0.04	0.04	–
Peach	15	7	0.01	0.02	0.02	0.0
Pineapple	20	25	0.07	0.02	0.09	0.1
Pomegranate	10	7	0.05	0.02	0.31	–
Strawberry	10	60	0.02	0.03	0.06	0.4

Table 4 Performance metrics of classifier

Training samples (%)	Testing samples (%)	Accuracy (%)	Precision (%)
40	60	91.26	89.83
50	50	91.82	97.87
60	40	94.10	98.74
70	30	96.58	98.06

class	1	2	3	4	5	6	7	8	9	10	11	12	13	14	15
1	152	2	1	0	0	2	1	3	0	1	0	1	1	0	0
2	0	164	0	0	0	0	0	0	0	0	0	0	0	0	0
3	2	3	140	0	4	0	1	4	1	1	2	2	1	0	3
4	0	2	0	156	0	1	0	0	0	2	0	0	0	2	1
5	0	0	3	0	164	0	0	0	0	0	0	0	0	0	0
6	1	1	0	0	3	147	1	0	2	4	0	0	0	0	3
7	1	0	0	0	0	1	160	0	0	1	0	0	1	0	0
8	1	0	0	2	0	0	1	158	0	0	0	2	0	0	0
9	4	4	0	8	0	3	3	0	135	2	0	2	0	1	2
10	0	2	4	0	0	2	0	0	0	158	0	2	0	0	0
11	0	0	1	0	0	0	0	1	0	0	162	0	0	0	0
12	3	0	4	2	2	4	2	0	3	1	1	140	5	0	1
13	0	0	4	0	2	0	2	0	3	0	0	0	148	5	0
14	2	0	0	0	3	0	0	0	1	0	0	2	0	156	0
15	3	6	1	0	4	0	6	1	0	5	1	3	0	8	126

Fig. 5 Confusion matrix indicating the rate of accuracy for individual class

In this experimentation, 15 classes are considered including apple Braeburn (class 1), apple golden 1 (class 2), apricot (class 3), avocado (class 4), banana (class 5), cherry (class 6), grape white (class 7), kiwi (class 8), mango (class 9), orange (class 10), papaya (class 11), peach (class 12), pineapple (class 13), pomegranate (class 14), strawberry (class 15).

In confusion matrix, the row corresponds to the predicted class (actual class) and the column represents the true class (target class). The diagonal cells are the observations that are classified correctly, and the off-diagonal cells are the incorrectly classified observations (Fig. 5).

5 Conclusion

The deep neural network is trained with around 15 varieties of fruits. The intention is to get a neural network that can identify a wider array of objects from the image. The segmentation of image algorithm is a very useful processing of image method, and it is extremely useful for post-processing. An overall accuracy of 93% is achieved. In the future work, we plan to create a mobile application which takes the picture of fruits and labels them accordingly. Another object is to expand the training and testing sets to include more items. This is time-consuming process since we want to include the items that were not used in most other examples.

References

1. Banot, S., Mahajan, P.M.: A fruit detecting and grading system based on image processing-review. Int. J. Innov. Res. Electr. Electron. Instrum. Control Eng. **4**(1), 47–53 (2016)
2. Suresha, M., Shilpa, N.A., Soumya, B.: Apples grading based on SVM classifier. In: National Conference on Advanced Computing and Communications, Apr 2012
3. Sagare, S.N.: Kore, fruits sorting and grading based on color and size. Int. J. Emerg. Technol. Comput. Appl. Sci. (IJETCAS), 12–333
4. Paulraj, M., Hema, C.R., Pranesh, R.K., SitiSofiah, M.R.: Color recognition algorithm using a neural network model in determining the ripeness of a banana (2009)
5. May, Z., Amaran, M.H.: Automated oil palm fruit grading system using artificial intelligence. Int. J. Eng. Sci. **11**(21), 30–35 (2011)
6. Yao, B.Z., Yang, X., Lin, L., Lee, M.W., Zhu, S.C.: I2t: image parsing to text description. Proc. IEEE **98**(8), 1485–1508 (2010)
7. Wang, H., Li, G., Ma, Z., Li, X.: Image recognition of plant diseases based on backpropagation networks. In: Image and Signal, Oct 2012
8. Donnelly, C.: Image caption generation with recursive neural networks
9. Sumi, P., Cse, F.Y.M.T.: A systematic approach for news caption generation
10. Vinyals, O., Toshev, A., Bengio, S., Erhan, D.: Show and tell: a neural image caption generator. In: Proceedings of the IEEE Conference on Computer Vision and Pattern Recognition, pp. 3156–3164 (2015)
11. Patel, H.N., Jain, R.K., Joshi, M.V.: Fruit detection using improved multiple features based algorithm. Int. J. Comput. Appl. **13**(2), 1–5 (2011)
12. Arivazhagan, S., Shebiah, R.N., Nidhyanandhan, S.S., Ganesan, L.: Fruit recognition using color and texture features. J. Emerg. Trends Comput. Inf. Sci. **1**(2), 90–94 (2010)

Modelling and Analysis of Volatility in Time Series Data

Siddarth Somarajan, Monica Shankar, Tanmay Sharma and R. Jeyanthi

Abstract The comprehension of volatility is a crucial concept in analysing data. It is of greater importance for financial data since it furnishes key aspects such as return on investments and helps with effective hedging. The unpredictable nature of volatility causes heteroskedasticity which leads to difficulty in modelling. Consequently, time series models are desirable to predict volatility. An illustration of the same has been shown through an example of fitting time series models on the volatility of a listing from the National Stock Exchange (NSE). This paper also attempts to treat heteroskedasticity using Box-Cox transformations to achieve equal error variances prior to the modelling.

Keywords Volatility · Time series models · Heteroskedasticity
Box-Cox transformations

1 Introduction

Investments are a key aspect of financial literacy. The objective of investing is to increase the value of assets compared to the liabilities. Volatility is the tendency of rapid change which is unpredictable. All investments have their own form of volatility. In econometrics, it is the standard deviation of returns or dispersion. In the context of the NSE, volatility is the dispersion of the received return from that of the

S. Somarajan · M. Shankar · T. Sharma · R. Jeyanthi (✉)
Department of Electronics and Communication Engineering, Amrita School of Engineering,
Amrita Vishwa Vidyapeetham, Bengaluru, India
e-mail: jeyanthi.ramasamy@gmail.com

S. Somarajan
e-mail: sid.info01@gmail.com

M. Shankar
e-mail: monicashankar21@gmail.com

T. Sharma
e-mail: tanmaysharma444@gmail.com

© Springer Nature Singapore Pte Ltd. 2019   609
J. Wang et al. (eds.), *Soft Computing and Signal Processing* ,
Advances in Intelligent Systems and Computing 898,
https://doi.org/10.1007/978-981-13-3393-4_62

expected return. Higher dispersion implies a higher volatility [1]. Most economic data does not have a constant mean, and many of them exhibit a trend. Moreover, the volatility in series is not a constant over time. There are many basic models to predict stochastic volatility such as exponentially weighted moving average volatility (EWMA) method, Heston model, constant elasticity of variance (CEV) model [2]. The challenge the modern econometrician faces is to develop simple and effective models capable of forecasting, interpreting and testing hypothesis concerning economic data. This is made simple and possible by the use of time series models as compared to the basic models.

Commonly used volatility models include Auto-Regressive Conditional Heteroskedasticity (ARCH) model developed by Engle (1982) and Generalized Auto-Regressive Conditional Heteroskedasticity (GARCH) model given by Bollerslev (1986). Both models use the error variance to forecast future values. Bala [3], studied co-movement in volatility between the Singapore Stock market and other stock markets around the world using univariate GARCH, Vector Auto Regression (VAR) and Asymmetric Multivariate GARCH model with Glosten, Jagannathan, and Runkle (GJR) extensions. In 2000, Sarkar [4] transformed the conditional variance (in regression analysis, it is called as the dependent variable) using Box-Cox transformations (BCT) to capture nonlinearity in financial data and found that a generalization of the linear ARCH model may perform much better than the linear ARCH model. AL-Najjar [5] studies the behaviour of ARCH and GARCH models on the return volatility in Amman Stock Exchange (ASE) and the Exponential Generalized Auto-Regressive Conditional Heteroskedasticity (EGARCH) model was used to capture the asymmetry in volatility. The results provide evidence to volatility clustering. As deduced by Nelson and Granger [6], Box-Cox transformed economic data does not always yield a consistent superior result and no value of λ may give a normally distributed data for a maximum likelihood procedure.

This paper explores the utilization of the symmetric GARCH process over the closing prices and the returns of a listing on the NSE [7]. It also compares the results with other time series models fit on BC transformed closing prices. The data used for modelling was collected over a period of 30 days and had a resolution of 2 min and was obtained from Google Finance. This paper has been organized as follows: The models used for fitting, method of transformation of the data and parameters to judge the goodness of fit are described under Sect. 1.1. The methodology of the experimental simulation is covered under Sect. 2. Analysis of the final results for the best fit models is discussed in Sect. 3. Section 4 discusses the conclusions resulting from the simulations which is followed by references.

1.1 Modelling of Volatility

In this paper, the volatility observed in the closing prices and their returns is modelled using parsimonious models. The data has further been transformed using Box-Cox transformations prior to volatility modelling in order to reduce irregularities such as non-normality and conditional heteroskedasticity. In finance, the term 'return' describes the amount of money gained or lost on a particular investment. The returns of the closing prices are defined as:

$$r(i) = \log[\frac{P(i+1)}{P(i)}]/[ticks(i+1) - ticks(i)] \tag{1}$$

where

$r(i)$ is the ith continuously compounded return,
$P(i)$ is the closing price and
ticks   is the time.

Prediction of volatility can be done using univariate ARCH, GARCH and asymmetrical GARCH models. ARCH is used when the error variance follows Auto-Regressive (AR) model. Bala and Premaratne [3] and Gamit et al. [8] provide further comprehensive information on ARCH models.

An Auto-Regressive–moving-average (ARMA) model [9] is one which consists of both AR and moving average (MA) components. AR process utilizes its own past values to predict the future values. MA process utilizes the history of errors to forecast future values. ARMA process was modelled by Peter Whittle et al. 1951, and is mathematically represented as:

$$y_t = a_0 + \sum_{i=1}^{P} u_i y_{t-1} + \sum_{i=0}^{q} \beta_i \varepsilon_{t-i} \tag{2}$$

where $\{a_i\}$ and $\{\beta_i\}$ are non-negative constants, 'p' is the order of AR model, 'q' is the order of MA model and $\{\varepsilon_t\}$ is white noise (zero mean and constant variance).

An ARCH (p) model [10] as formulated by Engle (1982) is represented as:

$$\hat{\varepsilon}_t^2 = \alpha_0 + \alpha_1 \hat{\varepsilon}_{t-1}^2 + \alpha_2 \hat{\varepsilon}_{t-2}^2 + \cdots + \alpha_q \hat{\varepsilon}_{t-q}^2 + v_t \tag{3}$$

where v_t refers to white noise, $\{a_i\}$ are non-negative coefficients that can be determined by maximum likelihood estimation (MLE).

In 1986, Bollerslev formulated the GARCH process [11] which is a parsimonious and generalized model for ARCH. GARCH utilizes past conditional variance in error for prediction of volatility [12]. The error is represented as,

$$\varepsilon_t = v_t \sqrt{\sigma_t} \tag{4}$$

Moreover, GARCH is a stationary model [13]. A GARCH (p, q) model is represented as:

$$\sigma_t = \omega + \sum_{i=1}^{q} \alpha_i \varepsilon_{t-i}^2 + \sum_{i=1}^{q} \beta_i \sigma_{t-i}^2 \tag{5}$$

where σ_t is a function of the previous values of ε_t^2, $\{a_i\}$ and $\{\beta_i\}$ are non-negative constants. Unlike ARCH, in case of GARCH, the error variance assumes an ARMA model. GARCH is incapable of capturing asymmetry in financial data. This implies that negative values of ε_t increases the conditional volatility higher than that of positive values of ε_t of equal magnitude. Thus, to apprehend both size and sign effects, asymmetrical GARCH models are preferred over symmetrical GARCH models.

1.2   Box-Cox Transformations

Box-Cox transformations (BCT) is a power transformation technique used to reduce irregularities such as non-normality and conditional heteroskedasticity. Ideally, transforming the data should linearize, normalize and treat its heteroskedasticity. Thus, not requiring the conditional heteroskedasticity (CH) component of ARCH and GARCH models, makes an AR model sufficient. As said by Maddala [14], it is important to take into consideration the regression assumptions before fitting an AR model. Further details about BCT can be found in [4]. However, [6] deduced that the use of transformations on financial data need not show any significant changes due to the presence of extreme non-normality. Also, there may exist no value of λ which gives normally distributed residuals. This can be tested visually by using normality plots.

The BC transformation formula is given by:

$$y(\lambda) = \begin{cases} \frac{y^\lambda - 1}{\lambda}, & \text{if } \lambda \neq 0; \\ \log y, & \text{if } \lambda = 0. \end{cases} \tag{6}$$

An extended form of BC transformation to accommodate negative values of y is as follows:

$$y(\lambda) = \begin{cases} \frac{(y+\lambda_2)^{\lambda_1} - 1}{\lambda_1}, & \text{if } \lambda \neq 0; \\ \log(y + \lambda_2), & \text{if } \lambda = 0. \end{cases} \tag{7}$$

where λ_2 satisfies the condition: $y + \lambda_2 > 0$ for any y.

1.3   Goodness of Fit

Goodness of fit is a measure of how well the statistical model matches for a given set of observations. It is the basis of model selection amongst a finite set of models. As suggested by Enders [1], there are several parameters to judge the goodness of fit for a model, such as R^2 (Residual squared), RSS (Residual sum square), AIC (Akaike information criteria), BIC (Bayesian information criteria).

1.3.1 R^2 and RSS

Residual squared is also known as the coefficient of determination. Higher the R^2, the better the model fit. R^2 is between 0 and 1.

$$R = \frac{Explained\ variable\ (of\ y)}{Total\ variable\ (of\ y)} \tag{8}$$

$R^2 = \frac{RSS}{TSS} < 0.9$ is acceptable for a volatile data (when collected from field or online). This is derived from a simple Analysis of Variance (ANOVA) [15] model.

ANOVA is a group of statistical procedures used to study the differences amongst group means. ANOVA tests, like t-test and ratios, are used for comparative analysis of experiments.

A simple ANOVA model is given below,

$$TSS = RSS + ESS \tag{9}$$

where TSS is the Total sum squared; RSS is the Residual sum squared; ESS is the Error sum squared.

1.3.2   Akaike Information Criterion and Bayesian Information Criterion

Akaike information criterion (AIC) and Bayesian information criterion (BIC) select the most suitable model, that is, one with the least number of parameters [6]. A model with lower AIC/BIC is preferred. They are given by the following equations:

$$AIC = 2\,ln(L) + 2k \tag{10}$$

$$BIC = -2\,ln(L) + k\,ln(n) \tag{11}$$

where

L   is the log-likelihood function;
K   is the number of parameters to be estimated;
n   is the number of observations.

1.3.3 t-Statistic

The t-statistic measures the number of standardized errors that the estimated coefficient is away from zero [16]. Usually, a t-statistic greater than 2 or lesser than -2 is acceptable. Higher the t-statistic, greater is the confidence in the estimated coefficient as a suitable predictor. Lower t-statistic values indicate low reliability.

2 Experimental Simulation

The data used for simulation is recorded market data of a listing on the NSE which was obtained from Google Finance through Excel Macros. The NSE [17] is India's leading stock exchange which began operating in 1994. In India, it is ranked as the largest exchange with respect to total and average daily turnover for equity. The resolution of the data acquired is 2 min. Data was collected for listings which were known to have high levels of volatility. This was done based on current affairs during the time of study. The factors contributing to the volatility in the listings consist of endogenous causes such as changes in top-level management, mergers and acquisitions and exogenous causes such as changes in governmental policies, taxation, condition of the economy. The data was collected for a period of 30 working days. Figure 1 shows the residuals of the closing prices, and Fig. 2 shows the returns of the closing prices of the listing chosen.

The experimental simulation followed is the algorithm developed by Box and Jenkins et al. [14]. The BJ methodology is an iterative process for model fitting. It involves model identification and parameter estimation followed by a diagnosis, which eventually leads to forecasting or reidentification. Prior to model identification, the ordinary least square (OLS) regression assumptions, stationarity and autocorrelation tests were conducted. Since GARCH accounts for conditional heteroskedasticity,

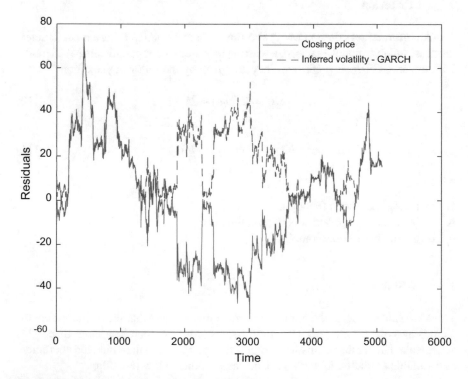

Fig. 1 Plot of closing returns versus inferred volatility (GARCH)

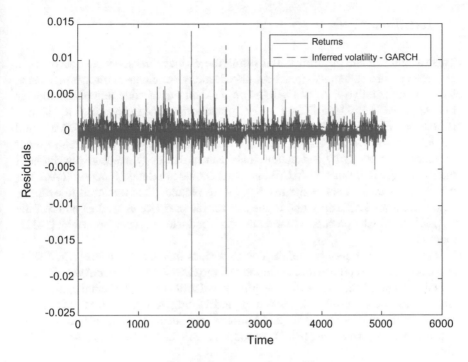

Fig. 2 Plot of actual returns versus inferred volatility (GARCH)

the model is fit on the non-stationary data. The resulting inferred volatilities are calculated from the conditional variances and plotted against the actual closing prices and returns.

This paper also ventures to linearize and treat the heteroskedasticity present in the dependent variable by treating the data with Box–Cox transformations. A comparative analysis of the time series models on the transformed data is represented in Table 1. The software chosen for the experimental simulation is MathWorks MATLAB 2017b which is equipped with an econometric toolbox.

Table 1 Goodness of fit for models over residuals and BCT data

Parameters	Residuals	BCT data		
	GARCH	AR	ARCH	GARCH
R^2	0.9928	0.2165	0.9906	0.9927
RSS	1,025,500	0.3833	0.1464	0.1461
AIC	16,338	Imaginary	−62,292	−63,583
BIC	16,345	Imaginary	−62,285	−63,577
t-statistic	>2	>2	>2	>2

3   Result Analysis

This section exhibits the results after fitting the various time series models over the chosen data. In an attempt to treat heteroskedasticity, the data was transformed using BCT. A comparative study of GARCH on the data versus other time series models like AR, ARCH and GARCH on the BC transformed data is illustrated in Table 1. Before proceeding with modelling, the Augmented Dickey–Fuller (ADF) test and Engle's test for residual heteroskedasticity or ARCH test were conducted in order to check for stationarity and conditional heteroskedasticity, respectively. Plots of the autocorrelation function (ACF) and partial Autocorrelation function (PACF) of the residuals and square of the residuals were plotted. This was used to examine autocorrelation, stationarity and to inspect for the presence as well as order of the AR, MA or ARMA process. These tests were conducted in accordance with [14] BJ method.

The residual variance is modelled using various time series models (AR, ARCH and GARCH). On comparison, we found that the GARCH (1, 1) model was the most suited fit. Figure 2 shows a plot of the inferred volatility versus the actual returns. The Ljung–Box Q-test and the ACF plots after model fitting show the presence of autocorrelation in the standardized residuals. However, the square of standardized residuals showed no autocorrelation [18]. The goodness of fit for the model is represented in Table 1.

Upon converting the closing prices to returns, the inferred volatility plot is shown to have a zero mean and both the ARCH and GARCH models have a competitive and tight fit. However, the GARCH model depicts a better fit visually (Fig. 2) and has better goodness of fit parameters (Table 2).

After transforming the data using BCT, it is noted that for values of $\lambda < 0$, the error variance amongst the transformed data is negligible. While these values of $\lambda < 0$ treat the presence of heteroscedasticity, they are not chosen since their residuals are a constant or 0. Also, comparing a plot of the standard deviation of the transformed data against various values of λ proves that for $\lambda = 0$, the standard deviation is the least and within the confidence interval of $\alpha = 0.05$. Therefore, $\lambda = 0$ is chosen for further modelling. However, the normality plots showed that the residuals were not perfectly normally distributed.

Table 2 Goodness of fit for returns

Parameters	Returns	
	ARCH	GARCH
R^2	0.3553	0.7092
RSS	0.0020	0.0023
AIC	−57,447	−64,264
BIC	−57,441	−64,255
t-statistic	>2	>2

After fitting time series models (AR, ARCH and GARCH) to the BC transformed data, estimation of the BCT-AR parameters, shows that $\alpha_1 \approx 1$. Hence, the roots of Eq. (2) are imaginary. This makes the BCT-AR model incompetent to predict volatility. The ARCH test on the BCT data shows the presence of heteroscedasticity. Therefore, ARCH and GARCH models were fit over the data. Amongst the two, BCT-GARCH shows the better fit (Table 1). The plot of the inferred volatility versus the actual returns showed to be similar to Fig. 1.

4 Conclusion

Consequent to examination of the various models fit over the residuals of the closing prices, the results show that the GARCH (1, 1) model was the most suitable fit. It was found that there is autocorrelation between the standardized residuals but no auto-correlation between the square of standardized residuals. This is due to the non-zero mean of the inferred volatility. From Table 1, it is observed that Box-Cox transformations of the data caused a highly significant drop in RSS and R^2 for BCT-GARCH. This implies a tighter and better fit for the model. Even though the heteroskedasticity was not treated completely, the Box-Cox transformations are useful in getting better model fit parameters. This is further confirmed by modelling for the returns which is of prime importance to the investor. Another important observation in modelling for the returns is, the standardized residuals and their squares for the GARCH (1, 1) were not correlated after model fitting which is a desirable result. Although the GARCH was the best fit model for the above-depicted scenarios, upon modelling for various other listings, it was found that the GARCH is not always the best fit model and modelling needs to be dynamic as per the data.

References

1. Enders, W.: Applied Econometric Time Series, 2nd edn. Wiley, Hoboken, New Jersey (2010)
2. Kim, J., Park, Y.J., Ryu, D.: Testing CEV stochastic volatility models using implied volatility index data. Phys. A Stat. Mech. Appl. **499**, 224–232 (2018)
3. Bala, L., Premaratne, G.: Stock market volatility: examining North America, Europe and Asia. SSRN Electron. J. Department of Economics, National University of Singapore (2004)
4. Sarkar, N.: Arch model with Box–Cox transformed dependent variable. Stat. Probab. Lett. Indian Statistical Institute, Economic Research Unit (2000)
5. AL-Najjar, D.: Modelling and estimation of volatility using ARCH/GARCH models in Jordan's stock market. Asian J. Financ Account. **8**(1) (2016). ISSN 1946-052X
6. Nelson Jr., H.L., Granger, C.W.J.: Experience with using the Box-Cox transformation when forecasting economic time series. J. Econ. (1979)
7. Nikita, B., Balasubramanian, P., Yermal, L.: Impact of key macroeconomic variables of India and USA on movement of the Indian Stock return in case of S&P CNX Nifty. In: International Conference on Data Management, Analytics and Innovation (ICDMAI), pp. 330–333, 18 Oct 2017. Article number 8073536

8. Gamit, P., Leua, A., Tandel V.: Modeling of sugar prices volatility in India using autoregressive conditional heteroskedasticity models. Indian J. Econ. Dev. **14**(1a) (2018)
9. Hiremath, N., Naveen Kumar, S., Surya Narayanan, N.S., Jeyanthi, R.: A study of dealing serially correlated data in GED techniques. In: IEEE International Conference on Signal Processing, Informatics, Communication and Energy Systems, SPICES, 31 Oct 2017. Article number 8091338
10. Engle, R.F.: Autoregressive conditional heteroscedasticity with estimates of the variance of UK inflation. Econometrica **50**, 987–1008 (1982)
11. Bollerslev, T.: Generalised autoregressive conditional heteroskedasticity. J. Econom. **31**, 307–327 (1986)
12. Li, X., Zhang, W.: Research on the Efficiency of Chinese Stock Index Future Market Based on the Test of GARCH Model. Management & Engineering, School of Business, University of Jinan (2017)
13. Jiratumpradub, N., Chavanasporn, N.: Forecasting option price by GARCH model. In: 8th International Conference on Information Technology and Electrical Engineering (ICITEE), Yogyakarta, Indonesia (2016)
14. Maddala, G.S.: Introduction to Econometrics, 3rd edn. Wiley (2003)
15. Bharathi, A., Natarajan, A.M.: Cancer classification of bioinformatics data using ANOVA. Int. J. Comput. Theory Eng. **2**(3) (2010)
16. Friedman, J.P.: Dictionary of Business and Economics Terms. 5th edn. Barron's Educational Series (2012)
17. The National Stock Exchange. https://www.nseindia.com/global/content/about_us
18. Thornton, T.D.: Least-squares regression cautions about correlation and regression. Lecture notes. https://goo.gl/djXF1B

Principal Component Analysis Based Data Reconciliation for a Steam Metering Circuit

C. Reddy Varshith, J. Reddy Rishika, S. Ganesh and R. Jeyanthi

Abstract Data reconciliation (DR) is playing an important role in reducing random errors usually occurred in measured data. Principal component analysis (PCA), on the other end, deals with the reduction of dimensions when there is large number of variables involved in a complex process. In this paper, we bring these two techniques together to deal with random errors in measured data of a steam metering circuit. The results prove that PCA-based DR is effective in dealing with random errors than DR alone. The study is also extended to work on a partially measured system where only partial information of the system is known.

Keywords PCA · DR · Steam metering circuit

1 Introduction

Data reconciliation (DR) came into existence several decades back which is used to provide accurate and efficient information regarding the data or records which are missing [1, 2]. It is also used for the reduction of random errors [3]. It finds its application in almost every fields ranging from the data sciences and nuclear power plants to medical and forecasting fields [4]. PCA is a multivariate data analysis method widely used in reducing the dimensions and de-noising the data and model identification. This paper deals with the application of PCA-based DR to reduce the

C. R. Varshith · J. R. Rishika · S. Ganesh · R. Jeyanthi (✉)
Department of Electronics and Communication Engineering, Amrita School of Engineering,
Amrita Vishwa Vidyapeetham, Bengaluru, India
e-mail: jeyanthi.ramasamy@gmail.com

C. R. Varshith
e-mail: c.varshithreddy@gmail.com

J. R. Rishika
e-mail: rishika.reddy03@gmail.com

S. Ganesh
e-mail: gaganarayan96@gmail.com

© Springer Nature Singapore Pte Ltd. 2019   619
J. Wang et al. (eds.), *Soft Computing and Signal Processing* ,
Advances in Intelligent Systems and Computing 898,
https://doi.org/10.1007/978-981-13-3393-4_63

random errors present in the measured data [5]. The study is implemented on steam metering circuit of methanol, and the results show its efficiency [6]. Investigation of the same is extended to partially measured systems where only part of the system information is known.

2 Methodology

In this section, simple DR and PCA-based DR are discussed along with the case of modified DR utilized for a partially measured system [5]. The steady-state measurement model of process variables is represented in Eq. (1).

$$Y = X + E \tag{1}$$

where

Y Measured variable vector ($n \times \mathcal{N}$)
X True value vector ($n \times \mathcal{N}$)
E Error vector ($n \times \mathcal{N}$)
n Number of process variables in the system
N Number of observations

The true value vector X is assumed to be fixed and does not vary due to process noise. The error vector contains random error with the following assumptions [6]:

i. $\in (j) \sim \mathcal{N}(0, \Sigma_\epsilon)$ (normally distributed and error covariance matrix Σ_ϵ is known)
ii. $E[\in (j) \in (k)^T] = 0, \ \forall \ j \neq k$ (errors are not correlated)
iii. $E[x(j) \in (j)^T] = 0$ (errors are not correlated with process variables)

where $\in (j)$ is a random error at jth position.

The data matrix is formed by the assumptions and the scenario taken as per any real-life industry-based situation.

2.1 Simple Data Reconciliation

The reconciled estimates of the measured data are calculated using Eq. (2), which are explained in [5].

$$\hat{x}(j) = y(j) - \sum_{\in} A^T (A \sum_{\in} A^T)^{-1} A y(j) = W y(j) \tag{2}$$

where

$\hat{x}(j)$ jth reconciled estimate of measured variable
A   Constraint matrix of this process network ($m \times n$) calculated through mass balance at each node (m).

2.2 PCA-Based Data Reconciliation

PCA is a multivariate analyst technique which is majorly used to transform possibly correlated process variables into a group of linearly uncorrelated variables which are also known as principal components (PCs) (this procedure mainly works on orthogonal transformation). In PCA, PCs are defined, arranged, and extracted in such a way that the first PC has highest variance and the second PC has the next highest variance among the remaining PCs, and so on. The PCs here are relatively smaller in number compared to the original process variables, and hence, we can see the drastic reduction in complexity of the system. PCA is also widely known to de-noise the given system. To perform PCA, the following steps are systematically performed. The eigenvectors of the covariance matrix are found, which are obtained by performing singular value decomposition (SVD) for the scaled data matrix of the system [7].

Let S_y be the covariance matrix of Y, and let Y be the (n × N)-dimensional data matrix defined by

$$S_y = \frac{1}{N} Y Y^T \tag{3}$$

The singular value decomposition is being given by the SVD function, [8, 2, 9]

$$svd\left(\frac{Y}{\sqrt{N}}\right) = U_1 S_1 V_1^T + U_2 S_2 V_2^T \tag{4}$$

where

U_1 and U_2 Corresponding orthonormal eigenvectors of largest 'p' eigenvalues and the remaining $(n - p)$ smallest of eigenvalues, respectively, of the covariance matrix S_y.

S_1 and S_2 Diagonal matrices where each diagonal element is the square root of eigenvalues of the covariance matrix S_y.

The PCs required to represent the system can be derived from the eigenvectors, and the first p PCs are given by $U_1^T y(j)$. The variance of these PCs is given by the corresponding eigenvectors. Based on a particular requirement of any situation, those PCs derived from eigenvectors are taken to represent the system which together represent the 95–98% range of the total variance [10, 11]. Using the orthonormal property of the eigenvectors, the reconciled variables are estimated using Eq. (5) [10]

$$\hat{X} = \sqrt{N} U_1 S_1 V_1^T \tag{5}$$

The derivation regarding the PCA-based DR, Eq. (5), can be looked up in [5].

2.3   DR for Partially Measured Systems

It is not always economically feasible to measure all the required variables in a system, and also sometimes, it is not physically possible to measure all variables in a process. In such cases, with available measurements all other unknown variables can be estimated through DR. This is very effective when only partial information of process variables of the system is known.

The constraint matrix of the partially measured system is represented by Eq. (6). Here, unknown variables and known variables are partitioned.

$$A_k x_k + A_u x_u = 0 \tag{6}$$

where

A_k and A_u are constraint matrices of known and unknown constraint matrices respectively.

x_k and x_u are true data vectors of known and unknown measurements.

Another problem arises in this technique due to the fact that x_u is unknown. Eliminating the term $A_u x_u$ is a must to perform simple DR. To facilitate this, projection matrix is found. It reduces the expression to eliminate the unknown terms ($A_u x_u$) and contains only the known constraints. Projection matrix is formed in such a way that it satisfies the condition: $PA_u = 0$, which reduces Eq. (6) to (by multiplying (6) with P):

$$PA_k X_k = 0 \tag{7}$$

where

P   projection matrix

QR   factorization of the unmeasured constraint matrix gives the projection matrix [12].

$$P = Q_2^T \tag{8}$$

$$A_u = [Q_1 Q_2][R_1 0] \begin{bmatrix} R_1 \\ 0 \end{bmatrix} \tag{9}$$

where

R_1 is an upper triangular and non-singular matrix

Q   is orthonormal column vector, ($Q_1 - m \times n$, $Q_2 - m \times m - n$)

The projection matrix P in Eq. (7) is obtained from Eq. (8). The reconciled estimates of known variables are arrived through Eq. (10)

$$\hat{x}_k(j) = y_k(j) - \sum_{\in} (PA_k)^T (PA_k \sum_{\in} (PA_k)^T)^{-1} (PA_k) y_k(j) \tag{10}$$

After estimating $\hat{x}_k$, unmeasured variables are estimated using Eq. (6).

3 Experimental Simulation and Results

As stated earlier, the study of the proposed analysis methods using steam metering circuit (Fig. 1) is implemented through MATLAB.

The above steam metering circuit (Fig. 1) of methanol which consists of 11 nodes and 28 process variables (flow rates) is taken for the purpose of studying the proposed DR methods. The notations of the inflows and outflows of the system are assigned through the direction of the flows at each node as observed in Fig. 1. The study is initialized with the objective of materializing a data matrix for the steam metering circuit, with the first step where the base value of the following flow rates (1, 5, 6, 8, 9) is fixed and the remaining are estimated through mass-balance equation.

Let $x(j) \in R^n$ be a n-dimensional vector holding the ideal values of the 'n' process variables. $x(j)$ represents the ideal flow rates expected at all the nodes, i.e., 'j' varying from 1 to 28. Let 'N' be the number of times the experiment is taking place, which is taken as 10,000. Hence, the updated flow rate vector is of dimension ($N \times n$) which is $(10,000 \times 28)$ represented as $X(i, j)$. Data matrix is subjected to errors.

Errors in the steam metering circuit are simulated with the aid of white Gaussian noise. Also, the process dynamics are taken into consideration and are combined with the ideal values of flow rate's matrix. A constraint matrix is a matrix that provides the information regarding the variables (flow rates) which are inputs or outputs or not associated with the nodes of the system [12].

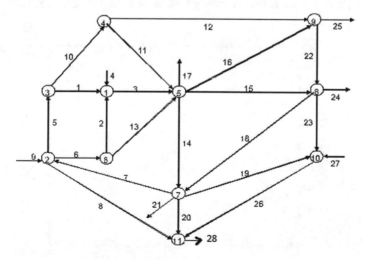

Fig. 1 Steam metering circuit of methanol

With the above data, DR can be performed using Eq. (3). Root-mean-square error (RMSE) is computed for each reconciled and the true value for the entire 10,000 samples simulated. RMSE is a measure of error which is commonly used to see the effect and how much of deviation from ideal values is present. RMSE is calculated as follows:

$$RMSE = \sqrt{\sum_{i=1}^{N} \left(r_{i,j} - \bar{r}\right)^2}$$

(11)

where

$\bar{r}$ Mean of ith variable

$r_{i,j}$ jth observation of ith variable.

Here, the same data matrix taken prior is used as it is the same system and PCA is performed in conjunction with DR to validate and verify its effectiveness. The Eigenvectors are found with Eq. (4), and then, PCs are derived. Depending on the computations performed, the variances are arranged in the descending order form, and later with the help of scree plot (Fig. 2), principal components are considered where one PC is deemed fit to represent the system (PC value is around 93, and rest of the variables are ignored which are around one to zero). The first PC contributes to about 95% of the variance, and hence, it is taken as the required principal component; utilizing it in the formula Eq. (5), PCA-based DR is performed and then the RMSE values are obtained.

For the study of DR for a partially measured system, the steam metering circuit considered in Fig. 1 is used. The unknown variables are considered to be (4, 9, 13, 21, 24, and 25) flow rates, and the remaining process variables are assumed to be known

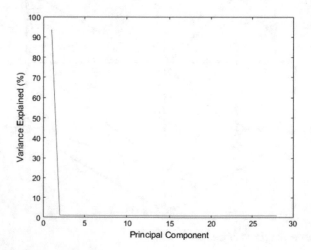

Fig. 2 Scree plot for the principal components

variables. With the data available, the new DR Eq. (8) is applied, and we obtain the reconciled values for the known process variables (flow rates); subsequently, the estimated values for the unknown process variables are obtained.

By the table (Table 1), we can clearly observe that the RMSE values of measurements are improved after performing data reconciliation and principal component analysis for complete and partially measured systems compared to the results before performing the methods. There are considerably 20%, 15%, and 75% improvement on application of DR, partial DR, and PCA, respectively.

Table 1 Performance of DR and PCA-based DR

Node number	Before DR	After DR	After PCA	After PDR
1	1.00348	0.7017	0.1622	0.7929
2	1.01227	0.7712	0.2213	0.9664
3	1.01589	0.8027	0.4125	0.9396
4	1.01146	0.7736	0.2351	0.8210
5	1.10017	0.7249	1.1868	0.9378
6	1.01596	0.7357	0.3929	1.0080
7	1.02442	0.8097	0.1601	0.8823
8	0.99659	0.7818	0.1359	0.7248
9	1.01115	0.8302	0.6952	0.7741
10	1.02023	0.6670	0.5369	0.8306
11	1.02374	0.7562	0.2919	0.9499
12	1.02899	0.7262	0.5313	0.9753
13	1.06845	0.7710	0.7592	0.9673
14	1.01374	0.8354	0.2560	0.9440
15	1.02943	0.8359	0.3470	1.0110
16	1.02314	0.8310	0.2997	0.8755
17	1.09571	0.9233	0.4874	0.8759
18	1.03375	0.8401	0.3199	1.0435
19	1.08885	0.8345	0.3742	0.9141
20	1.01485	0.8158	0.2254	0.7856
21	1.02696	0.8747	0.3451	0.9710
22	1.05951	0.8006	0.5369	0.9220
23	1.01705	0.8136	0.1779	1.4721
24	1.00253	0.8572	0.1050	1.5921
25	1.02963	0.8308	0.3762	1.2418
26	1.01263	0.7792	0.4840	1.2568
27	1.01573	0.8193	0.2519	1.3568
28	1.10945	0.8638	3.8714	1.0231

4　Conclusion

Data reconciliation is playing a key role in suppressing random errors present in measurement data. In this paper, we implemented the PCA-based DR proposed by Narasimhan and Bhatt [5] for the steam metering circuit which has large set of variables and nodes. The study has also been extended for the partially measured system. From the results, it is observed that PCA-based DR shows very good improvement in reducing RMSE. Apart from that, among 28 PCs, only one contributes more in overall variance. Hence, PCA helps to reduce the dimension and computational time and estimates the unknown variables. This highlights the importance of PCA in DR that it is not only reducing the complexity but also increasing the efficiency by drastic reduction of noise. This study is conducted under the assumption that there is no gross error in the measurement, and the errors are normally distributed and not correlated. Further investigation is required to analyze the performance of PCA-based DR for the measurement data which violates the above-said assumptions.

References

1. Romagnoli, J., Sanchez, M.: Data Processing and Reconciliation for Chemical Process Operation. Chemometric Monitoring: Product Quality Assessment, Process Fault Detection, and Applications (2000)
2. Narasimhan, S., Jordache, C.: The Importance of Data Reconciliation and Gross Error Detection, pp. 1–31. Gulf Professional Publishing, Burlington (1999)
3. Yoon, S., MacGregor, J.F.: Fault diagnosis with multivariate statistical models part I: using steady state fault signatures. J. Process Control 11(4), 387–400 (2001)
4. Yang, Y., Ten, R., Jao, L.: A study of gross error detection and data reconciliation in process industries. Comput. Chem. Eng. 19(Supplement 1), 217–222 (1995)
5. Narasimhan, S., Bhatt, N.: Deconstructing principal component analysis using a data reconciliation perspective. Comput. Chem. Eng. 77, 74–84 (2015)
6. Narasimhan, S., Shah, S.L.: Model identification and error covariance matrix estimation from noisy data using PCA. Control Eng. Pract. 16(1), 146–155 (2008)
7. Moore, B.: Principal component analysis in linear systems: controllability, observability, and model reduction. IEEE Trans. Autom. Control 26(1), 0018–9286 (1981)
8. Wentzell, P.D., Andrews, D.T., Hamilton, D.C., Faber, K., Kowalski, B.R.: Maximum likelihood principal component analysis. Chemometrics 11, 339–366 (1997)
9. Andersen, A.H., Gash, D.M., Avison, M.J.: Principal component analysis of the dynamic response measured by fMRI: a generalized linear systems framework. Magn. Reson. Imaging 17(6), 795–815 (1999)
10. Rao, C.R.: The use and interpretation of principal component analysis in applied research. Sankhyā Indian J. Stat. Ser. A (1961–2002) 26(4), 329–358 (1964)
11. Chan, N.N., Mak, T.K.: Estimation in multivariate errors-in-variables models. Linear Algebra Appl. 70, 197–207 (1985)
12. Hiremath, N., Kumar, S.N., Narayanan, N.S.S., Jeyanthi, R.: A study of dealing serially correlated data in GED techniques. In: International Conference on Signal Processing, Informatics, Communication and Energy Systems (SPICES), Kollam, 2017, pp. 1–6

Design of FIR Filter Architecture for Fixed and Reconfigurable Applications Using Highly Efficient Carry Select Adder

Shaurav Shah and Swaminadhan Rajula

Abstract With increased complexity in digital circuits, efficient performance of involved circuitry has become the part and parcel of the digital signal processors (DSPs). In this paper, we have designed an efficient FIR filter for fixed and reconfigurable applications by embedding an area, and delay efficient carry select adder (CSLA), implemented by optimizing the redundancies in the logical operations in conventional and BEC-based CSLA. The proposed CSLA involves less area and delay than BEC-based CSLA and conventional CSLA. Here the carry operation is scheduled before the ultimate sum unlike the traditional method. Having desirably less output area and delay this becomes the best choice for FIR filter of transpose form. For the reconfigurable filter design, it is seen that the delay is reduced by 26.66%, and for MCM-based filter, the delay is reduced by 20.23%. The efficacy of the proposed design is accompanied by 15% reduction in area.

Keywords Finite impulse response (FIR) filters · Carry select adder (CSLA)
Half summation generator (HSG) · Half carry generator (HCG)
Add carry generator (ACG) · Common sub-expression elimination (CSE)

1 Introduction

Nowadays low power, area and delay efficient design is used in almost all the communication devices, biomedical devices, and wireless receivers which makes the processing fast and efficient. A complex Digital Signal Processor (DSP) which is considered as a heart for signal processors requires an efficient adder which performs fast calculations. FIR is used in many signal processing, viz echo cancelation, noise

S. Shah (✉) · S. Rajula
Department of Electronics & Communication Engineering, Amrita School of Engineering,
Amrita Vishwa Vidyapeetham, Bengaluru, India
e-mail: shaurav.shah79@gmail.com

S. Rajula
e-mail: r_swaminadhan@blr.amrita.edu

© Springer Nature Singapore Pte Ltd. 2019
J. Wang et al. (eds.), *Soft Computing and Signal Processing* ,
Advances in Intelligent Systems and Computing 898,
https://doi.org/10.1007/978-981-13-3393-4_64

cancellation applications involving software-defined radio (SDR) [1]. The number of addition and multiplication requirement, however, increases with the increase in the number of taps and linearly with the filter order [2, 3] as a result the complexity increases. In order to avoid any redundant operations involved, many researchers used different techniques for the realization of the FIR filter such as DA (Distributed Arithmetic) and MCM (Multiple Constant Multiplication) techniques. The MCM technique decreases the number of adder needed for the implementation of products by using common sub-expression elimination algorithm when an input sample is multiplied by a group of constants. This technique is more efficient for the implementation of filters of higher order. For some applications, e.g., SDR channelizer requires a finite impulse response filter to be realized in a reconfigurable hardware to enable multi-standard wireless communication. For effective realization of reconfigurable filter (RFIR), multiple designs have been studied during the last decade based on generic multipliers and constant multiplication schemes. However, this paper significantly contributes to the points given beneath:

(1) Analysis of CSLA and deriving the optimized equation for each adder block.
(2) Design of an optimized CSLA and implementation of same in FIR.
(3) Optimization of both reconfigurable and fixed applications-based FIR filter design using the optimized CSLA.

2 Logic Formulation for CSLA

Basically, the CSLA consists of the following units viz;

- ACG, the add carry generator unit, and
- The add carry selection unit.

Among the above-mentioned units, mostly ACG unit consumes significant logic resources in CSLA contributing to the critical path. Therefore, it becomes imperative to study different logic designs for their functional implementation. Hence, study has been done on proposed logic designs for ACG unit of conventional, binary to excess code (BEC) and Common Boolean Logic (CBL) based-CSLAs of [4, 5] by suitable logic formulations. The chief goal here is to find redundancies in logic formulations and possible data dependencies. All redundant and sequence logic formulations are removed accordingly.

2.1 Comparison Between Add Carry Generator Unit of Conventional CSLA and BEC-Based CSLA

The logic formulation of the m-bit RCA consisted of different blocks, shown in Fig. 1, is given:

Fig. 1 Conventional CSLA

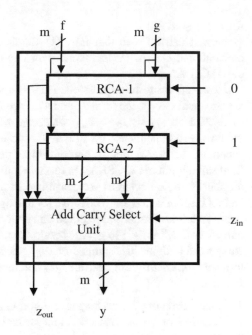

$$y_0^0(j) = f(j) \oplus g(j) z_0^0(j) = f(j) \cdot g(j). \tag{1}$$

$$y_1^0(j) = y_0^0(j) \oplus z_1^0(j-1). \tag{2}$$

$$z_1^0 = z_0^0(j) + y_0^0(j) \cdot z_1^0(j-1) \quad z_{out}^0 = z_1^0(m-1). \tag{3}$$

$$y_0^1(j) = f(j) \oplus g(j) z_0^1(j) = f(j) \cdot g(j). \tag{4}$$

$$y_1^1(j) = y_0^1(j) \oplus z_1^1(j-1). \tag{5}$$

$$z_1^1(j) = z_0^1(j) + y_0^1 \cdot z_1^1(j-1) \quad z_{out}^1 = z_1^1(m-1). \tag{6}$$

where $z_1^0(-1) = 0$, $z_1^1(-1) = 1$, and $0 \le j \le m-1$

As evident from the above logic expressions from (1), it is identical to that of (4). Therefore, the primary goal is to remove the redundant logical operations and to get a design which is optimized for the m-bit RCA-2, wherein the HSG and ACG of RCA-1 are shared for configuring RCA-2. In order to serve the same purpose, in [6, 7], an add-one circuit is used in place of RCA-2 in CSLA, whereas BEC circuit is made to use in [5]. Among existing CSLAs, the BEC-based design provides the best efficacy for area–delay–power. We further describe the logic equations of the add carry generator (ACG) unit:

$$y_1^1(0) = z_1^0(0)' \, z_1^1(0) = y_1^0(0) \tag{7}$$

$$y_1^1(j) = y_1^0(j) \oplus z_1^1(j-1). \tag{8}$$

$$z_1^1(j) = y_1^0(j) \cdot z_1^1(j-1). \tag{9}$$

$$z_{out}^1(j) = z_1^0(m-1) \oplus z_1^1(m-1). \tag{10}$$

for $1 \leq j \leq m - 1$

Now it can be seen that Eqs. (1)–(3) and (7)–(10), z_1^1 depends on y_1^0, unlike conventional CSLA. Hence, there is an increase in data dependency in the BEC-based CSLA.

However, from the expressions (1)–(3) and (4)–(6), we observe that y_1^0 and y_1^1 are identical except the term z_1^0 and z_1^1 since ($y_0^0 = y_0^1 = y_0$). Additionally, z_1^0 and z_1^1 depend on $\{y_0 = z_0 = z_{in}\}$, where $z_0 = z_0^0 = z_0^1$. Interestingly, since z_1^0 and z_1^1 have zero dependence on y_0^0 and y_1^1, the logic formulation of z_1^1 and z_1^0 can be scheduled before y_0^1 and y_1^1, and the select unit can select on from (y_0^1, y_1^1) for the final summation of the CSLA. On further analysis, it is found that a sufficiently great amount of logic resources are consumed in the calculation of $\{y_1^0, y_1^1\}$, and hence, cannot be considered as an effective logical approach to discard one sum word after the calculation. Rather the required carry words can be selected from the expected carry words $\{z^0$ and $z^1\}$ to get the resulting sum. Further, summation of the selected carry word with the half summation (y_0) generates the final summation (y). Now let us have a look at the design advantages obtained by the following the above described method.

1. Calculation of y_1^0 is no longer required in the add carry generator (ACG) unit.
2. Requires only m-bit selection unit instead of $m + 1$ bit.
3. No data dependence.
4. Final carry is obtained before the final sum operation.
5. Output carry delay is small.

We observe that the above features lead to an area-delay efficient design. Given below is the proposed logic formulation of the CSLA:

$$y_0(j) = f(j) \oplus g(j) \quad z_0(j) = f(j) \cdot g(j). \tag{11}$$

$$z_1^0(j) = z_1^0(j - 1) \cdot y_0(j) + z_0(j) \text{ for } (z_1^0(0) = 0). \tag{12}$$

$$z_1^1(j) = z_1^1(j - 1) \cdot y_0(j) + z_0(j) \text{ for } (z_1^1(0) = 1). \tag{13}$$

$$z(j) = z_1^0(j) \text{ if } (z_{in} = 0). \tag{14}$$

$$z(j) = z_1^1(j) \text{ if } (z_{in} = 1). \tag{15}$$

$$z_{out} = z(m - 1). \tag{16}$$

$$y(0) = y_0(0) \oplus z_{in} \quad y(j) = y_0(j) \oplus z(j - 1). \tag{17}$$

3   Proposed Design

The suggested CSLA configuration shown in Fig. 3 depends on the logic expression given in Eqs. ((11)–(17)) as evident from Fig. 2. It comprises of the following units: FSG, HSG, CS, and CG. The carry generator (CG) unit comprises two CGs (CG1 and CG0) for two input carry '1' and '0', respectively. The Half Summation Generator unit gets the m-bit inputs g and f and produces HS word y_0 and HC word z_0. Under CG unit, CG_0 and CG_1 take y_0 and z_0 in order to produce two m-bit full carry word.

Fig. 2 RCA showing the logic operations containing different add and carry generation blocks

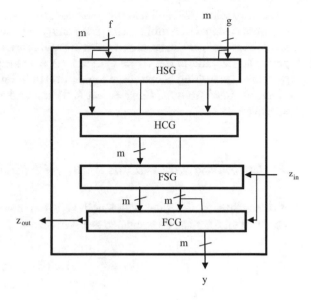

Fig. 3 Proposed design; m-bit-width of the input operand

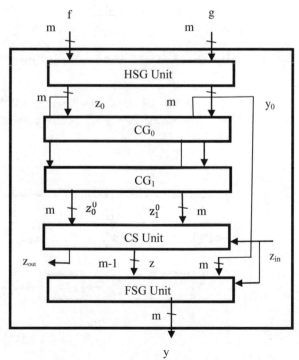

The carry select (CS) unit selects one of the carry words between the two generated carry words using the control signal z_{in}. When $z_{in} = 0$, z_1^0 is selected or z_1^1 is selected. The CS unit is constructed using an m-bit 2:1 multiplexer. However, z_1^0 and z_1^1 follow a particular bit pattern. If $z_1^0(j) = '1'$ then $z_1^1(j) = 1$ irrespective of $y_0(j)$ and $z_0(j)$, $\forall$ $0 \leq j \leq m-1$, resulting in optimized logic in CS unit. Finally, the output carry word is, thus, obtained from the Carry Select (CS) unit and the sum from the FSG unit, respectively.

3.1 Arithmetic and Analysis of the FIR Filter

Let us consider an FIR filter with length L, the input–output relation of the linear time-invariant filter can be computed as given below.

$$y(k) = \sum_{j=0}^{L-1} a(j) \times (k - j). \tag{18}$$

Above Eq. (18) can be expressed as given below:
[in Z domain]

$$y(z) = \left[z^{-1} \left(\ldots \left(z^{-1} \left(z^{-1}a(L-1) + a(L-2) \right) + a(L-3) \right) \cdots + a(1) \right) + a(0) \right] X(z). \tag{19}$$

The data flow graphs, DFG, of transpose form FIR filter (block size P=4) is shown in Fig. 4. These data flow graphs correspond to a block of output {(y(k)}, drawn from Eq. (19).

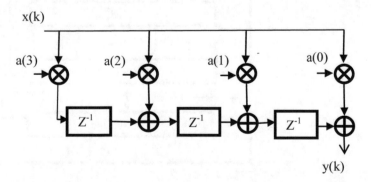

Fig. 4 DFG for y(k), for L=4

3.2   *Data Flow Graph Transformation*

In paper [8], we can see the realization of data flow graphs of non-overlapping blocks. On careful observation, it can be seen that Type-II configuration [8] contains less delay, while area remains the same and by the use of optimized CSLA, better performance in terms of delay and area is achieved.

4   Proposed Design of Transpose Form FIR Filter Using CSLA

In this segment, we describe two types of transpose form block FIR filter using highly efficient carry select adder (CSLA):

1. Reconfigurable FIR filter.
2. MCM-Based implementation.

1. Reconfigurable FIR filter.

For reconfigurable applications, we are presenting a new FIR filter architecture shown in Fig. 5, processing real-time inputs and coefficient monitoring circuit, as there are multiple applications where the coefficients are not constant and therefore needs to be tuned as per the requirements for different tap FIR filter. For example, application like software-defined radio (SDR)-based systems, the use of channelizers to extract the desired channel from received RF frequency bands require reconfigurable FIR filter design to support applications based on multi-standard wireless communication. In this section, design and processing of FIR filter (block transpose form) for reconfigurable applications are discussed having significantly lesser delay than [9].

The reconfigurable FIR filter consists of one CSU (coefficient storage unit), whose function is used to store multiple coefficients. It also consists of RU (register unit), W number of IMUs (Internal Multiplication Units), and one pipelined CSLA unit. The coefficient storage unit is implemented using N-ROM lookup tables (LUT) so that in a clock cycle, weight vectors on any specific channel filter are obtained. The register unit receives a block of P input samples (x_t) during the tth cycle and in turn gives P rows in parallel. These are transferred to W IMUs of the proposed design. Furthermore, the W numbers of Internal Multiplication Units receive W weight vectors coming from the CSU. Each IMU gives the internal product calculations of P rows of C_j^0 with the common vector d_m which is further added in the pipeline carry select adder unit to get a block of P filter outputs. Also, the critical path time period for every cycle is $T = T_W + T_A + T_{CSLA} \log_2 P$, where T_W corresponds to unit delay of multiplier, T_A is unit delay of adder, and T_{CSLA} is unit carry select adder (CSLA) delay.

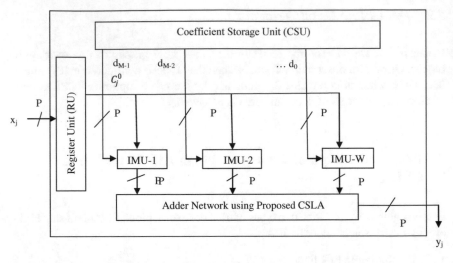

Fig. 5 Proposed FIR filter structure using CSLA jth block

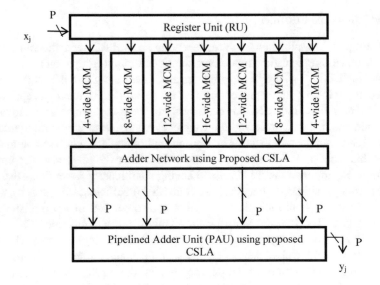

Fig. 6 Proposed MCM-based FIR filter structure using CSLA

2. MCM-Based Implementation.

Here we describe the design of the proposed structure of MCM-based block FIR filter using CSLA. The MCM-based structure is designed for filter tap $L = 8$. The MCM structure shown in Fig. 6 comprises seven MCM blocks with respect to seven input samples which are multiplied with different coefficients as detailed in Table 1.

Table 1 Input samples and coefficient groups in MCM

Inputs	Coefficients
x(4k)	{a(0), a(4), a(8), a(12)}
x(4k − 1)	{a(0), a(4), a(8), a(12)} {a(1), a(5), a(9), a(13)}
x(4k − 2)	{a(0), a(4), a(8), a(12)} {a(1), a(5), a(9), a(13)} {a(2), a(6), a(10), a(14)}
x(4k − 3)	{a(0), a(4), a(8), a(12)} {a(1), a(5), a(9), a(13)} {a(2), a(6), a(10), a(14)} {a(3), a(7), a(11), a(15)}
x(4k − 4)	{a(1), a(5), a(9), a(13)} {a(2), a(6), a(10), a(14)} {a(3), a(7), a(11), a(15)}
x(4k − 5)	{a(2), a(6), a(10), a(14)} {a(3), a(7), a(11), a(15)}
x(4k − 6)	{a(3), a(7), a(11), a(15)}

Since MCM deals with constant coefficients, there is no requirement of CSU. As evident from Fig. 6, each block gives the required product terms as given in Table 1. Interestingly, we can see that incorporating the proposed CSLA in MCM (Multiple Constant Multiplication)-based block FIR filter (based on common sub-expression elimination) leads to area and delay efficient configuration as compared to [10].

There are basically two main steps involved with CSE algorithm:

- To identify the occurrence of multiple patterns in the input coefficient matrix.
- To eliminate all occurrences of the selected pattern.

5   Results

We have implemented the proposed design in Verilog HDL for 8-tap filter, and the synthesis results are obtained using FPGA 3S1000lFT256-4 device which is shown below in the performance comparison Table 2 along with the device utilization report in Table 3. The proposed CSLA-based FIR structure both reconfigurable and MCM involves lesser area and delay as compared to the existing designs.

Table 2 Comparison between existing FIR [8] and proposed FIR

Method	Delay (ns)	Area (Total gate count)
Existing RFIR [8]	77.714	29,405
CSLA-based RFIR	57.061	29,093
Existing MCM FIR [8]	60.747	16,354
CSLA-based MCM FIR	48.454	13,937

Table 3 Device utilization summary of existing FIR [8] and proposed FIR

Method	No. of 4 input LUTs	No. of bonded IOBs	No. of occupied slices
Existing RFIR [8]	4814	43	2690
CSLA-based RFIR	3774	36	2049
Existing MCM FIR [8]	2189	39	1289
CSLA-based MCM FIR	1599	32	978

6   Conclusion

In this paper, we have made an analysis and described how the redundant logic involved with different blocks in CSLA can be optimized to get an area, delay efficient adder (CSLA). We have analyzed the chances of implementation of optimized transpose form block Finite Impulse Response (FIR) filter by leveraging the efficiency of optimized CSLA for both reconfigurable and MCM-based filters. The synthesis result of the proposed design shows lesser area and delay than the existing design. For an 8 tap reconfigurable filter, it is seen that the delay is reduced by 26.58%, and for MCM-based filter, the delay is reduced by 20.23%. Along with the aforementioned improvement in delay, the efficacy of the design is accompanied by 15% reduction (MCM) in area.

References

1. Vinod, A.P., Lai, E.M.: Low power and high-speed implementation of FIR filters for software defined radio receivers. IEEE Trans. Wirel. Commun. **7**(5), 1669–1675 (2006)
2. Sivaranjini, K., Jacob, N.S., Unnikrishnan, G., Heera, K.H.: Low power, high speed FIR filter design. Int. J. Appl. Eng. Res. **10**, 440–444 (2015)
3. Parhi, K.K.: VLSI Digital Signal Processing Systems: Design and Implementation. Wiley, New York, NY, USA (1999)
4. Manju, S., Sornagopal, V.: An efficient SQRT architecture of carry select adder design by common Boolean logic. In: Proceedings of VLSI ICEVENT, 2013, pp. 1–5 (2013)

5. Ramkumar, B., Kittur, H.M.: Low-power and area-efficient carry-select adder. IEEE Trans. Very Large Scale Integr. (VLSI) Syst. **20**(2), 371–375 (2012)
6. Kim, Y., Kim, L.-S.: 64-bit carry-select adder with reduced area. Electron. Lett. **37**(10), 614–615 (2001)
7. He, Y., Chang, C.H., Gu, J.: An area-efficient 64-bit square root carry-select adder for low power application. In: Proceedings of IEEE International Symposium on Circuits Systems, 2005, vol. 4, pp. 4082–4085
8. Mohanty, B.K., Meher, P.K.: A high-performance FIR filter architecture for fixed and reconfigurable applications. IEEE Trans. Very Large Scale Integr. (VLSI) Syst. **24**(2) (2016)
9. Karthick, S., Valarmathy, S., Prabhu E.: Reconfigurable FIR filter with radix-4 array multiplier. J. Theor. Appl. Inf. Technol. **57**, 326–336 (2013)
10. Mahesh, R., Vinod, A.P.: A new common subexpression elimination. Trans. Comput. Aided Des. Integr. Circuits Syst. **27**(2), 217–219 (2008)

Obstacle Detection and Distance Estimation for Autonomous Electric Vehicle Using Stereo Vision and DNN

Sarma Emani, K. P. Soman, V. V. Sajith Variyar and S. Adarsh

Abstract Automation—replacement of humans with technology—is everywhere. It is going to become far more widespread, as industries are continuing to adapt to new technologies and are trying to find novel ways to save time, money, and effort. Automation in automobiles aims at replacing human intervention during the run time of vehicle by perceiving the environment around automobile in real time. This can be achieved in multitude of ways such as using passive sensors like camera and applying vision algorithms on their data or using active sensors like RADAR, LIDAR, time of flight (TOF). Active sensors are costly and not suitable for use in academic and research purposes. Since we have advanced computational platforms and optimized vision algorithms, we can make use of low-cost vision sensors to capture images in real time and map the surroundings of an automobile. In this paper, we tried to implement stereo vision on autonomous electric vehicle for obstacle detection and distance estimation.

Keywords Radar · Lidar · TOF · Stereo vision · Object detection
Distance estimation

S. Emani · S. Adarsh
Department of Electronics and Communication Engineering, Amrita School of Engineering, Coimbatore, Amrita Vishwa Vidyapeetham, Coimbatore, India
e-mail: chandu9175@gmail.com

S. Adarsh
e-mail: s_adarsh@cb.amrita.edu

K. P. Soman · V. V. Sajith Variyar (✉)
Center for Computational Engineering and Networking (CEN), Amrita School of Engineering, Coimbatore, Amrita Vishwa Vidyapeetham, Coimbatore, India
e-mail: vv_sajithvariyar@cb.amrita.edu

K. P. Soman
e-mail: kp_soman@amrita.edu

© Springer Nature Singapore Pte Ltd. 2019   639
J. Wang et al. (eds.), *Soft Computing and Signal Processing* ,
Advances in Intelligent Systems and Computing 898,
https://doi.org/10.1007/978-981-13-3393-4_65

1 Introduction

Autonomous vehicles or Automated Driving System (ADS) as per Society of Automotive Engineers(SAE) J3106 are vehicles that are equipped with technology which aims to reduce vehicle crashes, congestion, energy consumption, and pollution, at the same time increasing transport accessibility. The challenges and future implications of ADS are discussed in [1]. ADS is also a widely researched topic in leading universities around the world, each university experimenting on various models of it. It is deployed as a transport service. Technologies like object detection developed as part ADS can be deployed in various other applications like waste management, disabled people assistance. The major demerits for this technology are reduction in man power may lead to more unemployment. Non-availability of this technology to below average income group is also a potential disadvantage. The long-term effects of ADS were discussed in [2]. ADS can be deployed using RADAR, LIDAR, TOF, Stereo Vision, Monocular vision and various other sensors. The active sensors like RADAR, LIDAR, and TOF are very costly and are not optimal for academic and research domains. Comparison of various imaging techniques using above-said sensors is carried out in [3]. Monocular camera application needs complex vision processing algorithms to process the images and identify the objects. Monocular versus stereo vision processing is done in [4]. Stereo vision can be implemented using low-cost webcams, and images captured can be processed using relatively less complex vision algorithms, thus making it an ideal choice for ADS systems and an accessible system for education and research domains. In this paper, we implemented real-time distance estimation to the obstacle in front of electric vehicle using stereo vision camera. The objective of the experiment is to detect a certain type of obstacles, in real time from a moving automobile. The main steps involved in this process are: creation of disparity map of the environment in real time using stereo vision, obstacle detection in both left and right images during run time, and estimation of the distance to the obstacle calculated from the disparity value. The initial phase of project was done using custom-made webcam stereo rig at the laboratory. Later, the same algorithms were applied to ZED stereo camera placed on electric golf cart moving at low speeds to capture real-time left and right images. Jetson TX1 is used for computations. OpenCV and python were used for algorithm development. Section 2 describes the state-of-the-art methodologies implemented, and Sect. 3 discusses the implementation details. Section 4 describes about distance estimation. Section 4 describes the experimental results.

2 Literature Survey

The state-of-the-art methods for various 3D imaging and ranging techniques were discussed by Aboali and Bandaru [5]. An efficient algorithm for disparity map generation using stereo vision was presented by Zhencheng and Uchimura [6]. Distance estimation using stereo vision is implemented by Patel et al. [7] in their paper. 360° depth perception around a vehicle using two cameras was proposed by Appiah and Bandaru [8]. Obstacle detection based on real-time ROI generation using FPGA and normal CPU was studied by Soon Kwon and Hyuk-Jae Lee in their work [9]. Real-time implementation of depth map generation was carried out indoors by Tianyu [10] in his thesis work. Distance estimation to an object using stereo cameras is discussed in detail by Jernej and Damir [11], which we used in our implementation.

3 Implementation

3.1 Stereo Vision

Depth information is lost when we capture a 3D object and convert it into 2D image. Stereo vision is the extraction of 3D information from 2D images. Retrieval of this depth information is possible if we capture the same object from different perspectives along common baseline. When a 3D object is captured from two cameras separated by a horizontal distance, the object lies at different positions in both images. This relative displacement of the object in both images can be used to calculate the distance of the object from either of the cameras. Relation between real-world point X, image plane UV, and camera center C (x; y; z) is shown in Fig. 1.

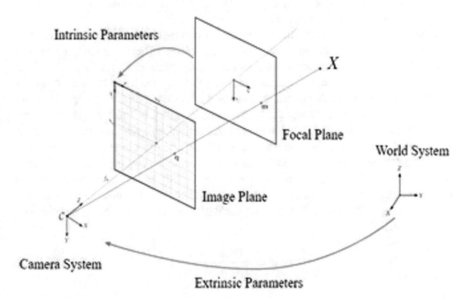

Fig. 1 Relation between real-world object and image coordinates

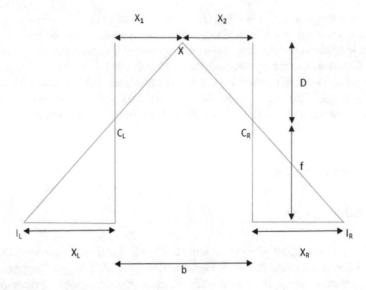

Fig. 2 Picture of an object taken from stereo cameras

The do it yourself (DIY) model includes two webcams mounted in a stereo rig. The images captured are then preprocessed. Then, disparity map of the rectified images is evaluated. In mounting, the cameras are placed such that there is no misalignment along the horizontal base of the cameras. Misalignment causes webcams to capture different fields of view of the scene.

In the preprocessing step, unknown camera parameters such as camera centers and rotational and translational matrices are calculated. These parameters are later utilized for the image rectification and disparity map evaluation. Rectification of the left and right images is done to align both images along the same horizontal line. If left and right images are not rectified, there will be errors in disparity map. Disparity map is calculated by measuring the relative displacement of the same object in left and right images. Same object when measured from stereo camera is shown in Fig. 2.

C_L and C_R represent the camera centers of left and right cameras, respectively. D is the distance of the object located at X from cameras, b is the baseline distance between the cameras, and f is the camera's focal length. I_L and I_R represent the left and right images on image plane of the cameras. Disparity between the two cameras is given by $X_L + X_R$. From Fig. 2 using similarity of triangles, we can get distance to the object as

$$D = \frac{(b * f)}{(X_L + X_R)} \tag{1}$$

Fig. 3 Stereo rig using webcams

Fig. 4 Hardware setup for the experiment

3.2 Experimental Setup

In the initial work, we placed two webcams in the stereo rig as shown in Fig. 3. Calibration was done using chessboard corners to find out the webcam intrinsic and extrinsic parameters. The images are then rectified and processed to calculate the disparity map. To test the algorithm with a standard input from an industrial standard stereo camera, the same algorithm is applied on ZED stereo camera. The final experimental setup is as shown in Fig. 4. We have mounted ZED camera and Jetson TX1 embedded board on an electric golf cart. ZED camera will capture the scene in front of the car, and vision processing algorithms are run on TX1 board. Images captured from ZED in real time are fed to deep neural network to identify objects in the image.

3.3   Object Detection

Object detection for autonomous driving in outdoor environment is discussed in [12, 13]. Similar object detection needs to be carried out at real time from a mobile electric vehicle. Major steps involved in object detection are object classification and localization. Object classification refers to identifying real-world object, and localization refers to drawing a region of interest around the classified object. CNNs are generally used for object detection but are very slow, complex, and used up a lot of space. An efficient alternative is using MobileNet architecture with single-shot detector as proposed in [14, 15], respectively.

3.3.1   MobileNet

MobileNet is an efficient architecture which can be used for resource-constrained devices, to get faster outputs. MobileNets differ from traditional CNNs through the usage of depth wise separable convolution as shown in Fig. 5, which are referred from [14]. From Fig. 5, we can see that a standard convolution layer has $D_F * D_F * M$ input feature map F and produces a $D_F * D_F * N$ output map G, where M is input channel depth, N is output channel depth, and D_F is spatial width and height of input map. The convolution kernel K of size $D_F * D_F * M * N$ is taken for the standard convolution layer. The output feature map G is calculated as

$$G_{k,l,n} = \sum_{i\,j\,m} K_{i,j,m,n} * F_{k+i-1,l+j-1,m} \tag{2}$$

And it has computation cost of $D_K * D_K * M * N * D_F * D_F$, whereas depthwise convolution is made up of two layers: depthwise convolution and pointwise convolution.

Depth wise convolutions apply a single filter per each input channel. Pointwise convolution, a simple 1×1 convolution, is then used to create a linear combination of the outputs of depthwise layer. The equation for output map G for depthwise convolution can be written as

$$G_{k,l,m} = \sum_{i\,j\,m} K_{i,j,m} * F_{k+i-1,l+j-1,m} \tag{3}$$

where K is the depthwise convolution kernel of size $D_K * D_K * M$ where the mth filter in K is applied to the mth channel in F to produce the mth channel of the filtered output feature map G. The computational cost of this is $D_K * D_K * M * D_F * D_F$. However, this is only input filter layer, and we need to add a 1×1 convolution layer (pointwise convolution) to get new features. Thus, total depthwise convolution cost is

Fig. 5 Standard and depthwise convolution

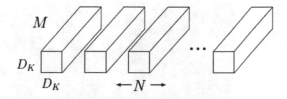

(a) Standard Convolution Filters

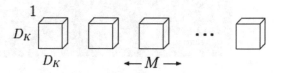

(b) Depthwise Convolutional Filters

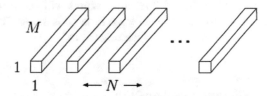

(c) 1×1 Convolutional Filters called Pointwise Convolution in the context of Depthwise Separable Convolution

$$\frac{(D_K * D_K * M * D_F * D_F + M * N * D_F * D_F)}{D_K * D_K * M * N * D_F * D_F} = \frac{1}{N} + \frac{1}{D^2 * K} \quad (4)$$

3.3.2　SSD

Single-shot detector detects objects in an image using single deep neural network (DNN). The model is based on feed-forward convolution neural network (CNN) which produces bounding boxes and scores of object classes present in that box. In this approach, various default bounding boxes of different aspect ratios and scales are taken per feature map for localizing the detected object. During training, SSD needs ground truth boxes of each object in the input image. Then, for every predicted default box, both offset from ground truth and confidence of object category are calculated. Default box and ground truth box are said to be matched only if calculated offset is higher than a threshold value. Thus, multiple boxes with high overlapping

Fig. 6 Object detection and distance prediction in left image

(greater than that of threshold) scores will be selected instead of single box with maximum overlapping which is time-consuming. This is followed by non-maximum suppression to produce final detections.

4 Distance Estimation

Object detection is carried out using the above-said deep learning algorithms in both left and right images. The object detection in left image is shown in Fig. 6. After same object is localized in both the images, a bounding box is drawn around the object. When this object lies along the same vertical line in both images, the horizontal coordinates of the object are noted. Difference in this horizontal position between the left and right image is calculated, which gives the disparity value. We already know baseline distance between the left and right cameras and focal length. Thus, plugging these values in Eq. 1, we get the distance of the object from camera.

5 Experimental Results

The experiment was carried out by running the above-said algorithms on custom-made webcam stereo rig as well as using ZED camera on Jetson TX1. The experimental results for various distances were given in Table 1.

Table 1 Experimental results

Serial No	Actual distance (cm)	Predicted distance using webcam rig (cm)	Predicted distance using ZED camera (cm)
1	100	90	92
2	140	130	135
3	190	190	193
4	240	232	235
5	290	292	289
6	390	395	405

6 Conclusion

In this paper, we have done real-time distance estimation using stereo vision. Various disparity maps are created for different block matching parameters and arrived at optimal values. The object distance estimation is done using both DIY stereo camera rig and industry standard ZED stereo camera. The results obtained from both the setups are almost similar. Thus, for college research purposes, stereo webcam setup can be used which is cost-effective and gives reasonable results for predicting the distance to obstacles.

References

1. Bagolee, S.A., Tvana, M., Asadi, M., Oliver, T.: Autonomous vehicles: challenges, opportunities, and future implications for transportation policies. J.M.T. **24**, 284–303 (2016)
2. Gruel, W., Stanford, J.M.: Assessing the long-term effects of autonomous vehicles: a speculative approach. Transp. Res. Procedia **13**, 18–29 (2016)
3. Aboali, M., Manap, N.A., Darsono, A.M., Yusof, Z.M.: Review on three-dimensional (3-D) acquisition and range imaging techniques. Int. J. Appl. Eng. Res. **12**, 2409–2421 (2017)
4. Aish, D.: Stereo vision facing the challenges and seeing the opportunities for ADAS applications (2016). http://www.ti.com/lit/wp/spry300/spry300.pdf
5. Magad, A., Nurulfajar, M., Majad, D., Zulkalnain, M.Y.: Review on 3-D imaging and range imaging techniques. Int. J. Appl. Eng. Res. **12**, 2409–2421 (2017)
6. Zhencheng, H., Uchimura, K.: U-V-disparity: an efficient algorithm for stereovision based scene analysis. In: IEEE Proceedings, Intelligent Vehicles Symposium, pp. 48–54 (2005)
7. Patel, D.K., Pankaj A.B., Nirav, R.S.: Distance measurement system using binocular stereo vision approach. IJERT **2**, 2409–2421 (2017)
8. Appiah, N., Bandaru, N.: Obstacle detection using stereo vision for self-driving cars (2015)
9. Kwon, S., Lee, H.: Dense stereo-based real-time ROI generation for on-road obstacle detection. In: International SoC Design Conference, pp. 179–180, Jeju (2016)
10. Tianyu, G.: Real time obstacle depth perception using stereo vision. Masters thesis, University of Florida, Florida, USA (2014)
11. Jernej, M., Damir, V.: Distance measuring based on stereoscopic pictures. In: 9th International PhD Workshop on Systems and Control: young Generation Viewpoint, Slovenia (2008)

12. Deepika, N., Sajith, V.V.: Obstacle classification and detection for vision based navigation for autonomous driving. In: Proceedings, ICACCI. pp. 2092–2097. IEEE Press, Udupi (2017)
13. Shimil, J., Sajith, V.V., Soman, K.P.: Effective utilization and analysis of ROS on embedded platform for implementing autonomous car vision and navigation modules. In: Proceedings, ICACCI. pp. 877–882. IEEE Press, Udupi (2017)
14. Howard, G.A., Menglong, Z., Bo, C., Dmitry, K., Weijun, W., Tobias, W., Marko, A., Hartwig, A.: MobileNets: efficient convolutional neural networks for mobile vision applications (2017). arXiv:1704.04861
15. Wei, L., Dragomir, A., Dumitru, E., Christian, S., Scott R., Fu, C.-Y., Berg, A.C.: SSD: Single shot multibox detector. arXiv:1512.02325 (2016)

Basic Framework of Vocoders for Speech Processing

R. Chinna Rao, D. Elizabath Rani and S. Srinivasa Rao

Abstract The main objective of this paper is to develop a basic framework of a vocoder for speech processing. The basic framework of a vocoder consists of voice preprocessing, encoder, decoder, and post-processing blocks and performing multithreading among these blocks. A GUI can be prepared which will take the input from the user and set the environment as required. The voice subsystem consists of transmission path (TX chain) and receiver path (RX chain) having various signal processing blocks. These blocks are executed in sequential order to process the voice signal. This basic framework allows user to completely analyze the effect of preprocessing blocks, post-processing blocks, and vocoders on the speech signal at different points. A thread is a process which takes samples as input, calls an appropriate block for processing, and gives processed samples as output.

Keywords Framework · Multithreading · Encoder · Decoder

1 Introduction

A vocoder is a system of coders and encoders which are used to reduce bandwidth over limited bandwidth requirements and limited capacity channels in real-time requirements. By analyzing the basic framework of vocoders, we can optimize the bandwidth requirements [1–3].

R. C. Rao (✉) · S. Srinivasa Rao
MallaReddy College of Engineering and Technology, Hyderabad, Telangana, India
e-mail: rauudu.chinnarao@gmail.com

S. Srinivasa Rao
e-mail: ssrao.atri@gmail.com

D. Elizabath Rani
Gandhi Institute of Technology and Management, Visakhapatnam, Andhra Pradesh, India
e-mail: elizabeth.varghese@gmail.com

© Springer Nature Singapore Pte Ltd. 2019
J. Wang et al. (eds.), *Soft Computing and Signal Processing* ,
Advances in Intelligent Systems and Computing 898,
https://doi.org/10.1007/978-981-13-3393-4_66

1.1 Basic Framework of Vocoders

Figure 1 represents the basic framework of vocoders which consists of the following basic blocks: (1) voice input/output management, (2) voice preprocessing, (3) encoding, (4) decoding, and (5) voice post-processing blocks. The basic framework of vocoders will be operated in two modes: (a) offline mode and (b) loopback mode. In offline mode, voice input/output management subsystem takes PCM samples and forward to voice preprocessing subsystem as shown in the diagram. The voice preprocessing subsystem processes PCM samples through different voice preprocessing algorithms (e.g., echo cancelation) and sends processed samples to encoder. Now the encoder encodes PCM samples and produces encoded packet [4, 5]. In loopback mode, encoded packet is given to decoder input. Decoder decodes the received PCM packet and produces PCM samples. These PCM samples are processed by different voice post-processing modules (e.g., automatic gain control, filter) and are forwarded to voice output management [6, 7]. The voice output management subsystem stores the output PCM samples in the form of raw file in offline mode or plays on with the connected speaker in real-time mode.

The basic framework of vocoders consists of six main subsystems as shown in Fig. 1, and the significance of each subsystem is mentioned below:

1. Voice Input/Output Management Subsystem: This subsystem is mainly responsible for collecting voice samples to the next preceding subsystem voice preprocessing block on the transmission path and accepting outputs from voice post-processing subsystem in the receiver path for every frame. In the transmission path, it collects input voice samples either from offline file or from real-time microphone device as per the condition prevailing and sends samples to voice preprocessing subsystem for every frame [8]. In the receiver path, it collects output samples from post-processing output subsystem buffer and stores either in offline file or plays it on speaker in real time as per the prevailing conditions [9].
2. Voice Preprocessing Subsystem: This subsystem is responsible for collecting the samples from the input buffer of voice input management subsystem and then

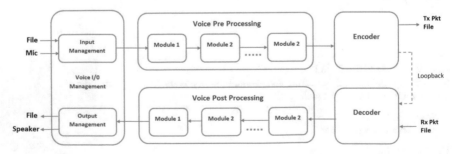

Fig. 1 Basic framework of vocoders

processing the collected samples by using different preprocessing modules as shown in Fig. 1.

3. Voice Encoding Subsystem: This subsystem is responsible for collecting samples from the voice preprocessing subsystem output buffer and then encoding these samples using different encoders (e.g., AMR-NB, FR, HR, AMR-WB) [10, 11]. After encoding, it writes the encoded packet in encoder output buffer as shown in Fig. 1.

4. Voice Decoding Subsystem: This subsystem is responsible for collecting encoded packet from the voice decoder subsystem input buffer and then decodes the packet using different decoders (e.g., AMR-NB, FR, HR, AMR-WB) and then writes decoded samples into decoder subsystem output buffer.

5. Voice Post-processing Subsystem: This subsystem is responsible for collecting samples from the input buffer and then processing the collected samples using different post-processing subsystem modules (e.g., filtering, volume control, automatic gain control). After processing, it writes the output samples in the post-processing subsystem output buffer as shown in Fig. 1.

6. Voice Timing Management Subsystem: This subsystem is the main and important module which will take care of all the timings required for all the subsystems such as voice input/output management subsystem, voice preprocessing subsystem, encoder subsystem, decoder subsystem, and voice post-processing subsystem.

Basic framework needs to be configured, and this will be done from the configuration file. The configuration file consists of multiple options for framework operations like

Sampling rate required for the framework,
Types of vocoders required for encoding and decoding,
Type of input file,
Type of output file.

In general, basic framework of vocoders is only 20 ms of frame size. We can design the basic framework of vocoders to operate both in narrowband (8000 kHz) and wideband (16000 kHz) sampling rates.

2 Algorithm Integration for Basic Framework of Vocoders

In general, there are two methods for algorithm integration. They are (1) symbolic algorithm integration and (2) numeric algorithm integration. Symbolic algorithm integration basically follows the same rules how a human would perform to integrate a function and then manipulate the function algebraically. On the other hand, numeric algorithm integration approximates algorithm integration by breaking the defined function into small modules, approximating those modules and then integrating them together [12].

By following numeric algorithm integration for basic framework of vocoders, the following sections highlight the significance of the algorithms for basic framework and the process of integrating them.

2.1 Adaptive Multirate Narrowband Vocoder

The adaptive multirate narrowband (AMR or AMR-NB or GSM-AMR) vocoder is an audio compression method used for optimizing speech. AMR-NB speech codec consists of a multirate narrowband speech codec which encodes narrowband (200–3400 Hz) signals at different bit rates ranging from 4.75 to 12.2 kbps with toll quality speech available from 7.4 kbps.

AMR was accepted as the standard speech codec by 3GPP in October 1999 and is now widely used in GSM and UMTS [13, 14]. It follows link adaptation method to select one of the eight different bit rates available based on the link conditions.

AMR also has a file format for storing human voice using the AMR codec. Today's modern mobile telephone handsets have the facility of storing small audio recordings in the AMR format. This can be done using free and proprietary programs. The common filename extension used is .amr. There exists another storage format for AMR which is suitable for applications which have more advanced requirements with respect to storage formats, like random access or synchronization with video. This format is specified by 3GPP standards referred as 3GP container format which is based on ISO base media file format.

Features of AMR-NB Vocoder

- The sampling frequency is 8 kHz/13-bit (160 samples for 20 ms frames), for a narrowband filter, filtered to 200–3400 Hz.
- The AMR-NB vocoder utilizes eight source codecs with bit rates of 12.2, 10.2, 7.95, 7.40, 6.70, 5.90, 5.15, and 4.75 kbps.
- AMR-NB vocoder generates frame lengths comprising of 95, 103, 118, 134, 148, 159, 204, or 244 samples for AMR-FR bit rates ranging from 4.75 kbps, 5.15 kbps, 5.90 kbps, 6.70 kbps, 7.40 kbps, 7.95 kbps, 10.2 kbps, or 12.2 kbps, respectively. AMR-HR frame lengths are different.
- AMR-NB uses a special technique called discontinuous transmission (DTX) with voice activity detection (VAD) and comfort noise generation (CNG) to reduce bandwidth usage during silence periods which in turn optimizes the bandwidth utilization.
- AMR-NB algorithmic delay is 20 ms per frame. For bit rates of 12.2 kbps, there is no 'algorithm' look-ahead delay as such. For other available rates, look-ahead delay comes to 5 ms.

2.1.1 Supported Application Program Interfaces for AMR-NB Vocoder

Application programming interface (API) consists of a set of subroutine definitions, protocols, and tools for building application software required for basic framework of vocoders. It establishes the communication between various software components [15]. In general, the documentation for the API is provided to facilitate usage and reimplementation of different subsystems that are used for building the basic framework of vocoders. In this process, the functionality of different APIs for encoder and decoders are given below:

Encoder

- **amrnb_enc_get_mem_size ()**
  This API returns the required size for the amrnb encoder library.
- **amrnb_enc_mem_alloc()**
  This API initializes amrnb lib pointers with allocated memory.
- **amrnb_enc_init()**
  This API initializes all parameters of amrnb encoder library.
- **amrnb_enc_set_rate ()**
  This API sets the amrnb encoder rate value and updates the amrnb encoder library. Supported rates: 12.2, 10.2, 7.95, 7.40, 6.70, 5.90, 5.15, and 4.75 kbps.
- **amrnb_enc_process()**
  This API processes 160 PCM samples and generates encoded packet for configured rate.
- **amrnb_enc_end()**
  This API initializes amrnb encoder memory pointers to NULL.

Decoder

- **amrnb_dec_get_mem_size ()**
  This API returns the required size for the amrnb decoder library.
- **amrnb_dec_mem_alloc()**
  This API initializes amrnb lib pointers with allocated memory.
- **amrnb_dec_init()**
  This API initializes all parameters of amrnb decoder library.
- **amrnb_dec_process()**
  This API decodes encoded packet and generates 160 PCM samples.
- **amrnb_dec_end()**
  This API initializes all pointers to NULL.

2.2 IIR Filter

In basic framework of vocoders, infinite impulse response (IIR) filters are used to modify the frequency response of voice and audio signals. The basic purpose of IIR filters in the transmission path is to compensate for the frequency response of electro-acoustic devices such as microphones, headsets, loudspeakers, receivers [16]. In general, IIR filters are mandatory and are applied on PCM data (voice signals) in the Transmission path to equalize the frequency responses and satisfy the requirements of 3GPP/3GPP2 standards.

IIR filter is integrated in basic framework of vocoders in voice preprocessing subsystem. Generally, IIR filter supports 8000 and 16,000 kHz sampling rates.

2.2.1 Supported APIs for IIR Algorithm

- **iir_set_config()**
  This API configures parameters required to calculate memory requirement, e.g., sampling rate information.
- **iir_get_mem_size()**
  This API returns the required size based on the configured sampling rate information.
- **iir_mem_alloc()**
  This API initializes IIR pointers with allocated memory.
- **iir_init()**
  This API initializes all parameters of IIR algorithm library.
- **iir_process()**
  This API processes input PCM samples filters signal and output PCM samples.
- **iir_end()**
  This API initializes all pointers to NULL.

2.3 FIR Filter

In basic framework of vocoders, finite impulse response (FIR) filters are used to modify the frequency response of voice and audio signals. In general, FIR filters are used in the receiver path to compensate the frequency response of electro-acoustic devices such as microphones, headsets, loudspeakers, receivers [17]. In basic framework of vocoders, FIR filters are used on PCM data (voice signals) to equalize the frequency responses and satisfy the requirements of 3GPP/3GPP2 standards.

FIR filter is integrated in basic framework of vocoders in voice post-processing subsystem. In general, FIR filter supports 8000 and 16,000 kHz sampling rates.

2.3.1 Supported APIs for FIR Filter

- **fir_set_config()**
  This API configures parameters required to calculate memory requirement, e.g., sampling rate information.

- **fir_get_mem_size()**
  This API returns the required size based on the configured sampling rate information.
- **fir_mem_alloc()**
  This API initializes IIR pointers with allocated memory.
- **fir_init()**
  This API initializes all parameters of IIR algorithm library.
- **fir_process()**
  This API processes input PCM samples filters signal and output PCM samples.
- **fir_end()**
  This API initializes all pointers to NULL.

3 GUI for Basic Framework of Vocoders

The graphical user interface (GUI) provided for basic framework of vocoders is a type of user interface that allows users to interact through graphical icons and visual indicators, instead of text-based user interfaces, typed command labels, or text navigation [18].

The actions in a GUI are usually performed through direct manipulation of the graphical elements. GUIs are mainly used in many handheld mobile devices such as MP3 players, portable media players, gaming devices, smartphones, and smaller household, office, and industrial controls [19].

3.1 Features in GUI for Basic Framework of Vocoders

- Input File
  This GUI for input file consists of PCM samples which need to be tested.
- Output File
  This GUI for output file represents PCM format after processing complete chain of voice system.
- Encoder Output File
  This GUI consists of a file used to dump encoded packet after encoding.
- Decoder Input File
  This GUI represents a file used to read encoded packet for decoding.
- Sampling Rate
  This GUI presents supported sampling rates of 8000 and 16000 kHz.
- Vocoder Type
  This GUI shows the supported vocoder type: AMR-NB.
- Vocoder Rate
  This GUI presents supported vocoder rates: 12.2, 10.2, 7.95, 7.40, 6.70, 5.90, 5.15, and 4.75 kbps.

- Duration of the Test
  This GUI shows the time to run test in milliseconds.
- Start Test
  This GUI represents the start button.

4 Conclusion

We successfully developed the basic framework of vocoders used for speech processing. The paper presented the three algorithms required for integration into basic
framework of vocoders, namely 1. AMR-NB vocoder subsystem, 2. IIR filter in voice
preprocess subsystem, and 3. FIR filter in voice post-processing subsystem. The
paper also presented different APIs which are frequently used for basic framework
of vocoders. The paper also presents the essence of GUI which will be user-friendly
and easy to operate the basic framework of vocoders.

References

1. Mishra, S., Singh, S.: A survey paper on different data compression techniques. Indian J. Res. Pap. **6**(5) (2016)
2. Holmes, J.N.: The JSRU channel vocoder. IEEE Proc. **127**, 53–60 (1980)
3. Lukasaik, J., McElory, C., Chang, E.: Compression transparent low-level description of audio signals. In: IEEE International Conference Multimedia and Expo, 6 July 2005
4. Akyol, E., Rose, K.: A necessary and sufficient condition for transform optimality in source coding. In: IEEE International Symposium on Information Theory Proceedings (2011)
5. Nanjundaswamy, T., Rose, K.: Cascaded long term prediction for coding polyphonic audio signals. In: IEEE w/s on Applications of Signal Processing to Audio and Acoustics, New Paltz, NY, Oct 2011
6. Raut, R., Kullat, K.: SDR design with advanced algorithms for cognitive radio. IJACS **1**(4), 134–141 (2011)
7. Srinivasan, P., Jameason, L.H.: High quality audio compression using an adaptive wavelet packet decomposition and psychoacoustic modelling. IEEE Trans. Signal Process. **XX**(Y) (1999)
8. Makhoul, J.: Linear prediction: a tutorial review. IEEE Proc. **63**(4), 561–580
9. Giacobello, D., Christensen, M.G., Murthi, M.N., Jenson, S.H., Moonen, M.: Speech coding based on sparse linear prediction. IEEE Trans. Acoust. Speech Signal Process. **15**(5) (2010)
10. Joseph, S.M., Shah, F., Babu Anto P.: comparing speech compression using waveform coding and parametric coding. Int. J. Electron. Eng. **3**(1), 35–38 (2011)
11. Giacobello, D., Christensen, M.G., Murthi, M.N., Jenson, S.H., Moonen, M.: Retrieving sparse patterns using a compressed sensing framework: applications to speech coding based on sparse linear prediction. IEEE Signal Proc. Lett. **17**(1)
12. Wichman, S.: Comparison of Speech Coding Algorithms ADPCM vs CELP. Department of Electrical Engineering. Texas University (1999)
13. Gersho, A.: Advances in speech and audio compression. IEEE Proc. **82**(6) (1994)
14. Kemper, G., Iano, Y.: An audio compression method based on wavelets subband coding. Lat. Am. Trans. IEEE (Revista IEEE Am.) **09**(05), 610–621 (2011)

15. Strahl, Stefan, Hansen, Heiko, Mertins, Alfred: A dynamic fine-grain scalable compression scheme with application to progressive audio coding. IEEE Trans. Audio Speech Lang. Process. **19**(1), 14–23 (2011)
16. Shirani, S.: Multimedia Communications Sub-band Coding. McMaster University
17. Zang, Y., Zeytinoglu, M.: Improved lifting scheme for block subband coding. In: The proceedings of IEEE SP. International Conference on Electronic Devices, Systems and Applications (ICEDSA1999), pp 487–490 (1999)
18. Zhang, Y., Zeytinoglu, M.: block subband coding and time-varying orthogonal filter banks. In: The Proceedings of 1999 IEEE PacRim'99 Conference on Communications, Computers and Signal Processing
19. Vetterli, M., Herley, C.: Wavelets and filter banks: theory and design. IEEE Trans. Signal Proc. **40**(9), 2207–2232 (2009)

Vein Detection System Using Quad-Core ARM Processor and Near-Infrared Light

Aashay Mhaske, Siddhant Doshi, Pranjal Chopade and Vibha Vyas

Abstract Venipuncture is the process of puncturing the vein to withdraw blood or to carry out an intravenous injection. It requires high level of expertise to achieve high rates of accuracy. In traditional ways, success depends heavily on the experience of the practitioner. Consequently, venipuncture has been reported as one of the leading causes of injury to patients. The estimation of failure ranges from 20 to 33% overall. Specifically in populations, which include children, obese and old people it ranges from 47 to 70%. To improve first stick accuracy, we propose a system which will help identify the suitable subcutaneous veins. With the help of near-infrared radiation (NIR), images are been captured. The captured images are processed to identify the veins. These identified veins can be further used by practitioner to carry out further analysis. These processed images of veins can be further projected directly on the limb.

Keywords Venipuncture · Intravenous · NIR

1 Introduction

Venipuncture or establishing access to venous bloodstream is one of the most commonly performed procedures and is carried out for one of the following reasons:

A. Mhaske · S. Doshi · P. Chopade (✉) · V. Vyas
College of Engineering Pune (COEP), Pune, India
e-mail: chopadep14.extc@coep.ac.in

A. Mhaske
e-mail: mhaskeyam14.extc@coep.ac.in

S. Doshi
e-mail: doshisr14.extc@coep.ac.in

V. Vyas
e-mail: vsv.extc@coep.ac.in

© Springer Nature Singapore Pte Ltd. 2019
J. Wang et al. (eds.), *Soft Computing and Signal Processing*,
Advances in Intelligent Systems and Computing 898,
https://doi.org/10.1007/978-981-13-3393-4_67

(1) To obtain blood for diagnostic purposes. (2) To monitor levels of blood components. (3) To administer therapeutic treatments including medications, nutrition or chemotherapy. (4) To collect blood for transfusions.

Although the procedure is extremely common, it is hardly straightforward. WHO has laid down some guidelines for the procedure which span over 100 pages. In most cases, the procedure is carried out using touch and vision and is dependent upon the experience and skill of the practitioner and on the patient's physiology. Rauch et al. [1] and Jacobson and Winslow [2] have estimated the failure rate at 20–33%. Failure is when the access to bloodstream is not established in the first try at venipuncture. It is also known as 'First-Stick Accuracy', which will then be 67–80%. In children, however, the failure rate is 47–70% or a first-stick accuracy of merely 30–50%, according to Black et al. [3]. The adverse effects of such failures are described by Stevenson M, Lloyd-Jones M, Morgan MY et al. in their work on non-invasive diagnostic assessment tools for the detection of liver fibrosis (2012) [4].

2 Concept

Vincent Pacquit et al. (2006) have conclusively shown that NIR light is the most effective way of vein detection [5]. Figure 1 shows the absorption of various components of blood for different wavelengths. The adipose tissue has minimum absorption in the window of 740–960 nm wavelength while oxyhaemoglobin has near to maximum. Deoxyhaemoglobin absorption is nearly the same as that of oxyhaemoglobin. Far-field infrared of 6–14 μm can pick up heat signature of veins. But there can be fluctuations that occur based on ambient temperature or if the target person is sick. A lot of random noise is brought into play. Therefore, by illuminating the target body part using NIR LEDs the veins appear darker than the surrounding tissue and this can be captured using an infrared camera.

3 Structure

The hardware system plays a very important role to obtain the subcutaneous vein images. The hardware system should meet the following considerations named below:

(1) The camera used should have near-infrared filter with sufficient resolution.
(2) The illumination by infrared LEDs should be capable of providing sufficient contrast to discriminate between surrounding tissue and veins. The proposed set-up is shown in Fig. 2.

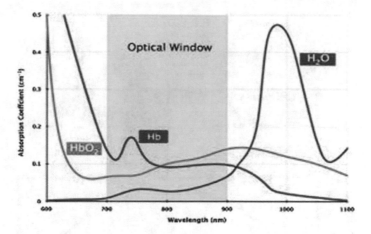

Fig. 1 Optical spectrum of NIR absorption

Fig. 2 Structure of entire system

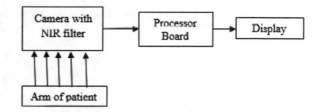

4 Radiation Optimization

To provide illumination of the hand surface, it is necessary to use an appropriate source of near-infrared light. This will be provided by LEDs arranged in a suitable array. One paper [6] tested several array geometries including double line, rectangular and concentric circles. All configurations had the camera centred in the centre of the array. It was found that the LED array giving most uniform illumination was the circular array. Figure 3 shows the irradiance pattern of the circular array, and Fig. 4 shows its variation with distance. Several papers aim to calculate the distance away from a geometric LED array where there is optimal uniformity in the developed light. Moreno et al. investigate, in detail, the illumination of several array patterns. For the concentric circle with number of LEDs, $N > 3$, the irradiance $E(x; y; z)$ at a point $(x; y; z)$ away from the origin which is at the centre of the LED array (in our case at the camera lens) is given by [7]:

$$E(x, y, z) = z^m A_{LED} L_{LED}$$

$$\sum_{n=1}^{N} \left\{ \left[x - \rho \cos\left(\frac{2\pi n}{N}\right) \right]^2 + \left[y - \rho \sin\left(\frac{2\pi n}{N}\right) \right]^2 + z^2 \right\}^{\frac{-(m+2)}{2}} \quad (1)$$

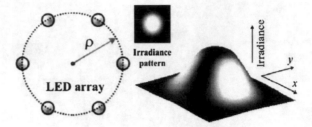

Fig. 3 LEDs pattern and irradiance pattern

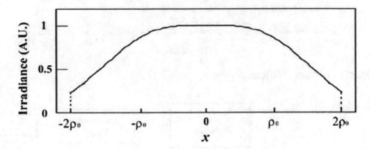

Fig. 4 Irradiance versus distance

where ρ_0 is the radius of LED array. The variable 'm' was defined as a function of viewing angle of the LED ($\theta_{1/2}$). By calculating second partial derivative, a relation was also found between the distance of optimal uniformity 'z' and radius 'ρ_o' of the array [7]:

$$m = \frac{-ln2}{\ln(cos\theta_{1/2})} \rho_0 = \sqrt{\frac{2}{m+2}} z \tag{2}$$

Near-IR LED sources consist of circular arrangement of N = 16 LEDs with 8 LEDs of 850 nm and other 8 LEDs of 940 nm placed alternately in a circle around camera in centre. The average size of hand for male is 84 and 74 mm for female. The value of $\theta_{1/2}$ is 10°. The calculated from above Eq. (2) value of is 'm' = 45.278. Taking value of ρ_0 = 4.86 cm to cover complete average hand size, i.e. of 8 cm × 8 cm and area the value of 'z' = 17.01 for uniform illumination. The battery of 12 V 1 amp is used to power LED array. The current flowing through the LEDs is 36 amp using a resistor of 68 Ω.

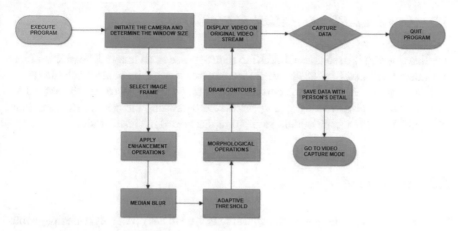

Fig. 5 Block diagram of software processing

5 Software

After capturing the image, it undergoes various operations like histogram equalization, low-pass filtering, thresholding, morphological operations. Figure 5 outlines the steps that involved in the software processing of the image.

The details of pre-processing of images are.

5.1 Contrast Limited Adaptive Histogram Equalization—CLAHE

Contrast limited adaptive histogram equalization (CLAHE) is used to improve contrast and enhance the vein pattern. Adaptive histogram equalization (AHE), a more generalized method than CLAHE, differs from ordinary histogram equalization as used by Chakravorty et al. [8]. The adaptive method divides the captured image into smaller blocks and computes regional histograms. These histograms are further equalized and used to redistribute the lightness values of the image. This aids in enhancing the edges of the image and improving regional contrast. But AHE tends to over-amplify noise. CLAHE takes care of this by limiting contrast enhancement [9].

5.2   Low-Pass Filtering Median Blur

Median filtering is used after CLAHE to reduce noise in an image. The median filter considers each pixel in the image and compares it with its nearby neighbours to decide if the chosen pixel represents its surroundings and replaces it with the median of those values [10]. The median is calculated by sorting the pixels into numerical order and then taking the middle value to replace pixel in consideration.

5.3   Segmentation

The segmentation technique of thresholding is used to convert a greyscale image into binary image. The number of pixels over the pixel value graph is plotted, and the pixel values above the threshold are settled to a foreground value, and the rest are settled to a background value [11].

(a) **Otsu's Method**

Otsu's method (Otsu [1979]) is an optimum alternative to AHE. In this method, a normalized histogram of the captured image is computed. Further, the cumulative sum, mean, global intensity mean and between-class variance are calculated using the components of the histogram. The component, for which the between-class variance is maximum, is equated to the Otsu threshold. If this value is not unique, the threshold is obtained by averaging out the values of various maxima detected. The computational time for this method is high which poses as one of the major drawbacks of this method.

(b) **Adaptive Thresholding**

In global thresholding, the pixel data of the entire image is considered while performing histogram equalization. The beauty of this method is that the threshold value changes dynamically. This more complex version of thresholding can account for the varying lighting conditions in the image which occur due to shadows or a strong illumination gradient.

In Fig. 6, the middle image is the sample image and the one on the left side is processed image by Otsu thresholding and the right side image by adaptive thresholding.

Various window sizes were tried, and optimum window size was decided depending on the time elapsed and the authenticity and quality of output image. Table 1 for the window size and time elapsed is given.

The above two methods were studied in detail, and it was decided that adaptive thresholding with window size of 16 is best for our usage. Figure 7 shows the graph for comparison between two images shown in Fig. 6. It is observed that if we increase the window size beyond certain window size, then we lose the quality of the output vein image.

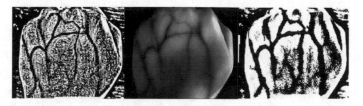

Fig. 6 Comparison between Otsu's method and adaptive thresholding

Table 1 Comparison between Otsu's method and adaptive thresholding

Window size	Otsu's method (ms)	Adaptive thresholding (ms)
9 × 9	1.71007	0.385885
16 × 16	0.840233	0.39515
32 × 32	0.539337	0.395786
64 × 64	0.457948	0.402231
128 × 128	0.453671	0.435384
256 × 256	0.44818	0.537373

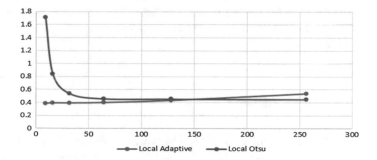

Fig. 7 Comparison graph *between Otsu's method and adaptive thresholding*

5.4 Morphological Operations

The form and structure of the image are used to remove the imperfections present in the binary image. Morphological image processing consists of a set of nonlinear operations associated with the shape or morphology of features in an image. The image is probed with a small shape or template called structuring element. The structuring element is moved across the image such that it covers all locations in the image. It is then compared with the corresponding neighbourhood of pixels. Some of the operations test if the element 'fits' within the neighbourhood, while the rest test whether it 'hits' or intersects with neighbourhood [12]. After the veins are distinguished, the contours are drawn on them with the help of libraries available in python.

6 Hardware Implementation of Model

6.1 Single Board Computer

The model was required to be lightweight and easy to access for these reasons, and the whole model was implemented on Raspberry Pi. Raspberry Pi has Broadcom chip BCM2837 which is a quad-core ARM Cortex A53 cluster with a dimension of 8.56 cm × 5.65 cm × 17 mm.

6.2 NoIR Camera

The Raspberry Pi Camera Module v2 features a fixed focus lens. It is a custom designed add-on board for Raspberry Pi with high-quality 8 MP Sony IMX219 image sensor. It does not have an IR filter, and therefore, it can capture images in near-field band. It can be connected to Raspberry Pi through one of the small sockets present on the upper surface of the board. The board has a dimension of 25 mm × 23 mm × 9 mm and a weight of 3 g. It connects to Raspberry Pi by way of a short flat ribbon cable (FRC).

6.3 Near-Infrared LEDs

We used two types of near-infrared LEDs having wavelength of 880 and 940 nm of emitted light with the emitting angle of 20°. The calculation for the geometric array is shown earlier in this paper.

6.4 Optical Filter

The efficiency of the camera can be increased if the spectrum other than the required one is blocked outside the camera. The solution to this problem of visible light interference with the camera was to use a band-pass IR filter in front of the camera that had a cut-off region below the useful 850–940 nm IR spectra emitted by the LEDs. The NIR filter was placed in front of the camera.

6.5 TFT Display

The touch screen uses Serial Peripheral Interface (SPI) to communicate with the main processor. The reason behind selection of TFT is fast response time, large viewing angle, lightweight and less power consumption.

7 Software Implementation of Model

The main goal of the software as outlined in this section is to capture images using the hardware discussed above and then process these images to a form that is suitable for vein detection.

7.1 Python

Python is a highly flexible and fast programming language that suits prototype. Here are some of the advantages:

a. Python is object orientated. This makes the system simple to code as the frequently used basic functions can be abstracted and used in a reiterative manner.
b. Python has mutable data types. This makes it easy to write and execute the functions without worrying much about the data types being passed to it.

7.2 OpenCV

OpenCV is an open-source library for computer vision applications. To use OpenCV, it is downloaded [13] onto the Raspberry Pi and compiled. The compiled files are then installed on the system. 'OpenCV contributions' serve as some extra and additional modules we need for the whole software procedure.

8 Result

The final result is shown in Fig. 8. The following results can be further projected or displayed on TFT screen.

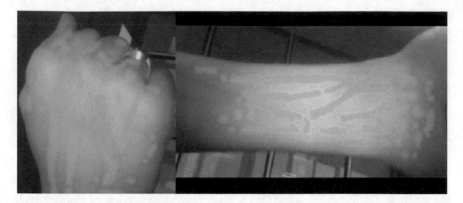

Fig. 8 Final obtained vein pattern

9 Conclusion

It was successfully shown that it is indeed possible to use near-infrared light as a method of detecting subcutaneous veins within the hand. It seems that with the improvements in the computational power of Raspberry Pi, it is becoming easier to implement some complex algorithms to process the acquired images on an embedded, system-on-chip kind of platform, greatly increasing the portability of the system at the same time as reducing cost and efforts. Software suite was developed for image enhancement which used CLAHE and median filtering. To extract the vein pattern, operations such as adaptive thresholding and morphological transforms were successfully used. Images were captured from a live stream, and then the stream was modified to show the vein pattern in an augmented reality type of application.

Acknowledgements We are grateful to Centre of Excellence in Image and Signal Processing of Electronics and Telecommunication Department, College of Engineering, Pune (COEP), for their support.

References

1. Rauch et al.: Peripheral difficult venous access in children (2009)
2. Jacobson, A.F., Winslow, E.H.: Variables influencing intravenous catheter insertion difficulty and failure (2005)
3. Black, K.J.L., Pusic, M.V., Harmidy, D., McGillivray, D.: Pediatric intravenous insertion in the emergency department: Bevel up or bevel down? Pediatr. Emerg. Care (2005)
4. Stevenson, M., Lloyd-Jones, M., Morgan, M.Y.: Non-invasive diagnostic assessment tools for the detection of liver fibrosis. Health Technol. Assess. **16**, 1 (2012)
5. Paquit, V., Price, J.R.: Near-Infrared Imaging and Structured Light Ranging for Automatic Catheter Insertion (2006)
6. Crisan, S., Tarnovan, I.G., Crisan, T.E.: Radiation Optimization and Image Processing Algorithms in the Identification of Hand Vein Patterns (2010)

7. Moreno, I., Avedano-Alejo, M., Tzonchev, R.I.: Designing Light-Emitting Diode Arrays for Uniform Near-Field Irradiance (2006)
8. Chakravorty, T., Sonawane, D.N., Sharma, S.D., Patil, T.: Low-Cost Subcutaneous Vein Detection System Using ARM9 Based Single Board Computer (2011)
9. Pizer, S.: Adaptive Histogram Equalization and Its Variations (1987)
10. https://www.cs.auckland.ac.nz/courses/compsci373s1c/PatricesLectures/Image%20Filtering_2up.pdf
11. Bradley, D., Roth, G.: Adaptive Thresholding Using the Integral Image (2001)
12. Gonzalez, R.C., Woods, R.E.: Digital Image Processing. Prentice Hall, New York (2008)
13. www.opencv.org

Hardware Trojan Detection Using Deep Learning Technique

K. Reshma, M. Priyatharishini and M. Nirmala Devi

Abstract A method to detect hardware Trojan in gate-level netlist is proposed using deep learning technique. The paper shows that it is easy to identify genuine nodes and Trojan-infected nodes based on controllability and transition probability values of a given Trojan-infected circuit. The controllability and transition probability characteristics of Trojan-infected nodes show large inter-cluster distance from the genuine nodes so that it is easy to cluster the nodes as Trojan-infected nodes and genuine nodes. From a given circuit, controllability and transition probability values are extracted as Trojan features using deep learning algorithm and clustering the data using k-means clustering. The technique is validated on ISCAS'85 benchmark circuits, and it does not require any golden model as reference. The proposed method can detect all Trojan-infected nodes in less than 6 s with zero false positive and zero false negative detection accuracy.

Keywords Controllability · Deep learning · Hardware Trojan
Transition probability

1 Introduction

Modern IC designs often involve intellectual property (IP) cores supplied by third-party vendors. Hardware Trojans (HTs) may be inserted or embedded into IP cores by some suppliers that cause its malfunction under very rare circumstances. Nowadays, lack of reference circuit is the main problem to detect a hardware Trojan. An emerging

K. Reshma (✉) · M. Priyatharishini · M. Nirmala Devi
Department of Electronics and Communication Engineering, Amrita School of Engineering,
Amrita Vishwa Vidyapeetham, Coimbatore, India
e-mail: k.reshma94@gmail.com

M. Priyatharishini
e-mail: m_priyatharishini@cb.amrita.edu

M. Nirmala Devi
e-mail: m_nirmala@cb.amrita.edu

© Springer Nature Singapore Pte Ltd. 2019
J. Wang et al. (eds.), *Soft Computing and Signal Processing* ,
Advances in Intelligent Systems and Computing 898,
https://doi.org/10.1007/978-981-13-3393-4_68

hardware security concern was emerged in which the hardware is often outsourced to the untrusted foundries for cost minimization. These untrusted foundries will access the hardware or modify the design during manufacturing, and there is a potential risk having Trojans intruded in the chip [1], which may enable cybersecurity attacks.

The Trojan detection approaches are developed in the hardware security community such as functional test [2], run-time monitoring [3], online checker [4], and side channel-based approaches [5]. A power profile-based approach is also developed for better refinement of the region in which Trojan is inserted. The power or the frequency analyses is also considered as side channel-based approach in which the power values are measured at different time instance and for different test patterns [6].

Hardware Trojans are mainly classified into three types [7] based on their physical, activation, and action characteristics. A classification of hardware Trojans and a survey of published techniques for Trojan detection are explained in [8]. The connectivity between the original circuit and the trigger input of the Trojan module will be independent, and that types of Trojans are known as always-on Trojans. 'Always on' means the Trojan is always active and can disrupt the chip's function at any time. It is very difficult to detect a Trojan like always-on Trojan. Trojan taxonomy is explained in [9] where the classification is based on the types of trigger and payload. Trojan analysis can be performed on various levels, viz. node selection, feature extraction, and the outcome is classified using different classifiers such as SVM, random Forest. Node selection is one among the techniques used in Trojan detection. The activation mechanisms include transition probability-based Trojan activation [10] and SCOAP measures [11]. This technique checks if any given node is Trojan infected or not depending on the values present.

In paper [12], a method has been proposed to detect the hardware Trojan-infected nodes using a support vector machine (SVM) technique. Using this technique, a low true positive value and high false positive value are obtained which occurs to be disadvantageous. Fifty-one gate-level Trojan features describe Trojan nets from a given netlist and then select the best 11 Trojan features to maximize the F-measures for the hardware Trojan classification [13]. This classification is done using random forest classifier. Hardware Trojan classification using multi-layer neural network is done in paper [14]. The 11 Trojan features are used to detect the hardware Trojan. Multi-layer neural network is used to learn the set of training data and classifies the nets as Trojan net and genuine nets. Ensembles of decision trees such as random forest are taken less time to train, but it takes more time to make predictions after the training. More accurate ensembles need additional trees, and it indicates that the model is slower. Deep learning techniques [15, 16] are proposed in order to overcome the above-specified disadvantages.

The proposed technique does not require any golden model as a reference, and it is validated on ISCAS'85 benchmark circuits with always-on Trojan. This paper is arranged such a way that proposed methodology is explaining in Sect. 2 and results, discussion in Sect. 3. This paper concludes in Sect. 4.

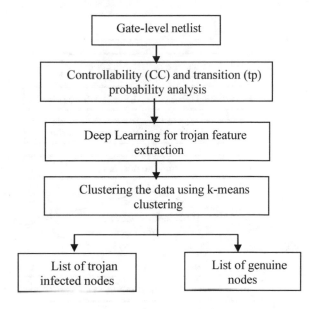

Fig. 1 Proposed hardware Trojan detection flow

2 Proposed Methodology

The flow of hardware Trojan detection using proposed method is given in Fig. 1. The gate-level netlist is taken as input. The netlist is obtained using Synopsys Tetra-MAX [17]. The determination of controllability and transition probability values corresponding to each node in the circuit is done using python language [18], and the controllability data are given to the deep learning algorithm. Then the data are clustered using k-means clustering. TensorFlow and Keras are used as deep learning models. The flow for obtaining the transition probability and controllability values is shown in Figs. 2 and 3.

2.1 Controllability Analysis

Combinational controllability measure can be used to identify nets with poor testability. A signal having very low switching activity may have high testability. The controllability values range from 1 to infinity. If 'i' is a primary input of a logic circuit, then the controllability values CC1(i) and CC0(i) are set to one. The controllability values are calculated from primary inputs toward primary outputs. The nodes with high controllability (CC) value are more suitable to insert a Trojan because the detection probability of such node is very low, and it is very difficult to control that particular node. Controllability (CC) is expressed as:

$$CC(i) = \sqrt{CC0(i)^2 + CC1(i)^2} \tag{1}$$

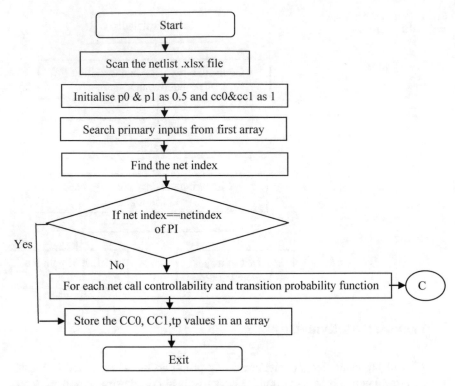

Fig. 2 Flow chart of controllability and transition probability measures

Fig. 3 Flow chart of controllability and transition probability function

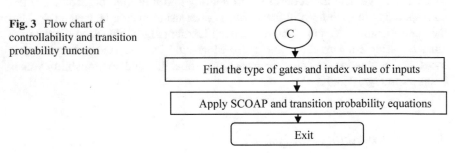

The relation between testability measures and the fault coverage of a logic circuit is explained in [19]. The combinational controllability values of different gates are explained in [20].

2.2 Transition Probability Analysis

Transition probability value also plays an important role for Trojan insertion. The node with low transition probability is susceptible for triggering the Trojan and Trojan payload. Low transition probability nodes indicate that it is rarely triggered and the attacker will always look for rarely triggered nodes for Trojan insertion. The transition probability (TP) is expressed as:

$$tp = (Logic\,0\,Probability) \times (Logic\,1\,Probability) \qquad (2)$$

2.3 Deep Learning Algorithm

The deep learning algorithm is formulated for extracting the Trojan features. Output of controllability and observability values is taken as an input of deep learning algorithm. In this proposed method, we are using unsupervised autoencoder algorithm that applies backpropagation [21, 22] and it will give an approximation to the input vector by minimizing the error value. Controllability values of all nodes in a circuit are taken as the input vectors of autoencoder, and it will produce normalized values corresponding to the given input vector. Then the data are clustering using k-means where the value of k is taken as 2 (1. high controllability Trojan-infected nodes, 2. low controllability genuine nodes), and we are assigning two cluster centers randomly. It calculates the inter-cluster distance between the data and clustering the data as Trojan-infected and Trojan-free. Deep learning algorithm is given below.

Algorithm 1: Deep learning

---

1. Read the training data as cctp.
2. Normalize the training data and set to X_train
3. Get the unsupervised feature representation of the input to the variable decode_data.   Decode_data = auto_encoder(X_train)
4. Implement k-means algorithm, where the   k value= 2
5. Visualize the clusters.

Algorithm 2: Auto_encoder(X_train):

---

1. Create a densely connected encoder neural nework with ReLU activation function.
2. Encode the training data into the dimension 32.
3. Create a densely connected decoder neural network with Sigmoid activation function.
4. Decode the encoded data into the dimension 2 using decoder network.
5. Obtain the feature representation for 100 epochs with batch size=5, optimization function   = adadelta and loss function as binarycrossentropy

Table 1 Processing time for hardware Trojan detection using machine learning technique

Benchmark circuit	No. of nets	Processing time for CC & TP (s)	Processing time for deep learning and clustering (s)
C17	17	0.0192	0.586
C499	499	0.1033	1.18
C880	880	0.3927	1.187
C1355	1355	0.12	2.49
C2670	2670	1.94	4.002

2.4 K-Means Clustering

Algorithm 3: K-means clustering:

---

1. Randomly assign two cluster centers
2. Assign data points to the cluster with closest cluster center
3. Compute new cluster centers by calculating the mean of the data points in each cluster
4. Repeat the steps 2 and 3 until convergence

3 Results and Discussions

The proposed technique is validated on ISCAS'85 combinational benchmark circuits with always-on Trojan and functionality modification Trojan. Here we are considering a ring oscillator as an always-on Trojan and the 'AND' gate as a trigger and 'XOR' gate as payload is considered as functionality modifying Trojan. Result shows that the node which has high controllability value is Trojan affected, and it is easy to cluster. The plot of k-means clustering of ISCAS'85 benchmark circuit with and without Trojan insertion is shown in Fig. 4. If the normalized controllability value corresponding to a node is above 0.5 then the node is a Trojan-infected node, and if the value is below 0.5, then the node is a Trojan-free node.

Table 1 shows the processing time for hardware Trojan detection using machine learning technique. From Tables 2 and 3, we can observe that if we insert a functionality modifying Trojan in C499, the controllability value does not have any change but the transition probability alone is changing. The result indicates that the Trojan detection is not only depending on the controllability value, but also the transition probability values.

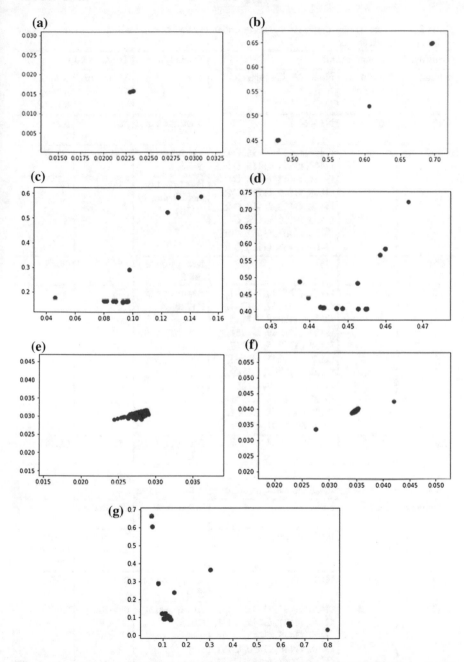

Fig. 4 **a** k-means clustering output of C17 circuit, **b** k-means clustering output of C17 circuit with always-on Trojan at N23, **c** k-means clustering output of C499 circuit with always-on Trojan, **d** k-means clustering output of C499 circuit with functionality modifying Trojan, **e** k-means clustering output of C880 circuit, **f** k-means clustering output of C880 circuit with always-on Trojan **g** k-means clustering output of C1355 circuit with always-on Trojan

Table 2 Hardware Trojan detection using controllability values and finding the Trojan-infected nodes using deep learning

Benchmark	Genuine circuit		Trojan type	Trojan circuit	
Circuit	Maximum CC value	Nodes having high CC		Maximum CC value	Nodes having high CC
C17	7.07	N10, N11	Always-on at N23	363	N23
C499	244	G436, G437, G438, G439, G440, G441, G442, G443, G444, G445, G446, G447, G448, G449, G450, G451, G452, G453, G454, G455, G456, G457, G458, G459, G460, G461, G462, G463, G464, G465, G466, G467	Always-on at G436	500	G436
			Functionality modification at G436	244	G436
C880	37.48	G813	Always-on at G813	363	G813
C1355	299.31	G1229, G1231, G1233, G1235, G1237, G1239, G1241, G1243, G1245, G1247, G1249, G1251, G1253, G1255, G1257, G1259, G1261, G1263, G1265, G1267, G1269, G1271, G1273, G1275, G1277, G1279, G1281, G1283, G1285, G1287, G1289, G1291, G1291	Always-on at G1229	555	G1229
C2670	135.72	G2214, G2243	Always-on at G2244	257	G2243

Table 3 Hardware Trojan detection using transition probability values

Benchmark circuit	Genuine circuit		Trojan type	Trojan circuit	
	Minimum TP value	Nodes having low TP		Minimum TP value	Nodes having low TP
C17	0.1875	N10, N11	Always-on at N23	0.15	N23
C499	0.00705	G436, G437, G438, G439, G440, G441, G442, G443, G444, G445, G446, G447, G448, G449, G450, G451, G452, G453, G454, G455, G456, G457, G458, G459, G460, G461, G462, G463, G464, G465, G466, G467	Always-on at G436	0.00705	G436

(continued)

Table 3 (continued)

Benchmark circuit	Genuine circuit		Trojan type	Trojan circuit	
	Minimum TP value	Nodes having low TP		Minimum TP value	Nodes having low TP
			Functionality modification at G436	0.00177	G436
C880	0.00095	G507, G508, G509, G510, G511, G512, G513, G514	Always-on at G507	0.00095	G507
C1355	0.00444	G1229, G1231, G1233, G1235, G1237, G1239, G1241, G1243, G1245, G1247, G1249, G1251, G1253, G1255, G1257, G1259, G1261, G1263, G1265, G1267, G1269, G1271, G1273, G1275, G1277, G1279, G1281, G1283, G1285, G1287, G1289, G1291, G1291	Always-on at G1229	0.00444	G1229
C2670	0.0024	G1814	Always-on at G2244	0.0024	G1814

4 Conclusion

Based on controllability and transition probability analysis, Trojan nodes are isolated using deep learning algorithm and the technique is validated on ISCAS'85 benchmark circuits, and it does not require any golden model as reference. The data are clustered as Trojan-infected nodes and Trojan-free nodes using k-means with $k = 2$. The analysis is based on static parameters; therefore, it does not require any test pattern generation. The proposed method can detect all Trojan-infected nodes in less than 6 s with zero false positive and zero false negative detection accuracy.

References

1. Tehranipoor, M., Koushanfar, F.: A survey of hardware trojan taxonomy and detection. In: IEEE Design & Test of Computers, pp. 10–25 (2010)
2. Wolff, F., Papachristou, C., Bhunia, S., Chakraborty, R.S.: Towards trojan-free trusted ICs: problem analysis and detection scheme. In: Design, Automation and Test in Europe, pp. 1362–1365 (2008)
3. Dubeuf, J., Hely, D., Karri, R.: Run-time detection of hardware trojans: the processor protection unit. In: IEEE European Test Symposium, ETS, pp. 1–6 (2013)

4. Chakraborty, R.S., Pagliarini, S., Mathew, J., Sree Ranjani, R., Nirmala Devi, M.: A flexible online checking technique to enhance hardware trojan horse detectability by reliability analysis. IEEE Trans. Emerg. Topics Comput. (2017)
5. Banga, M., Hsiao, M.: A region based approach for the identification of hardware trojans. In: IEEE International Symposium on Hardware Oriented Security and Trust (HOST), pp. 40–47 (2010)
6. Wei, S., Potkonjak, M.: Self-consistency and consistency-based detection and diagnosis of malicious circuitry. IEEE Trans. Very Large Scale Integr. (VLSI) Syst. **22**(9), 1845–1853 (2014)
7. Tehranipoor, M.: A survey of hardware trojan taxonomy and detection. In: IEEE Design, Test of Computers (2010)
8. Mohankumar, N., Periasamy, R., Sivaraj, P., Thakur, A.: Hardware trojan detection-a survey. In: 4th National Conference on Recent Trends in Communication Computation and Signal Processing (RTCSP-2013), pp. 99–102 (2013)
9. Kamran Haider, S., Jin, C., Ahmad, M., Manikantan Shila, D.: Hardware trojan: Threats and Emerging Solutions. ePrint Arch., Tech. Rep. (2009)
10. Popat, J., Mehta, U.: Transition probabilistic approach for detection and diagnosis hardware trojan in combinational circuits. In: IEEE (2016)
11. Salmani, H.: COTD: reference-free hardware trojan detection and recovery based on controllability and observability in gate-level netlist. IEEE Trans. Inf. Forensics Secur. **12** (2017)
12. Hasegawa, K., Oya, M., Yanagisawa, M., Togawa, N.: Hardware trojans classification for gate-level netlists based on machine learning. In: Proceedings of the IEEE/ACM International Conference (2016)
13. Hasegawa, K., Yanagisawa, M., Togawa, N.: Trojan-feature extraction at gate-level netlists and its application to hardware-trojan detection using random forest classifier. In: IEEE (2017)
14. Hasegawa, K., Yanagisawa, M., Togawa, N.: Hardware trojans classification for gate-level netlists using multi-layer neural networks. In: IEEE (2017)
15. LeCun, Y., Bengio, Y., Hinton, G.: Deep Learning. Macmillan Publishers Limited (2015)
16. Introduction to Deep Learning. http://introtodeeplearning.com
17. Synopsys TetraMAX. http://www.synopsys.com/Tools/Pages/default.aspx
18. Dive Into Python. http://diveintopython.org. O'Reilly, Associates
19. Hasegawa, K., Oya, M., Yanagisawa, M., Togawa, N.: Controllability and observability. In: Proceedings of the IEEE/ACM International Conference (2016)
20. Bushnell, M., Agrawal, V.: Essentials of Electronic Testing. Springer, Berlin (2015)
21. Autoencoder. http://ufldl.stanford.edu/tutorial/unsupervised/Autoencoders/
22. Autoencoder. https://github.com/Rentier/keras-autoen-coder

Design of Parallel Coupled Line Band-Pass Filter Using DGS for WiMAX Applications

Karunesh Srivastava and Rajeev Singh

Abstract Filter constitutes an integral part of microwave wireless communication systems. In this work, a parallel coupled line band-pass filter is proposed using defected ground structure (DGS) for WiMAX applications, which not only reduces the size but also the mutual coupling between the resonators is enhanced. The parallel coupled band-pass filter is implemented by introducing DGS in the ground plane. Effective inductance and capacitance of a transmission line are dependent on DGS as observed from results. The entire proposed filter is designed and fabricated on 1.6-mm-thick FR-4 substrate with tangent loss of 0.02 by using 50 Ω resistance. For validation of fabricated design, fabricated filter is measured using vector network analyzer and also compared with simulated results.

Keywords Dumbbell-shaped DGS · Insertion loss
Parallel coupled band-pass filter · Return loss · VSWR

1 Introduction

Impact of microwave engineering on modern communication systems such as WiMAX/WLAN, satellite, civil and military radar communication systems, etc. is huge. In microwave communication systems or even otherwise, filters play a dominant role to select desirable and pass undesirable frequencies [1]. Typically a filter with low insertion loss, large return loss with better impedance matching is desirable for good selectivity and minimum interference between two signals [1, 2]. Parallel coupled band-pass filters utilize resonators or model resonators as transmission lines. A typical parallel coupled line band-pass filter uses half wavelength line res-

K. Srivastava (✉) · R. Singh
Department of Electronics & Communication, University
of Allahabad, Allahabad, Uttar Pradesh, India
e-mail: karunesh.ec@gmail.com

R. Singh
e-mail: rsingh68@gmail.com

© Springer Nature Singapore Pte Ltd. 2019   681
J. Wang et al. (eds.), *Soft Computing and Signal Processing* ,
Advances in Intelligent Systems and Computing 898,
https://doi.org/10.1007/978-981-13-3393-4_69

onators, wherein the resonators are collateral to each other along their length. By providing suitable gaping between resonators, we observe a good coupling as well as sufficient bandwidth as compared to other filter structures [3, 4]. Due to the rapid growth of electronic devices and systems in wireless communication systems, we need small devices to miniaturize the overall system [5, 6]. The parallel configuration is widely used for the design of band-pass filters. Even though the designs of these filters reported in [3–6] are simpler but the main disadvantages are that they are not very compact and suffer from poor stop-band performance [7]. Nowadays, we often see that defected ground structures (DGSs) are used to design filters, oscillators, and antennas. The defected ground structures act as parallel LC resonators and their resonating frequencies are determined by its shape and size. Dumbbell shape DGS with embedded capacitor is reported [8] in designing low pass filter. For getting high-performance microwave filters, one of the design techniques is using defected ground structure resonators and these defected ground structures are typically etched for obtaining various shapes in the ground plane [9]. A circuit model using parallel coupled dual mode resonator filter operating as dual mode resonator is reported [10]. The parallel coupled dual mode resonator is advantageous as they are compact in size and demonstrate better Chebyshev and Butterworth response. A compact band-pass filter with multiple harmonics suppression by means of a folded coupled micro-strip line is reported and the filter is designed using low impedance feeding system which enables to attain sharp lower and upper cutoff frequencies. The filter so designed achieves 2.45 GHz center frequency with 0.1 dB insertion loss [11]. A simple compact wideband band-pass filter using DGS and spur-line provides 35% fractional bandwidth with 5 GHz central frequency is reported [12]. Effective inductance and capacitance of a transmission line can be altered by placing DGS of suitable shape and size [13]. Numerous design techniques of band-pass filter are reported for WiMAX applications [14–16].

In this paper, the band-stop characteristics of defected ground structure (DGS) of different shapes are studied in terms of inductance, capacitance, and their sharpness factor. Various DGS shapes are etched on the ground plane of parallel coupled micro-strip line. All the structures are simulated using CST EM simulator by keeping the cutoff frequency constant for all the structures.

2 Theoretical Details

2.1 Parallel Coupled Micro-strip Line

Characterization of a parallel coupled micro-strip line can be done by determining its characteristic impedance and effective permittivity of even and odd modes. A typical parallel coupled line band-pass filter can be designed using half wave line resonators. Resonators are so designed and placed so that they are parallel to each other. Good coupling and bandwidth in parallel coupled line structure resonators can be achieved with appropriate spacing between resonators. The coupling gaps resemble the admittance inverter in the band-pass circuit. Even and odd mode characteristics impedances of parallel coupled half wave resonators can be computed by using admittance inverters [17].

2.2 Defected Ground Structure (DGS)

The concept of DGS (defected ground structure) borrowed from photonic gap structure (PBG) is a useful technique to reduce the size of the microwave passive elements. Figure 1 depicts various designs of DGSs, and the dimensions are tabulated in Table 1. Designs shown in Fig. 1 are simulated by means of Computer Simulation Technology tool. Simulation results show that the dumbbell-shaped defected ground structure (cf. Fig. 1a) has better selectivity and sharpness as compared to other geometries of Fig. 1 (cf. Fig. 1b, c).

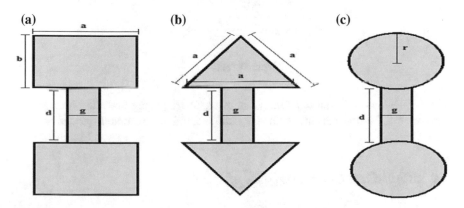

Fig. 1 Various DGS pattern: **a** dumbbell shape, **b** triangular shape, **c** circular shape

Table 1 Dimensions of various DGSs at cutoff frequency

Design configuration of DGS	a (mm)	b (mm)	d (mm)	g (mm)	r (mm)
Square	2	2	3.6	0.5	–
Triangular	2.2	–	3.6	0.5	–
Circular	–	–	3.6	0.5	1.5

Fig. 2 Circuit equivalent of
DGS

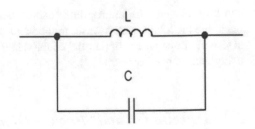

2.3 LC Circuit Equivalent of DGS

LC circuit equivalent of DGS shapes (cf. Fig. 1) is represented by Fig. 2. The value
of effective inductance (L), effective capacitance (C), and sharpness of the filter can
be computed using Eqs. (1)–(3) [18].

$$C = \frac{5 f_c}{\pi \left(f_o^2 - f_c^2 \right)} pF \tag{1}$$

$$L = \frac{250}{C (\pi f_o)^2} nH \tag{2}$$

$$\text{Sharpness factor} = \frac{f_c}{f_o} \tag{3}$$

where f_c is cutoff frequency and f_o is resonant frequency and the simulated S-
parameter of the designs can be used to extract cutoff and resonant frequency.

2.4 Band-Pass Characteristics of DGS

Figure 3 shows the simulated S-parameter (S_{21}) of DGS geometries of Fig. 1. We
observe different resonant and cutoff frequencies for different shapes of DGS. To
optimize different structures, different extracted cutoff frequencies are chosen. Sharpness factor for circular and triangular structure is less as compared to dumbbell shape
defected ground structure. Effective capacitance (C) plays major role in deciding the
sharpness of filter. It is clear from Fig. 4 that effective capacitance of dumbbell shape
structure is highest among three shapes.

Fig. 3 S_{21} of different DGS

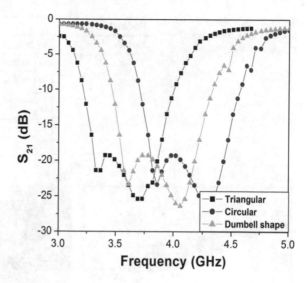

Fig. 4 Parameter comparison of DGSs

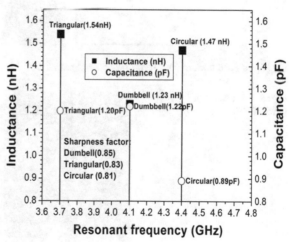

3 Results and Discussion

Design of the filter is conceptualized in a way that neighboring resonators are parallel to each other along half of their length. Such type of setup provides good coupling for a specified gaping between resonators. Filter structure thus becomes easier to implement and on other hand we observe a wider bandwidth as compared to other reported filters. The final dimensions optimized by CST are given in Table 2 and shown in Fig. 5a.

Table 2 Optimized geometrical parameters of proposed band-pass filter

Parameters		Values (mm)
Length	L_1 to L_{10}	10
Width	$W_1 = W_9$	3.137
	$W_2 = W_3 = W_7 = W_8$	1
	W_4	3.363
	W_5	3.037
	W_6	2.967

(a)

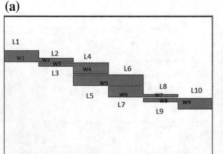

(b)

Fig. 5 Top and bottom view of proposed band-pass filter

Fig. 6 Front view of fabricated design

The proposed dumbbell shape DGS unit is developed to obtain good sharpness. This DGS provides reduction in size in the ground plane of the filter. To improve the pass-band match, width of compensated micro-strip on the top of the DGS is broadened as shown in Fig. 5a. Large pass-band, low insertion loss, high return loss, VSWR<2 and flat group delay is provided by the designed filter. The filter was designed on FR-4 substrate ($\varepsilon = 4.4$, thickness-1.6 mm), and the performance of the designed band-pass filter was evaluated using simulation technique by means of CST Microwave Studio. Figures 6 and 7 present the layout of the fabricated filter design.

Fig. 7 Bottom view of
fabricated design

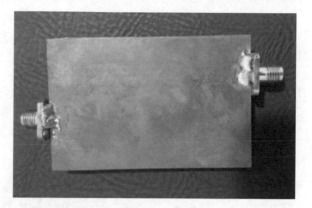

Fig. 8 S-Parameter versus
frequency

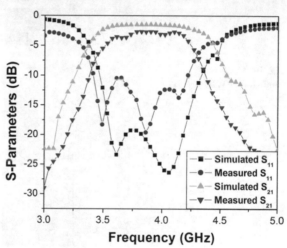

Figures 8 and 9 represent the simulated and measured insertion loss, return loss, and VSWR (Voltage Standing Wave Ratio), respectively. It is observed that simulated return loss, insertion loss, and VSWR are −26.8, −1.7 dB, and 1.7395, respectively between 3.48 and 4.31 GHz.

It is clear from Figs. 10 and 11 that percentage fractional bandwidth and Q_{loaded} decreases when the return loss and fractional bandwidth increases, respectively. From Fig. 12, it can be observed that load impedance decreases when VSWR crosses its perfect matching condition. Simulated and measured group delay for the proposed filter is shown in Fig. 13, which is below 0.5 ns over the whole pass-band which is suggestive of the fact that there is minimum distortion between transmitted and received pulses, and thus, filter performance is stable.

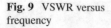

Fig. 9 VSWR versus frequency

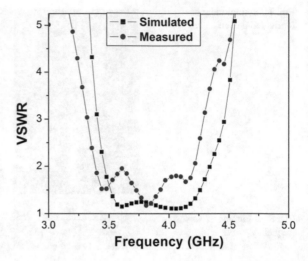

Fig. 10 Fractional bandwidth (FWB) versus $|S_{11}|$

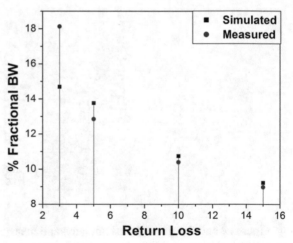

A marginal shift between the measured and simulated $|S_{11}|$ and $|S_{21}|$ and VSWR is observed (cf. Figs. 8 and 9) at mid-band frequency range. The difference observed in the simulated and measured $|S_{11}|$ and $|S_{21}|$ is attributed to change in overall circuit impedance, capacitance, and inductance caused by soldering and joints. The change in overall circuit impedance causes small change in operating frequencies of pass-band as well as in stop band.

The proposed method covers large bandwidth of 1.25 GHz with better return loss, insertion loss and the design is cost-effective as well. The proposed design satisfies the design goals of the band-pass filter and is economically viable.

Fig. 11 Quality factor versus FWB

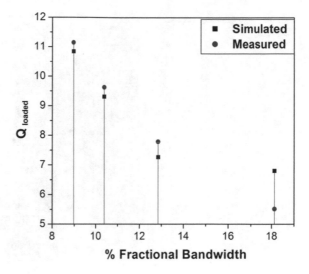

Fig. 12 Load impedance versus VSWR

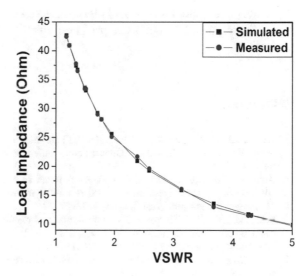

4 Conclusions

Synthesis, design, and analysis of parallel coupled line band-pass filter using DGS for WiMAX applications are demonstrated in this work. Defected ground structure is used for reducing the size as well as for improving the sharpness of the band-pass filter. The proposed design was simulated and experimentally validated. Simulated return loss, insertion loss, and VSWR are observed as −26.8, −1.7 dB, and 1.7395, respectively, between 3.48 and 4.31 GHz. Maximum measured insertion loss

Fig. 13 Group delay versus frequency

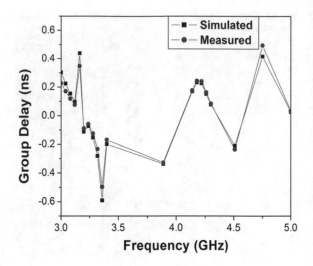

(−2.48 dB) and return loss (−19.75 dB) is observed with a group delay of less than 0.5 ns. The proposed design is compact, cost-effective and is suitable for WiMAX applications.

References

1. Rabahallah, D., Challal, M., Talaharis, N.: Tri-band microstrip bandpass filters for GSM and WiMAX applications. In: 4th IEEE International Conference on Electrical Engineering (ICEE), pp. 1–4 (2015)
2. Akra, M., Pistono, E., Ferrari, P., Jrad, A.: A novel accurate method for synthesizing parallel coupled line bandpass filter. In: Microwave Symposium (MMS), 13th Mediterranean Microwave Symposium (MMS), pp. 1–4 (2013)
3. Saha, P.B., Roy, S., Bhowmik, M.: Improvement of parallel coupled bandpass filter using coupled closed loop square resonator structure. In: 2nd IEEE International Conference Signal Processing and Integrated Networks (SPIN), pp. 65–68 (2015)
4. Chen, C.J.: A four-pole parallel-coupled dual-mode resonator bandpass filter. In: IEEE MTT-S International Microwave Symposium (IMS), pp. 1–3 (2016)
5. Lee, S., Lee, Y.: Generalized miniaturization method for coupled-line bandpass filters by reactive loading. IEEE Trans. Microw. Theory Tech. **58**(9), 2383–2391 (2010)
6. Yuan, D.M., Zhao, S.W., Zhang, H.X.: Design and analysis of a structure-based microstrip bandpass filter. J. China Univ. Posts Telecomm. **21**(4) 6495–6497 (2014)
7. Sánchez-Soriano, M.Á., Bronchalo, E., Torregrosa-Penalva, G.: Parallel-coupled line filter design from an energetic coupling approach. IET Microwaves Antennas Propag. **5**(5), 568–575 (2011)
8. Chen, Z.-Y., Li, L., Chen, S.-S.: A novel dumbbell-shaped defected ground structure with embedded capacitor and its application in low-pass filter design. Prog. Electromagn. Res. Lett. **53**, 121–126 (2015)
9. Weng, L.H., Guo, Y.C., Shi, X.W., Chen, X.Q.: An overview on defected ground structure. Prog. Electromagn. Res. B **7**, 173–189 (2008)

10. Chen, C.J.: Design of parallel-coupled dual-mode resonator bandpass filters. IEEE Trans. Compon. Packag. Manuf. Technol. **6**(10), 1542–1548 (2016)
11. Marimuthu, J., Bialkowski, K.S., Abbosh, A.M.: Compact bandpass filter with multiple harmonics suppression using folded parallel-coupled microstrip lines. In: IEEE Microwave Conference (APMC), Asia-Pacific, vol. 2, pp. 1–3 (2015)
12. Tang, C., Lin, X., Liu, W., Fan, Y., Song, K.: A compact wideband bandpass filter based on parallel-coupled stub-loaded resonator. In: IEEE MTT-S International Microwave Workshop Series on Advanced Materials and Processes for RF and THz Applications (IMWS-AMP), pp. 1–3 (2016)
13. Khan, T.M., Zakariya, A.M., Saad, M.N.M., Baharudin, Z., Rehman, M.U.: Analysis and realization of defected ground structure (DGS) on bandpass filter. In: 5th IEEE International Conference on Intelligent and Advanced Systems (ICIAS), pp. 1–4 (2014)
14. Sandhya, R.K., Monisha, B.: A novel hair pin line band pass filter design for WIMAX applications. Int. J. Adv. Res. Electron. Commun. Eng. **3**(4), 457–460 (2014)
15. Shaik, A., Vinoy, K.J.: Design of bandpass filter using branch-line directional couple. In: Radio Frequency Integrated Circuits and Systems, pp. 1–4 (2011)
16. Mudrik, A.: Designing microstrip bandpass filter at 3.2 GHz. Int. J. Electr. Eng. Informatics **2**, 71–83 (2010)
17. Hong, J.-S., Lancaster, M.J.: Microstrip filters for RF/microwave applications. Wiley, New York (2001)
18. Ahn, D., Park, J.S., Park, C.S., Kim, J., Kim, Y., Qian, T.: Design of the low-pass filter using the novel microstrip defected ground structure. IEEE Trans. Microw. Theory Tech. **49**(1), 86–93 (2001)

LSTM- and GRU-Based Time Series Models for Market Clearing Price Forecasting of Indian Deregulated Electricity Markets

Ashish Ubrani and Simran Motwani

Abstract In a deregulated electricity market scenario, formulation of bidding strategies and investment decisions depends majorly on forecasting of Market Clearing Price (MCP). This research proposes and compares models based on Long Short-Term Memory (LSTM) and Gated Recurrent Unit (GRU) networks to achieve the same. Data for training and testing of the proposed models is collected from Indian Energy Exchange (IEX). Trained models are used to perform day-ahead and week-ahead predictions. Mean Absolute Percentage Error (MAPE) for each model in each case is calculated. Results show that both LSTM- and GRU-based models deliver a reasonably good overall performance with LSTM performing slightly better.

Keywords Market clearing price (MCP) · Deregulated electricity markets
Long short-term memory (LSTM) networks
Gated recurrent unit (GRU) networks

1 Introduction

Price of electricity is of prime concern for buyers and sellers in today's deregulated electricity market. In order to devise an effective bidding strategy, a concrete model for price forecasting is necessary [1]. While a lot of work is done in load forecasting, price forecasting is still in its primitive stages. According to [2], in this scenario, structure of price curve is richer than that of load curve.

Electricity price forecasting methods are classified into three major classes [3]:

A. Ubrani (✉)
Electrical Department, Veermata Jijabai Technological Institute, Mumbai, India
e-mail: ubraniav@gmail.com

S. Motwani
Computer Department, Veermata Jijabai Technological Institute, Mumbai, India
e-mail: simran.motwani29@gmail.com

© Springer Nature Singapore Pte Ltd. 2019
J. Wang et al. (eds.), *Soft Computing and Signal Processing* ,
Advances in Intelligent Systems and Computing 898,
https://doi.org/10.1007/978-981-13-3393-4_70

1. Game theory—Situation is modeled as a power transaction game, and strategies are devised accordingly. Here, the equilibrium point is analyzed by the participants to maximize the profits. Different models like Nash equilibrium and Bertrand model are available.
2. Time series—Price at a future time is predicted by the model which takes past prices into account [4].
3. Simulation—System is modeled and simulated considering the physical parameters. Price is predicted on the basis of solution of mathematical model.

Pindoriya et al. [5] proposes an adaptive wavelet neural network (AWNN) for short-term price forecasting (STPF). They have tested their model by performing a day-ahead prediction of Market Clearing Price (MCP) of Spain's electricity market. In [6], two models based on Neuro-Evolutionary Cartesian Genetic Programming Evolved Artificial Neural Network (CGPANN) algorithm are proposed. To predict the future price, they take previous 12 and 24 h data, respectively. This system is also tested on Spanish market data. An Asymmetric Gaussian Fuzzy Inference Neural Network (AGFINN) based on Takagi–Sugeno–Kang (TSK) structure is proposed in [7]. It is tested on the electricity data of ISO New England market. Yang and Duan [8] presents a forecasting model based on recurrent neural network (RNN). It uses chaos theory to reconstruct phase space, from time series, in which RNN is used to perform global and local forecasting. Model is tested on energy prices on New England market. Anamika and Kumar [9] presents a simple artificial neural network (ANN), i.e., a multilayer perceptron (MLP) feedforward network-based system to forecast MCP for Indian electricity markets. Its performance is compared with a regression model. In their study, regression model is found to perform better than ANN.

From the above discussion, it can be observed that comparatively lesser research is done, in this area, on Indian electricity markets. It is also clear that not a lot of work is done, in this area, on Long Short-Term Memory (LSTM) networks and Gated Recurrent Unit (GRU) networks. LSTMs and GRUs, which are advancements of RNN, are known to perform extremely well in the problems of time series forecasting. In the view of that, we present models based on LSTM and GRU networks, respectively. Proposed models are trained and tested on data of Indian electricity markets obtained from Indian Energy Exchange (IEX) [10].

Section 2 presents the fundamental concepts/methodologies of LSTM and GRU networks. Section 3 proposes the system for solving the problem. Further, Sect. 4 presents the results and comparative analysis of the performance of the proposed system with LSTM and GRU, respectively. Both the models are compared on their performance on weekly and daily forecasts.

2   Fundamental Concepts

2.1   Recurrent Neural Networks (RNNs)

Recurrent neural networks are a type of neural networks having loops in their hidden layers. That is, at any time, they take into account current input and recently computed hidden unit values to determine the current value of hidden unit and eventually the output. This means that they use new information and previously known knowledge to make a decision. This replicates our human's thought process in real life, especially when dealing with sequences. It does excellent work modeling long-term dependencies. Hence, RNNs work really well with time series and textual input.

To put this mathematically, the value of hidden layer unit at time t, h_t, depends on input at time t, x_t, and the value of hidden layer unit at time $t - 1$, h_{t-1}.

$$h_t = f(x_t, h_{t-1}) \tag{1}$$

The value of h_t is calculated according to Eq. (2):

$$h_t = \emptyset(Wx_t + Uh_{t-1}) \tag{2}$$

Here, W is the weight matrix for input units and U is a matrix mapping from one hidden state to another. $\emptyset$ is any activation function like sigmoid or ReLU. Because of this form of equation, the current value of hidden layer unit not only depends on immediately previous value, but also on the values of hidden layer unit before that.

2.2   Problem of Vanishing and Exploding Gradients

The problem of vanishing and exploding gradients occurs when RNNs are trained with a gradient-based optimization method. The gradient of weights with respect to error is used to update all weights through time. If the gradient is less than 1, the compound effect of backpropagating through time results in a value very less than 1. Hence, the network learns slowly. This becomes more problematic as various hidden layers are stacked upon each other. Exploding gradients are similar, except value of the gradient is greater than 1. Hence, weights become saturated. However, they are easily solved in comparison with vanishing gradient by truncation.

2.3   Long Short-Term Memory (LSTM) Networks

LSTM networks model long-term dependencies very well by taking care of vanishing gradients problem [11]. Instead of a single activation function, LSTMs maintain a cell state and perform updates by considering its previous cell state. LSTM does this by using three gates: forget gate, input gate, and output gate.

Forget gates f decide whether to keep or forget the previous information. If the value of forget gate is 1, previous cell state, C_{t-1}, is passed unchanged. If the value is 0, it means to completely forget the old cell state. The equation for forget gate is:

$$f_t = \sigma\left(W_f h_{t-1} + U_f x_t + b_f\right) \tag{3}$$

Input gate i decides what new information to be added to the memory or the cell state.

$$i_t = \sigma(W_i h_{t-1} + U_i x_t + b_i) \tag{4}$$

Now, the cell state is updated taking into account new values and previous cell state.

$$C\sim_t = \tanh(W_c h_{t-1} + U_c x_t + b_c) \tag{5}$$

$$C_t = f_t * C_{t-1} + i_t * C\sim_t \tag{6}$$

In the next and final step, output gate o decides which values to add to hidden layer output using Eqs. (7) and (8):

$$o_t = \sigma(W_o h_{t-1} + U_o x_t + b_o) \tag{7}$$

$$h_t = o_t * \tanh(C_t) \tag{8}$$

2.4 Gated Recurrent Unit (GRU) Networks

GRU is a variant of LSTM [12]. It has two gates: update z and reset r. Update gate decides how much previous state memory to keep, and reset gate decides how to combine new input values with old cell state. It does not have an output gate. As GRU has less parameters, it trains faster and is little more efficient than LSTM. Equations (9)–(12) are used in a GRU:

$$z_t = \sigma(W_z h_{t-1} + U_z x_t) \tag{9}$$

$$r_t = \sigma(W_r h_{t-1} + U_r x_t) \tag{10}$$

$$h\sim_t = \tanh(W_h(h_{t-1} * r) + U_h x_t) \tag{11}$$

$$h_t = (1 - z) * h\sim_t + z * h_{t-1} \tag{12}$$

Fig. 1 Flowchart of the process

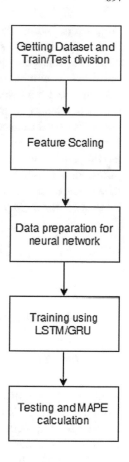

3 Proposed System

Figure 1 shows the flowchart of methodology. The various steps followed in detail are:

Getting Dataset and Train/Test Division. Electricity data containing MCPs in desired interval is obtained from the energy exchange. For this system, data from two intervals is used: weekly and daily. Data from each interval is then divided into training set and test set.

Feature Scaling. Scaling of data is performed before inputting it to neural network model. Min–Max normalization is used for the process. Data is then scaled according to Eq. (13):

$$v' = \frac{v - \min(x)}{\max(x) - \min(x)} \tag{13}$$

where x is the list of data values of electrical unit prices.

Data Preparation. To bring the data to the format that the network expects, some extra preparation steps are needed. LSTM and GRU need timesteps, which is the number of times the hidden layer will be looped back. For this, the input sequence of timesteps' number of values is made rather than a single input value.

Training. For both weekly and daily data, one LSTM and GRU network is trained for each. That is, four networks are trained for comparison of LSTM and GRU networks' performance on data. The optimizer used is Adam optimizer with a learning rate of 0.001. The loss to be minimized is the mean squared loss.

Testing and Calculation of MAPE. Same data preprocessing in way of scaling and preparation is used for test set, and values are predicted for it. Data is scaled back to original scale to calculate error. The error metric we have used is Mean Absolute Percentage Error (MAPE). The equation for MAPE is:

$$MAPE = \frac{1}{N} \sum_{i=1}^{N} \frac{|Actual_i - Predicted_i|}{Actual_i} \times 100 \qquad (14)$$

4 Implementation and Results

4.1 Dataset

Required data is collected from the Web site of IEX [10]. For weekly interval, data from first week of 2012 to last week of 2016, i.e., a total of 261 weeks, is used as training set and model is tested on first 20 weeks of 2017. For daily interval, data from first day of 2016 to last day of 2017, i.e., a total of 731 days, is used as training set and data from first 3 months, i.e., 90 days of 2018, is used as test set.

4.2 Training

In each network, four stacked layers of LSTM/GRU with hidden layer size of 50 units are used. Dropout rate is set to 0.3 to prevent overfitting of data to training set. The final output layer has 1 neuron which gives the predicted value. For weekly data, timestep is taken as 25 and for daily data as 60. The data is trained with a batch size of 32 and in 200 epochs.

4.3 Results

MAPEs of the test set for the four networks described above are calculated. Results are presented in Table 1.

Table 1 MAPE (%) of LSTM and GRU on weekly and daily data

	LSTM	GRU
Weekly	4	4.5
Daily	6.9	7.1

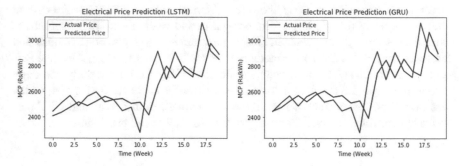

Fig. 2 Weekly prediction graphs of LSTM and GRU

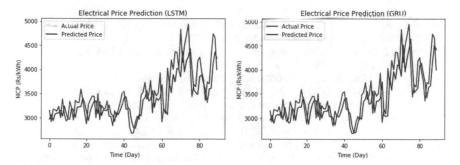

Fig. 3 Daily prediction graphs of LSTM and GRU

As can be seen from the table, LSTM performs better than GRU for both weekly and daily data. LSTM networks optimize more parameters. Additionally, they are known to perform well for longer timesteps. These both factors contribute toward their good performance on our data. Also, accuracy achieved by both LSTM and GRU on weekly data is more than that on daily data. The weekly data is more generalized and also more stable than daily data, which tend to have unprecedented highs and lows. Hence, both the networks do a superior job while performing week-ahead MCP predictions. Finally, LSTM for weekly data manages to get the best performance of all.

Figures 2 and 3 show how well each network predicts the trends in electrical unit prices.

5 Conclusion

In this paper, the problem of prediction of Market Clearing Price of Indian deregulated electricity markets was studied using time series analysis. Two advancements of recurrent neural networks, Long Short-Term Memory Network and Gated Recurrent Unit Network, were used to perform day-ahead and week-ahead MCP predictions using data over the years from Indian Energy Exchange. The evaluation demonstrated the superiority of LSTM for the prediction task with respect to Mean Absolute Percentage Error. LSTM attained better results than GRU for both weekly and daily forecasting tasks. LSTM for week-ahead predictions achieved a MAPE of 4, which was overall best performance from the four networks. Both LSTM and GRU performed better for week-ahead predictions than for day-ahead predictions.

References

1. Rodriguez, C., Anders, G.: Bidding strategy design for different types of electric power market participants. IEEE Trans. Power Syst. **19**, 964–971 (2004)
2. Bunn, D., Karakatsani, N.: Forecasting electricity prices. Review paper. London Business School Working Paper (2003)
3. Gonzalez, A., Roque, A., Garcia-Gonzalez, J.: Modeling and forecasting electricity prices with input/output hidden Markov models. IEEE Trans. Power Syst. **20**, 13–24 (2005)
4. Box, G., Jenkins, G., Reinsel, G.: Time Series Analysis. Prentice Hall, Englewood Cliffs (1994)
5. Pindoriya, N., Singh, S., Singh, S.: An adaptive wavelet neural network-based energy price forecasting in electricity markets. IEEE Trans. Power Syst. **23**, 1423–1432 (2008)
6. Khan, G., Arshad, R., Khan, N.: Efficient prediction of dynamic tariff in smart grid using CGP evolved artificial neural networks. In: 16th IEEE International Conference on Machine Learning and Applications (ICMLA), pp. 493–498. IEEE, Cancun (2017)
7. Alshejari, A., Kodogiannis, V.: Electricity price forecasting using asymmetric fuzzy neural network systems. In: IEEE International Conference on Fuzzy Systems (FUZZ-IEEE), pp. 1–6. IEEE, Naples (2017)
8. Yang, H., Duan, X.: Chaotic characteristics of electricity price and its forecasting model. In: Canadian Conference on Electrical and Computer Engineering—CCECE 2003, pp. 659–662. IEEE (2003)
9. Anamika, Kumar, N.: Market clearing price prediction using ANN in Indian electricity markets. In: International Conference on Energy Efficient Technologies for Sustainability (ICEETS), pp. 454–458. IEEE, Nagercoil, India (2016)
10. Indian Energy Exchange. https://www.iexindia.com/
11. Hochreiter, S., Schmidhuber, J.: Long short-term memory. Neural Comput. **9**, 1735–1780 (1997)
12. Chung, J., Gulcehre, C., Cho, K., Bengio, Y.: Empirical evaluation of gated recurrent neural networks on sequence modeling. NIPS 2014 Workshop on Deep Learning, December 2014 (2014)

MRI Brain Tumor Segmentation Using Automatic 3D Blob Method

B. Jyothi

Abstract Brain tumor surgery totally depends on accurate detection and segmentation of a brain tumor. For this, 3D detection is very necessary to know the actual depth of a tumor. This paper presents the segmentation and detection of brain tumors based on 3D blobs. We detect and segment the single and multiple tumors from 2 to 31,738 mm^3 in volume. The proposed method uses Laplacian of Gaussian (LoG) filtering, skull removal, affine adaptation and shape pruning. The LoG finds 3D blobs and extremely sensitive to minute abnormalities per scan. The proposed approach has 90.70 and 96.40% detection rates and a typical end running time of less than 2 min. The results exhibit that it is possible to categorize normal and abnormal.

Keywords Brain tumor · MRI brain asymmetry · 3D independent LoG

1 Introduction

Image segmentation is a crucial process in observing typicality in the human body, mostly through clinical investigation. In the latest years, research worker's interest in cancer or tumor identification has inflated quickly, as tumors have become the most important explanation for death. Tumors are accumulation of mass caused by the irregular expansion of cells. Normally, brain tumors' square measures are benign and malignant. A nonmalignant tumor does not unfold during a speedy manner and does not have an effect on healthy adjacent cells. These neoplasm cells, rather than spreading all through the full body grow during a restricted area and Type A lump of moles square measure samples, are unfold all over the body and touch organs. Ultimately, it ends in the death of the patient. A malignance is additionally referred to as cancer [1].

B. Jyothi (✉)
Department of ECE, Malla Reddy College of Engineering and Technology, Hyderabad, India
e-mail: bjyothi815@gmail.com

© Springer Nature Singapore Pte Ltd. 2019
J. Wang et al. (eds.), *Soft Computing and Signal Processing* ,
Advances in Intelligent Systems and Computing 898,
https://doi.org/10.1007/978-981-13-3393-4_71

Fundamentally, two scan modalities are used for brain tumors: magnetic resonance imaging (MRI) and computed tomography (CT) scans. CT scans utilize X-ray emission, while MRIs use tough magnetic field to create a high-eminent scan as against CT [2]. The brain contains complicated 3D structure. An MRI scan represents a 3D brain analysis addicted to a huge quantity of 2D slices altogether directions; however, this ends up in essential data being missed. In medical imaging, this loss of knowledge ends up in the incorrect identification of imperfection [3]. The accessibility of 3D imaging has created imperfection identification and surgical designing eminent that leads to saving patient lives [4]. In several hospitals, the tumor segmentation from associate MRI is completed by radiologist hand because the variety of doctors concerned will increase, thus will the price of treatment. Owing to this, it is terribly essential to style a full computer-aided diagnosis (CAD) system [5, 6]. The MRI image segmentation is that the initial inspiration to build up a CAD system, as labor-intensive segmentation could result in erroneous recognition of disease. The reform of a 3D from 2D MRI slices is that the most recent works are presented in [7, 8].

2 Proposed Method

The proposed algorithm for 3D blob is presented here. This method extracts the region of interest for tumor detection. Image is read into the system, and numerous features are extricated from each slice. Architectures of the proposed procedure is shown in Fig. 1.

Architecture of 3D segmentation mainly includes many layers, namely Laplacian of Gaussian filtering, skull removal, affine adaptation and shape pruning as explained in next sections. We compute saliency response of a blob (A), shape compactness score (B), and bilateral brain asymmetry (C). These feature scores are joined to decide a tumor strength C:

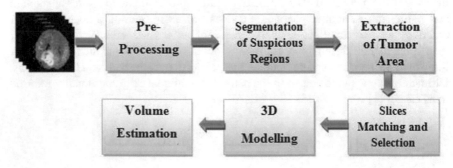

Fig. 1 Architecture of 3D segmentation

$$C(A, B, C) = (A + C)/B \tag{1}$$

Algorithm

```
X=Read image
  Kernel= [Ix2 Ix Iy IxIz
           Iy2 IyIz IyIx
           Iz2 IxIy IyIz];
  B=gaussianfilter(X, kernel);
  Binary=Convert to a binary image
  E=erosion (binary);
  D=dilation (E);
  B=binarylabe (D);
  Calculate are of an each region of B
  If area< threshold
  Set zeros in B
  Otherwise
  Set ones in B
  End
  [x, y, z]Z=Convert image into 3-Dimension
  A=affine interpolation (B, x, y, z);
  Affine=morphological (A, kernel)>threshold
  Filter= bilateral3 (affine);
```

2.1 Laplacian of Gaussian (LOG) Filtering

The LoG filter proposes to use blob detection from tumor. It is a universally used 3D blob detector for quick calculation:

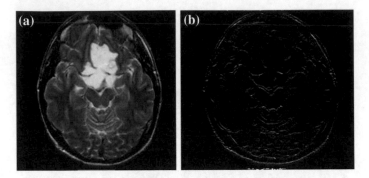

Fig. 2 **a** Input MRI slice and **b** MRI slice after Gaussian high-pass filtering

$$[(f \otimes \Delta gx) \otimes gy] \otimes gz+$$
$$p(x, y, z|\sigma) = [(f \otimes \Delta gy) \otimes gx] \otimes gz+ \qquad (2)$$
$$[(f \otimes \Delta gz) \otimes gx] \otimes gz+$$

where p is the LoG-filtered image, Δg and g are 1D LoG and Gaussian filters, respectively. This formulation extensively minimizes the voxel of computing from n3 to 9n. Using three difference masks (*Gx, Gy,* and *Gz*) of Gaussian-weighted moment vector operator in the *x, y,* and *z* directions, respectively, shown in Fig. 2.

$$B = \sigma^2(x, y, z/\sigma) \qquad (3)$$

2.2 Skull Removal

The second step in several MRI examination sequences is the removal of extra tissues on the complete MRI volume. The skull removal is obtained with the morphological operations of erosion and dilation to get rid of skull. This algorithmic program will manufacture needed results terribly with efficiency if it might distinguish between the original and also the false background [9].

2.3 Affine Adaptation and Shape Pruning

The 3D LoG might develop structure besides tumors. We have a tendency to prune the identified 3D blobs by using affine method and eliminating with extremely elliptical for each identified 3D blob, we calculated the gradient capsulated victimization its 3D formation M, where M = I 0 I and I = [Ix, Iy, Iz].

$$M = \begin{matrix} Ix^2 & IxIy & IxIz \\ IxIy & Iy^2 & IyIz \\ IxIz & IyIz & Iz^2 \end{matrix} \qquad (4)$$

The eigenelements signify the 3D elliptical outline of every 3D blob which provides the orientation of elliptic axis in 3D. These values are used to find the structural matrix M and the fraction of the direct and highest axis.

2.4 Bilateral Symmetry-Based Pruning

Approximately, human brains demonstrate a bilateral symmetry, whereas nonbrain-stem tumors frequently split this symmetry. To find out the asymmetry formed via a 3D blob, we use Earth mover's distance (EMD) as a measurement to evaluate the similarity of the closed collective intensity distribution.

EMD is defined as:

$$EMD(x, y) = \frac{1}{n} \sum i = 1|xi-yi| \qquad (5)$$

The circle is created around the tumor by having four methods to find and segment the tumor from image and calculate the tumor area to detect the edge for overall the area, and finally, it draws the circle based on tumor in MATLAB using rectangle comment.

3 Implementation Results

The performance of the projected method is tested by using precision and recall rates as measures. The precision rate is the fraction of tumor-detected blobs versus all determined blobs, whereas the recall rate is the fraction of determined tumors toward all of the exact tumors. Precision is defined as $Tp = (Tp + Fp)$, and recall is defined as $Tp = (Tp + Fn)$. We experimented our algorithm on four clinical 3D MR images, and it is possible to detect total 21 tumors that are 2–8 and 39 mm in diameter and 2- 31,738 mm^3 in volume shown in Figs. 3, 4, 5, 6 and 7 and Table 1.

4 Conclusion

The proposed unsupervised statistical asymmetry-based method for 3D brain tumor segregation by using 3D blobs technique uses the LoG to identify minute abnormal-

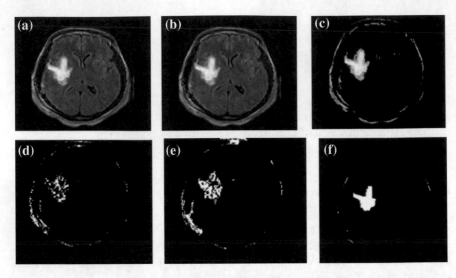

Fig. 3 **a** Selected MRI image, **b** LoG transformation, **c** skull removal, **d** affine transform, **e** bilateral filter, and **f** segmentation of proposed method

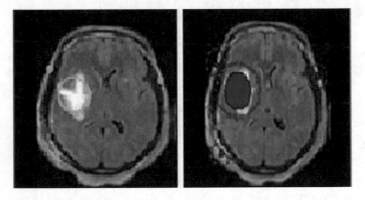

Fig. 4 Final tumor detection result

Table 1 Measurements of proposed method

Measurements	Values
Time	1.2 s
Precision	0.6
C	1.1
Tumor area	839

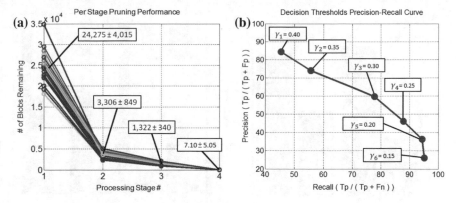

Fig. 5 **a** Precision curve and **b** recall curve

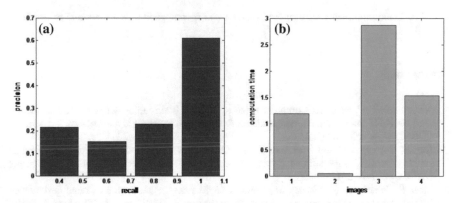

Fig. 6 **a** Precision and recall of brain tumor and **b** computation time of proposed method

ities. Using MATLAB 108 R2010a software, our average 96.4% detection rate and under 2 min run-time are upgrading over the algorithm. Finally, the proposed method finding results is utilized 110 as forefront seeds for regular tumor description.

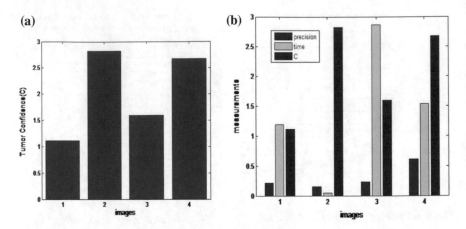

Fig. 7 **a** Tumor confidence of proposed method and **b** comparative measurements of proposed method

References

1. Akram, M.U., Usman, A.: Computer aided system for brain tumor detection and segmentation. In: IEEE International Conference on Computer Networks and Information Technology (ICCNIT), pp. 299–302 (2011)
2. Kumar, S.A.: Robust and automated lung nodule diagnosis from CT images based on fuzzy systems. In: IEEE International Conference on Process Automation, Control and Computing (PACC), pp. 1–6 (2011)
3. Shen, S., Sandham, W., Granat, M., Sterr, A.: MRI fuzzy segmentation of brain tissue using neighborhood attraction with neural-network optimization. IEEE Trans. Inf. Technol. Biomed. **9**(3), 459–467 (2005)
4. Mustaqeem, A., Javed, A., Fatima, T.: An efficient brain tumor detection algorithm using watershed & thresholding based segmentation. Int. J. Image Graph. Signal Process. (IJIGSP) **4**(10), 34 (2012)
5. Haider, W., Sharif, M., Raza, M.: Achieving accuracy in early stage tumor identification systems based on image segmentation and 3D structure analysis. Comput. Eng. Intell. Syst. **2**(6), 96–102 (2011)
6. Badran, E.F., Mahmoud, E.G., Hamdy, N.: An algorithm for detecting brain tumors in MRI images. In: International Conference on Computer Engineering and Systems (ICCES), pp. 368–373
7. Gopal, N.N., Karnan, M.: Diagnose brain tumor through MRI using image processing clustering algorithms such as Fuzzy C Means along with intelligent optimization techniques. In: IEEE International Conference on Computational Intelligence and Computing Research (ICCIC), pp. 1–4 (2010)
8. Ambrosini, R., Wang, P.: Computer-aided detection of metastatic brain tumors using automated three-dimensional template matching. J. MRI (2010)
9. Mahmoud, T.A., Marshall, S.: Edge-detected guided morphological filter for image sharpening. EURASIP J. Image Video Process. (2008)

Reconfigured VLSI Architecture for Discrete Wavelet Transform

Pramod Naik, Hansraj Guhilot, Arun Tigadi and Prakash Ganesh

Abstract This paper presents reconfigured dual-memory controller-based VLSI architecture for discrete wavelet transform to meet the wide variety of diverse computing requirements of the future generation system on chip designs. The proposed architecture mainly consists of a reconfigured DWT processor with dedicated memory which enhances the overall performance of the design. Generally any image and video processing algorithm memory plays are major criteria in determining the overall performance of the design. We have designed and implemented a dual-memory controller DWT processor on ZedBoard. This created dual-memory controlled DWT IP can be reconfigured as per designer requirement for high-end application. The DWT is applied to various images, and the compression ratio is obtained. The above architecture is implemented in SoC XC7Z020 FPGA, and the results show that architecture has reduced memory, low power consumption, and high throughput over several designs. The performance is compared with the available architectures showing good agreements. The area utilized by the DWT architecture is 13%, the delay is 11.577 ns, and the total on-chip power consumption is 23.8 mW. The number of slice LUTs and slice registers utilized for the design is 912 and 1469, respectively.

Keywords DWT · SoC · Dual-port memory · IP

P. Naik (✉) · P. Ganesh
CoreEL Technologies Pvt Ltd, Bengaluru, Karnataka, India
e-mail: pramodnaik40@gmail.com

P. Ganesh
e-mail: prakash.g@gmail.com

H. Guhilot
EC, K C College of Engineering, Thane, Mumbai, Maharashtra, India
e-mail: hansraj.g@gmail.com

A. Tigadi
EC, KLE Dr. M.S.S.CET, Belagavi, Karnataka, India
e-mail: arun.t@gmail.com

© Springer Nature Singapore Pte Ltd. 2019
J. Wang et al. (eds.), *Soft Computing and Signal Processing* ,
Advances in Intelligent Systems and Computing 898,
https://doi.org/10.1007/978-981-13-3393-4_72

1 Introduction

Image and video compressions decompose image and frames into multiple subbands of low- and high-frequency components. The compression ratio of the video depends on the algorithm and the area of application [1]. Generally, DWT video compression enables the compression in spatial as well as temporal domain suited for video compression algorithm [2, 3]. Moreover, DWT compression provides us with the scalability with the different levels of decomposition [4]. Thus, the goal of the video compression algorithm in our design is to reduce the amount of visual information data which would help in reducing the memory size required for storage, transmission, and display [5, 6]. The DWT processor architecture computes on larger data and performs intensive mathematical operations. To increase the efficiency of the processor, we have designed a dual-port memory controller. The designed DWT processor IP core is analyzed for its performance in term of area, speed, and power consumption.

The efficient design would be widely used for image and video processing application [7]. The designed DWT architecture as an IP core is analyzed as the requirement of the designer to analyze the overall performance by deploying the algorithm on seven series FPGA [8, 9]. This video compression architecture is a major research area with the advent and wide usage of multimedia devices [10]. The success of the present-day compression architecture greatly influences the overall performance of multimedia devices [11]. The importance of the search is that they should have a high speed of operation and consume low power [12]. In this paper, we have chosen a dual-memory controller-based DWT processor for compression and analyzed the performance metrics of the architecture.

Organization of the rest of the paper is as follows. Section 2 discusses the detailed description different DWT architectures. In Sect. 3, the proposed architecture is projected. In Sect. 4, we have given a detailed overview of a real-time hardware implementation of DWT processor. In Sect. 5, we have given the results and performance comparisons. Finally, the concluding remarks are given in Sect. 6.

2 Discrete Wavelet Transform

The basic DWT architecture has low-pass and high-pass filters. There are various DWT architectures which can be broadly classified into the regular structure and irregular structure as shown in Fig. 1. In regular structures, we find repeatability of the common blocks in regular order so that the circuit complexity in design and implementation is reduced [13]. In irregular structure, the redundancies in regular structures are exploited and the data flow is controlled, and irregular structures also require the same set of building blocks as that of regular structure [14].

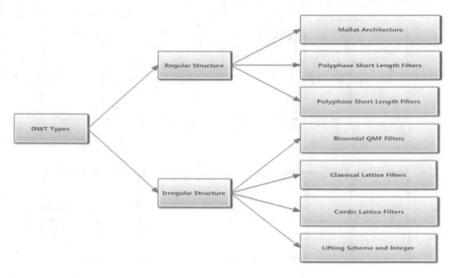

Fig. 1 Various structures of DWT types

The convolution-based DWT has a number of architectures which have been proposed based on various calculations in terms of performance, processing speed, or efficiency [15]. The lifting scheme-based DWT architecture capable of performing input decomposition using predicts and update logic is developed [16]. The added advantage of lifting scheme is that we can obtain forward and inverse transform of same architecture [17, 18]. There are two types of lifting schemes as shown in Fig. 2. In the lifting scheme, we factorize the polyphase matrix into alternative upper and lower triangular matrices and followed by a diagonal matrix [19].

The low-pass and high-pass analysis filters are represented by $\overline{h(Z)}$ and $\overline{g(z)}$; low-pass and high-pass synthesis filters are represented by $h(z)$ and $g(z)$. Thus poly-phase matrices are defined as mentioned in Eq. 1 and Eq. 2

$$\overline{P}(z) = \begin{bmatrix} \overline{h}_e(Z) & \overline{h}_o(Z) \\ \overline{g}_e(Z) & \overline{g}_o(Z) \end{bmatrix} \tag{1}$$

$$P(z) = \begin{bmatrix} h_e(Z) & h_o(Z) \\ g_e(Z) & g_o(Z) \end{bmatrix} \tag{2}$$

The lifting schemes architectures are shown in Fig. 2. There are two schemes shown in the figure through which one can perform DWT- and IDWT-based compressions. The two schemes can be mathematically represented as follows:

$$\overline{P}_1(z) = \begin{bmatrix} K & 0 \\ 0 & 1/K \end{bmatrix} \prod_{t=1}^{m} \begin{bmatrix} 1 & \overline{S}_t(z) \\ 0 & 1 \end{bmatrix} \begin{bmatrix} 1 & 0 \\ \overline{t}_i(z) & 1 \end{bmatrix} \tag{3}$$

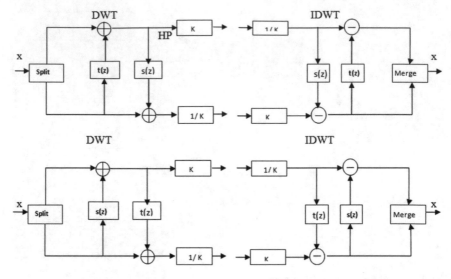

Fig. 2 Lifting schemes [20]

$$\overline{P}_2(z) = \begin{bmatrix} K & 0 \\ 0 & 1/K \end{bmatrix} \prod_{t=1}^{m} \begin{bmatrix} 1 & 0 \\ \overline{t}_i(z) & 1 \end{bmatrix} \begin{bmatrix} 1 & \overline{S}_t(z) \\ 0 & 1 \end{bmatrix} \tag{4}$$

Figure 2 corresponds to the two types of lifting scheme. Scheme one corresponding to Eq. 3 consists of steps mentioned below:

Step 1: The first step is the predicting step where the even samples of the input are multiplied by that is the time domain and added to the odd samples.

Step 2: The second step is the update step where the updated odd samples are multiplied by that is the time domain and added to the even samples.

Step3: In the third and final step, the updated that is 1/K multiplied to even samples and K is multiplied by odd samples [20].

Scheme two corresponding to Eq. 2 consists of following steps:

Step 1: In the first step, the even samples are calculated
Step 2: In the second step, the odd samples are calculated
Step 3: In the third step, the inverse is obtained by traversing in the reverse direction.

3 Proposed Architecture

The proposed block diagram of the design is shown in Fig. 3. As shown, the architecture consists of a DWT processing unit and dual-memory controller. The internal buses take care of read and write operations between the processor. In this architecture, the read and write actions occur in parallel. This architecture increases the

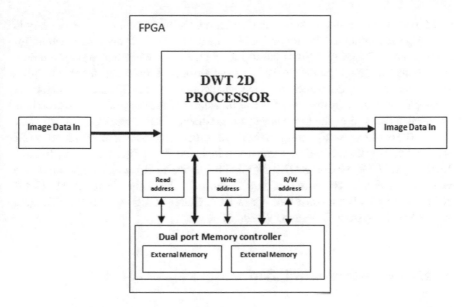

Fig. 3 Block diagram of DWT processor

processing time and overall performance of the design. In the design, the row/column processing is performed and is processed by the controller. The necessary parallel and pipelining processing is taken care in the design which makes the control design simpler and efficient. The test bench of the DWT processor would consist of the processor and the behavioral model of the dual-port memories. The dual-port memory is initialized with the pixel values of the grayscale image. The pixel value of the input image is passed as input to the DWT processor. This input is interleaved into the two banks in a column first indexing strategy as per our design specification.

If the image contains a checkerboard of black and white pixels, then all of the black pixels will end up in the one bank, and all of the white pixels in the other. The processor includes a memory controller that schedules the reads and writes to memory and an image processing filter kernel that computes the 2D DWT decomposition of the image [21]. The processor has been configured to compute two levels of decomposition on a 64×64-pixel image. The computation is done in place in the single image memory space, and the initial pixel ordering is preserved throughout. A full-color version of the design would easily fit in the same part as only the DWT data path would require replication; therefore, memory control could remain unchanged. This is a behavioral model for a dual-port memory. It is an obtuse model. Basically, there is a vector register for storage in the feedback loop. Vectors are constructed in front of this loop to allow one word to be inserted into the storage at the appropriate spot if the write enable is asserted. Following the storage, the loop is the read section that picks off the appropriate word from the memory vector. In the case of overlapping address, the read is done before the write changes the stored word. At every time step of simulink simulation, the full contents of the memory in the design are output as a vector of double data type. This is a very handy feature for analysis and debugging. This is the top level of the wavelet filter implementation. It contains the design gateways, which have all been double registered. The filter kernel, the memory controller, pipeline balancing delay blocks and a crossbar for every pair of signals in the design is interfaced to the memory bank. The crossbars are needed because of the interleaving of the image pixels across the two memory banks. The hardware setup configuration in the proposed model of reconfigured VLSI architecture for discrete wavelet transform included Zynq 7000 series (ZedBoard) Xilinx Zynq-7000 AP SoC XC7Z020-CLG484 series FPGA with a high-end laptop as shown in Fig. 8. The software requirements for the design included MATLAB 2017b including all the latest tool boxes-HDL Coder, XILINX Vivado 2017.4 and Xilinx Vivado System Generator, Vivado SDK.

4 Hardware Implementation

We have designed the above design in MATLAB and Simulink, and the same design is modeled in Xilinx system generator [22]. During simulation we generate the Xilinx equivalent DWT processor block and we perform co-simulation of the design. Finally implement the design on ZedBoard. After implementation of the design the equivalent netlist files are also generated. This process reduces the time to market of any design. Generally, the hardware description language have become extremely popular because the same language is used by design engineers to test, verify and validate algorithms on CPLD and FPGA's.

We have tested and simulated the design at every level of the design process to check the functionality before deploying the algorithm on FPGA. The register transistor logic is described in hardware description language and tested by simulation of the corresponding test bench. Based on the obtained results, the designer can synthesize the design and map the obtained netlist to be converted to gate-level design. Multiple test benches are applied to the design to see that the design is error free. This design is finally implemented on FPGA, and various functionalities of the design are verified. Finally, the memory controller-based DWT processor IP is obtained is shown in Fig. 4. This IP of DWT processor can be reconfigured further as per requirement of the designer. The IP level design of memory controller-based DWT processor architecture is shown in Fig. 4.

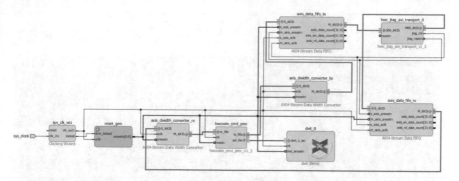

Fig. 4 IP of memory controller-based DWT processor

5 Results

In this section, we describe the figures of merit obtained for our proposed architecture and design when implemented on FPGA. The functionally verified netlist is synthesized using Vivado 2014.7 and implemented on ZedBoard. The adopted hardware was ZedBoard with board specification XC7Z020 and IC specification clg400 with grade speed of −1 which is from the Zynq family. The source estimation, power analysis and timing reports of the corresponding design is obtained from Xilinx Vivado software. As the per the design specification, the design is synthesis and implemented on ZedBoard. The area utilization is mentioned in Table 1.

The memory controller-based DWT processor model is designed and developed, and the simulation result of the entire design is shown in Figs. 5 and 6, respectively.

The laptop is interfaced with the development ZedBoard FPGA for programming. Once the interfacing is done, the programming file of the design is generated. The device is then configured so that the generated programming file can be successfully implemented on ZedBoard. The RTL schematics and block-level internal routing of the implemented design of the proposed design with interconnects internal routing of the implemented design are shown in Fig. 7.

Table 1 Logical resource utilization

Parameters	Available	Used number	Utilization (%)
Slices	53200	1124	2
Slice registers	105400	1607	2
F7 muxes	13300	5	<1
BUFG CTRL	32	3	1
Bonded IOBs	200	1	1
Block RAM tile	140	2	1

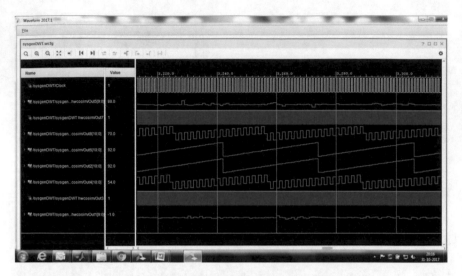

Fig. 5 Simulation results of memory controller-based DWT processor

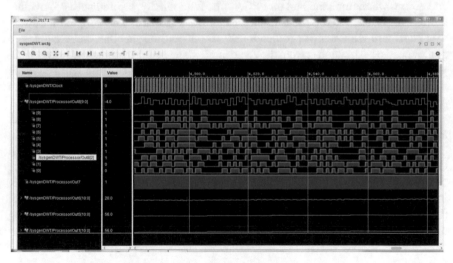

Fig. 6 Simulation results of a memory controller-based DWT processor

The power and area of the corresponding design can be further optimized based on the HDL algorithm. Xilinx Synthesis tool will give the Estimated Power and Implementation tool will give actual power. Based on the Utilization Report of Logical resources like FF's, BRAM, LUT's and Slice Registers which in turn dependent on Design Implementation. The power analysis parameters of synthesized and implemented design are captured and listed below in Table 2.

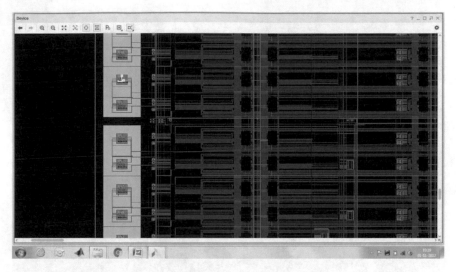

Fig. 7 Results of RTL of memory controller-based DWT processor

Table 2 Experimental results of power consumption

Parameters	Synthesized design	Implemented design
Total on-chip power	0.237 W	0.306 W
Junction temperature	27.7 C	28.5 C
Thermal margin	57.3 C (4.8 W)	71.5 C (5.5 W)
Power supplied to off-chip devices	0 W	0 W
Dynamic on-chip power	0.118 W (50%)	0.118 W (39%)
Static on-chip power	0.120 W (50%)	0.188 W (61%)

The verified model is synthesized using Vivado targeting ZedBoard FPGA consisting of 250 thousand gates. The results obtained show that the proposed design operates at a maximum frequency of 300 MHz and consumes power less than 0.306 W, with less than 1% resource utilization. In Fig. 8 we can see the hardware setup of the design.

Fig. 8 Hardware setup of memory controller-based DWT processor

6 Conclusion

In this paper, we have designed and implemented a memory controller-based DWT processor. This design is not only implemented on FPGA but also been synthesized on Mentor Graphics for ASIC implementation. The performance of the design in terms of area utilization, timing, and power is shown in the previous section. The obtained results of the architecture are efficient to produce accurate results. Based on the results, the obtained DWT processor IP can be further reconfigured for high-end video compression applications. At the same time, the memory controller increases the controllability. The flow control parameters increase the performance by passing the inputs based on priority and scheduling

Acknowledgements I would thank the CoreEL Technologies Pvt Ltd for providing the various facilities and resources available. I would also thank KLE Dr. M.S.S.CET, Belagavi, for their support in completing this work.

References

1. Kumar, P., Tigadi, A., Guhilot, H., Vyasaraj, T.: Design and analysis of real-time video processing based on DWT architecture for mobile robots. Procedia Comput. Sci. (2016)
2. Mehrseresht, N., Taubam, D.: An efficient content adaptive motion-compensated 3D-DWT with enhanced spatial and temporal scalability. IEEE Trans. Image Process. 15(6) (2006)
3. Srinivasarao, B.K.N., Chakrabarti, I.: High-performance VLSI architecture for 3-D discrete wavelet transform. In: International Symposium on VLSI Design, Automation and Test (VLSI-DAT) (2016)
4. Das, A., Hazra, A., Banerjee, S.: An efficient architecture for 3-D discrete wavelet transform. IEEE Trans. Circuits Syst. Video Technol. (2014)
5. Said, A., Pearlman, W.A.: A new fast and efficient image codec based on set partitioning in hierarchical trees. IEEE Trans. Circuits Syst. Video Technol. 6(3) (1996)
6. Naveen, Ch., Satpute, V.R., Keskar, A.G., Kulat, K.D.: Fast and memory efficient 3D-DWT based video encoding techniques. In: Proceedings of the International Multiconference of Engineers and Computer Scientists, vol. 1, pp. 427–433 (2014)
7. He, C., Dong, J., Zheng, Y.F., Gao, Z.: Optimal 3-D coefficient tree structure for 3-D wavelet video coding. IEEE Trans. Circuits Syst. Video Technol. 13(10), 961–972 (2003)
8. Potluri, U.S., Madanayake, A.: Improved 8-point approximate DCT for image and video compression requiring only 14 additions. IEEE Trans. Circuit Syst. 61(6) (2014)
9. Kalia, K., Pandey, B., Nanda, K., Malhotra, S., Kaur, A., Hussain, D.M.A.: Pseudo open drain IO standards-based energy efficient solar charge sensor design on 20 nm FPGA. In: IEEE 11th International Conference on Power Electronics and Drive Systems (2015)
10. Naveen, Ch., Satpute, V.R., Kulat, K.D., Keskar, A.G.: Comparison of 3D-DWT based video pre-post processing techniques. In: Selected at World Congress on Engineering and Computer Science, San Francisco, pp. 22–24 (2014)
11. Sureshraju, K., Satpute, V.R., Keskar, A.G., Kulat, K.D.: Image compression using wavelet transform compression ratio and Psnr calculation. In: Proceedings of the National Conference on Computer Society and Informatics (2012)
12. Pilrisot, C., Antonini, M., Barlaud, M.: 3-D scan based wavelet transform and quality control for video coding. Eur. Assoc. Signal Process. J. Appl. Signal Process., 56–65 (2003)
13. Daubechies, I., Sweldens, W.: Factoring wavelet transforms into lifting steps. J. Fourier Anal. Appl. 4(3), 247–269 (1998)
14. El Gamal, A., Eltoukhy, H.: CMOS image sensors. IEEE Circuits Devices Mag. 21(3), 6–20 (2005)
15. Mohanty, B.K., Meher, P.K.: Memory-efficient high-speed convolution-based generic structure for multilevel 2-D DWT. IEEE Trans. Circuits Syst. Video Technol. 23(2), 353–363 (2013)
16. Das, B., Banerjee, S.: Data-folded architecture for running 3-D DWT using 4-tap Daubechies filters. Proc. Circuits Devices Syst. 152(1), 17–24 (2005)
17. Naveen, Ch., Satpute, V.R., Kulat, K.D., Keskar, A.G.: Video encoding techniques based on 3D- DWT. In: IEEE Students' Conference on Electrical, Electronics and Computer Science, SCEECS (2014)
18. Barua, S., Carletta, J.E., Kotteri, K.A., Bell, A.E.: An efficient architecture for lifting-based two-dimensional discrete wavelet transforms VLSI. J. Integr. 38(3), 341–352 (2005)
19. Taghavi, Z., Kasaei, S.: A memory efficient algorithm for multidimensional wavelet transform based on lifting. Proc. IEEE Int. Conf. Acoust. Speech Signal Process. (ICASSP) 6, 401–404 (2003)
20. Mammeri, A., Hadjou, B., Khoumsi, A.: On the selection of appropriate wavelet filters for visual sensor networks. In: 3rd International Symposium on Applied Sciences in Biomedical and Communication Technologies (2010)

21. Das, B., Banerjee, S.: Low power architecture of running 3-D wavelet transform for medical imaging application. Proc. Eng. Med. Biol. Soc./Biomed. Eng. Soc. Conf. **2**, 1062–1063 (2002)
22. Darji, A., Shukla, S., Merchant, S.N., Chandorkar, A.N.: Hardware efficient VLSI architecture for 3-D discrete wavelet transform. In: Proceedings of 27th International Conference on VLSI Design and 13th International Conference on Embedded Systems, pp. 348–352 (2014)

Design and Analysis of Universal-Filtered Multi-carrier (UFMC) Waveform for 5G Cellular Communications Using Kaiser Filter

Ravi Sekhar Yarrabothu, Narmada Kanchukommula
and Usha Rani Nelakuditi

Abstract Currently, the 5G research is trying to address the existing OFDM-based LTE problems such as high peak-to-average power ratio (PAPR) and spectral loss. Universal-filtered multi-carrier (UFMC) technique can be considered as a candidate waveform for 5G communications because it provides robustness against inter-symbol interference (ISI) and inter-carrier interference (ICI) and suitable for low-latency scenarios. In this paper, Kaiser filter is used instead of standard Dolph–Chebyshev for UFMC-based waveform to provide better power spectrum and to avoid spectral leakage and is simulated using MATLAB software. A comparative study for Dolph–Chebyshev and Kaiser filters is performed, and the results are presented in this paper. The results show improvement in PSD and reduction in PAPR for UFMC with Kaiser-based window, in contrast to UFMC with Dolph–Chebyshev filter and conventional OFDM.

Keywords UFMC · 5G · Dolph–Chebyshev filter · Kaiser filter and OFDM · LTE

1 Introduction

The appetite of human society for more bandwidth-based applications forcing the cellular industry to work towards better technologies and is fuelling the development of 5G research. By 2020 frame, fifth-generation (5G) cellular access will be a reality and already trails are going on across the world.

R. S. Yarrabothu (✉) · N. Kanchukommula · U. R. Nelakuditi
Depatment of Electronics and Communication Engineering, VFSTR (Deem to be University),
Guntur, India
e-mail: ykravi@gmail.com

N. Kanchukommula
e-mail: kknarmada81@gmail.com

U. R. Nelakuditi
e-mail: usharani.nsai@gmail.com

© Springer Nature Singapore Pte Ltd. 2019
J. Wang et al. (eds.), *Soft Computing and Signal Processing* ,
Advances in Intelligent Systems and Computing 898,
https://doi.org/10.1007/978-981-13-3393-4_73

721

Currently, orthogonal frequency-division multiplexing (OFDM) technique is extensively used in wireless communications and also in many recent communication systems due to its efficiency. However, it has limitations such as larger PAPR and ICI. In OFDM, cyclic prefix (CP) causes inter-symbol interference (ISI) which is caused due to delay in distribution of the channel is higher than CP length [1].

With the aim to overcome these drawbacks, new techniques are introduced in 5G. Universal-filtered multi-carrier (UFMC) was one of the proposed techniques. This waveform groups number of subcarriers into sub-bands. In this way, UFMC procedure increases good spectral efficiency as a result of lack of cyclic prefix and diminishing out-of-band emission.

UFMC with Dolph–Chebyshev filter is not the best possible filter, since the side-lobe fall rate of Dolph–Chebyshev filter [2] is not very quick, which causes an increase in the spectrum leakage. Hence, in this paper, UFMC with Kaiser filter is implemented. The Kaiser filter or Kaiser-Bessel window is a flexible smoothing window whose shape can be by adjusted by modifying **beta**. Depending on the application, the shape of the window can be changed to control the amount of spectral leakage [3]. It provides a sub-optimal solution for the out-of-band emissions. The most important factor of Kaiser filter is to provide the flexibility to control stop band and the side-lobe attenuation, by using β input.

This paper is divided into four sections: Section 2 describes the architecture of UFMC system; Sect. 3 talks about the Kaiser window for UFMC; simulation setup parameters and results are presented in Sect. 4; and finally in Sect. 5, the conclusions are explained.

2 UFMC Architecture

UFMC is a technique where blocks of subcarriers are filtered before transmission and reception for eliminating inter-carrier and inter-symbol interferences. UFMC is a modulation technique that can be considered as a generalized filter bank multi-carrier technique (FBMC) and filtered OFDM. FBMC applies a filtering to each subcarrier while in a single shot and has many advantages like lesser ICI, in case of frequency offsets or jitter. However, it increases computational complexity of the signal [4]. In filtered OFDM, filtering is applied to the total band. Hence, the bandwidth of the filter is large and filter length is much lesser than FBMC. In UFMC, subsets are filtered in the whole band instead of total band or single subcarriers [5].

A. **UFMC Transmitter Design**:

UFMC transmitter [6] structure with B sub-bands is as shown in Fig. 1. UFMC's ith sub-module, where i ∈ {1, 2, …, B}, generates the time-domain baseband vector x_i with (N + Nfilter-1) dimension for the sub-bands, respectively, which carries the QAM symbol vector Si with dimension $n_i x_i$ where N is the number of samples required per symbol which represents all sub-bands without aliasing, and Nfilter indicates the length of the filter. From Fig. 1, transmit vector x is synthesized by

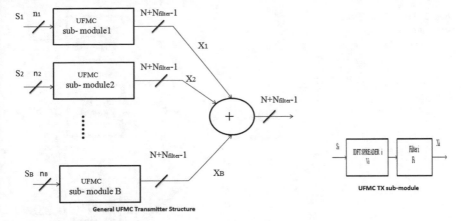

Fig. 1 UFMC transmitter structure and Tx sub-module

combining single sub-band signals. In downlink, multiple users get the transported data on single sub-modules over the all available frequency bands. In uplink (UL), the user uses the assigned frequency portion.

By using IDFT spreader, the complex QAM symbol ni is changed to time domain, and then sub-band filtering is performed. For a given multi-carrier symbol, the time-domain transmit vector is the superposition of the sub-bandwise filtered components is given in Eq. (1)

$$X = \sum_{i=1}^{B} F_i V_i S_i \tag{1}$$

Vi indicates dimension $N \times n_i$, includes the relevant columns of the IFT matrix as per the position of sub-band; Fi is a Toeplitz matrix whose dimension is (N+Nfilter − 1) × N and is composed of the FIR, which enables the convolution.

The rewritten signal, without the summation, is defined as follows:

$$\overline{F} = [F_1 F_2, \ldots, F_B] \tag{2}$$

$$\overline{V} = \text{diag}(V_1, V_2, \ldots, V_3) \tag{3}$$

$$\overline{S} = [S_1^T, S_2^T, \ldots, S_B^T]^T \tag{4}$$

This enables columnwise piling of filter matrices, generation of an IDFT matrix and pooling up all the data symbols into a single column.

This results into:

$$X = \overline{F} \, \overline{V} \, \overline{S} \tag{5}$$

The following design parameters are defined for the UFMC system:

Fig. 2 UFMC receiver
structure

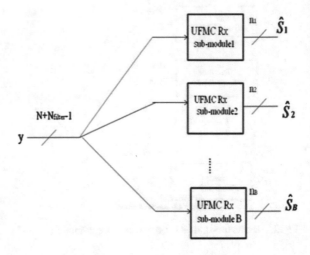

- B: No. of sub-bands
- n_i: No. of subcarriers in sub-band i
- N: Overall no. of subcarriers
- Filter i (F_i): Bandwidth or length.

Here, B varies according to the UFMC transmitter spectral settings. If the system is applied to fragmented spectrum, B should be selected depending on the available number of spectral sub-bands. Alternatively, for overall system stream lining and controlling the more fine-grained spectral characteristics, the single sub-band can be divided into minute chunks of same size in every sub-band. These particular spectral chunks are called as physical resource blocks (PRBs), as in the LTE [7].

B. UFMC Receiver Design:

UFMC receiver processing is done as depicted in Fig. 2. In reception, processing is done on the elements for simplicity which deals with blocks like channel estimation/equalization and error correction in the UFMC signal processing. From Fig. 2, y is the received signal vector through the channel after propagation, with Toeplitz structure by convolution matrix H, including noise n is given in Eq. (6)

$$y = Hx + n = H\bar{F}\bar{V}\bar{S} + n \tag{6}$$

Each Rx sub-module of UFMC transmits the output symbol vectors to the respective sub-bands along with distortions. In uplink (UL), the Rx sub-modules are used for data transmissions to cover the complete frequency range, where as in downlink (DL) the receiver acts as a part of the user device. Active single sub-modules or the PRBs carry the data and control messages relevant to an assigned user. Different methods for the design of UFMC receiver are still under research.

In an ideal linear receiver case under an AWGN channel, zero-forcing filter (ZF) and minimum mean square error (MMSE) can be given as:

$$W_{ZF} = \left(\overline{FV}\right)^+ = T^+ \tag{7}$$

$$W_{MMSE} = \left(T^H T + \sigma^2 I\right)^{-1} T^H \tag{8}$$

where T^+ is the Moore-Penrose inverse of a matrix, T^H is the hermitian transpose, σ^2 is the noise variance and I is identity matrix. The receiver operation is given in Eqs. (7) and (8) and can be viewed as a concatenation of inverse filtering and DFT despreading.

3　Proposed UFMC with KAISER Window

In the UFMC transmitter section, a filter is used for each sub-module as shown in Fig. 1. In this paper, in order to control the spectral leakage an FIR filter, Kaiser window is used. The DPSS window based upon Bessel functions is well-known as the Kaiser window (or Kaiser-Bessel window).
Definition:

$$w_k(n) \triangleq \begin{cases} \dfrac{I_0\left(\beta\sqrt{1-\left(\frac{n}{M/2}\right)^2}\right)}{I_0(\beta)}, & -\dfrac{M-1}{2} \le n \le \dfrac{M-1}{2} \\ 0, & elsewhere \end{cases} \tag{9}$$

Window transforms:
The Fourier transform of the Kaiser window Wk(t) is given in Eq. (10). Here, t is a continuous signal.

$$W(w) = \frac{M}{I_0(\beta)} \frac{\sinh\left(\sqrt{\beta^2 - \left(\frac{M_w}{2}\right)^2}\right)}{\sqrt{\beta^2 - \left(\frac{M_w}{2}\right)^2}} \tag{10}$$

$$W(w) = \frac{M}{I_0(\beta)} \frac{\sin\left(\sqrt{\left(\frac{M_w}{2}\right)^2 - \beta^2}\right)}{\sqrt{\left(\frac{M_w}{2}\right)^2 - \beta^2}} \tag{11}$$

where I_0 is the zero-order-modified Bessel function of the first kind.

$$I_0(x) \triangleq \sum_{k=0}^{\infty} \left[\frac{\left(\frac{x}{2}\right)^k}{k!}\right]^2 \tag{12}$$

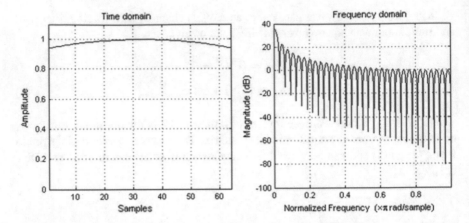

Fig. 3 Kaiser filter characteristics in time domain and frequency domain ($\beta = 1.0$, filter length: 64)

From Fig. 3, it can be observed that the beta input in the Kaiser window affects the side-lobe attenuation parameter of the Fourier transform. The β parameter provides continuous control over the window exchange between main-lobe breadth and side-lobe level. Larger β values with lower side-lobe levels provide a wider main lobe.

4 Simulation Results

UFMC system is simulated using MATLAB software with both Kaiser window and Dolph–Chebyshev filter. The general input parameters used for the simulations are as shown in Table 1. For Kaiser filter, side-band attenuation is 1 db and remaining other parameters are same as mentioned in Table 1.

A. **Waveform Filters**:

The simulated UFMC filter characteristics in frequency domain for both Dolph–Chebyshev and Kaiser filters are depicted in Fig. 4. From Fig. 4, it is observed

Table 1 UFMC and OFDM simulation parameters

Total number of used carriers	256
Number of symbols in a frame	20
Modulation	64 QAM
UFMC estimation	MMSE estimation
UFMC block size	16
UFMC number of resource blocks	16
Filter length	7
Filter side-band attenuation	40 dB
Channel	Perfect, no noise

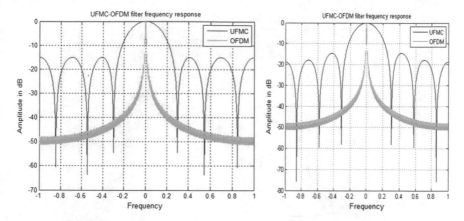

Fig. 4 UFMC with Dolph–Chebyshev, Kaiser and OFDM filter characteristics—frequency domain

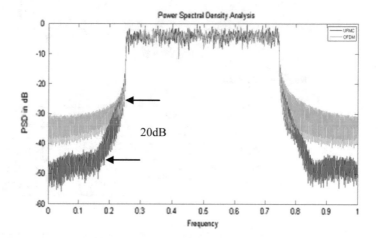

Fig. 5 PSD analysis using Dolph–Chebyshev window

that the UFMC filter side-band attenuation with Kaiser is slightly better than both the UFMC (Dolph–Chebyshev) and OFDM. The frequency responses show that the Kaiser-based UFMC filter provides small main-lobe width which indicates steepness than Dolph–Chebyshev filter; hence, it provides better power spectrum.

B. Power Spectral Density Analysis:

Figures 5 and 6 show the PSD analysis results for the 60-dB SNR situations when both OFDM and UFMC utilize the MMSE estimation for both Dolph–Chebyshev and Kaiser filters, respectively. As shown in Fig. 5, UFMC has around 20 dB better performance on the stop band because of Dolph–Chebyshev filter.

As seen on Fig. 6, UFMC has around 24 dB better performance on the stop band because of Kaiser filter. From Table 3 UFMC with Kaiser filter performs about 22 dB

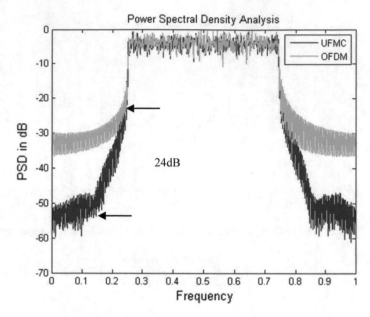

Fig. 6 PSD analysis using Kaiser window

Table 2 Channel power measuring values for Dolph–Chebyshev filter

System	SNR	AWGN	ACP
OFDM	60	YES	−62.9805
OFDM	–	NO	−64.7169
UFMC (DC)	60	YES	−80.3931
UFMC (DC)	–	NO	−83.5258

Table 3 Channel power measuring values for Kaiser filter

System	SNR	AWGN	ACPR
OFDM	60	YES	−63.5876
OFDM	–	NO	−64.1924
UFMC (Kaiser)	60	YES	−87.5264
UFMC (Kaiser)	–	NO	−85.8745

better than OFDM. Adjacent channel power ratio (ACPR) analysis for OFDMA and UFMC is performed by using both the Dolph–Chebyshev and Kaiser filters is captured in Tables 2 and 3.

From Tables 2 and 3, we can notice that the UFMC with Kaiser filter is slightly better than ACPR compared to UFMC with Dolph–Chebyshev and superior to OFDM system.

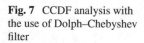

Fig. 7 CCDF analysis with the use of Dolph–Chebyshev filter

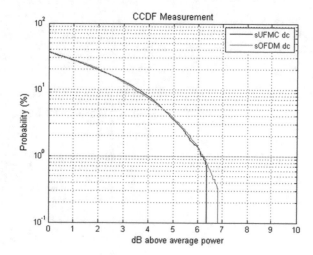

C. CCDF Analysis

Complementary cumulative distribution function (CCDF) shows about peak-to-average power ratio (PAR). As the name indicates it shows the distribution of power in dB above average power. In communication systems small PAR is better, where as high PAR values causes saturation in both transmitter and receiver.

Average power in dB is represented by the horizontal axis and probability of the existence of the signal power on vertical axis. Figures 7 and 8 illustrate the CCDF analysis for the OFDM and UFMC transceivers for both Dolph–Chebyshev and Kaiser filters, respectively. One can infer from the figures that OFDM has slightly higher probability of carrying signals with higher PAPR, from 0.6 dB to 1 dB, when compared to UFMC. And also, it can be observed that the UFMC system with Kaiser filter has low PAR by 0.4 dB, when compared to UFMC with D–C filter.

5 Conclusions

The simulation results show that the UFMC with Kaiser window can provide slightly improved PAPR characteristics by 0.4 dB less and the side-lobe suppression of 4 dB compared to UFMC with Dolph–Chebyshev filter and conventional OFDM.

Fig. 8 CCDF analysis with
the use of Kaiser filter

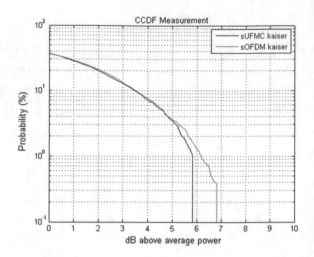

References

1. Hamiti, E., Sallahu, F.: Spectrum comparison between GFDM, OFDM and GFDM behavior in a noise and fading channel. Int. J. Electr. Comput. Eng. Syst. **6**(2), 39–43 (2015)
2. Mukherjee, M., Shu, L., Kumar, V., Kumar, P., Matam, R.: Reduced out-of-band radiation-based filter optimization for UFMC systems in 5G. In: IEEE 81st Vehicular Technology Conference (2015)
3. Kaiser, J., Schafer, R.: On the use of the I0-sinh window for spectrum analysis, IEEE Trans. Acoust. Speech Signal Proc. **28**(1), 105–107 (1980)
4. Farhang-Boroujeny B.: OFDM versus filter bank multicarrier. IEEE Signal Process. Mag. **28**, 92–112 (2011)
5. G Waveform Candidate selection: D3.1, Version 1.0. http://www.5gnow.eu/wpcontent/uploads/2015/04/5GNOW_D3.1_v1.1_1.pdf
6. Knopp, R., Kaltenberger, F., Vitiello, C., Luise, M.: Universal filtered multicarrier for machine type communications in 5G: European conference on networks and communications, Athens, Greece (2016)
7. Vakilian, V., Wild, T., Schaich, F., Ten Brink, S., Frigon, J.-F.: Universal-filtered multi-carrier technique for wireless systems beyond LTE. In: Proceedings IEEE Globecom Workshops, pp. 223–228 (2013)

Dynamic Spectrum Allocation and RF Energy Harvesting in Cognitive Radio Network

Pallavi Shetkar and Sushil Ronghe

Abstract RF energy harvesting (EH) is a new paradigm constituted by the wireless sensor network which enables it to recharge through the directed electromagnetic energy transfer. In a practical scenario, secondary users (SUs) are unacquainted with the traffic statistics of primary users (PUs). Thus, maximizing bandwidth utilization is one of the objectives of this paper. In order to obtain reasonably accurate estimations of spectrum opportunities (channels), the modified myopic scheme is implemented in this paper. In addition, energy detection algorithm is implemented for spectrum sensing. A scheme which proficiently executes channel selection, channel allocation and energy harvesting for the system model of multiple PUs and SUs in cognitive radio network (CRN) is also proposed in this paperwork. The outcome of the paper proves that the proposed scheme maintains a satisfactory balance between accessing the spectrum and harvesting energy while maintaining the fairness among SUs.

Keywords Cognitive radio network · Channel selection · Spectrum sensing
Spectrum allocation · RF energy harvesting

1 Introduction

Federal Communications Commission disclosed that the problem of spectrum shortage came into view due to inefficient use of available spectrum [1, 2]. To overcome this problem, dynamic spectrum access (DSA) technique is put forth to enable spectrum sharing among SUs (users with no authority to use channels). DSA technique permits SUs to dynamically access idle frequencies or primary channels without interrupting PUs (users with authority to use channel) [3].

P. Shetkar (✉) · S. Ronghe
College of Engineering, Pune, Pune 411005, India
e-mail: shetkarpa16.extc@coep.ac.in; pallavishetkar7588@gmail.com

S. Ronghe
e-mail: sbr.extc@coep.ac.in

© Springer Nature Singapore Pte Ltd. 2019   731
J. Wang et al. (eds.), *Soft Computing and Signal Processing* ,
Advances in Intelligent Systems and Computing 898,
https://doi.org/10.1007/978-981-13-3393-4_74

As the statistics of the PUs are unknown, it is necessary for the SUs to predict nearly accurate spectrum opportunities in coordination with spectrum sensing to increase the spectrum usage. For spectrum sensing, the channel selection schemes play a vital role in order to increase the throughput of the SUs in CRN [4]. If the channel sensed by the SU is found to be idle then that channel can be used for packet transmission. Hence, channel selection is an important step to be executed prior to channel sensing. In this paper, the modified myopic scheme (MMS) is implemented for predicting channels to sense for a given time slot. With the help of MMS, SUs can learn the occupancy status of the PUs [4–6].

The SUs in CRN require energy for packet transmission. All SUs are equipped with a battery. So, sustainability of a CRN can be improved by harvesting energy from immediate surrounding sources such as radio frequency (RF) signals, thermal energy, solar energy [7, 8]. The RF energy harvesting is a proficient technique to widen the lifetime of CR networks without using external energy sources [7]. At a particular time slot, SUs can either transmit data on a channel or harvest energy; hence, EH should be integrated with DSA.

The motivation behind this paperwork is to design an efficient system consists of multiple SUs and PUs on more than one channel. This paperwork proposes a flexible and efficient framework to integrate dynamic spectrum allocation and energy harvesting along with channel selection and spectrum sensing. Particularly, the mode selection policy is proposed which enables every SU to decide its favoured mode of operation, i.e. either data transmission or energy harvesting. Our aim is to improve overall system efficiency, i.e. optimizing channel allocation while maintaining energy efficiency at the satisfactory level.

2 System Model

This paperwork considers time-slotted cognitive radio network which consists of PUs, SUs and a central node called Supernode. Let M number of SUs is waiting for data transmission and L number of PUs is present in the network. Every channel is divided into T number of time slots by assuming N number of duplex channels in the network. Supernode broadcasts an encoded channel allocation matrix which contains the pairs of the channel and the corresponding SU. Each SU decodes its allocated channel; if the channel is found to be idle then packet transmission is performed otherwise energy harvesting is performed. Then the acknowledgement is sent back to Supernode from each SU after the execution of the mutually exclusive events (i.e. packet transmission or EH).

2.1 Channel Selection

Modified myopic scheme (MMS) is implemented to get successful channel selection. MMS is implemented in two phases: learning phase and channel selection phase. Learning phase consists of initialization period in which channel statistics (either busy or idle) are learned. Initialization period is $\alpha ln T$ time slots per channel where $\alpha > 1$. Let i and j be the indices for the channel and the time slot, respectively. $\sigma_i(j)$ represents the count when the jth time slot of the channel i is sensed and $\rho_i(j)$ represents the count when a channel i is sensed and found to be idle till jth time slot. The channel idle probability $(\acute{\varepsilon}_i(j))$ can be calculated as per Eq. 1. Let β is the number of packets transmitted by the multiple SUs in a given time slot for different channels.

$$\acute{\varepsilon}_i(j) = \frac{\rho_i(j)}{\sigma_i(j)}. \tag{1}$$

The steps for MMS are as follows:

1. Sense $\alpha ln T$ time slots per channel in the initialization period.
2. Compute $\sigma_i(j)$, $\rho_i(j)$ for $i = 1, 2, \ldots, N$.
3. After the initialization period, update $\acute{\varepsilon}_i(j)$ of the ith channel till time slot j.
4. The channel $p = argmax_i \acute{\varepsilon}_i(j)$ has maximum channel idle probability; hence it can be utilized for data transmission and the channel $q = argmin_i \acute{c}_i(j)$ has minimum channel idle probability which implies that it has a strong possibility of getting occupied by PU. Hence, it can be used for energy harvesting, i.e. the channel q can be allocated to that SU whose energy level in the battery is minimum.

2.2 Spectrum Sensing

In this paper, spectrum sensing is performed by using energy detection algorithm [9]. Spectrum sensing means inspecting the presence of PU on a channel by monitoring the energy level at that channel. If the energy level is exceeding the predefined threshold then primary signal is present otherwise channel is said to be idle. The threshold is computed using Monte Carlo simulation model [10]. The false alarm probability and detection probability are the appraisals of sensing performance. The theoretical Eq. 2 indicates predefined threshold (γ) as given as follows:

$$\gamma = \frac{Q^{-1}(p_f)}{\sqrt{l+1}}. \tag{2}$$

where signal transmitted by primary user is QPSK signal with zero mean, noise considered as a white Gaussian noise, p_f is the false alarm probability and n is the number of samples taken to average out energy. Steps of energy detection algorithm:

1. Each SU computes average energy of l samples in the time slot of corresponding channel.
2. Compare the calculated value with the threshold (γ).
3. If the value is below the threshold then mark the given channel as idle.

2.3   Encryption and Decryption

Instead of using single channel per time slot with the help of CSMA/CA protocol [11], this paper put forth a new scheme in which all channels are utilized per time slot without increasing hardware complexity at the SU's end. In order to achieve this goal, encryption and decryption scheme is developed to access multiple channels per time slot.

In the system, Supernode develops the channel allocation matrix on the basis of channel idle probabilities, transmission count and energy count from each SU. Supernode broadcasts the encrypted channel allocation matrix (A) of order M × N, where rows correspond to SU and columns correspond to channel as shown in Eq. 3. If $A_{m,n} = 1$ then it implies that the nth channel is allocated to the nth user. In order to decrypt the matrix, each secondary user has given a unique matrix (B) for its identification as shown in Eq. 4. The matrix will have one row and M columns where M represents SUs. So for a particular SU, the corresponding element in the matrix B will be 1 and remaining all matrix elements are 0. For example, if $B_{1,m} = 1$ then it implies that it is the mth user identification code. Each SU needs to multiply that matrix with the encrypted matrix broadcasted by the Supernode to find out which channel is allocated to that particular SU, i.e. decryption matrix (D) of order 1 × N as per Eq. 5. If $D_{1,n} = 1$ then it implies that the nth channel is allocated to the corresponding user.

$$A = \begin{bmatrix} A_{11} & A_{12} & & A_{1N} \\ A_{21} & A_{22} & \cdots & A_{2N} \\ & \vdots & \ddots & \vdots \\ A_{M1} & A_{M2} & \cdots & A_{MN} \end{bmatrix}_{M \times N.} \tag{3}$$

$$B = [\, B_{11} \ B_{12} \ \ldots \ B_{1M} \,]_{1 \times M.} \tag{4}$$

$$D = B * A. \tag{5}$$

2.4 Criterion for Secondary User Selection

In CRN, Supernode will decide whether to allot idle channel or busy channel to a particular SU depending upon its transmission count and energy level count. Note that Supernode possesses the database of transmission count and energy level count of all SUs present in the network. In addition, it will update the database depending upon the acknowledgements given by SUs. It is necessary that every SU should get fair opportunity to transmit data. The channels having considerably greater channel idle probability are allocated to those SUs which are waiting for data transmission and having minimum data transmission count. The channels with lower channel idle probability are allocated to those users which are having minimum battery levels. The SUs who did not get chance for data transmission will get it in the next time slot provided it has nonzero battery levels.

Steps for SU selection are as follows: Initialize the arrays $[F]_{1 \times M}$ and $[E]_{1 \times M}$, where each element in these arrays maintains the count of data transmission and energy level count per SU, respectively.

1. Update F and E when data is transmitted in the given time slot by the corresponding SU. After packet transmission, the energy level will decrease, and transmission count will increase. Note that, the database will be updated after receiving the acknowledgement at the Supernode.
2. Update values of $\sigma_i(j)$, $\rho_i(j)$, $\acute{\varepsilon}_i(j)$, and ζ of the ith channel in the jth time slot.
3. Supernode will sort the estimated channel availability database in descending order and packet transmission count database in ascending order. Now, with reference to these databases, one-to-one mapping will be performed by the Supernode. Note that if the SU has less packet transmission count with zero energy levels available in the battery then to such SUs busy channel will be allotted by the Supernode unless it successfully harvests energy. In this way, the channel allocation matrix will be formed.
4. Supernode will broadcast the encrypted channel allocation matrix to all SUs as explained in Sect. 2.3.

Note that if channel is unfortunately found busy after allotment of idle channel by the Supernode to a SU then corresponding SU will perform energy harvesting and the values of $\sigma_i(j)$, $\rho_i(j)$ and $\acute{\varepsilon}_i(j)$ are modified accordingly and in the next time slot the channel idle probability of the ith channel will be lowered. Note that the SU sends only the status of the channel (idle or busy) as an acknowledgement to the Supernode.

2.5 Energy Harvesting

In the proposed scheme, the channels with considerably low channel idle probability are allocated to those SUs who possess the lowest battery level. The energy count

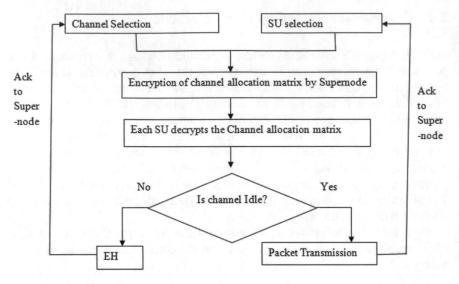

Fig. 1 Overall flow chart of proposed scheme

is maintained in the energy buffer. In this paper, let Z be the energy level in the battery, where $Z = 0, 1, 2, \ldots, e_{max}$. When SU transmits the packet then energy level decreases by 1. Steps for EH algorithm are as follows:

1. Update energy buffer, i.e. increment the count if energy is harvested or decrement the energy count of a particular SU in energy buffer when packet is transmitted.
2. Send the acknowledgement (status of the channel) to the Supernode.

Note that unfortunately if the channel is found idle after sensing then corresponding SU will transmit packet on that channel provided it has nonzero energy levels in the energy buffer and the values of $\sigma_i(j)$, $\rho_i(j)$ and $\grave{\varepsilon}(j)$ are updated accordingly and in the next time slot the channel idle probability of the ith channel will be increased. In this way, SUs are selected and channel idle probabilities are updated. In addition, acknowledgement sent to Supernode. Overall flow chart of the proposed scheme is shown in Fig. 1.

3 Simulation Results

For all the simulations $N = 10$, $\alpha = 1.5$, $T = 10{,}000$ are considered. In case of spectrum sensing, value of p_f is chosen to be 0.01 and the generated value of Monte Carlo threshold (γ) is 0.6163. Maximum energy level (e_{max}) owed by each secondary user at the initial phase is 10.

Figure 2 shows the time-averaged throughput (ζ) of the proposed system for two cases mainly for one SU and 100 SUs. By keeping all other parameters unchanged,

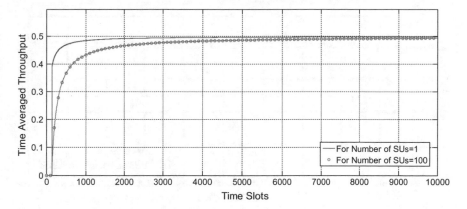

Fig. 2 Time-averaged throughput of the system

the number of secondary users is varied from 1 to 100. Learning time required for this system consisted of 10 channels is $10 * (\alpha \ln T)$ which is 140. Hence, throughput is zero till 140th time slot. The throughput of the system saturates around 0.5 refer to Eq. 6. It is because half of the time slots are utilized for data transmission and another half is invested for energy harvesting. Observations show that as the number of SUs increases the time required for each SU to reach the maximum throughput also increases. The upper graph with solid line represents the throughput of system with one SU reaching the saturation earlier as it has no competitors for data transmission.

$$\zeta = \frac{\beta}{T * N}. \tag{6}$$

Figure 3 shows the throughput of the system with EH scenario. Moreover, this throughput plot is also compared with non-EH scenario for single user. It can be observed from the graph that the throughput saturates around 0.5 for the SUs with limited source of energy. Also, the maximum throughput around 0.85 is observed for those SUs which are equipped with undying battery. In the non-EH scenario, it is possible to achieve this extraordinary threshold mark because the MMS always selects the channel with highest idle probability.

Figure 4 shows the throughput of each secondary user (η) in the system. System consists of 20 SUs. Each bar shows the normalized count of data transmission by each user which is around 0.25 refers to Eq. 7. All users are having same normalized count that means each secondary user is getting fair chance of data transmission. The objective of maintaining fairness among SUs via round robin algorithm is achieved.

$$\eta = \frac{\text{Transmission count}}{T}. \tag{7}$$

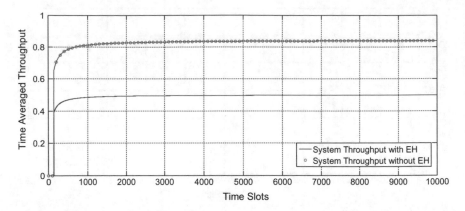

Fig. 3 System throughput with EH and without EH

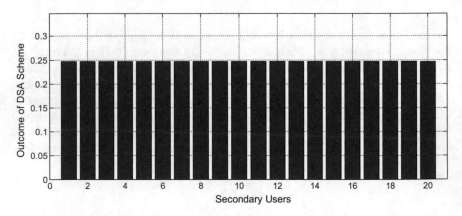

Fig. 4 Throughput of DSA maintaining fairness among SUs

4 Conclusion

This paper put forth an effective scheme for channel selection, dynamic spectrum allocation and RF energy harvesting. In this paperwork, the channel selection, spectrum sensing, dynamic spectrum allocation and RF energy harvesting are successfully implemented for multiple SUs and PUs coordinated by the Supernode. This proposed encryption and decryption scheme helps multiple SUs to transmit packet by utilizing the same time slots of the different channels. The objective of efficient bandwidth utilization and maintaining fairness among SUs is also achieved.

References

1. Andrews, J.G., Buzzi, S., Choi, W., Hanly, S.V., Lozano, A., Soong, A.C., Zhang, J.C.: What will 5G be? IEEE J. Sel. Areas Commun. **32**(6), 1065–1082 (2014)
2. Jacob, P., Sirigina, R.P., Madhukumar, A.S., Prasad, V.A.: Cognitive radio for aeronautical communications: a survey. IEEE Access **4**, 3417–3443 (2016)
3. Bhardwaj, P., Panwar, A., Ozdemir, O., Masazade, E., Kasperovich, I., Drozd, A.L., Varshney, P.K.: Enhanced dynamic spectrum access in multiband cognitive radio networks via optimized resource allocation. IEEE Trans. Wireless Commun. **15**(12), 8093–8106 (2016)
4. Rastegardoost, N., Jabbari, B.: On channel selection schemes for spectrum sensing in cognitive radio networks. In: Wireless Communications and Networking Conference (WCNC), pp. 955–959. IEEE (2015)
5. Ronghe, S.B., Kulkarni, V.P.: Modelling and performance analysis of RF energy harvesting cognitive radio networks. In: International Conference on Communication and Electronics Systems (ICCES), pp. 1–6. IEEE (2016)
6. Ali Ahmad, H., Liu, M., Javidi, T., Zhao, Q., Krishnamachari, B.: Optimality of myopic sensing in multichannel opportunistic access. IEEE Trans. Inf. Theory **55**(9) (2009)
7. Lu, X., Wang, P., Niyato, D., Kim, D.I., Han, Z.: Wireless networks with RF energy harvesting: a contemporary survey. IEEE Commun. Surv. Tutor. **17**(2), 757–789 (2015)
8. Chalasani, S., Conrad, J.M.: A survey of energy harvesting sources for embedded systems. In: Southeastcon, 2008, pp. 442–447. IEEE (2008)
9. Zhang, W., Mallik, R.K., Letaief, K.B.: Optimization of cooperative spectrum sensing with energy detection in cognitive radio networks. IEEE Trans. Wireless Commun. **8**(12) (2009)
10. Cabric, D., Tkachenko, A., Brodersen, R.W.: Experimental study of spectrum sensing based on energy detection and network cooperation. In: First International Workshop on Technology and Policy for Accessing Spectrum, p. 12 (2006)
11. Jones, S.D., Merheb, N., Wang, I.J.: An experiment for sensing-based opportunistic spectrum access in CSMA/CA networks. In: New Frontiers in Dynamic Spectrum Access Networks, 2005. DySPAN 2005, pp. 593–596. IEEE (2005)

A Study on Image Segmentation in Curvelet Domain Using Snakes and Fractals for Cancer Detection in Mammograms

R. Roopa Chandrika, S. Karthik and N. Karthikeyan

Abstract Breast cancer is one of the common diseases that affect the quality of life in women. To improve the mortality rate, screening procedures are recommended by medical practitioners. The procedure involves detecting and diagnosing the suspected regions. Detecting the suspected regions in the digital mammograms is a challenging task. In this work, multiresolution analysis is done to resolve the curve discontinuities and procedures like fractals and snakules are applied for segmenting the suspected region. The algorithms have been tested on mini-MIAS and observed that snakes have identified the regions better than fractals. Snakes are able to converge within few iterations determining the circumference of the suspected region and also able to differentiate tumorous and non-tumorous regions.

Keywords Curvelet · Fractals · Snakes · Mammogram · Breast cancer

1 Introduction

Cancer cells are abnormal cells that multiply rapidly invading the normal cells of human body. The parts of the human body affected by cancer are lung, breast, stomach, ovary, bone, etc. Breast cancer is the cancerous cells developed in the breast region, and more new cases are reported in women than men. According to the cancer

R. Roopa Chandrika (✉)
Department of Information Technology, Malla Reddy College
of Engineering and Technology, Hyderabad, Telangana, India
e-mail: mroopachandrika@gmail.com

S. Karthik (✉)
Department of Computer Applications, SNS College of Engineering and Technology,
Coimbatore, Tamil Nadu, India
e-mail: profskarthik@gmail.com

N. Karthikeyan (✉)
Department of Computer Applications, SNS College of Engineering, Coimbatore, Tamil Nadu,
India
e-mail: profkarthikeyann@gmail.com

© Springer Nature Singapore Pte Ltd. 2019
J. Wang et al. (eds.), *Soft Computing and Signal Processing* ,
Advances in Intelligent Systems and Computing 898,
https://doi.org/10.1007/978-981-13-3393-4_75

statistics [1], the estimated new cases are 249,260 and among them the estimated deaths are 40,890. To reduce the death rate in individuals affected by the disease, screening of the breast is recommended to women from the age of 40. The screening procedure is a simple task to detect cancers at an early stage. Mammography is the procedure that passes low dose of X-rays to detect the abnormal growth in the breast region. The X-ray image from the procedure is digitized mammogram. Review shows [2, 3] that due to low dosage of X-rays, from the digitized images, the medical practitioners find it difficult to interpret the abnormal cells. For a better interpretation, an automated system like computer-aided detection (CAD) is used for detecting and classifying abnormal cells.

The initial step of a CAD system is preprocessing involving image denoising filters and transforms to enhance features in the given image. Natural images show point or line singularities, i.e. lines or curves having discontinuities. Therefore, procedures are followed to handle these discontinuities for further processing. In this research, curvelet transform, variant of wavelets is applied on mammogram as a preprocessing tool to filter noise and to resolve the discontinuities. Fractals and snakules are applied to segment the suspected ROI from the original image.

2 Literature Review

Liu et al. have shown that multiresolution analysis [4] have improved the detection accuracy in mammograms. Wavelet transforms had limitations like increasing number of coefficients while reconstruction of edges in the image, increasing the complexity [5]. To ease the case, curvelet transforms were developed that includes orientation, scaling and locations. To isolate the image at different scales, multiscale ridgets [7] along with bandpass filtering are incorporated. The region of interest (ROI) is extracted from the image, and curvelet is applied on the region and the moment features are calculated. The results had shown that the dimension of the feature set has been reduced after applying curvelets [6, 7]. Different approaches like box-counting method and ruler set method for calculating fractal dimension in order to classify breast masses in mammograms [8]. The experimental analysis of fractal methods have shown they are good in characterizing the shape of breast tumour. The contour and shape are important parameters for classification of tumours.

In [9], curvelets and B-spline are combined to improve the performance of defining the shape of a ROI in medical images. Apart from fractal methods, active contour models are being applied to determine the shape of the breast mass. Snakes have the ability to grow, deform and adapt to the ROI with spiculations [10], and the performance has improved the CAD system specificity. Researchers have experimented and analysed fractals and snakules. In the following sections, experiments have been performed using the above methods and a comparative study on their performance is analysed. The study is about image preprocessing in curvelet domain and segmentation using fractal method and snakules. A comparative study is done on both the contour models in mammograms.

3 Materials and Methods

Mini-MIAS database is used for the comparative study. The image size is 1024×1024 with the images centre of matrix [11]. The database provides the details of severity of abnormality-normal, benign and malignant tumours, class of abnormality-dense, circumscribed, spiculated, fatty, glandular, normal, asymmetry and ill-defined mass. The radii enclosing the suspected region are given in the database. The images are in portable greyscale format and each given a reference number. The normal images do not include the radii measurements showing no tumour.

4 Curvelet Transform

Multiresolution analysis is involved in medical image processing and computer vision applications. Wavelets, ridgelets, curvelets are multiresolution or multiscale methods. Wavelets applied in images do not exhibit directional sensitivity. Directional wavelets were modelled but with limitations. To overcome this, Wirth and Stapinski [12] defined the variant of wavelets, the curvelet transform to handle the curve singularities and efficient for representing curve-like edges. Curvelet allows to analyse the image in different sizes of blocks with only a single transformation. It is a multiscale pyramid, each level representing different subbands. The subband decomposition

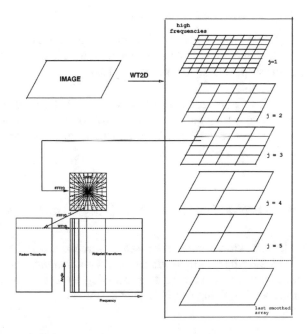

Fig. 1 Image decomposed into subbands and ridglet applied in each subband [13]

follows anisotropic parabolic scaling [13, 14] with the parameters width and length of the pyramid equivalent to *width* ≈*length*^{2}.

From Fig. 1, the discrete curvelet algorithm decomposes given n × n image *Im* in the form

$$Im(x, y) = C_{RJ}(x, y) + \sum_{j=1}^{RJ} \left(w_j(x, y) \right) \tag{1}$$

CRJ is the smoothened image of the original and j is the scaling parameter with j = 1 the band with higher frequencies.

Discrete Curvelet Transform (DCT) Algorithm

Step 1: á trous procedure is applied with RJ scales.
Step 2: set the block sizes B$_1$= B$_{min}$
Step 3: for (j from 1 to RJ scales) repeat
              The subbands are decomposed such that w$_j$ of block size B$_j$;
              To each block Ridgelet transform is applied;
      For analysing the jthsubband- [2^j,2^{j+1}], the side length of the local window is
      doubled at every second subband or dyadic subband
                     if (j mod 2 = 1) then
B$_{j+1}$ = 2B$_j$
              else
B$_{j+1}$ = B$_j$

Step 4: thresholding is computed

The curvelet algorithm denoises the given image and Fig. 2a, b shown below is the mammogram before and after applying the transform. The intensity is enhanced in the suspected region in a dense glandular mammogram after applying the curvelet transform.

In the earlier research works, wavelets [15] have shown better performance compared to other image transforms but had complex decomposition properties. Curvelets have proven to exhibit better performance in curve edges that is essential to segment a region of irregular shape.

4.1 Image Segmentation: Fractals and Snakes

The following section discusses the procedures to implement fractals and snakes model. Both these methods are tested on mammograms to locate the breast masses and segment them for further processing. Before segmentation, the image is contrast enhanced applying GWFEV [16] technique.

(a) Before Curvelet Transform   **(b)** After Curvelet Transform

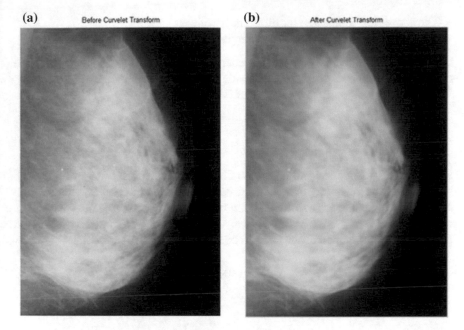

Fig. 2 **a** mdb001 before DCT **b** mdb001 after DCT

4.1.1 Fractals Using Box-Counting Method

Considering,

- the image Im of size n × n array,
- the fractal dimension (FD) of each pixel is stored in an array.
- For each pixel in Im,

  - a small window Wnof size m × m is formed surrounding the pixel under consideration.
  - FD is calculated for each window.
  - The value is assigned in the array.

- Histogram is calculated based on the fractal dimension values and each bin is the segmented image.

  Fractal Dimension [17] is calculated by implementing the following method:

- A coarse grid is visualized of size mr × mr where r is ratio in the range [0, 1].
- The grid will lie on the top of the small window Wn.
- The number of cells in the coarse grid that occupies the underlying pixel in image Im is counted.

- The above step is repeated for different 'r' values.
- Linear regression is performed to compute the slope of the line ln(N) and ln(1/r).

where N is the number of pixels in the coarse grid that has a value greater than the threshold. The computed fractal dimension for each pixel is assigned to the array. Thus, a fractal is formed on the region of interest.

4.1.2 Snakes Contour Model

Snakes are also called as local minimum finding contours placed at a given desired region. As the name of the model, the curve slithers around the ROI with minimum energy forces. Snakes are capable to bend or stretch yet preserving the smoothness of the curve. The objective of snakes is to draw contour on the desired region. The parametric representation or the contour of the model is given in Eq. 2

$$v(s) = (x(s), y(s)) \tag{2}$$

$$E_n s_n = \int_0^0 E_{in}(v(s)) + E_{im}(v(s)) + E_{con}(v(s)) ds) \tag{3}$$

and the energy measures are given in Eq. 3, where E_{in} is the internal energy required for the bending operation, E_{im} the energy to force the contour to move towards the image features like edges and Econ to drive the contour to the local minimum. The bending function depends on the intensity of the image. The snake will tend to move to the lighter or darker intensity region based on the neighbouring values. Snake is considered to be a function of time and on each iteration, control point is updated only if new position has a lower external energy. The snakes terminate on edges or smoothed regions.

5 Results and Discussion

The study is to experiment denoising in mammogram with curvelet transforms and to enhance the image for further ease in segmentation. Multiresolution methods have also proven for better segmentation of images where there are occurrences of curve discontinuities. Discontinuities cannot be interpolated with random values, and hence, curvelets were chosen. Image segmentation is a challenging task in medical images in which the suspected regions are to be identified correctly for diagnosing the suspected area as cancerous or non-cancerous. Fractals are used to model natural images and hence experimented with the medical dataset. Active contour model is

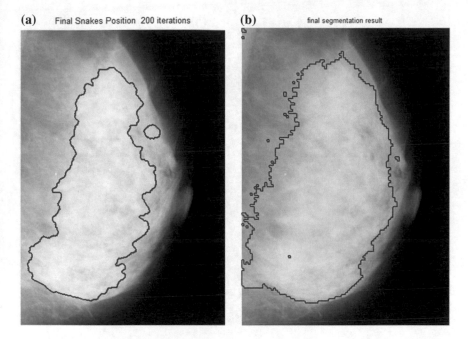

(a) Final Snakes Position 200 iterations   **(b)** final segmentation result

Fig. 3 **a** Im01—area covered 173 pixels **b** Im01—area covered 205 pixels

also applied for a comparative study with the fractals. From Fig. 3a, b it can be seen that the dense glandular tissue is segmented using both snake and fractals models. The area covered by fractals is 205 pixels and snakes 173 pixels. Snakes represented in blue shows a better segmentation than the fractals.

From MIAS dataset, the actual coverage area is 190 pixels and fractals have segmented an area that covers non-cancerous tissue. From Fig. 4a, b an image that has microcalcifications is segmented. The area covered by fractals is 13 pixels and snakes 46 pixels. Snakes have segmented the image showing the regions with microcalcifications and that fractals segmented only few areas.

From Fig. 5a, b, a normal image is chosen with snakes showing no segmentation and fractals showing few areas of calcifications. Snakes performed better than fractals in segmenting the suspected tissues. Snakes have shown better convergence to the suspected regions. The segmented results will support further diagnosing process in a CAD system.

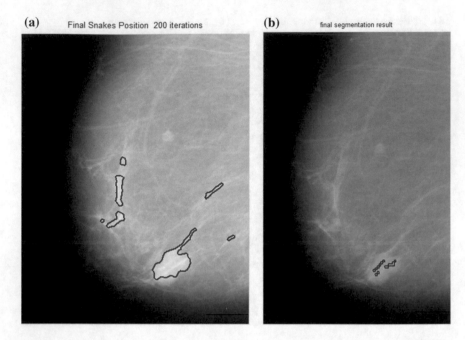

Fig. 4 **a** Im02—area covered 46 pixels **b** Im02—area covered 13 pixels

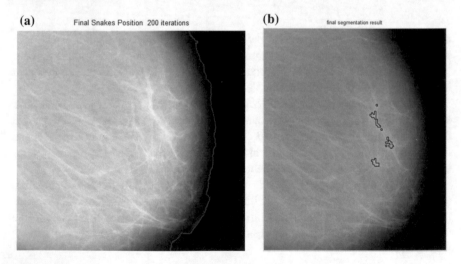

Fig. 5 **a** Im03—area covered 46 pixels **b** Im03—area covered 13 pixels

References

1. Miller, K.D., Jemal, A., Siegel, R.L.: Cancer statistics CA. A Cancer J. Clin. **66**(7–30). https://doi.org/10.3322/caac.21332 (2016)
2. Cheng, H.D., Shi, X.J., Min, R., Hu, L.M., Cai, X.P., Du, H.N.: Approaches for automated detection and classification of masses in mammograms. Pattern Recogn. **39**(4), 646–668 (2006)
3. Pal, S.K.: Fuzzy image processing and recognition: uncertainty handling and applications. Int. J. Image Graph. **1**(02), 169–195 (2001)
4. Liu, S., Babbs, C.F., Delp, E.J.: Multiresolution detection of spiculated lesions in digital mammograms. IEEE Trans. Image Process. **10**(6), 874–884 (2001)
5. Donoho, D.L., Duncan, M.R.: Digital curvelet transform: strategy, implementation, and experiments. In: Proceedings of SPIE 4056, Wavelet Applications VII (2000)
6. Eltoukhy, M.M., Samir, B.B., Faye, I.: Breast cancer diagnosis in digital mammogram using multiscalecurvelet transform. Comput. Med. Imaging Graph. **34**(4), 269–276 (2010)
7. Candes, E.J., Donoho, D.L.: Curvelets: A Surprisingly Effective Nonadaptive Representation for Objects with Edges. Stanford Univ Ca Dept of Statistics (2000)
8. Rangayyan, R.M., Nguyen, T.M.: Fractal analysis of contours of breast masses in mammograms. J. Digit. Imaging **20**(3), 223–237 (2007)
9. Tuyet V.T.H.: Active contour based on curvelet domain in medical images. In: Vinh, P., Barolli, L. (eds.) Nature of Computation and Communication. ICTCC 2016. Lecture Notes of the Institute for Computer Sciences, Social Informatics and Telecommunications Engineering, vol 168. Springer, Cham (2016)
10. Muralidhar, G.S., et al.: Snakules: a model-based active contour algorithm for the annotation of spicules on mammography. IEEE Trans. Med. Imaging **29**(10), 1768–1780 (2010)
11. Suckling, J.: The mammographic image analysis society digital mammogram database. In: Exerpta Medica International Congress Series 1069, pp. 375–378 (1994)
12. Wirth, M.A., Stapinski, A.: Segmentation of the breast region in mammograms using active contours. In: VCIP, pp. 1995–2006 (2003)
13. Ma, J., Plonka, G.: The curvelet transform. IEEE Signal Process. Mag. **27**(2), 118–133 (2010)
14. Starck, J.-L., Candès, E.J., Donoho, D.L.: The curvelet transform for image denoising. IEEE Trans. Image Process. **11**(6), 670–684 (2002)
15. Graps, A.: An introduction to wavelets. IEEE Comput. Sci. Eng. **2**(2), 50–61 (1995)
16. Chandrika, R.R., Karthikeyan, N., Karthik, S.: Simplified contrast enhancement fuzzy technique in digital mammograms for detecting suspicious cells. J. Med. Imaging Health Inf. **7**(2), 316–322 April (2017).
17. Vuduc, R.: Image segmentation using fractal dimension. Report on GEOL 634 (1997)

Key Frame Extraction Using Rough Set Theory for Video Retrieval

G. S. Naveen Kumar and V. S. K. Reddy

Abstract Key frame is a representative frame which contains the entire information of the shot. It is used for indexing, classification, analysis and retrieval of video. The existing algorithms generate relevant key frames but they also generate a few redundant key frames. Some of them are not able to represent the entire shot since relevant key frames are not extracted. We have proposed a more effective algorithm based on DC coefficients and Rough Sets to prevail over the rest. It extracts the most relevant key frames by eliminating the vagueness of the selection of key frames. It can be applied for compressed MPEG videos hence decompression is not required. The performance of this algorithm shows its effectiveness.

Keyword MPEG · DC coefficients · Rough Set Theory · Key frame extraction Content based video retrieval

1 Introduction

The size of the video information base is escalating rapidly as video recording devices have become cheap and the broadband speed has enhanced. So, it has made the text based retrieval of the video from the database cumbersome. Content based video retrieval (CBVR) can solve this problem. Key frame extraction is prerequisite for CBVR [1]. In video content analysis, key frame extraction follows shot boundary detection [2]. Key frames that represent the entire shot, remove redundant and repeated frames are called representative frames. In conventional methods, key frames are taken out from the video shot that is based on the position of the frames such as the initial frame, center frame or the final frame. The execution of this method is simple, but some video content may be missed in these key frames [3]. In some

G. S. Naveen Kumar (✉) · V. S. K. Reddy
Malla Reddy College of Engineering and Technology, Hyderabad, Telangana, India
e-mail: gsrinivasanaveen@gmail.com

V. S. K. Reddy
e-mail: vskreddy2003@gmail.com

© Springer Nature Singapore Pte Ltd. 2019
J. Wang et al. (eds.), *Soft Computing and Signal Processing* ,
Advances in Intelligent Systems and Computing 898,
https://doi.org/10.1007/978-981-13-3393-4_76

other methods, key frames are extracted based on the video feature extraction such as color, shape, size and texture [4]. They fail when we apply them on the compressed video since decompression is required. In the proposed method, key frames are extracted based on DC coefficients and Rough Set Theory [5, 6]. This algorithm can directly be applied on compressed videos and therefore decompression is not required. This results in the reduction in the number of computations and the amount of time taken is reduced. The Rough set is an effective tool that can handle vagueness, ambiguity and uncertainty in the selection of data. It helps in the selection of relevant key frames. The remaining paper has been structured as under. Section 2 refers to the proposed System. Experimental results and conclusion have been focused in Sects. 3 and 4 respectively.

2   The Proposed System

2.1   Key Frame

A video sequence from which key frames are extracted is a set of frames. These frames can summarize the whole content of the video. The Key frames highly reduce a large amount of data into a compressed one. Since the key frames can compress the data, they assist in video indexing, classification and especially retrieval of video. Key frame extraction follows shot boundary recognition in the video structure analysis. It plays a significant role in the video retrieval [7].

2.2   Compressed MPEG Video Stream

MPEG stands for Moving Photographic Expert Group and it is the most popular video compression standard. It is a sequence of the frames which contains I frames, P frames and B frames. I frames are intra-coded pictures, P frames are predictive-coded pictures and B frames are bidirectional predictive-coded pictures. I frame is taken into account since it is the first frame that contains original video frame information whereas the P and B frames have only motion information [8]. The criteria for taking out the Key Frames for MPEG videos are DC coefficients. The steps for the extraction of DC coefficients are as follows.

1. Each I frame of MPEG video is divided into a number of blocks without over lapping and the individual block size of which is 8×8.
2. Discrete Cosine Transform (DCT) is to be applied on each block [9]. DCT is defined in the Eq. 1.

$$Q(u, v) = p(u)p(v) \sum_{x=0}^{N-1} \sum_{y=0}^{N-1} f(x, y)\cos\frac{(2x + 1)u\pi}{2N}\cos\frac{(2y + 1)v\pi}{2N} \tag{1}$$

For u, v = 0, 1, 2, ..., N − 1 and p(u), p(v) are defined following

$$p(u) = \begin{cases} \sqrt{\frac{1}{N}}, & u = 0 \\ \sqrt{\frac{2}{N}}, & u = 1, 2, 3, \ldots N - 1 \end{cases}$$

3. Every DCT block has DCT coefficients. The coefficients whose intervals is (0, 0) is named DC coefficient; other coefficients are named AC coefficients.
4. Since DC coefficients have the most significant information and the average of DCT block, these are used for the extraction of key frames [10].

2.3 Rough Set Theory

Existing systems can decide key frames from the sequence of frames in shot. In some cases, they take out key frames from shot video imperfectly resulting in vagueness, uncertainty, ambiguity and imperfection. In the existing systems, n-dimensional space frames are depended on for similarity matching. There is a possibility of irrelevant frames in extracted key frames. In the proposed system, the Rough Set Theory has solutions for all these problems. The Rough Set Theory clears vagueness and rectifies the imperfect data [11]. Rough set performs feature selection by distinguishing relevant features from irrelevant ones. Hence, it results in the reduction of dimensionality. When the selected frame is matched with the sequence of frames in shot, there is a similarity or dissimilarity between the two. When there is ambiguity whether there is similarity or dissimilarity between the two, the situation is called roughness.

A set with the elements having precise information is called crisp set. A set with one or more elements having imprecise (vague) information is called the rough set. Let U be Universal set have a non empty set and X be a subset of U. An equivalence relation E, categorizes U into a set of subsets. U/E = {X1, X2, ..., Xn} in which the subsequent conditions are satisfied:

(i) Xi ⊆ U, Xi ≠ Ø for any i
(ii) Xi ∩ Xj ≠ Ø For any i, j

Rough set is classified into lower approximation and upper approximation sets. In the lower approximation set, elements surely belong to the given set satisfying conditional information. Elements that surely and possibly belong to the given set are included in the upper approximation set.

$$\underline{E}X = \{x \in U : [x]_R \subseteq X\} \tag{2}$$

$$\bar{E}X = \{x \in U : [x]_R \cap X \neq \emptyset\} \tag{3}$$

In the upper approximation set, the elements surely and possibly belong to the given set satisfying conditional information. The border region of the Rough set is obtained when the lower approximation set is subtracted from the upper approximation set.

$$\text{BNR}(X) = \bar{E}X - \underline{E}X \tag{4}$$

The Information System (IS) table is a function of (U, A) here U represents the finite set of objects that are in the rows and A represents attributes that are in the column. In the Information system table the attributes are conditional information. If the decision attributes are added to the Information table, it is called Decision system (DS) table. The lower approximation and upper approximation sets are derived from Decision system table based on certainty and uncertainty of decision rules. In the Decision system table, some conditional attributes are redundant and some attributes have sufficient information to represent the decision attribute(s). The set of conditional attributes that gives sufficient information about the decision table is called the Reduct set. This set removes redundant conditional attributes. The intersection of reduct attributes is called the core set [12]. The Core set is a salient attribute in the Information system. The I frames in the core set are the representative frames that are extracted as key frames.

2.4   Proposed Algorithm Framework

Extraction of key frames using DC coefficients and Rough Set Theory is the proposed algorithm and it is applied on compressed MPEG video. Initially, an MPEG video is partitioned into shots using shot boundary detection techniques [13]. The step by step method for taking out the key frames from a shot using the proposed algorithm is as follows.

1. The Extraction of I frames from the compressed video is the first step.
2. The second step is an extraction of DCT coefficients from each of the I frames.
3. Thirdly, DC coefficients, that contain essential information, are extracted from DCT blocks.
4. An Information system table is created in which the rows contains DC coefficients and columns contains I frames. Here, the rows represent objects and the columns represent attributes.
5. The number of redundant I frames is reduced by using attribute reduced theory of Rough Set Theory. The remaining I frames constitute a core set.
6. Core set is a salient attribute in the Information system. The I frames in the core set are the representative frames that are extracted as key frames.

The implementation of this algorithm is shown in the Fig. 1.

Fig. 1 The framework for proposed System

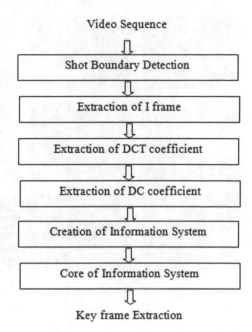

Video Sequence

⇩

Shot Boundary Detection

⇩

Extraction of I frame

⇩

Extraction of DCT coefficient

⇩

Extraction of DC coefficient

⇩

Creation of Information System

⇩

Core of Information System

⇩

Key frame Extraction

3 Experimental Results

We test the efficiency of the proposed technique by giving different categories of compressed MPEG videos as inputs. The input video has been segmented into a number of shots. Next, I frames have been extracted from each shot. Later, DCT coefficients have been calculated for 8×8 blocks of I frame. Finally, we have taken DC coefficient from each DCT block. After extraction of DC coefficients, we have created an Information System table. The rows that are known as objects represent DC coefficients and the columns that are also called attributes represent I frames. We have eliminated the redundant I frames by using the attribute reduced theory of Rough Set Theory and the Core set has been extracted. Lastly, we have extracted the key frames that are in the core set. Figure 2 shows the extracted key frames for the given input video.

The efficacy of this algorithm has been measured by two performance metrics. The first metric is the compression ratio that calculates the rate of data compression [14]. The compression ratio is the total number of frames in the video shot to the quantity of key frames selected. The compression ratio is defined in the Eq. 5.

$$Compression\,Ratio = \frac{N1}{N2} \tag{5}$$

where N1 is the total number of frames in shot and N2 is the total number of extracted key frames.

Fig. 2 Key frame extraction for flower video

Fidelity that is the second metric is defined as semi- Hausdorff distance among the I frames of the shot and the key frames [15].

$$\text{Dist}_{SH} = \max(\min(d(\text{KFj, SFi})));\, i = 1, 2, \ldots, N \text{ and } j = 1, 2, \ldots, k \quad (6)$$

The fidelity measure is given as

$$fidelity = 1 - \frac{\text{DistSH}}{\max(\max(dij))} \quad (7)$$

where dij represents the dissimilarity matrix of the shot set S.

Table 1 shows the compression ratio and the fidelity of different categories of videos. The compression ratio indicates that the proposed algorithm is effective in data compression. The fidelity shows that the resultant key frames have most significant content of the video. Hence, the proposed algorithm is the most effective Key frame extraction and it is useful in content based video retrieval.

Table 1 Performance measurement for different categories of videos

Video	No. of shots	Total frames	Key frames	Compression ratio (%)	Fidelity
Flower	10	1562	18	86.77	0.75
Sports	17	2345	28	83.75	0.74
News	9	1463	16	91.43	0.79
Cartoon	13	1812	19	95.36	0.82

4 Conclusion

In this paper, a new algorithm of key frame extraction using DC coefficient using Rough Set Theory has been proposed. There is ambiguity in the selection of key frames in the existing algorithms and the proposed algorithm has overcome it with Rough Set Theory. We have applied this system on the compressed video directly and hence the decompression process has been avoided. The performance measurements of compression ratio and fidelity have shown that a small fraction of key frame data can represent the whole content of the video.

References

1. Hu, W., et al.: A survey on visual content-based video indexing and retrieval. IEEE Trans. Syst. Man Cybern. Part C (Applications and Reviews) **41**(6), 797–819 (2011)
2. Mendi, E., Bayrak, C.: Shot boundary detection and key frame extraction using salient region detection and structural similarity. In: Proceedings of the 48th Annual Southeast Regional Conference, p. 66. ACM (2010)
3. Sun, Z., Jia, K., Chen, H.: Video key frame extraction based on spatial-temporal color distribution. In: International Conference on Intelligent Information Hiding and Multimedia Signal Processing, IIHMSP'08, pp. 196–199. IEEE (2008)
4. Ejaz, N., Mehmood, I., Baik, S.W.: Efficient visual attention based framework for extracting key frames from videos. Signal Process. Image Commun. **28**(1), 34–44 (2013)
5. Hua, G., Chen, C.W.: Distributed video coding with zero motion skip and efficient DCT coefficient encoding. In: 2008 IEEE International Conference on Multimedia and Expo, pp. 777–780. IEEE (2008)
6. Wang, T., Yu, W., Chen, L.: An approach to video key-frame extraction based on rough set. In: International Conference on Multimedia and Ubiquitous Engineering, MUE'07, pp. 590–596. IEEE (2007)
7. Gianluigi, C., Raimondo, S.: An innovative algorithm for key frame extraction in video summarization. J. Real-Time Image Proc. **1**(1), 69–88 (2006)
8. Liu, G., Zhao, J.: Key frame extraction from MPEG video stream. In: 2010 Third International Symposium on Information Processing (ISIP), pp. 423–427. IEEE (2010)
9. Xu, J., Yuting, S., Liu, Q.: Detection of double MPEG-2 compression based on distributions of DCT coefficients. Int. J. Pattern Recognit. Artif. Intell. **27**(01), 1354001 (2013)
10. Uehara, T., Safavi-Naini, R., Ogunbona, P.: Recovering DC coefficients in block-based DCT. IEEE Trans. Image Process. **15**(11), 3592–3596 (2006)
11. Pawlak, Z.: Rough set theory and its applications to data analysis. Cybern. Syst. **29**(7), 661–688 (1998)
12. Shirahama, K., Matsuoka, Y., Uehara, K.: Event retrieval in video archives using rough set theory and partially supervised learning. Multimed. Tools Appl. **57**(1), 145–173 (2012)
13. Wu, Z., Xu, P.: Shot boundary detection in video retrieval. In: 2013 IEEE 4th International Conference on Electronics Information and Emergency Communication (ICEIEC), pp. 86–89. IEEE (2013)
14. Borth, D., Ulges, A., Schulze, C., Breuel, T.M.: Keyframe extraction for video tagging & summarization. Informatiktage **2008**, 45–48 (2008)
15. Nutanong, S., Jacox, E.H., Samet, H.: An incremental Hausdorff distance calculation algorithm. Proc. VLDB Endow. **4**(8), 506–517 (2011)

Information Retrieval Using Image Attribute Possessions

D. Saravanan

Abstract Extracting defined information from the huge data set really challenging task for many researchers, especially this data set like image data's process is too complex. As image data consist of motion, time, text, audio, pixel difference and more. From this complex data set, extracting the domain knowledge takes more time. This process differs from traditional text mining, because the nature of the data sets. Extracting information from image data, user needs additional knowledge; i.e., users required domain knowledge. This attracts many users concentrate on this field. Currently, many research works carried on this particular domain. Advancement of technology more and more image data is created and uses, for this urgent attention required in the field of image mining. This paper focuses on image mining help of clustering technique. First video data are grouped into frames, from the cleaned frameset process are done client- and server-side operations. The proposed technique works well, and experimental results also verified this.

Keywords Video data mining · Key frame analysis · Clustering technique
Image mining · Frame comparison · Knowledge extraction

1 Introduction

Technology brings image-based extraction so simple and easy to the users. This process brings the output with greater accuracy. Today most of the applications which are used in this process effectively some of them are in traffic control, criminal analysis, weather forecasting, finance, marketing, operations and more. In the above domain, vast amount of images are collected and constructed image data based.

D. Saravanan (✉)
Faculty of Operations & IT, ICFAI Business School (IBS), Hyderabad, India
e-mail: Sa_roin@yahoo.com

D. Saravanan
The ICFAI Foundation for Higher Education (IFHE) (Deemed-to-be-University u/s 3
of the UGC Act 1956), Hyderabad, India

© Springer Nature Singapore Pte Ltd. 2019   759
J. Wang et al. (eds.), *Soft Computing and Signal Processing* ,
Advances in Intelligent Systems and Computing 898,
https://doi.org/10.1007/978-981-13-3393-4_77

From these huge data sets, extraction of needed knowledge is one of the challenging tasks for many researchers [1]. Reason today most of the research work is carried in this particular domain. It not only brings the need information it also brings the greater accuracy. In video data mining, the most challenging task aims to perform these operations automatically and extracts the needed knowledge effectively. Many image mining algorithms perform this operation effectively. Increasing the content of the image day by day, it required advance searching operations on this field [2]. It helps the researchers and users extract the needed cone tent and improve the searching operation in the field of education, training, and many industrial functions. But this process is not easy to develop because of nature of the input data sets [3]. Creation of image data sets is easy, retrieve the needed content from this data sets are challenging tasks to the users [4]. Video data mining performs its operation on the following category (1) understand the input data sets (2) content-based extraction (3) knowledge extraction (4) knowledge representation (5) store the relevant contents. This process is need domain knowledge and proper mining techniques. Most of the image data sets are containing sequential and multidimensional properties this property to need to treat properly [5]. The proposed method can find the best matching sequence in many messy match results, which effectively excludes false "high similarity" noise and compensate the limited description of image low-level visual features [6]. These functions are mostly treated separately, and these attributes are not able to mix with other functions. Based on this work motive in this paper the input video files are converted to static frames, using image properties those frames are analyzed and noisy frames are separated. Using RGB-based image pixel property, images are stored and retrieved based on the users input image query [7, 8].

2 Related Work

A new approach is carried out for deep concept-based multimedia information retrieval, which focuses on high-level human knowledge, perception, incorporating subtle nuances and emotional impression on the multimedia resources [9]. It provides a critical evaluation of the most common current multimedia information retrieval approaches and proposes an innovative adaptive method for multimedia information search that overcomes the current limitations. The main focus of this approach is concerned with image discovery and recovery by collaborative semantic indexing and user relevance feedback analysis.

2.1 Negative Aspects of Related Work

1. No proper clustering algorithm.
2. Cluster formation takes more time.
3. System works only specific set of files only.
4. One or two image attributes only consider.
5. Efficient and effectiveness regarding image retrieval is low.

3 Proposed System

The proposed system works based on image RGB pixel evaluation. First, video files are converted into static frames from this frame sets unwanted i.e. duplicate frames are eliminated using image threshold value, after this refined frames are stored for further operation. Using hieratical clustering technique, frames are trained client and server process [10, 11]. This technique image extraction gets improved and the same more number of outputs is extracted for given input frame. Experimental design explains this process more detail.

3.1 Advantage of Proposed Technique

1. Works well in all types of video files.
2. Input data sets consist of cleaned frames.
3. Cluster formation is quick compared to existing image mining technique.
4. Retrieved more number of outputs.
5. Efficiency improved.

4 Experimental Methodology

4.1 Data Mining Using Image Data

Data mining defines extract the new knowledge from the stored data set. In the same video, data mining defines finding interesting pattern that is unknown before. Every mining process undergone the preprocessing operations because of noisy data need to be cleaned before it starts extracted. Here the preprocessing operation of video mining is shown in Fig. 1. In stage 1, raw video data are converted into frames, noisy frames are eliminated; cleaned frames are stored in the target database for further operations. Those data sets are transformed into operations database for finding knowledge. This extracted knowledge represents the user community.

4.1.1 The Most Commonly Used Techniques in Data Mining Are

Artificial neural networks: This is similar to biological network structure. The process consists of input layer, hidden layer and output layer. Based on the input, hidden layer may add and form multilayer architecture. This technique consists of

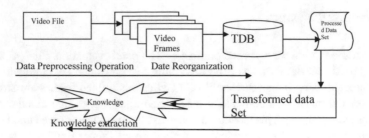

Fig. 1 Data mining interactive process system

back rogation helps the users to derive the need output. This technique works well when other data mining classifications not produced satisfied result.

Decision trees: Data sets are represented in hierarchical structure. Each structure characterizes decision and outcome of the decisions. Many decision tree algorithms are available, pre-pruning and post-pruning techniques help to eliminate noise in the give data set.

Nearest neighbor method: One of the best clustering technique. Information's are processed based on the values assigned. This technique never required and background knowledge. Most similar data are combined and form one unique big data set.

Rule induction: Decisions are derived based on rule-based operation. After constructing rules, finding the best rule are needed.

Many of these technologies have been in use for more than a decade in specialized analysis tools that work with relatively small volumes of data.

4.2 Image Categorization

After frames are cleaned, frames are stored in the operational data set. This video classification defined assign proper labels among the groups. It helps the users to extract the needed information accordingly [12]. Data's are more closer are put into one group, and not more closer are put in another group. Every group resolution limits is identified. This process also helps to avoid time taken for data extraction this process output shown in Fig. 2. Information here properly indexed, so extraction is easy. Every group one key frame are selected, using the key frame groups are easily identified.

Fig. 2 Model input frames of different types of test videos

4.3 Image Clustering

Clustering is the technique to form the data sets in a group. If the items are most similar are grouped to one and items are dissimilar, they are grouped separately. It helps the identify the similarities between the data sets. It also improved the searching speed of the user query. Items are grouped together, so extraction is very easy. Information extraction based on the image frame count is shown in Table 2. Number of techniques is used to forming the cluster, most of the existing clustering technique based on distance between the data points. This process output is shown in Figs. 3 and 4 and Table 1. Based on the pixel difference items are grouped. Difference between the data points inside the clusters is very high compared to the items which far away from

Table 1 Cluster formation versus time taken

Id	Frame	No.of clusters	Time
1	0	35,204	3:17:55 pm
2	1	35,175	3:18:01 pm
3	2	29,162	3:18:07 pm
4	3	29,762	3:18:14 pm
5	4	30,475	3:18:20 pm
6	5	30,975	3:18:26 pm
7	6	31,351	3:18:33 pm
8	7	31,867	3:18:39 pm
9	8	32,339	3:18:45 pm
10	9	32,427	3:18:52 pm
11	10	32,482	3:18:58 pm
12	11	32,440	3:19:04 pm
13	12	32,842	3:19:11 pm
14	13	32,964	3:19:17 pm
▶ (Auto number)	0		

Table 2 Frame count versus time taken

Frame count	Milliseconds	Category
25	2586	News
50	2640	News
75	2625	News
100	3062	News
125	3375	News
150	3421	News

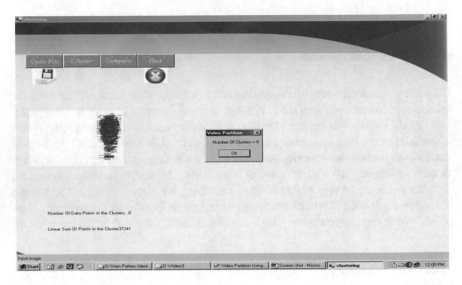

Fig. 3 Cluster the image

the group. Among this, image mining and video-based cluster are differ from normal cluster formations. Video is the combination of audio, text, motion, time and pixel quality. Based on the above, video clustering is the most challenging task for any users. Existing clustering technique falls on either top-down approach or bottom-up approach. Either single data sets are separated with different groups or different items are grouped together and form a one big clustering group.

5 Image Extraction

There are two major issues that will affect the image data mining process. One is the notion of similarity matching and the other is the generality of the application area. For a specific application area [7], associated domain knowledge can be used to improve the data mining task. Since data mining relies on the underlying querying capability of the CBIR system, which is based on similarity matching, user interaction

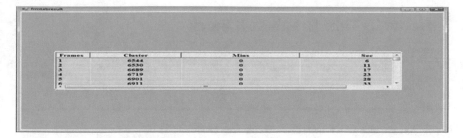

Fig. 4 Different image cluster results

will be necessary to refine the data mining process. With image, mining process will consider the four broad problem areas associated with data mining: finding associations, classification, sequential pattern and time series pattern. With all these, the essential component in image mining is identifying similar objects in different images. This process output is shown in Fig. 5. Performance comparison of various image extractions using different clustering is also shown in Fig. 6.

5.1 Image Mining Algorithm Step

The algorithms needed to perform the mining of associations within the context of image. The four major image mining steps are as follows:

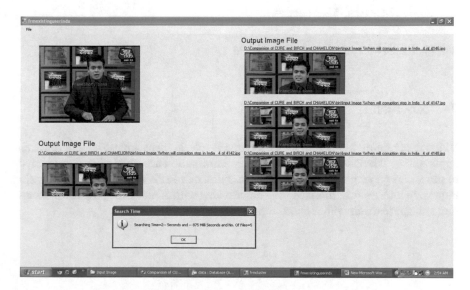

Fig. 5 Image retrieval using RGB technique 1 input and 5 outputs

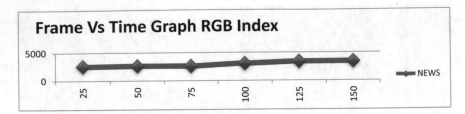

Fig. 6 Performance graph frame count versus time taken

1. Feature Extracting: Image mining starts with image feature extraction, after successfully separates the frames, every frames attribute are extracted and stored separately. This process done on both client and server sides, using this value user can compare or extract needed frames. This is the important initial step for any image mining.
2. Object identification and record creation: After successful creation of frame identifier immediately compares this values it helps to remove the duplication frames. For camper the frame values image threshold values are used. Compare objects in one image to objects in every other image. It is one of the image preprocessing steps.
3. Create auxiliary images: After successfully remove the duplication, rest of the frame is stored under unquiet class for further operation.
4. Repeat the step 1–3 unit all image data's are segregated.

6 Experimental Outcomes

See Figs. 2, 3, 4, 5 and 6, and Tables 1 and 2.

7 Conclusion

Video data mining today plays very important role in many industrial and research applications. It helps the users with needed information accurately in a short period of time. This paper proofs information extraction more effectively than existing techniques. In future works this study can be extended with other image properties, brings more accurate result with reduced time period.

References

1. Pan, J.-Y., Faloutsos, C.: VideoCube: a new tool for video mining and classification. In: ICADL, Dec 2002, Singapore
2. Saravanan, D.: Design and implementation of feature matching procedure for video frame retrieval. Int. J. Control Theory Appl. **9**(7), 3283–3293 (2016)
3. Fayyad, U.M., Piatetsky-Shapiro, G., Smyth, P., Uthurusamy, R.: Advanced in Knowledge Discovery and Data Mining. AAAI/MIT press (1996)
4. Ng, R.T., Han, J.: CLARANS: a method for clustering objects for spatial data mining. IEEE Trans. Knowl. Data Eng. **14**(5) (2002)
5. Zhao, W., Wang, J., Bhat, D., Sakiewicz, K., Nandhakumar, N., Chang, W.: Improving color based video shot detection. In: IEEE International Conference on Multimedia Computing and Systems, vol. 2, pp. 752–756 (1999)
6. Zabih, R., Miller, J., Mai, K.: A feature-based algorithm for detecting and classifying scene breaks. In: Proceedings of the ACM Multimedia 95, San Francisco, CA, pp. 189–200, November 1995
7. Saravanan, D.: Image frame mining using indexing technique. In: Data Engineering and Intelligent Computing, pp. 127–137. Springer (2017). ISBN:978-981-10-3223-3
8. Saravanan, D.: Video content retrieval using image feature selection. Pak. J. Biotechnol. **13**(3), 215–219 (2016)
9. Shyu, M., Xie, Z., Chen, M., Chen, S.: Video semantic event/concept detection using a subspace-based multimedia data mining framework. IEEE Trans. Multimed. **10**, 252–259 (2008)
10. Saravanan, D.: Information retrieval using: hierarchical clustering algorithm. Int. J. Pharm. Technol. **8**(4), 22793–22803
11. Saravanan, D.: Effective video data retrieval using image key frame selection. Adv. Intell. Syst. Comput. 145–155
12. Zhang, L., Lin, F., Zhang, B.: A CBIR method based on color-spatial feature. In: The IEEE Region 10 Conference Proceedings, pp. 166–169, Sept 1999

Blind DCT-CS Watermarking System Using Subsampling

S. M. Renuka Devi and D. Susmitha

Abstract Digital image watermarking is the technology used to protect copyright information of multimedia objects. In this paper, a novel blind image watermarking in transform domain is proposed. The cover image is subsampled into four images, and the correlation between the subimage DCT coefficients is exploited for selecting the embedding locations for watermark. The CS measurements of the watermark are embedded into the selected DCT coefficients of the subsampled image. The watermark is compressively sensed for providing higher embedding capacity than the traditional transform domain watermarking techniques. The outcome of this method is improvement in the security of the system, as the watermark can be recovered only if the receiver has the knowledge of the measurement matrix which serves as a key. Experimental results demonstrate that the extracted watermark and watermarked image is better in terms of PSNR, SSIM, NC, and RMSE under with and without noise attack conditions.

Keywords Discrete cosine transform · Watermarking · Compressive sampling
Subsampling

1 Introduction

With the advent of the Internet, the transmission of digital media over the Internet had increased rapidly over the last decade. This has resulted in a serious threat on the security of digital information being transmitted. One solution to this risk is integrating some kind of ownership information into the digital media. This process of embedding authentication data into the digital media is known as watermarking [1]. Depending on the watermark visibility in the digital media, watermarking can

S. M. Renuka Devi (✉) · D. Susmitha
G. Narayanamma Institute of Technology and Science (for Women), Hyderabad, India
e-mail: renuka.devi.sm@gmail.com

D. Susmitha
e-mail: susmitha.daparty18@gmail.com

© Springer Nature Singapore Pte Ltd. 2019
J. Wang et al. (eds.), *Soft Computing and Signal Processing* ,
Advances in Intelligent Systems and Computing 898,
https://doi.org/10.1007/978-981-13-3393-4_78

be visible or invisible [1]. Based on the watermark extraction, it can be classified into full reference and blind image watermarking [2]. Full reference watermarking requires cover image to be available at the receiver for extracting the watermark while blind image watermarking does not require cover image for extracting the watermark. Our proposed method deals with the blind image watermarking. Watermark can be embedded into the cover image either in transform domain (DCT, DWT, SVD, PCA, DCT-DWT, etc.) [1] or in spatial domain [2]. Though the embedding capacity is large in spatial domain techniques, it is also seen that the robustness against attacks is poor [2]. So we prefer the transform domain technique in our approach and DCT is used. Many researchers [3–5] have discussed watermarking performance evaluation in terms of embedding capacity, transparency, and algorithm robustness against various attacks. It is required to keep a fair balance between all these factors. Few of the authors have proposed watermarking based on wavelets [4, 6], DCT [7], compressive sampling [3], and subsampling approach [8].

Our main contribution to this paper is in proposing blind image watermarking by combining the concepts of subsampling and compressive sampling. This approach derives the advantages of subsampling, i.e., it eliminates the requirement of cover image for extracting the watermark at the receiver, and compressive sampling, i.e., it enhances the security due the use of measurement matrix. The results observed are compared with two other approaches of watermarking—the basic DCT method [9] and DCT-CS method [10]. Our method outperforms the basic DCT method and the DCT-CS method in terms of performance characteristics of similarity measure, correlation, signal-to-noise ratio, and mean square error.

This paper is organized as follows: Sect. 1 deals with the Introduction to watermarking, Sects. 2 and 3 brief about the compressive sampling and the orthogonal matching pursuit algorithm, respectively, Sect. 4 describes our proposed system, Sect. 5 is about the experimental analysis of the proposed system, and Sect. 6 concludes the paper.

2 Compressive Sampling

The traditional approach of sample and then compress is actually a cumbersome and also expensive process. Therefore, there arises a question of obtaining the samples directly in the compressed domain, to which compressive sampling is an answer [10]. Compressive sampling uses the fact that all natural signals are sparse in some domain [10]. Considering a one-dimensional signal F of finite length N,

$$F = \sum_{i=1}^{N} K_i \emptyset_i \text{ or } F = \emptyset K \qquad (1)$$

where K is column vector of dimension N × 1. K and F are equivalent representations of the same vector except for in different domains. K is called the sparse representation of F since the number of non-zero samples in K is much lesser than the total number samples in F.

Recovering the signal F at the receiver [11, 12] can be done by computing dot product between F and ϕ_j where $j = 1$ to L as

$$y_j = \langle F, \phi_j \rangle \tag{2}$$

$$Y = \phi F = \phi \emptyset K = \theta K \tag{3}$$

θ is the dimensionally reduced space of F. Recovery algorithms like greedy algorithms and basic pursuit can be used for recovering the original signal F.

3 Orthogonal Matching Pursuit

Greedy algorithms like OMP an iterative process which uses only N^3 operations [12] where N is the total number of samples. The algorithm for realizing the OMP is as follows. Given the matrix A, the vector Y, and the error threshold T.

Initialization: Initialize k=0, and set
1. The Primary solution $x^0 = 0$.
2. The Primary residue $r^0 = Y - Ax^0 = Y$.
3. The Primary solution support $s^0 = \text{support}\{x^0\} = \text{NULL}$.
Main Iteration : INCREASE k by 1 and perform the following steps :
1. Sweep: Compute the errors $e(j) = \min_{z_j}||a_j z_j - r^{k-1}||_2^2$ for all j using the optimal choice $z_j^* = a_j^T r^{k-1}/||a_j||_2^2$.

2. Update Support : Find a minimizer,j_0 of e(j) : for all $j \notin s^{k-1}$, $e(j_0) \leq e(j)$, and update $s^k = s^{k-1} \cup \{j_0\}$.
3. Update Provisional Solution : Compute x^k, the minimizer of $||Ax - Y||_2^2$ subject to $\text{support}\{x\} = s^k$.
4. Update Residual : Compute $r^k = Y - Ax^k$.
5. Stopping Rule : If $||r^k||_2 < T$, stop. Else, apply another iteration.
Output : The proposed solution is x^k obtained after k iterations.

4 Proposed System

Watermark embedding: Figure 1 depicts the watermark embedding process. Given the cover image V (p_1, p_2) where $p_1 = 1, 2, \ldots, N_1$ and $p_2 = 1, 2, \ldots, N_2$ and $N_1 \times N_2$ is the size of the cover image. Performing subsampling operation,

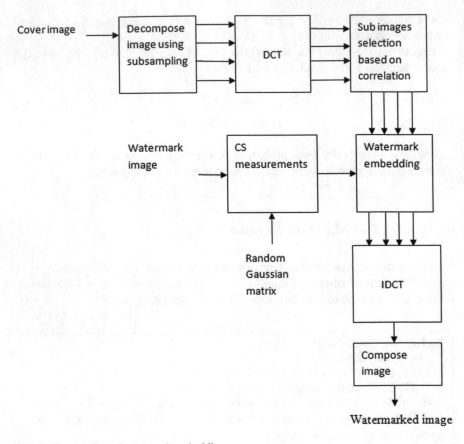

Fig. 1 Process flow of watermark embedding

$$V_1(m_1, m_2) = V(2m_1, 2m_2),$$
$$V_2(m_1, m_2) = V(2m_1 - 1, 2m_2)$$
$$V_3(m_1, m_2) = V(2m_1, 2m_2 - 1),$$
$$V_4(m_1, m_2) = V(2m_1 - 1, 2m_2 - 1)$$

(4)

where $m_1 = 1, 2, \ldots, \frac{N_1}{2}$ and $m_2 = 1, 2, \ldots, \frac{N_2}{2}$ are the subimages obtained by subsampling. Take DCT to obtain the sets of coefficients $V_i(L_1, L_2)$ where i = 1, 2, 3, 4 and $L_1 = 1, 2, \ldots, N_1/2$ and $L_2 = 1, 2, \ldots, N_2/2$. The subimages are highly correlated [13] since the intensities are approximately same in any particular region. Given the watermark sequence Y of N samples. Apply the compressive sampling measurements by using a predefined random Gaussian matrix A of size M × N, then with X = A * Y, X will have M samples. The scheme described below uses only the first and second subimages:

$$V_1'(L_n) = V_1(L_n)(1 + g * X(n)),$$

$$V_2'(L_n) = V_2(L_n)(1 - g * X(n)) \tag{5}$$

where n = 1, 2,..., N. In the above equations, $L_n = (L_{1n}, L_{2n})^T$ is the nth location vector. DCT coefficients of the watermarked subimages are denoted V_1' and V_2'. The constant g (0 < g ≤ 1) is known as watermark strength [2], the choice of g depends on balance concerning image distortion and detection accuracy.

Watermark Extraction: Figure 2 shows the watermark extraction process. For extracting the watermark, it is required to share the random Gaussian matrix or measurement matrix as the key to the authorized receiver. In order to recover the image, the following equations are used.

$$V_1'(L_n) - V_2'(L_n) = V_1(L_n) - V_2(L_n) + g * X(n)[V_1(L_n) + V_2(L_n)] \tag{6}$$

Using the fact that V_1 is approximately equal to V_2, the above equation can be approximated to $V_1'(L_n) - V_2'(L_n) \approx 2 g * V_1(L_n) * X(n)$. Now using the watermark embedding equations, and substituting $V_1(K_n)$,

$$X(n) \approx \frac{1}{g}\left(\frac{V_1'(L_n) - V_2'(L_n)}{V_1'(L) + V_2'(L_n)}\right) \tag{7}$$

Thus the watermark can be recovered by simply calculating the difference between V_1' and V_2'. The success of our proposed system relies on the fact that how well is the approximation of above equations [13].

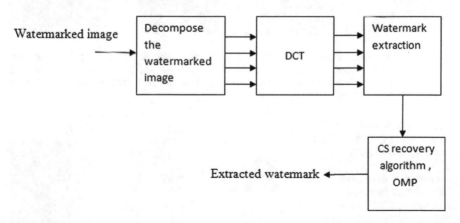

Fig. 2 Process flow of watermark extraction

5 Results and Discussion

When matched to the spatial watermarking, frequency domain watermarking offers less embedding capacity due to the fact that not all the transform coefficients are suitable for embedding the watermark. In DCT watermarking system, around 30% of the center frequencies are apt for modification with low visual degradation [4]. This experiment is carried on cover image of size 256×256 and watermark of size 64×64 (alphabet A, logo) as shown in Figs. 3 and 4. The quality of watermark is evaluated in terms of SSIM, PSNR, RMSE, and NC [14].

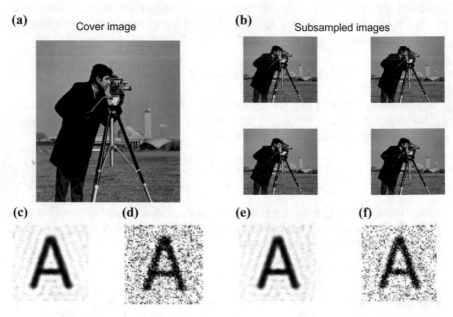

Fig. 3 **a** Cover image of size 256×256, **b** subsampled images of size 128×128, **c** extracted watermark without noise attack, **d** extracted watermark under salt and pepper noise attack (0.025), **e** extracted watermark under Gaussian filtering attack (5×5), and **f** extracted watermark under AWGN noise attack (SNR = 45 dB)

Fig. 4 **a** Embedded watermark of size 64×64, **b** extracted watermark by using right key, i.e., measurement matrix and NC = 0.9127 for alphabet A and NC = 0.7376 for logo, and **c** extracted watermark by using wrong key, and NC = 0.0012 and NC = 0.0025

Tables 1, 2, 3, and 4 showcase the performance of watermarked image and extracted watermark. We have compared two schemes: "Basic DCT watermarking system [9]", and the proposed "Blind DCT-CS watermarking system based on subsampling". Basic DCT watermarking is a full reference watermarking system, whereas subsampling-based DCT-CS watermarking is a blind image watermarking system. From Tables 1, 2, 3, and 4, it is seen that this method of blind DCT-CS watermarking system based on subsampling performs far well than the conventional DCT-watermarking in terms of NC, SSIM, PSNR, and RMSE for the watermarked image and also extracted watermark.

In Fig. 3c, watermark extracted is shown without any noise attack and also Fig. 3d–f shows extracted watermark when watermarked image is subjected to different noise attacks. Table 5 depicts the better performance of our proposed method of blind DCT-CS watermarking system based on subsampling under different noise attacks.

Embedding capacity of our algorithm is far better than the DCT-CS [5] algorithm, since our algorithm supports 4000 bits for embedding, whereas the DCT-CS algorithm supports a maximum of 600 bits. Further our method had an added advantage of using monochrome watermark image and binary watermark image, whereas the DCT-CS [5] approach is applicable only for binary watermark. Due to this reason,

Table 1 Comparison of PSNR (dB) and NC of the watermarked image of size 256×256 under no noise attack

Cover image	Watermarked image (256×256)			
	PSNR (dB)		NC	
	Basic DCT	Blind DCT-CS watermarking system based on subsampling	Basic DCT	Blind DCT-CS watermarking system based on subsampling
Baboon	31.9633	36.0103	0.9663	0.9724
Cameraman	33.9203	35.0294	0.9534	0.9605
Couple	33.0139	35.0193	0.9590	0.9623

Table 2 Comparison of SSIM and RMSE of the watermarked image of size 256×256 under no noise attack

Cover image	Watermarked image (256×256)			
	SSIM		RMSE	
	Basic DCT	Blind DCT-CS watermarking system based on subsampling	Basic DCT	Blind DCT-CS watermarking system based on subsampling
Baboon	0.9711	0.9756	0.1195	0.1062
Cameraman	0.9523	0.9621	0.1254	0.1079
Couple	0.9602	0.9698	0.1208	0.1155

Table 3 Comparison of PSNR (dB) and NC of extracted watermark of size 64×64 under no noise attack

Cover image	Extracted watermark (64×64)			
	PSNR (dB)		NC	
	Basic DCT	Blind DCT-CS watermarking system based on subsampling	Basic DCT	Blind DCT-CS watermarking system based on subsampling
Baboon	12.0329	15.1933	0.7892	0.9018
Cameraman	13.1938	15.2043	0.8131	0.9127
Couple	12.9283	15.0093	0.7994	0.9421

Table 4 Comparison of SSIM and RMSE of extracted watermark of size 64×64 under no noise attack

Cover image	Extracted watermark (64×64)			
	SSIM		RMSE	
	Basic DCT	Blind DCT-CS watermarking system based on subsampling	Basic DCT	Blind DCT-CS watermarking system based on subsampling
Baboon	0.7238	0.8074	0.2019	0.1543
Cameraman	0.5204	0.7906	0.2031	0.1509
Couple	0.6075	0.7990	0.2074	0.1532

Table 5 Comparison of PSNR (dB) and SSIM of the extracted watermark image of size 64×64 under noise attacks

Noise attack	Extracted watermark (64×64)			
	PSNR (dB)		SSIM	
	Basic DCT	Blind DCT-CS watermarking system based on subsampling	Basic DCT	Blind DCT-CS watermarking system based on subsampling
Salt and Pepper (0.025)	5.1983	8.7521	0.1201	0.1743
Gaussian filter (5×5)	9.2942	21.3233	0.1502	0.2019
AWGN (SNR = 45 dB)	9.8973	20.0122	0.1532	0.2310

Fig. 5 Quality of extracted watermark (alphabet A) after salt and pepper noise attack

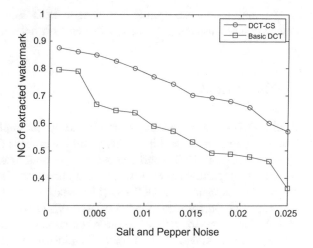

Fig. 6 Quality of extracted watermark (alphabet A) after AWGN channel attack

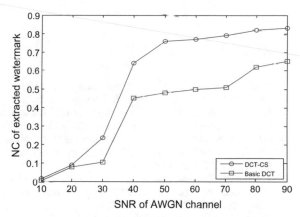

comparing both the algorithms in terms of NC, PSNR, RMSE, and SSIM is not possible, and so, we compared our method with the basic DCT [9].

Figure 4 demonstrates the security of the system when a different random Gaussian matrix was used for recovering the watermark. It is clear that the extraction of the watermark will not be successful in this case, and random Gaussian matrix serves as key during watermark extraction.

The plot in Fig. 5 indicates the NC of extracted watermark of alphabet A (shown in Fig. 3), in blind DCT-CS watermarking, is better than DCT at all values of noise density. In particular, comparing at specific value of salt and pepper noise density of 0.005, NC of blind DCT-CS and DCT are of 0.8534 and 0.6676, respectively.

Figure 6 shows that blind DCT-CS watermarking performs best under AWGN attack. Successful extraction of the watermark is still possible when the AWGN SNR is greater than 45 dB. Successful extraction implies 70% of perceptual quality which is measured in terms of NC.

Similar analyses are seen if logo watermark is used with salt and pepper noise and Gaussian noise attacks. The results and plots are not shown for want of space.

6 Conclusion

A blind DCT-CS watermarking system based on subsampling is presented with experimental results and capabilities. The algorithm embeds CS measurements of watermark in mid-band DCT coefficients corresponding to two subimages of the cover image. The experimental results show that our method has greater embedding capability when compared to DCT-CS [5] method. Also the results demonstrate that this approach outperforms the basic DCT [9], in terms of PSNR, SSIM, NC, and RMSE, for extracted watermark and watermarked image. Further, the performance of proposed method is superior for with and without noise attacks.

References

1. Pandey, S., Gupta, R.: A comparative analysis on digital watermarking with techniques and attacks. Int. J. Adv. Res. Comput. Sci. Softw. Eng. **6**(2), (2016)
2. Potdar, V., Han, S., Chang, E.: A survey of digital image watermarking techniques. In: 3rd IEEE International Conference on Industrial Informatics. IEEE (2005)
3. Neetha, K.K., Koya, A.M.: A compressive sensing approach to DCT watermarking system. In: International Conference on Control, Communication and Computing India (ICCC). IEEE Conference Publications (2015)
4. Korrai, P.K., Deergha Rao, K.: Compressive sensing and wavelets based image watermarking and compression. In: 2014 IEEE Region 10 Conference TENCON 2014. IEEE Conference Publications (2014)
5. Huang, H.-C., Chang, F.-C.: Robust image watermarking based on compressed sensing techniques. J. Inf. Hid. Multimed. Signal Proc. **5**(2) (2014)
6. Tsai, M.-J., Hung, H.-Y.: DCT and DWT-based image watermarking by using subsampling. In: Proceedings 24th International Conference on Distributed Computing Systems Workshops. IEEE (2004)
7. Rachmawanto, E.H., Sari, C.A.: Secure image steganography algorithm based on DCT with OTP encryption. J. Appl. Intell. Syst. **2**(1) (2017)
8. Luo, H., et al.: Blind image watermarking based on discrete fractional random transform and subsampling. Optik-Int. J. Light Electr. Opt. **122**(4) (2011)
9. Lin, S.D., Chen, C.-F.: A robust DCT-based watermarking for copyright protection. IEEE Trans. Consum. Electron. **46**(3) (2000)
10. Baraniuk, R.G.: Compressive sensing. IEEE Signal Proc. Mag. (2007)
11. Pavithra, V., Renuka Devi, S.M.: An image representation scheme by hybrid compressive sensing. In: 2013 IEEE Asia Pacific Conference on Postgraduate Research in Microelectronics and Electronics (PrimeAsia), Visakhapatnam, pp. 114–119 (2013)
12. Tropp, J.A., Gilbert, A.C.: Signal recovery from random measurements via orthogonal matching pursuit. IEEE Trans. Inf. Theory **53**(12) (2007)

13. Chu, W.C.: DCT-based image watermarking using subsampling. IEEE Trans. Multimed. **5**(1) (2003)
14. Sharma, A., Singh, A.K., Ghrera, S.P.: Robust and secure multiple watermarking for medical images. Wirel. Pers. Commun. **92**(4) (2017)

Author Index

© Springer Nature Singapore Pte Ltd. 2019
J. Wang et al. (eds.), *Soft Computing and Signal Processing*,
Advances in Intelligent Systems and Computing 898,
https://doi.org/10.1007/978-981-13-3393-4